PERSPECTIVES ON WATER

PERSPECTIVES
ON
WATER
USES
AND ABUSES

EDITED BY

DAVID H. SPEIDEL, *Queens College of the City University of New York*

LON C. RUEDISILI, *University of Toledo*

ALLEN F. AGNEW, *Oregon State University*

New York · OXFORD UNIVERSITY PRESS · Oxford
1988

Oxford University Press

Oxford   New York   Toronto
Delhi   Bombay   Calcutta   Madras   Karachi
Petaling Jaya   Singapore   Hong Kong   Tokyo
Nairobi   Dar es Salaam   Cape Town
Melbourne   Auckland

and associated companies in

Beirut   Berlin   Ibadan   Nicosia

Library of Congress Cataloging-in-Publication Data
Speidel, David H., 1938–
Perspectives on water.
Bibliography: p.   1. Water-supply. 2. Water use. 3. Hydrologic cycle.
I. Ruedisili, Lon C. II. Agnew, Allen
Francis, 1918–   III. Title.
TD345.S67 1988   333.91   86-12792
ISBN 0-19-504247-6
ISBN 0-19-504248-4 (pbk.)

9 8 7 6 5 4 3 2 1

Printed in the United States of America
on acid-free paper

This book is dedicated to the directors of the Water Research Institutes and Departments of Natural Resources of the various states, whose continuing efforts are essential for the development and protection of our water resource.

PREFACE

The noblest of elements is water.—Pindar

The earth is a water world. From outer space the oceans are visible as great swirls of blue, green, and grey covering over 71 percent of the surface. The blue-white of ice and snow covers another 3.5 percent. Below the patches and linear streaks of clouds are lakes and rivers, ponds and streams; unseen beneath the surface are vast reservoirs of ground water.

Earth is the only water world we know, the only planet where the compound water exists as liquid, vapor, and solid. There appears to be only a very small set of conditions in any solar system where the three states of water can so exist. Earth's size and location relative to those of the sun fit within that small set. The change of water from one physical state to another in the hydrologic cycle is a major factor influencing the geological, chemical, physical, and biological processes operating on the surface of the earth, including the development and maintenance of all life.

Water is a resource. We drink it. We use it for bathing and cleaning, cooling and heating, growing food and fishing, recreation, transporting goods, disposal of waste, and decoration. Water is so useful that conflicts over its use have existed for thousands of years.

Until recently, concern about water usually occurred only when it was not available to us in the amounts we wanted (as in droughts), when it was present in overabundance (as in floods), or when its usefulness was limited by pollution. Today there is a growing awareness of the importance of water, of the key role it plays in supporting our society, and of the many problems other than drought and floods involved in its use. We have assembled this textbook to provide an overview of the full range of water as a resource.

Part I, "Water: The Compound, the Resource," first describes the unique physical and chemical properties of water and then reviews the many aspects involved in its use as a resource. Part II, "Water in the Environment," discusses the hydrologic cycle and illustrates the physical limitations of water as a resource.

The range of uses of water is illustrated in Part III, "Water Use," by chapters on irrigation, drinking water, energy and water, and industrial uses of water. Every use of a natural resource produces environmental problems; water is no exception. Part IV, "Problems and Hazards," examines issues of water quantity and quality, land-use and hydrologic hazards, and other natural and human-instigated problems.

Finally, Part V, "Law, Economics, and Management of Water," describes how society manages water use. Many of the problems associated with water are not about quality, quantity, or location of water but arise from legal, economic, and management procedures.

Our aim is to present a reasoned look at the problems of water, being neither apocalyptic (overly pessimistic) nor Pollyannish (blindly optimistic). Severe problems do exist, but if attention is paid to them, we feel that they can be handled.

Who should pay that attention? First, students of hydrology must be made aware that the hydrologic cycle is more than a geologic process; it is also a human process. This book provides a companion to the standard hydrology text by stressing the human interaction with the water cycle.

Second, students in arts, business, humanities, and social sciences from whose ranks will come the politicians, economists, and managers making future decisions about water should have a knowledgeable base for such decisions. The study of resources must include discussions of physical limits and systemic problems as well as economic arguments. We feel that this collection of articles provides the basis for a course on water resources for such nonscientists and nonengineers.

Third, individuals concerned with problems of resources and the environment should have an appreciation for this critical resource. History is spotted with the stories of cities, regions, and even civilizations that are believed to have failed because of the misuse of water. Everyone should be aware of how hearty, yet, at the same time, how fragile this resource is.

CONTENTS

? = retrieval

PERSPECTIVES ON WATER

PART I

WATER: THE COMPOUND, THE RESOURCE

The earth is the only planet we know where water freely exists. Water is a critical factor in chemical, physical, and biological processes, and hence all aspects of this unusual compound deserve examination. In Chapter 1, "Water, Something Peculiar," J. Lyklema and T. E. A. Van Hylckama present a lighthearted discussion of some anomalous physical-chemical properties of water and their environmental and biological significance.

The structure of the water molecule, the "Mickey Mouse" of Figure 1-2b, is the key to many of the properties of water. What the diagram does not indicate is that while the water molecule as a whole is electrically neutral, the distribution of positive and negative charges is not uniform. The "ears" are slightly positive and the "chin" is slightly negative. Molecules of liquid water thus have a tendency to line up ears to chin, to ears to chin, and so on, held together by the attraction between a positive charge and a negative one. This "hydrogen bonding" is used to explain the anomalous stability of liquid water and its many properties.

In addition to the properties discussed in Chapter 1, several others are of environmental interest. One such set is the latent heats of melting (fusion) and vaporization (evaporation), illustrated in Figure I-1. The heat that is added to 1 g of material at its melting point sufficient to cause a change of state from solid to liquid, thus melting the material, is called the *latent heat of melting*. Similarly, the amount of heat necessary to change 1 g of liquid at its boiling point to vapor is called the *latent heat of vaporization*. For water the latent heat of melting is 80 cal, and the latent heat of vaporization is 540 cal. The term *latent* is used because the heat that is added to a given mass of ice to turn it into liquid or to a given mass of water to turn it into vapor is held by that mass of water or vapor and then released when the process is reversed. There are some practical effects to these latent heats. Cans of beer set in ice will cool. Heat energy drawn from the cans to the ice will provide sufficient energy to change water from the solid state to the liquid state as the ice melts. Water-cooled air-conditioning works the same way. Hot, dry air is passed through or over water and will lose heat to the water. The water will absorb a large amount of heat before it itself is vaporized.

The process works the other way also. Did you ever feel warmer when it started to snow than when you were in a cold rain? You were indeed warmed, because the latent heat of melting was released when the rain was turned into snow. Evaporation of water transfers solar energy from the liquid in the oceans to the vapor in the atmosphere. When the water falls as precipitation, the latent heat of vaporization is released. The rainfall releases heat into the atmosphere. Thus coastal regions generally have milder winters than inland, drier continental regions.

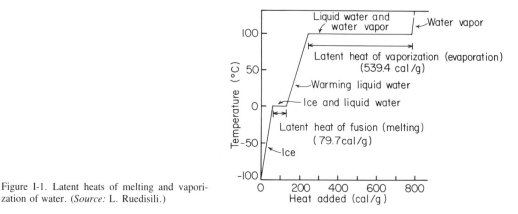

Figure I-1. Latent heats of melting and vaporization of water. (*Source:* L. Ruedisili.)

Density is another valuable property of water. Under normal pressure conditions liquid water at 4°C is the most dense form. Ice is less dense and will float. This means that water bodies will freeze from the surface downward and thus allow circulation to continue under the frozen surface, so that fish can survive.

Water as a resource has several characteristics that were described in the recent report *Global 2000:*

1. *Water is ubiquitous.* No place on earth is wholly without water. In general, most locales of human activity have vast quantities of water available nearby. While the means by which water is moved to the point of use may be of concern, and while the quality of available water may not be of the standard desired, the presence of water is constant.

2. *Water is a heterogeneous resource.* Water exists in three different physical states, each of which is commonly found. The liquid form, the state that is most often used as a resource, varies widely in chemical composition. The beneficial use of water requires that the characteristics of the water supplied must match the requirements of the use for which the water is demanded.

3. *Water is a renewable resource.* The water cycle was briefly discussed in the Preface and will be more fully discussed in Part II. The natural process is not completely independent of humans. Modern technology permits the exchange of water between surface sources and ground water sources, can restore quality to contaminated water, can remove salts from ocean water to provide new sources of fresh water, and can even alter the pattern of precipitation. These actions are capable of modifying the renewable characteristic of water.

4. *Water may be a common property.* Not only is water usually found everywhere, but it is nonstationary, with typically ill-defined property rights. If water is available to all without direct charge for use, any opportunity costs that might be associated with such use are not paid by the user. Without such charge no mechanism for allocation of water exists. Water can thus be treated as a free commodity, even during times of scarcity, when many potential users may be excluded. The costs of capturing, treating, and transporting water are clearly recognized, but the cost of water itself is not.

     An important exception to common property occurs in the western United States where ownership of water does occur, usually on a first-come, first-served basis. The drought in the western United States during the 1985 summer was so severe in some places that only those individuals with water claims on the water made prior to 1886 still were able to obtain a supply.

5. *Water is used in vast quantities.* The quantity of water used far exceeds the total quantity used of any other single resource. For example, the world's production of minerals, including coal, petroleum, metal ores, and nonmetals, has been estimated to be about 8 billion metric tons per year. Total water use has been estimated to be about 3000 billion metric tons worldwide, or approximately 800 metric tons per person per year.

6. *Water is very inexpensive.* Water's common-property aspect, the nature of water supply technology, and economies of scale combine to make water very inexpensive. For example, municipal water supplied to its point of use usually costs less than 30 cents a metric ton, and irrigation water can cost as little as 3 cents per metric ton. Sand and gravel, probably the cheapest mineral commodity, costs approximately $3 per ton, and iron ore is approximately $30 per ton.

Of the six characteristics of water, the first three—ubiquity, heterogeneity, and renewability—make it difficult to characterize the available water supply, either now or in the future. The quantity and quality of water available at a particular time and place constitute the relevant supply; aggregate or summary statistics, therefore, are nearly meaningless.

The common-property characteristic of water, the large quantities used, and the low user costs all act as deterrents to accurate forcasting of future water use.

Chapter 2 in this part, "Water Resource Adequacy: Illusion and Reality" by Gilbert W. White, examines recent shifts in technical thinking on how to assess the adequacy of fresh water supplies for the future. White's paper was one of the inspirations for this book.

1

WATER, SOMETHING PECULIAR

J. LYKLEMA and T. E. A. VAN HYLCKAMA

Water would be an excellent topic for an after-dinner talk. Not only do we have it in our drinks, (which are mostly water anyway—whether you like it or not), but even our solid food is nearly 50 percent water. It should not amaze us that water is the most abundant chemical substance in the earth's crust. More than half of the 5-km (3 mi) deep outer shell of the earth is water. And that is just one peculiarity of water.

Nearly all living things also consist mostly of water. You and I are about 65 percent water. You may consider this undignified, but that's the way we're built. A 1-month-old fetus is 90 percent water, and babies are more watery than grown-ups. Most mothers are aware of this. It is also the reason that babies can suck more out of their thumbs than we can.

The word for "water" occurs in any language. Table 1-1 gives just a small example. Contrast this with the word for "ice," which does not occur in all languages. In Indonesian, for instance, ice is described as *air batu,* or "stone water," although in modern Indonesian one finds *es,* a transliteration of the Dutch *ijs.*

We are so completely familiar with water that even if the word is not mentioned it is taken for granted. If somebody says: "My cup runneth over", nobody thinks the overflow to consist of syrup or gravel (although alcohol may be considered). If we talk about irrigation, nobody thinks we mean whisky, and when I mention drainage, nobody refers to tequila. When we say that a fish or a swimmer is in his element, again we mean water, even though water is not an element but a chemical combination, notwithstanding the fact that the Greek and Chinese had different opinions about this.

Water and watery solutions play an all-important

Table 1-1. Water in 30 Languages

Algonquin/Cree	nibi	Hungarian	viz
Arabic	mayah	Indonesian	air
Chinese	shuĭ	Italian	acqua
Danish	vand	Japanese	mizu'k
Dutch	water	Latin	aqua
English	water	Norwegian	vann
Esperanto	akvo	Polish	woda
Finnish	vettä	Portuguese	agua
French	eau	Russian	woda
Frisian	wetter	Sanskrit	udan
German	wasser	Spanish	agua
Greek	hydōr	Swahili	maji
Hawaiian	wai	Swedish	vatten
Hebrew	măyĭm	Tchech	woda
Hindi	păni	Turkish	su

role in the physical and biological processes on the earth. Of the physical ones I'll mention only the effects of rainfall on riverflow and of erosion on the landscape.

In living organisms water is active (1) in the transportation of nutrients and other chemicals; (2) in determining the structure and properties of large molecules such as proteins and nucleic acids; (3) in the biochemical reactions taking place in cells and in tissues; and (4) in heat-regulating processes such as moisture absorption and transpiration.

All these factors are essential for the processes in living organisms, and it is therefore most likely that life on this earth originated in the oceans. So we may pose this question: What are the unique properties of water, and is it possible to comprehend them on the basis of molecular interpretation? To answer this question, we must first assemble

Dr. Lyklema was professor of physics, Agricultural University, Wageningen, The Netherlands. Mr. van Hylckama, now retired, was a hydrologist with the U.S. Geological Survey, Tucson, Arizona 85701. This material was originally presented by Dr. Lyklema on the occasion of the 56th Dies Natalis of the University. This article was freely translated, with Dr. Lyklema's approval, by Mr. van Hylckama and published in *WRD Bulletin* (April–June 1975): 64–68. It was also published in *Hydrological Science Bulletin,* 24 (4) (1979): 499. Reprinted by permission of Mr. van Hylckama.

Table 1-2. A Rundown on Water

Chemical formula: H_2O
Molecular weight: 18
Physical characteristics (at room temperature and
    standard atmosphere)
  Freezes at 0°C
  Boils at 100°C
  Colorless, odorless
  Expands on freezing
  Maximum density at 4°C
Specific heat: 1 cal/(g/°C) = 75.25 J/mole·°C
Heat of vaporization at 100°C: 538 cal/g = 40.6 kJ/mole
Solubility for selected substances
  Kitchen salt (NaCl)   360 g/L
  1-Butanol (C_4H_9OH)   80 g/L
  Ethanol (alcohol)(C_2H_5OH)   All proportions
  Fatty substances   Very small
Viscosity at 20°C: 1 centipoise (cP = mPa·s); diminishes
    slightly with temperature
Conductivity at 20°C: 4×10^{-8} mho/cm = 4×10^{-6}
    siemen (S)/m
Dielectric constant at 25°C: 78.5 times that in vacuum
Relaxation time: 10^{-19} to 10^{-11} s

Table 1-3. Boiling Points of Fluids (°C at Sea Level)

Fluid	Formula	Molecular Weight	°C
Water	H_2O	18	100.0
Alcohol	C_2H_5OH	46	78.5
Ether	$C_2H_5OC_2H_5$	74	34.6
Propane	C_3H_8	44	−44.5
Carbon dioxide	CO_2	44	−78.5
Methane	CH_4	16	−161.0

the facts, facts that are usually expressed in form of numbers.

Table 1-2 is a small collection of facts. When you look at this list, I would not be surprised if some of the items sound familiar to you, while others are incomprehensible. *Familiar* means that it fits into one's frame of reference. If I say that Florida bananas are 12 to 15 cm (or 5 to 6 in.) long, then such a statement has meaning to you because it fits in your frame of reference. But if I asked you if Mr. X in Kandahar, who earns 100,000 puls per month, has a decent income, then most of you won't be able to answer that question unless you happen to be a specialist in Afghanistan economy. Even if I tell you that 100,000 puls is about equivalent to $25, that won't be of much help because you have to know what the average income of an Afghan is and what he can do with $25. Well, the average income is about $5 per month, so you see Mr. X has a princely income and should be taxed much heavier.

Well, back to water. It boils at 100°C and it freezes at 0°C. We are used to that; we consider it self-evident. But in our daily lives we usually do not talk about viscosities of 1 centipoise (cP) or relaxation times of a millionth of a millionth of a second. Therefore we consider such numbers uncommon or unusual. Here truly lurks a danger of jumping to conclusions, for we might consider that something we are familiar with is also scientifically self-evident. And in the case of water this just is

not so. There is hardly another fluid with such a small molecule that has such a very high boiling point. On the other hand, viscosities of 1 cP are not at all uncommon among fluids.

To compare water with other fluids in an objective manner, we'll have to look for other frames of reference. We'll have to compare the boiling point of water with the boiling point of other fluids and forget about making tea.

A few examples are given in Table 1-3. First of all, keep in mind that the boiling point of fluids generally increases with the weight of the molecule. The best frame of reference is therefore a series of fluids with a molecular weight comparable to that of water. As you can see, the boiling point goes up as the molecular weight goes up, but alcohol is out of line and water completely so.

Figure 1-1 presents another way of illustrating how peculiar water is. Here are three substances

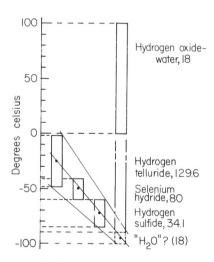

Figure 1-1. Boiling and freezing temperatures of water and waterlike substances. The numbers following the chemical names are molecular weights. (*Source:* After K. S. Davis and J. A. Day, 1961, *Water, the mirror of science.* New York: Doubleday.)

Table 1-4. Specific Heat of Fluids at 25°C

Fluid	Formula	J/g·°C
Water	H_2O	4.18
Alcohol	C_2H_5OH	2.49
Carbon tetrachloride	CCl_4	0.98
Chloroform	$CHCl_3$	0.90
Ether	$C_2H_5OC_2H_5$	2.31
Mercury	Hg	0.14

with a molecule similar to that of water. As the molecular weight goes down, so do boiling and freezing points. Water should be boiling at −91 and freezing at −100—but it is peculiar and behaves differently.

Why is water so exceptional, and once we know that, is it possible to make other fluids that also take such an exceptional position? We can at least say this: During the process of evaporation, molecules of a fluid have to dissociate themselves from each other. A high boiling point means that the molecules can, so to speak, resist such a dissociation. In this case we get the impression that there are very strong attractive forces between molecules of water, much stronger than in any other fluid.

Another interesting comparison is given in Table 1-4. (Don't bother about the units; we're just comparing numbers.) It deals with the specific heat, that is, the amount of heat a fluid needs to raise the temperature of one unit volume by one degree. As you can see, mercury needs a very small quantity, and that is fortunate, because it accounts for the fact that mercury is such a fast indicator of the rise and fall of temperatures. By contrast, water needs about thirty times as much heat. That is why we speak also of heat capacity.

Now this large heat capacity of water has an equalizing effect on our climate. Long after the summer is gone, the oceans are still warmer than the continent, and areas close to the ocean enjoy a much milder climate than those remote from the seas. As you can see, alcohol also has a pretty high heat capacity. That is not completely accidental, but it has nothing to do with diluting wine with water and other diverting chemical pastimes, such as making cocktails. (Incidentally, the units used here are those of the international system. Remember that we used to talk about calories or British thermal units.)

The relatively constant temperature of the ocean has possibly played a role in the formation of life on earth, although it may not have been the most important one. Whatever the history, specific heat does play a role in the heat-regulating mechanism of higher developed organisms and especially of warm-blooded animals. One can imagine that somebody with chloroform instead of water in his or her blood, and suffering from influenza, would have a much higher fever than ordinary, we might say, watery patients have. From a physical-chemical point of view the high specific heat means that water, notwithstanding the fact it is a fluid, must be highly structured or, you might say, organized. It needs a lot of heat to break that structure down and make the molecules move around individually, which happens when temperature rises.

Table 1-5 illustrates another property of water. It deals with viscosity. (Don't worry about the unit centipoise; as before, we're only comparing numbers.) Viscosity indicates how easily a fluid runs. The higher the number of centipoise, the more the stuff runs like molasses in January. For a solid we could say that the viscosity is infinitely high. Gases, of course, have very low viscosities, and fluids are in between. As you can see in this case, water is not particularly exceptional compared with, say, alcohol or carbon tetrachloride. But the interesting part is that this is completely unexpected. We just mentioned that water is structured, and when the molecules are so strongly attached to each other, one would expect that it also would be difficult for them to move around the way molecules do when a fluid is running. What we have here is a strong coherence without loss of mobility. This is most important for living organisms, because they do not need extra sources of energy to keep their blood moving around.

Viscosity depends on temperature and pressure. During influenza epidemics people ask themselves why they get fever. I do that, too. But an advantage is, anyway, that for any degree of fever the viscosity goes down by about 2½ percent and so your heart has to do comparatively less work to pump the blood around. Now this should make you happier when you come down with influenza.

In nearly all fluids viscosity increases with increasing pressure. That is not the case with water. The viscosity diminishes with increasing pressure,

Table 1-5. Viscosity of Fluids at 20°C

Fluid	cP = Pa·s
Water	1.00
Alcohol	1.20
Chloroform	0.58
Carbon tetrachloride	0.97
Ether	0.23
Glycerine	1.50
Sulfuric acid	25.4
Olive oil	84.0

at least when pressure doesn't get too high. You may never have thought of it, but if viscosity would increase significantly with pressure, you would get water out of your kitchen tap only a drop at a time, and deep-sea fishes would have much less fun chasing each other around.

The dissolving power of water, mentioned in Table 1-2, is also interesting. We know that kitchen salt, sugar, alcohol, and aspirin dissolve very easily, but oils and fats do not. To dissolve those, you need detergents, of which the housewife nowadays knows a great number. Animal organisms also use a kind of a detergent to "dissolve" the fats from our food into the large surplus of water in the body. It is of interest to realize that many biologically important molecules such as proteins and nucleic acids have large molecules, some parts of which dissolve easily in water and other parts less easily so or not at all. This is the result of the very characteristic structure of these molecules.

Another fascinating peculiarity of water is the dielectric constant given in Table 1-6. If you imagine two parallel plates, one having a positive and the other a negative electric charge, then there exists an attracting force between these two plates, which is proportional to the product of the charges and inversely so to the square of the distance between the plates. This is called Coulomb's law. Now if we put some fluid or solid, for instance paper, in between those plates instead of air or vacuum, the attractive force will be diminished by a factor depending on the dielectric constant of the substance. As you can see from Table 1-6, water, compared with more or less familiar chemicals, has a very high dielectric constant, with alcohol a good second. There are some substances that are even higher, such as hydrocyanic acid and formamide. Both have constants well over 100, but they are rare exceptions.

Because water, unlike oil or carbon tetrachloride, has such a high dielectric constant, the positive and negative ions can remain as loose, isolated units. This is very significant for plants, animals, and soil—for instance, for the transportation of

Table 1-6. Dielectric Constants for Some Fluids at 25°C

Water	78.5
Alcohol	24.3
Chloroform	4.8
Ether	4.3
Carbon tetrachloride	2.2
Formamide (HCONH$_2$)	109
Hydrocyanic acid (HCN)	114

Note: Constants are relative to a dielectric constant in a vacuum equal to 1.

nutrients and for the effectiveness of electrical pulses in the transmission of nerve signals. It is a sobering thought that if you want to electrocute somebody, it would save a lot of electricity if you could replace the water in his blood by carbon tetrachloride or ether. Even a slug of alcohol would help! Some of the proponents of the death penalty should be interested in this.

Finally, consider that relaxation time mentioned at the bottom of Table 1-2. How do we visualize that? We use the verb *relax* quite frequently. "Relax, for Pete's sake," we say, meaning "slow down" or "take it easy." In physics we give it another meaning. Relaxation time is the time a substance needs to yield to or recover from a deforming pressure or treatment. Solids have a very long relaxation time—long, but not infinitely long. Glaciers move due to the deforming forces of gravity, and the Himalaya Mountains have been rising for millions of years (and still are rising) due to the deforming pressure of the Indian land mass against the Asian continent. We say "Faith can move mountains," but that should really read, "Faith and patience can move mountains."

Fluids do not yield to deforming forces; they rather run away immediately. Tar is sort of an in-between. You can build a tar baby, but it will slowly relax into a tar puddle. Water has a relaxation time of 10^{-13} to 10^{-11}s, depending on temperature and pressure. But we cannot visualize such a short time. It is certainly faster than the building of the Taj Mahal, and it is faster than turning over your hand. It's even much faster than the fastest film, but we cannot "feel" the difference between 10^{-13} and 10^{-11}, although 10^{-11} is 100 times larger than 10^{-13}. But we do "feel" the difference between 10 percent and 11 percent inflation, although that factor is only 1.1.

The differences in relaxation time are important, though, in molecular thermodynamic studies, but this is beyond today's subject. Let me say only that water in this respect does not differ very much from other fluids, and this has to do with the fact that viscosity is not much different either.

So far, then, we have seen quite a few interesting peculiarities of water. We saw that the molecules hang together very strongly. There seems to be a certain amount of organization (whatever that is in a fluid), and the molecules are capable of cutting down electrical attractive forces.

Can this be explained on the basis of the geometric structure of a water molecule? A water molecule consists of a large oxygen atom and two very much smaller hydrogen atoms. It is nearly, but not completely, round, with a radius of 0.14 nm. If you put 150 million in one row, you can cover them with the width of your thumb.

Hydrogen

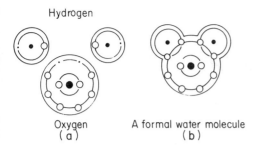

Oxygen   A formal water molecule
(a)   (b)

Figure 1-2. (a) Two hydrogen molecules and one oxygen, showing the electron orbits and the spots where
electrons are missing. A black dot is a nucleus and a cir-
cle an electron (schematic and not to scale.) (b) The water
"mouse." *(Source: After Davis and Day, Water.)*

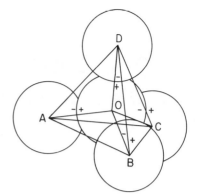

Figure 1-3. Five water molecules in a tetrahedral group-
ing of four around a central one. (*Source:* After E. C.
Childs, 1969, *Soil water phenomena.* New York, Wiley.)

Figure 1-2(a) shows an oxygen atom with a
nucleus, two electrons in the inner orbit, and six
in the outer orbit, but there is space for two more.
There are also two hydrogen atoms, each with a
nucleus and one electron, but there is space for
one more. The hydrogen electrons will fit in the
open spaces of the oxygen atom, and the result is
a stable molecule, schematically pictured in Figure
1-2(b).

In three-dimensional form we could, again very
schematically, visualize a simple configuration as
in Figure 1-3, which shows a slightly irregular
tetrahedron, four molecules attached to the two
positive and two negative poles of a center molecule. Of course this building goes on and on, and
it does so in ice, which has an open structure and
therefore floats. Water expands on freezing, and
its structure is much less regular. Groupings of

molecules continuously break apart and re-form,
quite nimbly so, if you think how fast water runs.

Thus we deduce that there are two basic factors
that cause many of the peculiarities of water. The
electric charge in a molecule is not evenly divided,
and the charge is divided tetrahedrally.

The knowledge of these peculiarities of water
(and there are many more) can be applied in understanding and developing processes in fields of
agriculture, of transport through membranes, of
hydration of foods and of making sweet water out
of the sea even without diluting it with wine—in
short, in the fields of many biological and technological actions in us and around us that govern
our being and our well-being.

2

WATER RESOURCE ADEQUACY: ILLUSION AND REALITY

GILBERT W. WHITE

Technical thinking about how to appraise the adequacy of fresh water to satisfy expected demands has been changing in ways that are strongly affecting prevailing views as to the resource and its wise management. These shifts are well illustrated by the evolving efforts to take the measure of the water situation on a world scale, including the recent *Global 2000 Report to the President*[1] and the United Nations Environment Programme assessment of *The World Environment, 1972–1982*.[2] In those documents and in the body of national and international studies from which they draw are found at least six concepts of water resources and its uses that are undergoing significant change. To the extent these changes are accepted, the methods and policies for dealing with water problems also are altered.

None of the shifts in thinking is wholly new: Each marks a different emphasis to take account of accumulating experience and to move away from certain widespread illusions about what determines the availability of water to serve human needs. They have to do with the principal constraints on drinking water adequacy, the alternatives to increasing physical supply, the relationship of population growth to industrial and agricultural water use, the meaning of flood and drought disasters, the risks to aquatic ecosystems, and the implications of a possible climate change. I call attention to them because I believe they ought to be in mind whenever water plans are in preparation. They are taken for granted wholly or in part, in much planning today, but they also are ignored in many instances.

As background for a discussion of each shift, the major attempts at appraisal of global water availability are noted. Each change in thinking then is examined briefly. The discussion is in no sense comprehensive but is intended to state the change concisely and to outline the realities and illusions it confronts.

EVOLVING EFFORTS TO ASSESS THE WORLD WATER SITUATION

At the first United Nations conference on natural resources at Lake Success in 1947, the hydrologic data were so incomplete and scattered that there seemed little point in attempting a worldwide estimate of water supply and use.[3] Instead, the discussions hinged on modes and effects of various types of water management within river basin, national, and subnational units. The chief topics were hydrologic appraisal, water supply and pollution, water conservation strategies, flood control, navigation, irrigation and drainage, and hydroelectric power. The conference heightened interest in strengthening hydrological networks, in basin surveys, and in technical assistance for water planning at national levels.

By the time of the Water for Peace conference in 1967 a variety of international programs to take stock of water resources had been launched, including the UNESCO arid zones research program, the International Hydrological Decade (IHD), and a large number of surveys and studies funded by the United Nations Development Programme and

Dr. White is Gustavson distinguished professor emeritus of geography in the Institute of Behavioral Science, University of Colorado, Boulder, Colorado 80302. He has taken part in a variety of national and international water resource studies ranging from those of the National Resources Planning Board in the 1930s to those of the United Nations Environment Programme in the 1980s. This material is taken from *Natural Resources Forum*, 4 (1) (1983): 11. Reprinted by permission of the author and *Natural Resources Forum*, Graham and Trotman, London. Copyright © 1983 by the United Nations.

bilateral assistance agencies in cooperation with developing counties. The papers at the Water for Peace conference canvased much of this rapidly growing body of information on precipitation, stream flow, ground water, unconventional sources, water technology, water quality, water use, education, training, organization, and financing.[4] In a major paper Doxiadis ventured a global estimate and produced a projection of water supply and use by sectors for 1975, 2000, and up to 2090 without breaking it down by area or showing how it was computed.[5]

During the next 15 years the calculations of the volume of fresh water available to the human race tended to converge. More data were collected, and the uncertainty as to how much surface water and ground water is available and its distribution in time and space narrowed. Knowledge as to ground water location remained rudimentary in many areas, and the estimates as to stream discharge were handicapped by the highly probabilistic evidence as to the magnitude and frequency of low and peak flows. However, the main physical parameters of surface supply came to be recognized with a moderate degree of accuracy for large continental areas. Information on ground water and water quality and its trends is far less satisfactory.

In 1974 L'vovich completed a calculation of total supplies and withdrawal of fresh water by continental units.[6] The USSR Committee for IHD[7] also published a worldwide computation of water balance, as did Baumgartner and Reichel a year later.[8]

It remained for the United Nations Water Conference in 1977 to bring together the immense body of information that had been accumulating in national and international records and had been analyzed by a number of investigators.[9] The review paper for that conference, "Resources and Needs," compared the estimates of global water balance by Baumgartner and Reichel, L'vovich, and the USSR Committee for the IHD,[10] and reviewed the information in hand as to uses.

These reviews showed that the prevailing scientific consensus placed the total volume of fresh water at less than one-thirtieth of all water on the globe. The water moving in streams in a year was estimated at 40 to 47 trillion m³. Altogether, at any one time about 22 percent of the fresh water is in the soil or in the ground, 0.35 percent in lakes and wetlands, 0.04 percent in the atmosphere, less than 0.001 percent is in the streams, and about 77 percent in snow and ice.[11] The point of giving these figures is to show that the great reserve of liquid fresh water, in contrast to snow and ice, is underground, while the running streams and lakes are a very small, rapidly circulating proportion of

Table 2-1. Estimates (Mid 1970s) of Volume of Fresh Water Withdrawn (Billions m³)

	Reference		
	Doxiadis[a] (1975)	Global 2000[b]	USSR[c] (1970s)
Domestic and municipal	228	201	150
Industrial	184	305	630
Irrigation	2000	1830	2100
Total	2412	2838	2880

[a]C. A. Doxiadis, 1967, Water and environment, in *Water for peace*, vol. 1 (Washington, D.C.: U.S. Government Printing Office), 33–60.
[b]U.S. Council on Environmental Quality and the Department of State, Gerald O. Barney (director), 1980, *The global 2000 report to the president: Entering the twenty-first century*, vol. II (Washington, D.C.: U.S. Government Printing Office), 147.
[c]M. W. Holdgate, M. Kassas, and G. F. White, eds., 1982, *The world environment 1972–1982: A report by the United National Environment Programme* (Dublin: Tycooly International, for the United Nations Environment Programme), 131.

the total resource. There is doubt as to how much of the ground water is within economic pumping depth: perhaps only one-third at most.

The calculations of water use roughly agreed as to total but differed as to sectoral use. As of the mid 1970s, the volume of fresh water withdrawn annually for human uses was believed to be of the order of the figures given in Table 2-1. Withdrawals for other uses were not included in these computations, and no figures were attached to in-stream uses such as for navigation, wildlife conservation, and waste dilution. The estimates were drawn from scattered national statistics or were estimated by assigning assumed per capita use to total population: The volumes for countries compiling national data were much more nearly accurate. In few cases was there a distinction between the volume of water withdrawn and the volume consumed in evaporation or transpiration and therefore not returned to the surface or underground water body. The proportion consumed has a profound effect upon subsequent availability: A power plant consuming less than 1 percent of the water it takes out of a stream for cooling has quite a different effect on stream flow than an irrigation project returning to the stream only 40 percent of the water diverted onto its fields. To arrive at the volume of water consumed would require taking 1 to 15 percent of domestic and industrial uses and 40 to 80 percent of irrigation uses.

The United Nations Conference on Desertification in 1978 focused attention on resource problems of the arid and semiarid third of the continental land areas.[12] More details thereby were provided on the water-land situation in dry areas and on management strategies appropriate to the high variability and prolonged droughts in those regions. One of the estimates produced at that time was the Swedish report on water in dry lands, which carried forward the calculations of water availability per unit of land area.[13]

At the end of 1980 the *Global 2000 Report* in the United States built upon the previously published data to make projections of water availability by country for all purposes from 1971 or 1975 to 2000. The projections were made by taking the L'vovich calculations of the amount of annual surface runoff "plus ground water flows" for each country and dividing this by 1971 populations.[14] The resulting per capita figures then were reduced in volume for the year 2000 in exact proportion to the projected increase in population made under the *Global 2000* population model.[15]

From this and from other GNP and resource projections it was concluded that "increases of at least 200 to 300 percent in world water withdrawals are expected over the period 1975–2000."[16] As indicated in two of its maps showing per capita availability, the report asserted that "population growth alone will cause demands for water to at least double relative to 1971 in nearly half the countries of the world. Still greater increases would be needed to improve standards of living."[17] Referring to Food and Agriculture Organization's (FAO) 1977 estimates, it noted that by far the largest absolute increase in demand would be for irrigation. The study stressed the problems of meeting the needs for potable water and of using irrigation water effectively without causing soil and water deterioration.

A principal recommendation of the United Nations Water Conference was that the secretary general should "make appropriate arrangements for organizing meetings of existing international river commissions, with a view to developing a dialogue between the different organizations on potential ways of promoting the exchange of their experience."[18] The first such meeting took place in Dakar in May 1981 and examined a wide range of legal, institutional, cooperative, and economic topics associated with river basin development.[19] In these discussions the growing complexity of carrying out effective water planning was recognized, and the fruits of exchanging experience were specified, but the comparison of total supply and projected use did not play an important role.

When United Nations Environment Programme (UNEP) undertook its review of changes in the world environment during 1972–1982, it did not attempt any new projections of water use. Instead, *World Environment* examined the principal ways in which the workings of the hydrological cycles were known to have been affected by human activities in the decade following the Stockholm Conference on the Human Environment, and identified trends that either enhanced or degraded the supplies of surface water and ground water. The emphasis thus shifted away from total supply and total withdrawal and consumption over large areas toward the effects that changes in land and water use in particular areas were exercising on water quality and water distribution.

My thinking about the highly significant aspects of water adequacy on the world scene has been influenced by having shared in that UNEP review. It revealed numerous points at which gaps in scientific knowledge or basic monitoring inhibit wise resource management. It also showed how irrelevant any broad generalizations about water supply and use over large areas may be to understanding what, in fact, is happening to water resources as they affect human well-being. If these are to be meaningful, the descriptions of those features need to be stated in terms of the geographic realities of combinations of physical and biological factors in local areas, and with an eye to the effects of technology, social organization, and economic productivity upon the uses made of the physical resource.

SIX MAJOR SHIFTS IN THINKING

After reviewing advances in evaluating water resources and in marshalling and forecasting techniques, the *World Environment* report presented the global water balance in terms of annual surface runoff and the ground water situation, noted efforts to augment supplies, and commented on changes in water use in major sectors. Further attention was given to alterations in water quality and aquatic ecosystems and to developments in water planning and legislation. It attempted finally to sum up the 19 major changes of the decade as falling on either the positive or negative side. Its concluding observation was as follows:

Efforts toward international co-operation accelerated during the 1970s, but the record was not one of unmixed improvement. In some places and in some ways the water resource was worse off in 1980 than when the Stockholm Conference called for a reversal of the then prevailing trends. Although water quality and accessibility improved in some regions, the absolute numbers of people without access to safe water grew. A rough balance of gains and losses suggests that the trends were generally—albeit slightly—favorable to the sustained use of the basic resource (p. 161).[20]

The trends and changes reported there were chiefly in information techniques, social organization, and the quality of the resource. In part implicit in them and in part independent of them were the shifts in thinking that promise to have large impact upon the way in which water is managed in future. I have selected six of those.

Drinking Water and Sanitation for All

The task taken on as a goal in 1980 of providing safe drinking water and sanitation for all the human family marked a new recognition of public responsibility to meet those needs and called for fresh approaches. It was immediately clear that a continuation of the level of improvement activity prevailing in the 1970s would increase the absolute numbers served with potable water but would leave the growing world population relatively no better off. A different type or magnitude of effort was needed.

Beginning on a revised program with a comparison of water availability and increasing use reveals nothing of significance. The physical supply-use ratio per capita is unimportant in most areas; other factors govern the reliability and safety of supply. Somewhere in the neighborhood of 150 to 250 billion m^3 of water is withdrawn for domestic purposes in a year. The mean annual flow of all of Italy's rivers is approximately the same magnitude.[21] Withdrawals for domestic purposes are less than 1 percent of the total flow of all continental rivers. This does not subtract the heavy drafts made on ground water to supply numerous urban places.

Obviously, not all the water occurs where and when it is needed, and the availability is set basically by the social cost of transporting and storing water so as to deliver it at the needed place and time in a quality suitable for drinking. Only about 43 percent of the population of developing nations were reported to have had reasonable access to safe supplies in 1980.[22] Physical supply is not the prime limiting factor in these areas. Nor is cost necessarily the chief constraint in many instances. The experience with programs to speed up the provision of improved supplies suggests that the rate and the extent of improvement are affected by combinations of cost, financial capacity, trained personnel, and community organization that vary from country to country. The challenge is to find methods of mobilizing those resources appropriate for the particular natural and social environment of the country.

The populations of the industrialized nations are generally provided with potable supplies, but two conditions throw some doubt on their adequacy. The two conditions relate to the state of older systems and to the possible health effects of increasing loads of organic chemicals in surface and ground sources. In the industrialized countries many of the water and sewer facilities have deteriorated so that heavy expenditures are required for rehabilitation, quite aside from investment to extend storage or conveyance facilities to meet the needs of enlarged populations.

At the same time serious questions have been raised as to the possible public health effects of new chemical compounds, chiefly organic compounds, reaching surface and ground sources of supply for municipal users.[23] These leave uncertainty as to whether or not the conventional methods of treatment will be adequate in future to meet health standards. In neither case is the volume of supply a major constraint, although here and there the technical and economic problems of enlarging supply are serious.

Increased Supply and the Alternatives

Where available supplies are judged inadequate to meet current or prospective demands, the next step usually has been to explore the possibility of enlarging the supply, and here the illusions that may work against sound long-term solutions are that there are readily available technical means, especially new techniques, of increasing supply, and that management of demand is of secondary significance.

The need for seeking to enlarge supply may spring from degradation of the quality of supply, from competitive use of surface and ground supplies, or from pumping aquifers at a rate greater than recharge. As ground water mining has expanded in some areas, pumping levels have increased, and it is accurate to speak of exhaustion, oftentimes irreversible. The past decade saw heightened public concern with such depletion near a few cities and irrigation areas.

Among the unconventional technological fixes attracting attention in recent decades as a remedy for threatened water shortage have been cloud seeding, desalination, and towing icebergs. Weather modification efforts to augment precipitation after more than thirty years of trial still have not been shown to have consistently positive results, although their advocates continue to find them promising in a few areas. Desalting operations have grown only modestly. They have proven economically feasible principally in arid regions where heavy investment seems warranted for supplementing existing urban and industrial supplies or, as in the case of United States delivery of Colorado River water for irrigation in Mexico, where it is the most feasible means to meet a political commitment. Icebergs are still in the realm of science fiction. There should be no hesitation in exploiting these unconventional measures whenever they might help. However, they have been all too often an excuse to pass over other means of coping with rising demand.

In favorable circumstances such new technology may offer a promising local solution, but two sets of advanced techniques of greater impact and broader application are those relating to ground water exploration and extraction and to reuse of water. Improved seismic and geological surveys, well-drilling, and pumping methods are opening up a huge volume of water previously ignored or inaccessible. This has been of major importance in developing countries, which could gain access to previously untapped supplies without building elaborate storage and conveyance works. Technical assistance has played an influential role in diffusing the new methods.[24] Rising energy costs, in contrast, have inhibited some new pumping projects.

As a result of advances in treatment methods and in system planning, the reuse of water is beginning to be viewed as a practical measure in both urban and agricultural settings. With techniques that meet health requirements and are otherwise publicly acceptable, the diversion of a stream may lead to a sequence of uses without seeking a new source for each use.

A much wider range of previously tested alternatives is receiving appraisal in the search for adequate water. Pricing as a device for managing demand is increasingly recognized as a means of holding withdrawals to economically warranted levels.[25] Methods of locating leaks in urban water systems may identify losses claiming as much as one-half the total withdrawal. Water-conserving devices for households and for lawn watering are reducing per capita use by as much as 20 to 40 percent in some cities.[26] Water conservation in irrigation encompasses canal lining, water application scheduling, drip irrigation, and choice of water-efficient crops.

Reduction of water use in industrial processes has been pursued with fresh vigor in response to increased water costs and public concern with stream pollution. Supply is rarely a limiting factor for industrial development. In a few cases water supply dictates choice of a site, but more often it is obtained as necessary to serve the site favored by other factors such as transport, raw materials, and labor. Industrial location may, however, be strongly influenced by considerations of waste disposal, and the questions of how much and what kind of burden

the effluent places on receiving water may be pivotal in the design of manufacturing processes and in setting the required withdrawal.

One of the profound changes in thinking about water adequacy during the past decade has to do with industrial waste disposal. For many years the prevailing approach to effluents from manufacturing was to discharge them into water bodies capable of assimilating them without undue harm downstream or to treat the waste sufficiently to meet some minimum standard of ambient quality in the receiving waters. The cost of maintaining water quality was seen as the construction and maintenance of waste treatment works. According to this view, projections of water use could assume that industrial demand would increase as rapidly or several times as rapidly as the population. Slowly, more emphasis was placed on altering the plant production process so as to reduce the volume of waste and of water used for cooling and material transport. Less waste thereby required dilution or treatment, and the plants withdrew less water. Reports from France, Japan, the United States, and the Soviet Union show decreasing use of water per capita for industry.[27]

Population Growth and Industrial and Agricultural Water Use

The ideal of minimal water consumption and minimal production of waste material is far from achievement in many areas, but as it is sought in new housing and industrial installations and in rehabilitation of old facilities, per capita use may decrease rather than increase. In these circumstances the notion that municipal and industrial water use will mount in direct proportion to population becomes illusory. The second United States national water assessment projected a decrease of as much as 60 percent in industrial withdrawals in the 25 years after 1975.[28]

In a strict sense the availability of surface or ground water sets the outside limits for agricultural development in the roughly 13 percent of cultivated lands dependent upon irrigation, just as precipitation is a major limit to cultivation in more humid lands. In a relatively few semiarid and arid areas such as the Indus and Rio Grande basins, the water supply is already fully utilized for irrigation. Elsewhere, the prospect for agricultural development, where sparse and variable precipitation is a significant constraint, rests more heavily in the short run upon making full and effective use of available water supply than upon expansion of basic irrigation sources.

The indicative plans by the FAO for agriculture to 2000 give one assessment of the part that irrigation improvements are expected to play in enhancing the capacity of the growing population in 90 developing countries to feed themselves at a minimal level of nutrition.[29] One of that study's conservative projections estimated that only about 28 percent of the desired increase in agricultural output would come from expansion of cropped area and that irrigation would be a major factor in increasing crop yields and cropping intensity. It judged that about one-half of the land in irrigation schemes in those countries is fully irrigated. Some of the unused lands will never be suitable for irrigation, but others may become highly productive. And new lands may be developed. In Egypt, which depends almost exclusively upon irrigation for its agriculture, there is proposed a modest expansion in cropped area, but the primary increase in production is planned to derive from intensification of cropping in lands now watered from the Nile. The emphasis is less upon more water than upon making efficient use of water now within call.

Although there are no thorough statistics on the state of irrigation and drainage on lands already irrigated, the scattered and somewhat anecdotal evidence in hand suggests that a very substantial proportion of those lands are subject to different degrees of deterioration due to excessive water applications or inadequate drainage. Water logging, salinization, and alkalinization are widespread, but causes and remedies are known in many areas.[30] In those circumstances the critical needs are for improved management and for applying lessons from the presently cultivated lands to design and management of new projects.

The opportunities for conservation of agricultural water through improved conveyance and farm distribution systems, application methods, scheduling, crop selection, and cropping practices are large. So, too, are the opportunities to halt and reverse the degradation of irrigated lands beset by waterlogging and salinization. Such efforts are far less dramatic than the construction of a huge storage reservoir or a network of canals on a parched landscape. Yet they offer solid economic returns in numerous projects and would warrant being undertaken regardless of the physical limits of supply.

Flood, Drought, and Physical Remedies

The occurrence of major floods or widespread droughts sometimes is cited as evidence that water resources are out of adjustment to human needs. Floods are seen as indications of accelerated runoff rendering the supply less accessible in time, and droughts as evidence of dwindling water supplies. To varying degrees this is true, but often the disastrous social consequences are the products of changes in land and water use rather than of alterations in physical supply, and ideas are changing as to how to best cope with the threat of such disaster.

Flood frequency in numerous urban areas undoubtedly has increased as a result of buildings, paving, and channel encroachment. Likewise, in some mountain terrain the devastation of torrents and mud flows has expanded in the trail of deforestation and unsuitable cropping practices. In the greater number of areas reporting expanding flood disasters, the growth in damages is to be attributed to human invasion of vulnerable areas. To some extent the intensified use of floodplains may be warranted by the prospective economic returns, but elsewhere the remedy may be in revised land use, floodproofing of buildings, insurance, or improved warnings rather than water control.

Drought disasters follow in the train of prolonged periods of unusually dry weather in areas where the pattern of land use and agricultural production is ill prepared for such dry periods, and as those patterns shift, the magnitude of human suffering may grow. Thus far there is not evidence that droughts themselves have increased in frequency. The appropriate preventive measures are seen as including such activities as land use regulation, limits on herd size, provision of alternative employment, and market improvement, as well as water conservation in suitable areas.

The accumulating experience with damaging floods and droughts suggests a basic observation about the course of water management. Reliance upon technological solutions to the exclusion of complementary changes in social and economic institutions and process may be counterproductive. A flood control structure may lead to increased flood damages if land use in vulnerable areas is not managed at the same time. Providing more water wells to a drought-stricken pastureland may enlarge the destruction of both herds and vegetation. Thinking is moving away from primary reliance upon water control toward a combination of water development, land use management, and economic and social adjustments. When dry periods come, water is not seen as the panacea.

In recent years the spectacular alterations in river regime effected by large artificial lakes have probably attracted more popular attention than any other type of water management. However, the interest often has focused on a few detrimental or allegedly detrimental consequences for human health or agricultural productivity and has neglected some less dramatic but possibly more far-reaching consequences of water control activities for terrestrial and aquatic ecosystems.

In Western Europe, Australia, and North America the era of building large storage works and canals for power, irrigation, and related uses has nearly run its course; a large proportion of favorable dam sites and of unregulated flows have been claimed. A few promising sites remain. In certain regions of Africa, Asia, and South America the physical conditions exist for large new developments. Over the past two decades a series of lessons have been learned the hard way as to how to plan and carry out a large water impoundment without provoking serious side effects.[31] While some of those lessons are still not followed adequately, the glaring social and environmental costs are recognized and minimized in many new projects. Net benefits from power production and irrigation thereby are increased.

On the other hand, unfortunately, little attention has been paid to the maintenance of aquatic ecosystems in the path of water diversion, water storage, channel works, and inputs of fertilizer, pesticides, and other waste flows from non–point sources. The effects of altered stream flow patterns upon aquatic life are still not well understood; desirable in-channel flows are difficult to establish. The contributions of non–point pollutants to river pollution are now seen along many reaches to be more difficult than waste from point sources of city or factory.

The most dramatic of these changes in Western Europe and eastern North America are in acidification of streams and lakes attributed to sulfur and nitrogen in precipitation.[32] While the processes of atmospheric transport and deposition in water bodies and soils are in controversy, it is apparent that this type of change is widespread.[33] The concept of availability of water for human use is being revised to recognize that the quality of water may be influenced by land uses in nearby or distant areas and that these may be more troublesome to handle than simple deficiencies in physical supply.

Speculation on Climate Change and Its Effects

Amid the welter of speculation that has spread during the past decade on the likelihood of climate change, the possible consequences for water resources have received their share of attention.[34] *Global 2000* used an array of opinions solicited from competent meteorologists.[35] The speculation is in two stages. The first asks what shifts in climate pattern might be expected from anticipated changes in atmospheric CO_2, ozone, sulfur, and trace elements generated by human activity. The second, to the extent those alterations in temperature, pre-

cipitation, and other parameters are indeed realized, inquires into what difference it would make to water resources distribution in time and place.[36]

Because of the dilemma attaching to the first stage of examination, it is difficult to judge how seriously the hypothetical future climate should be considered in looking into future water availability. If the change is sufficiently large to be measured by normal observations of temperature and precipitation, it is likely then to be far enough advanced so that efforts to reverse the causes in energy generation and forest management would at best take many years to achieve. In view of the large uncertainties water planners are cautious about taking explicit measures to cope with climatic alterations. Nevertheless, they are becoming sensitive to the possibility that, if and when identified, the effects on distribution of water, temperature, and evapotranspiration in place and time might trigger severe reactions in economy and society.

The speculation, while causing anxieties in the next few years or decades, may turn out to foster readjustments in water management that would be beneficial in any event. It may be that the same kinds of measures—such as provision for additional carryover storage or the encouragement of water conservation—that would be desirable in making current water uses less vulnerable to the vagaries of weather would be those that would be sought were climate change to render a region more arid. Some scientists think shifts in climate may be observed within a few years or a decade; others believe the processes are so complex it may be much longer. Whatever the time scale, this consideration now is entering into water-planning activity. Its pursuit is likely to have a large influence in elevating the attention and the priority given to water conservation measures.

NEW ORIENTATIONS

In view of these shifts in thinking the type of estimate of water availability presented in *Global 2000* for 1971 and 2000 may have little relevance to the actual situation and, indeed, may promote action in unproductive directions. For example, the *Global 2000* calculations of "water availability per capita" for 1971 and again for 2000 show Egypt and Saudi Arabia in the class of nations having the least amounts and assert that "the greatest pressure will be on those countries with low per capita water availability and high population growth, especially in parts of Africa, South Asia, the Middle East, and Latin America."[37] In both countries water is not likely to be the limiting factor in food production in the next few years: Suitable land, agricultural services, capital, and other factors may be determinative.

Global 2000 recognizes that population growth is only one relatively crude measure among many important factors that could affect a country's water situation, but it concludes that "population growth will be the single most significant cause of increased future demand."[38] As population grows, to be sure, the aggregate demand enlarges. Yet for reasons given previously, this need not imply that aggregate withdrawal will mount in the same degree.

The social complications of managing water do seem likely to increase. It is revealing that in the United States, where water science and technology are relatively sophisticated, a recent review by the National Research Council of the outlook for science and technology affecting water development did not dwell on issues as to the volume of potential supply.[39] After pointing out the need for future research on the process of deterioration in ground water quality and on the response of aquatic ecosystems to toxic inputs and non–point pollution, the National Research Council group directed urgent attention to the importance of exploring and appraising innovative ways of organizing water management at the subnational level.

One of the observations emerging strongly in *The World Environment* was that the political stability of nations in the international economic system was a basic and often determinative factor in the application of scientific and technical knowledge to environmental management. In place after place the primary, guiding constraint on use of water for the public good is not information on stream flow or ability to pump ground water or knowledge of soil-water-plant relations, although there are and will continue to be glaring gaps in that knowledge. The practical use of what already is known and the wily search for what it would be most important to learn in addition about water are dependent upon ability to marshal people and ideas in a consistent, effective fashion.

To learn how adequate the water resource is for a present or future society, one must look into how the physically available waters are being used and what physical, social, economic, and political constraints apply to further use. In some places the physical limits are severe, and changes in technol-

ogy, as with central-pivot irrigation, can alter the estimates drastically. More frequently, the crucial factors are social, economic, and political. The capacity of a society to manage demand, to ensure continuity in management policy, and to place values on environmental effects and the maintenance of the quality of surface water and ground water is more likely to influence the course of development.

CONCLUSIONS

There is bound to be continuing appeal in attempts to sum up the world water prospect in terms of the total volume of water available for the globe as a whole or for major regions as related to growing human uses. The arithmetic of fixed supply and growing demand is simple and therefore attractive. Yet understanding of the current situation or skill in forecasting the future will be advanced only when the calculations are based on the realities of the available alternatives for changing the location and form of use in relation to supply.

The contrast in approach is exemplified by the manner in which the *Global 2000* and *The World Environment* deal with water. Both canvas the basic evidence as to water supply and use, and both discuss some of the difficulties in promoting economic use and environmental quality. *Global 2000* emphasizes the ratio between physical supply and growing use as a curb on development. *The World Environment* report dwells on problems and trends in management of the supply. One finds grounds for concern in the diminishing volume of water per capita. The other derives cautious hope from improved methods of management.

Technicians and administrators looking to the future will do well to canvas the basic hydrologic facts as far as those can be ascertained, but their efforts will be prescient if they give equally discerning examination to the social and economic conditions within which fundamental choices as to technology and water management strategies are made. The likelihood of the world running out of water for sustaining its life is zero; the likelihood grows of its grossly mismanaging its water resource unless the proper political and technological decisions are made.

NOTES

1. U.S. Council on Environmental Quality and the Department of State, Gerold O. Barney (director), 1980, *The global 2000 report to the president: Entering the twenty-first century,* vols. I–III (Washington, D.C.: U.S. Government Printing Office).

2. M. W. Holdgate, M. Kassas, and G. F. White, eds., 1982, *The world environment 1972–1982: A report by the United Nations Environment Programme* (Dublin: Tycooly International, for the United Nations Environment Programme).

3. United Nations Department of Economic Affairs, 1950, *Water resources*, vol. IV, Proceedings of the United Nations Scientific Conference on Conservation and Utilization of Resources, August 17–September 6, 1949, Lake Success, N.Y.

4. U.S. Government, 1967, *Water for peace*, 8 vols., International Conference on Water for Peace, May 23–31, 1967 (Washington, D.C.: U.S. Government Printing Office).

5. C. A. Doxiadis, 1967, Water and environment, in *Water for peace*, vol. 1, 33–60.

6. M. I. L'vovich, 1974, *Global water resources and their future* (Moscow: Micl, in Russian).

7. USSR National Committee for the International Hydrological Decade, 1974, *World water balance and water resources of the earth* (Leningrad: Hydrometeoizdat, in Russian).

8. A. Baumgartner and E. Reichel, 1975, *The world water balance: Mean annual global, continental and maritime precipitation, evaporation and runoff* (Amsterdam: Elsevier Scientific).

9. United Nations, 1972, *Water development and management*, vols. 1–4, Proceedings of the United Nations Water Conference, March 1977, Mar del Plata, Argentina (Oxford: Pergamon (for the United Nations).

10. Ibid., vol. 1, 14.

11. Ibid., 5.

12. United Nations, 1977, *Desertification: Its causes and consequences*, Proceedings of the United Nations Conference on Desertification (Oxford: Pergamon Press).

13. M. Falkenmark and G. Lindh, 1976, *Water for a starving world* (Boulder, Colo.: Westview Press).

14. *Global 2000*, vol. II, 152–3.

15. Ibid., 7–28.

16. Ibid., vol. I, 26.

17. Ibid.

18. United Nations, 1977, *Report of the United Nations water conference* (New York: United Nations E.77.II.A.12).

19. United Nations, 1982, *Report of the interregional meeting of international river organizations, Dakar, Senegal, 5–14 May 1981*. (New York: United Nations).

20. Holdgate, Kassas, and White, *World Environment, 1972–1982*.

21. F. Van der Leeden, 1975, *Water resources of the world: Selected statistics* (Port Washington, N.Y.: Water Information Center) 2.

22. United Nations, 1980, *International drinking water supply and sanitation decade; present situation and prospects* (New York: United Nations A/35/367).

23. U.S. Council on Environmental Quality, 1981, *Contamination of ground water by toxic organic chemicals*, (Washington, D.C.: U.S. Government Printing Office).

24. United Nations Department of Technical Co-operation for Development, 1979, *A review of the United Nations ground-water exploration and development programme in the developing countries, 1962–1977*, Natural Resources/Water Series, no. 7 (New York: United Nations 79.II.A.4).

25. United Nations Department of Technical Co-operation for Development, 1980, *Efficiency and distributional equity in the use and treatment of water: Guidelines for pricing and regulations*, Natural Resources/Water Series, no. 8 (New York: United Nations 80.II.A.11).

26. D. D. Baumann, et al., 1979, *The role of conservation in water supply planning*, prepared for United States Army Corps of Engineers Institute for Water Resources, National Technical Information Service, Springfield, Va. U.S. General Accounting Office, 1979, *Water resources and the nation's water supply: Issues and concerns*. (Washington, D.C.: U.S. Government Printing Office, Report CED–79–69).

27. Holdgate, Kassas, and White, *World Environment, 1972–1982*, 415–428.

28. Water Resources Council, 1979, *Second national water assessment*. (Washington, D.C.: U.S. Government Printing Office).

29. Food and Agriculture Organization of the United Nations, 1979, *Agriculture: Toward 2000*. Proceedings of the Conference, Twentieth Session, November 10–29, 1979, Rome.

30. FAO UNESCO, 1973, *Irrigation, drainage and salinity: An international source book* (Paris: UNESCO; London: Hutchinson).

31. W. C. Ackerman et al., eds., 1973, *Man-made lakes: Their problems and environmental effects*. (Washington, D.C.: American Geophysical Union).

32. Holdgate, Kassas, and White, *World Environment, 142–144*.

33. Committee on the Atmosphere and the Biosphere, 1981, *Atmosphere-biosphere interactions: Toward a better understanding of the ecological consequences of fossil fuel combustion* (Washington, D.C.: National Academy Press).

34. Panel on Water and Climate, Geophysics Study Committee, 1977, *Climate, climatic change, and water supply*, Studies in Geophysics (Washington, D.C.: National Academy of Sciences).

35. *Global 2000*, vol. II, p. 51–65.

36. For example, see: Kellogg, W. W. and Schware, R., 1981, *Climate Change and Society: Consequences of Increasing Atmospheric Carbon Dioxide*. Westview Press, Boulder, CO.

37. *Global 2000*, vol. II, 153.

38. Ibid.

39. National Research Council, 1982, *Outlook for science and technology: The next five years* (San Francisco: Freeman), 255–285.

PART II

WATER IN THE ENVIRONMENT

Availability of water is controlled by the hydrologic cycle. The *hydrologic cycle* is a natural distillation process powered by solar energy. Water evaporates directly or transpires from plant surfaces; the resulting vapor is transported to the atmosphere and returns to the earth as precipitation. Precipitation that falls on the ocean starts the cycle again; precipitation that reaches land can infiltrate the ground, accumulate on the surface, or move on the surface as runoff to streams and eventually the sea.

Surface accumulations of water ultimately evaporate, returning the water as vapor to the atmosphere. Infiltrating water enters openings in the soil or rock until the pores are saturated. This reservoir of accumulated ground water slowly feeds springs and streams. Part of the infiltrating water is taken up by plant roots and, moving through the plant, is eventually released to the atmosphere from the leaves through transpiration during photosynthesis. Runoff carries the dissolved and detrital results of rock weathering. The many millions of years of geologic history can be followed through the varying cycles of erosion, transportation, deposition, lithification, uplift, and erosion again.

In Chapter 3, "The World Water Budget," David H. Speidel and Allen F. Agnew point out that the amounts of water that occur in various places and in various forms are not known with any great precision. Even with these uncertainties, general trends can be established and wet and dry locations identified.

One way of examining the behavior of the water cycle for a particular place is to construct a water budget diagram. An idealized water budget is shown in Figure II-1, with a yearly plot of precipitation and evapotranspiration. There is no difference between water that is evaporated directly from the surface and water that is transpired through the leaves of plants, so the sum is indicated as evapotranspiration. The amount of transpiration that is plotted is the *potential* evaporation, that is, the amount of water that would be evaporated if sufficient water were available. This is not always the case. Early in the year precipitation (curve 1) exceeds potential evapotranspiration (curve 2), and the actual amount evaporated is equal to curve 2. Thus the area between the curves, region *A*, indicates those times when water is available for runoff to streams and can accumulate in downstream reservoirs. It also is the time when the precipitation can occur as snow, accumulating in packs to be released later in the year during melting.

Water can also percolate below the surface into the soil. This soil moisture can be considered as a savings account for the water budget. It is not a typical savings account in that there is a limit to soil moisture capacity. Instead, when that limit is reached, the remaining water percolates deeper and recharges ground water.

During the time period indicated by region *B*, the demand for evapotranspiration exceeds the amount of precipitation available. What happens? The savings account of soil moisture is used

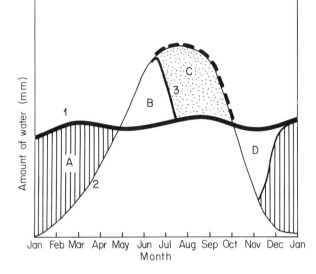

Figure II-1. An idealized water budget. The year starts in January. Precipitation is indicated by curve 1, and potential evapotranspiration is shown by curve 2. The striped area of region *A* represents times of water budget surplus, and the stippled area of region *C* represents times of water budget deficit. Region *B* represents times when soil moisture is used to balance demand, and region *D* represents times when soil moisture is recharged by the excess of precipitation over evapotranspiration. Compare the idealized water budget with the real ones shown in Figure 4-4.

to make up the difference between the "bill" of evapotranspiration and the "cash available" of the precipitation. The boundary between regions *B* and *C,* curve 3, defines the time of year when soil moisture is no longer available to make up this difference.

Region *C* shows that even though there is significant precipitation during that time, there is a water shortage because of the very high evapotranspiration. In region *A,* for example, no irrigation is needed, but by the time of region *C* the amount of irrigation needed to bring demand and availability of water into balance is indicated by the difference between curves 1 and 2. The actual evapotranspiration (AE) follows the potential evaporation demand unless there is insufficient moisture to satisfy the demand. Thus the AE follows curve 2, bounding regions *A* and *D*. The actual evapotranspiration continues to follow curve 2 along region *B* until the soil moisture bank is depleted. As the depletion occurs, the AE will leave curve 2 and follow curve 3 to curve 1, the precipitation curve.

Finally, in region *D* precipitation once again exceeds the potential evapotranspiration. The soil moisture savings account begins to be replenished. By the time this account is once again balanced, water is available for runoff and ground water recharge.

We have already mentioned that generalizations can not tell us what is happening in a particular region. Chapter 4, "Water Supply: The Hydrologic Cycle," by the Office of Technology Assessment of the U.S. Congress, discusses the local variations of the hydrologic cycle and water budgets in the western United States. Figure 4-4 illustrates the water budgets for selected stations in the western United States. It is instructive to compare the idealized version of Figure II-1 with the real examples.

Together, the Great Lakes of North America are the largest body of fresh water in the world,

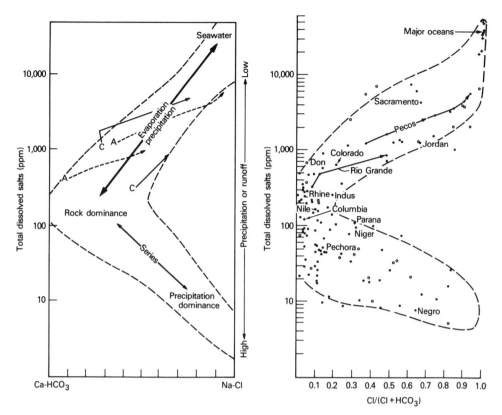

Figure II-2. Mechanisms controlling salinity and chemical compositions of world surface water; variation of salinity and Cl/(Cl + HCO₃). (*Source:* R. J. Gibbs, 1970, Mechanisms controlling world water chemistry, *Science,* 170:1088–1090. Copyright © 1970 by the AAAS.)

accounting for a full 20 percent of the total surface water. In Chapter 5, "The Great Lakes Rediscovered," Robert A. Ragotzkie gives an indication of the scientific and technological problems of trying to study such bodies of water.

How does water get into lakes and streams? We mentioned that after precipitation water can return to the atmosphere through evapotranspiration, infiltrate into the ground, or run off over the surface. Chapter 6, "Description of Runoff Processes," by Thomas Dunne, discusses the conditions under which overland flow and subsurface flow take place.

Water can also percolate downward through voids and cracks in soil and rock until it reaches a level at which all the available spaces are saturated. The top of this zone of saturation is the *water table,* and the water in the zone is commonly referred to as *ground water.* Chapter 7, "Ground Water," by Ralph C. Heath, provides a general discussion of ground water, including the variation in host rock to be expected and the variation in physical properties. He further discusses the major ground water regions of the United States.

The chemistry of river water can be described as influenced in one of three ways: by rock control, by precipitation control, or by evaporation and crystallization control. All are a result of the interaction of topographic relief and climate. In rock-dominated areas the waters are in

Table II-1. Comparison of Water Chemistry

	Rain Water[a]		River Water	Ground Water[b]			Ocean Water
	A	B		A	B	C	
Ca/Na (weight)	0.34	8.3	2.2	10	4.9	5.2	0.038
HCO₃/Cl (weight)	1.3	16.2	7.0	48	32	24	0.0074
Salinity (ppm)	12.5	3.7	130	420	430	470	34,500

Source: D. H. Speidel and A. F. Agnew, 1982, *The Natural Geochemistry of Our Environment* (Boulder, Colo: Westview Press), Table 3-4.
[a]Rain water: A = analysis from California; B = analysis from Wyoming.
[b]Ground Water: A = carbonate host rock; B = sandstone host rock; C = unconsolidated sands and gravels.

partial equilibrium with the materials in their basins and have high (Ca–HCO₃)/(Na–Cl) ratios, as shown in Figure II-2. With increased rainfall the dilution caused by increased runoff decreases the salinity. Also, with increased rainfall the elements added thereby begin to dominate; that is, the proportions of Na and Cl increase relative to those of Ca and HCO₃. The other major mechanisms of concentration of Na and Cl are evaporation, which increases salinity by removing water, and crystallization (precipitation) of CaCO₃ from solution, which increases the Na–Cl proportions. It is interesting to note that high Na–Cl proportions occur in warm to hot climates, high salinities in dry climates, and low salinities in humid climates.

Table II-1 illustrates some of the chemical differences among rain water, river water, ground water associated with different rock types, and ocean water. Wide variations in the chemistry of all water types are found. The chemistry of ground water varies with the type of host rock, a result to be expected if the ground water reacts slowly with the host material.

The sources of commonly found dissolved constituents of ground water are indicated in Table II-2. In addition, some of the effects of these dissolved constituents are indicated. Ground water is highly variable locally for total dissolved solids and for minor elements as well.

Table II-2. Sources and Effects of Dissolved Materials in Ground Water

Constituent or Physical Property	Source or Cause	Significance
Silica (SiO₂)	Dissolved from practically all rocks and soils, usually 1 to 30 ppm[a]	Forms hard scale in pipes and boilers and on blades of steam turbines
Iron (Fe)	Dissolved from most rocks and soils; also derived from iron pipes. More than 1 or 2 ppm of soluble iron in surface water usually indicates acid wastes from mine drainage or other sources	On exposure to air, iron in ground water oxidizes to reddish brown sediment. More than about 0.3 ppm stains laundry and utensils. Objectionable for food processing. Federal drinking water standards state that iron and manganese together should not exceed 0.3 ppm. Larger quantities cause unpleasant taste and favor growth of iron bacteria
Calcium (Ca) and magnesium (Mg)	Dissolved from moist soils and rocks, but especially from limestone, dolomite, and gypsum	Cause most of the hardness and scale-forming properties of water. Waters low in calcium and magnesium are desired in electroplating, tanning, dyeing, and textile manufacturing

Constituent or Physical Property	Source or Cause	Significance
Sodium (Na) and potassium (K)	Dissolved from most rocks and soils	Large amounts, in combination with chloride, give a salty taste. Sodium salts may cause foaming in steam boilers, and a high sodium ratio may limit the use of water for irrigation
Bicarbonate (HCO$_3$) and carbonate (CO$_3$)	Action of carbon dioxide in water on carbonate rocks	Produce alkilinity. Bicarbonates of calcium and magnesium decompose in steam boilers and hot water facilities to form scale and release corrosive carbon dioxide gas. In combination with calcium and magnesium they cause carbonate hardness
Sulfate (SO$_2$)	Dissolved from many rocks and soils	In water containing calcium forms hard scale in steam boilers. Federal drinking water standards recommend that the sulfate content not exceed 250 ppm
Chloride (Cl)	Dissolved from rocks and soils; present in sewage and found in large amounts in ancient brines, seawater, and industrial brines	In large amounts in combination with sodium gives salty taste. In large quantities increases the corrosiveness of water. Federal drinking water standards recommend that the chloride content not exceed 250 ppm
Dissolved solids	Chiefly mineral constituents dissolved from rocks and soils, but includes organic matter	Federal drinking water standards recommend that the dissolved solids not exceed 500 ppm. Waters containing more than 1000 ppm dissolved solids are unsuitable for many purposes
Hardness as CaCO$_3$ calcium carbonate	In most water nearly all the hardness is due to calcium and magnesium	Consumes soap before a lather will form. Deposits soap curd on bathtubs. Hard water forms scale in boilers, water heaters, and pipes. Hardness equivalent to the bicarbonate and carbonate is called *carbonate hardness*. Any hardness in excess of that is called *noncarbonate hardness*.
Acidity or alkalinity (hydrogen ion concentration, pH)	Acids, acid-generating salts, and free carbon dioxide lower pH. Carbonates, bicarbonates, hydroxides and phosphates, silicates, and borates raise pH	A pH of 7.0 indicates neutrality in a solution. Values greater than 7.0 denote increasing alkalinity; values less than 7.0 indicate increasing acidity. Corrosiveness of water generally increases with decreasing pH
Dissolved oxygen (O$_2$)	Dissolved in water from air and from oxygen given off in photosynthesis by aquatic plants	Increases the palatability of water. Under average stream conditions 4 ppm is usually necessary to maintain a varied fish fauna in good condition. For industrial uses zero dissolved oxygen is desirable to inhibit corrosion

Source: Office of Technology Assessment, 1983, *Water-related technologies for substantial agriculture in U.S. arid and semiarid lands,* (Washington, D.C.: U.S. Government Printing Office), Table 70.
[a]ppm = parts per million.

3

THE WORLD WATER BUDGET

DAVID H. SPEIDEL and ALLEN F. AGNEW

RESERVOIRS

The surface of the earth is dominated by water, the exact amount of which is difficult to measure. The range of recent estimates of these amounts is presented in Table 3-1. There is tremendous difficulty in determining the volume of the ocean basins and in estimating the amount of ground water. There is general agreement about the total area covered by water, but there is not always agreement as to which ocean a particular portion belongs to. Because ocean volumes are calculated by using an average depth multiplied by the area, variations in calculated ocean volume caused by uncertainties in the area can be huge—up to 100 percent for the Arctic in one comparison. The areas used for the world oceans are indicated in Table 3-2.

Individual seas outlined by land masses account for 5.4 percent of the volume and 13.3 percent of the area of the world ocean. No systematic comparison of the chemical behavior of elements within the individual seas with the open ocean is feasible because of wide differences in size, depth, and relationship. Clearly, seas such as the Baltic, Mediterranean, and Gulf of Mexico deserve close examination for the environmental impact of humans—and in some instances they have received such attention.

The movement of water within the oceans is easy to visualize on a large scale. Circular motions, called *gyres,* are centered about 30°N and 30°S latitude, with west-flowing equatorial currents that turn poleward at the western boundaries of the oceans. Examples of such currents are the Japan Current in the North Pacific, the Gulf Stream in

the North Atlantic, and the Brazil Current in the South Atlantic. Through the midlatitudes there is a slow movement of water eastward that is deflected toward the equator near the eastern edges of the ocean. This equatorial flow is associated with upwelling of cold bottom water, resulting in the Canaries Current and the Benguela Current in the Atlantic, and the Peru Current and the California Current in the Pacific. The Antarctic Current is not intercepted by land but continues to flow to the east.

In the Northern Hemisphere the position of the land masses causes cold Arctic waters to move southward through the Bering Strait, Denmark Strait, and Davis Strait, their currents being called, respectively, Kamchatka, Greenland, and Labrador. A profile through the ocean would show surface movement of warm water from the equator toward the poles and deep movement of cold water from the poles toward the equator.

Atmospheric Water

The amount of water in the atmosphere is estimated to be 13,000 km^3, an amount that would correspond to a thickness of 25 mm if distributed uniformly on the surface of the earth. However, the amount of water in the atmosphere is locally variable, ranging from estimates of 1.5 mm over Antarctica to 70 mm in a typhoon over Japan.

A 50-km/h wind moving in a belt 1 km wide and containing 60 mm of water would move 800 metric tons in 1 s. Because moving air masses are hundreds of kilometers across, it is evident that huge amounts of moisture are moved in this man-

D. H. Speidel is a professor of geochemistry and chairman of the Department of Geology, Queens College, City University of New York, Flushing, New York 11367. A. F. Agnew is a professor of geology, Oregon State University, Corvallis, Oregon 97330. Dr. Agnew was environmental policy specialist for the Congressional Research Service and has served as director of water research centers at Washington State University and Indiana University, state geologist of South Dakota, and geologist with the U.S. Geological Survey. This material is excerpted from Water, in *The natural geochemistry of our environment,* (Boulder, Colo.: Westview Press, 1982), Chapter 3. Reprinted by permission of the authors and Westview Press, Inc. Copyright © 1982 by Westview Press, Inc.

Table 3-1. Water Reservoirs and Fluxes

	Values (km³)	Range of Values in Recent Literature (km³)
Reservoirs		
Ocean	1,350,000,000	$1.32–1.37 \times 10^9$
Atmosphere	13,000	10,500–14,000
Land		
Rivers	1,700	1020–2120
Freshwater lakes	100,000	30,000–150,000
Inland seas, saline	105,000	85,400–125,000
Soil moisture	70,000	16,500–150,000
Ground water	8,200,000	$7–330 \times 10^6$
Ice caps/glaciers	27,500,000	$16.5–29.2 \times 10^6$
Biota	1,100	1000–50,000
Flux		
Evaporation	496,000	446,000–577,000
Ocean	425,000	383,000–505,000
Land	71,000	63,000–73,000
Precipitation	496,000	446,000–577,000
Ocean	385,000	320,000–458,000
Land	111,000	99,000–119,000
Runoff to oceans	39,700	33,500–47,000
Streams	27,000	27,000–45,000
Ground feed	12,000	0–12,000
Glacial ice	2,500	1700–4500

Source: Refs. 1, 2, 3, 4, 5, 6, 7, 8, 9, 10.

ner. For example, 311×10^6 kg/s of water crosses the Pacific coast of North America during the three autumn months and about 250×10^6 kg/s comes across the Gulf coast during the winter months (11). A yearly average of 250×10^6 kg/s of water moves across the Atlantic coast, with about 110×10^6 kg/s moving over the Arctic Ocean.

The major global wind systems are similar to the ocean currents: Trade winds move from east to west on either side of the equator, separated by doldrums where the air is rising, and midlatitude westerlies move from west to east between 30° and 70° N and S, with possible high speeds (100 m/s). Air moving toward the surface beyond 30° moves poleward; particles introduced at 30° or beyond will therefore tend to drift poleward, whereas particles introduced at the equator could drift either north or south.

Residence Time

The preceding discussion of evaporation and precipitation implies a *flux*, a transfer of material from one reservoir (the atmosphere) to another (the surface ocean). The total amount of precipitation distributed across the whole earth is estimated to be 2.7 mm/day, which suggests that in less than 10 days the atmosphere has had its water removed, or that water resides in the atmosphere for less than 10 days. Residence time is the expression used to describe the behavior of an element in a particular reservoir. It is defined as

$$\tau = \frac{Q}{dq/dt}$$

where Q is the total amount of element present in the reservoir and dq/dt is the flux into or out of the reservoir at any specific time. If the influx and outflux are not the same value, the residence time of the elements will depend on what value is chosen. If the flux changes, the residence time of all of the elements present in the reservoir changes according to this definition. It can be seen by this discussion that residence time is a very useful device for a steady-state system: influx =

Table 3-2. Geographic Distribution of the World Water Cycle

	Area (10³ × km²) A	Precipitation (km³) B	Evaporation (km³) C	Runoff (km³) D	Runoff (mm) E	Land with Interior Drainage (%) F	Precipitation on Land with Interior Drainage (%) G	Continental Runoff (km³) H	Continental Runoff (mm)	Ocean Flow Compensation (km³) I
Sea	361,110	385,000	424,700	−39,700				−39,700		
North Polar	8,509	826	452	374				2,611	(344)	2,985
Atlantic	98,013	74,626	111,085	−36,459				19,351	(310)	−17,108
Indian	77,770	81,024	100,508	−19,483				5,601	(397)	−13,882
Pacific	176,888	228,523	212,655	15,868				12,137	(194)	+28,002
Land	148,904	111,100	71,400	39,700	(266)	22.3	7.4	39,700		
Europe	10,025	6,587	3,761	2,826	(282)	17.5	12.2			
Asia	44,133	30,724	18,519	12,205	(276)	28.7	8.7			
Africa	29,785	20,743	17,334	3,409	(114)	41.0	13.3			
Australia	8,895	7,144	4,750	2,394	(269)	47.2	14.0			
North America	24,120	15,561	9,721	5,840	(242)	3.7	2.0			
South America	17,884	27,965	16,926	11,039	(618)	8.2	2.2			
Antarctica	14,062	2,376	389	1,987	(141)		6.0			
Global	510,014	496,100	496,100							

Source: Data mainly taken from Ref. 3.

Note: Runoff (column D) is measured by (Precipitation − Evaporation) (D = B − C) for volume and by (Precipitation − Evaporation)/Area (E = (B − C)/A) for mm values (column E). Columns F and G indicate the amount of runoff that is drained to the interior and column H indicates the intensity of continental runoff in those areas that have runoff to the oceans. The volume to the different oceans is also given in column H. Column I (D + H) indicates the necessary movement of oceanic water to balance the world water cycle.

outflux = constant. The problem is that steady-state conditions occur only under strictly defined and temporarily limited natural conditions.

A variety of residence times for water in the atmosphere and ocean can be calculated with differences between output and input. For example, water added to the atmosphere by evaporation from the ocean will have a residence time of $(13,000/425,000) \times 365 = 11$ days, whereas water that is precipitated from the atmosphere has a residence time of only $(13,000/496,000) \times 365 = 9.5$ days. Water in the whole ocean has a residence time of 255 years if it is assumed that all the water evaporated comes from the surface reservoir, the top 500 cm. Residence time varies as the choice of flux varies; when total precipitation (output) and total evaporation (input) are considered, the calculated residence times are approximately the same, 10 days. The short residence time of water in the atmosphere emphasizes and focuses attention on the possible movement of pollutants.

Land Reservoir

By far the greatest surface reservoir of water other than the oceans is that of ice caps and glaciers, with estimates varying from 16,500,000 km^3 to 29,200,000 km^3. More than 80 percent of the ice is in the south polar region, and another 10 percent is in Greenland. Part of the variation is due to inaccurate (or lack of) conversion of ice volume to equivalent water volume, and part to increased knowledge of the thickness of the ice packs. This amount of ice is equivalent to the runoff of all the world streams for 900 years. Glaciers can be considered a special kind of river, and ice packs can be considered a special kind of lake (8). The discharge of water from glaciers to the oceans is about 2500 km^3/year (less than 10 percent that of streams), and the residence time for water that is present as ice in Antarctica is approximately 9500 years. Because of minimal mixing of the ice, ice that is millions of years old has been identified in Antarctica and Greenland and used to show changes in chemistry of the atmosphere and oceans during past glacial ages. Residence times in valley glaciers can vary from tens of years to hundreds of years.

In addition, permanently frozen soil and ground water are present throughout much of North America and Siberia—the permafrost. Permafrost water generally does not enter into the cycle unless it is thawed, and its exact behavior is not yet well known.

Streams, lakes, and inland seas contain approximately 207,000 km^3 of water, or about 16 times as much as the atmosphere. Inland seas and saline lakes account for about half of that amount, with the Caspian Sea making up 75 percent of that half. Evidence shows that different climates in the past have caused major runoff into areas having no ocean drainage and that the Great Basin of the western United States and similar areas in Australia once had major inland seas. Because virtually all water leaves these lakes only through evaporation, they are good objects for studying the volatility of elements.

The bulk of the fresh water is concentrated in a few large lakes. Lake Baikal in Asia, the Great Lakes in North America, and the East African lakes account for almost 90 percent of the fresh lake water. Lakes, along with reservoirs, perform a stabilizing function on runoff. The total amount of water present in streams at one time is estimated to be 1700 km^3, or about 15 percent of the total instantaneous volume of water in the atmosphere. This amount is equivalent to the runoff received by the ocean in about three weeks. The variation of amount present in any particular river at different times is striking and will be addressed in later sections on rivers as an agent of flux.

Water under the surface of the ground is of three types: soil water, vadose water, and ground water. *Soil moisture* is of immediate biological concern, as it is the prime source of water for plants—although not all of it is available. Soil moisture is the linkage between precipitation and ground water. *Vadose water* lies in the zone of nonsaturation; that is, not all the pores in the earth material are filled with water. *Ground water* is defined as water in the zone of saturation, where all pores are saturated. Estimates of the amount of water in each of the three types are only rough. The depth to which soil moisture is affected by precipitation averages about 105 cm, which might be used to define the boundary between soil water and vadose water (12). On the other hand, some plant roots extend 5 m into the ground, indicating that the moisture at that level is eventually transpired to the atmosphere. The amount of soil water is estimated to be about 25,000 km^3 (8)—significantly greater than the 1700 km^3 in streams and double the 13,000 km^3 in the atmosphere. Estimates of the volume of vadose water are about 45,000 km^3. By far the largest underground reservoir is that of ground water, with estimates ranging from 7,000,000 (8) to 330,000,000 km^3 (5), including the pore water of sediments. Generally, the value for all three types of underground reservoirs is estimated at 8,100,000 to 8,400,000 km^3 and given here as 8,270,000 km^3.

Discharge of ground water to the sea has been estimated to comprise about 5 percent of the stream flow. Thus the residence time of water in ground water that discharges to the sea is approximately

4000 years. Most ground water enters stream channels within the continents and accounts for 30 percent of the stream runoff (6). This "nonflood," or stable, stream runoff is raised to 36 percent with the addition of surface water behind dams and lakes. The problem of evaluating ground water contributions to streams is discussed in a series of chapters of U.S. Geological Survey Professional Paper 813 (13). The water in streams during periods of high flow is derived from both surface runoff and ground water discharge. As stream flow decreases, the percentage of stream flow derived from direct surface runoff also decreases. The flow that is equaled or exceeded 90 percent of the time is generally assumed to be totally supplied by ground water. Seasonal variation in the base flow is assumed to be dependent on vapor loss through evapotranspiration. For example, vegetation on floodplains is situated between ground water recharge areas and stream channels and therefore intercepts the groundwater before it is discharged. The cycle of vegetation through the year controls this interception, which is more important in temperate than tropical climates. The 90 percent flow is an indication of base flow in a stream during summer, when vapor discharge is at a maximum, and the 40 percent flow duration value is an indicator during winter, when evapotranspiration is at a minimum. However, as this has nothing to do with the mean annual flow, it is difficult to construct from past data collections. For example, if the 90 percent flow (discharge of liquid is low) is 650 million gallons per day (mgd) and the 60 percent flow (vapor discharge is low) is 2460 mgd, then the seasonal variation in ground water discharge is 1810 mgd (60 percent flow minus 90 percent flow). If the 60 percent flow is a legitimate measure of ground water discharge, as proposed by the U.S. Geological Survey, it is also a measure of ground water recharge. In some areas mining of ground water is now used as a prime source instead of a supplementary source, so that both new and fossil ground waters are being added to the water used.

Land Biota

Animal tissues are largely water, and plants contain much water—the estimate used here is 1100 km³. Transpiration water cannot be distinguished from evaporation water, so there is no way to be certain of the amount of water that cycles through the biota. However, evaporation from land area is 71,400 km³, and if we assume that one-third moves through vegetation, the residence time is about $(1100/23,800) \times 365 = 17$ days, an amount almost equal to the stable stream mentioned previously.

FLUXES

River

It is important to repeat that the water budget, while balanced over the whole earth, is highly variable locally. The continents have an excess of precipitation over evaporation of 39,700 km³/year (Table 3-2). This runoff balances the deficit of the ocean reservoir but very unevenly, because the Northern Hemisphere is about 50 percent land area and the Southern is only 25 percent land area. Each continent is also different. The intensity of runoff can be calculated by dividing runoff by area, giving a depth of runoff per year per square kilometer (Table 3-2). This indicates that Africa and Antarctica are very dry (114 mm and 141 mm); South America is very moist (618 mm); and Europe (282 mm), Asia (276 mm), Australia (269 mm), and North America (242 mm) have about the same intensity of overall runoff. However, this is misleading, because a significant portion of the land (22 percent) does not drain to the ocean. Land with interior drainage is also distributed unevenly, with North America having about 5 percent, South America about 10 percent, Europe about 20 percent, Asia about 30 percent, Africa about 40 percent, and Australia about 50 percent. In addition, rainfall is distributed unevenly, with less than 8 percent of the total precipitation occurring on land having interior drainage. When the intensity for the ocean runoff areas is calculated, it shows that Australia (509 mm) and Asia (397 mm) are the continents that seem to vary the most, ranging from areas that receive practically no water to areas of high runoff. The bulk of the European interior runoff goes to the Caspian Sea.

Because runoff to the oceans is a critical parameter of the natural cycles, it is important to determine which oceans the runoff feeds and from which continents it comes (Table 3-3). About half of the total runoff goes to the Atlantic (48.7 percent), and about half of that (49.9 percent) comes from South American runoff. It is interesting to note that both the Atlantic and Pacific oceans receive about 70 percent of their inflow from their west coasts, whereas the Indian Ocean receives about 75 percent from its east coast. Table 3-4 lists the major rivers by discharge. During the year rivers have fluctuations in discharge, which affect the transport of sediment. Note that the Amazon River alone accounts for 15 percent of the total discharge of water (3767.8 km³/year) to the world oceans.

The total land area drained by the major rivers amounts to 42 percent of the continental masses. The intensity of the discharge is calculated by dividing discharge by basin area. For the major

Table 3-3. Distribution of Continental Runoff: Source of Runoff to Oceans

Distribution of Continental Runoff	Amount of Runoff (km³)	Percent to Ocean from Continent			
		North Polar	Atlantic	Pacific	Indian
Europe	2,564	1.4	98.6		
Asia	12,467	18.1	1.7	51.8	28.3
Africa	3,409		82.7		17.5
Australia	2,394			75.7	24.3
North America	5,840	5.4	61.8	32.8	
South America	11,039		87.5	12.5	
Antarctica	1,987		36.6	28.8	44.6
Total	39,700	6.6	48.7	30.6	14.1

Sources of Runoff to Oceans	Amount of Runoff Added (km³)	Percent from Continent to Ocean						
		Europe	Asia	Africa	Australia	North America	South America	Antarctica
Atlantic	19,351	13.1		14.5		18.7	49.9	2.7
Pacific	12,137		53.2		14.9	15.8	11.4	4.7
North Polar	2,611	1.4	86.6			12.0		
Indian	5,601		63.1	10.7	10.4			15.8
Total	39,700	6.5	31.4	8.6	6.0	14.7	27.8	5.0

Source: Information from Ref. 3.

river basins it is as follows: Irrawaddy, 1029 mm; Ganges, 899 mm; Mekong, 664 mm; Orinoco, 594 mm; Amazon, 534 mm; Brahmapputra, 509 mm; Fraser, 512 mm; Yangtze, 353 mm; and Congo, 340 mm. The first six are in monsoon areas, showing the inefficiency of water distribution there. The Fraser River is dependent on glacier melting.

The uneven annual distribution has not been factored into the calculation. If, for example, 50 percent of the annual runoff for the Mekong occurs within two months, during that time the intensity is increased from the annual average of 664 to almost 2000 mm. Any understanding of river water as a flux must consider these variations. Unfortunately, adequate data are not available.

Ocean-to-Ocean Movement and Residence Time

The movement of water as a function of the evaporation-precipitation differential has already been discussed. We emphasized that runoff from the land balances the excess evaporation, as indicated in Tables 3-1 and 3-2. These tables and Table 3-3 indicate that in the North Polar Ocean there is a surplus of 400 km³/year of excess precipitation, which, when coupled with Asian river runoff (75 percent), Mackenzie River runoff (11 percent), and

runoff from northern Europe (19 percent), amounts to an addition of 3000 km³ annually. In the Atlantic Ocean the water deficit of 36,500 km³ is not balanced by the river discharge (19,300 km³), so there is a deficit of 17,200 km³. This could be balanced by the excess 3000 km³ from the Arctic and 14,200 km³ from the Pacific, where there is a surplus. Although actual amounts must be known in order to determine the speed of mixing and the residence times, calculations based on the net effect should give maximum residence times.

Table 3-5 indicates the order of magnitude of residence times of water in the different parts of the world ocean. Line D indicates the net water movement compensation discussed previously, where the minimal flux provides the maximum residence time (time = amount/flux). The uncertainty of exact oceanic volumes does not change the orders of magnitude. Note that the total compensation time for water in the Indian, Atlantic, and Pacific oceans is 20,000 to 25,000 years; but it is only 3000 years for the North Polar Ocean, thus showing very rapid movement of water through the North Polar seas. The residence time, assuming that all compensation takes place within the surface reservoir of the ocean (upper 200 m), has also been calculated (line E). The rationale for using this upper zone is that the precipitation-evaporation interchange takes place

Table 3-4. General Features of the World's Biggest Rivers

River A	Length (miles) B	Basin Area (km³) C	Discharge (km³/yr) D	Intensity (mm/yr) E	Dissolved Transport, Td (t/km²/yr) F	Solid Transport, Ta (t/km²/yr) G	Ta/Td H	Amount Transported (t × 10⁶/yr) I
1. Amazon	3,915	7,049,980	3,767.8	534	46.4	79.0	1.7	290.0
2. Congo	2,716	3,690,750	1,255.9	340	11.7	13.2	1.1	47.0
3. Yangtze	3,434	1,959,375	690.8	353	NA	490.0	NA	NA
4. Mississippi- Missouri	3,860	3,221,183	556.2	173	40.0	94.0	2.3	131.0
5. Yenisei	3,100	2,597,700	550.8	212	28.0	5.1	.2	73.0
6. Mekong	2,600	810,670	538.3	664	75.0	435.0	5.8	59.0
7. Orinoco	1,283	906,500	538.2	594	52.0	91.0	1.7	50.0
8. Parana	2,406	3,102,820	493.3	159	20.0	40.0	2	56.0
9. Lena	3,636	2,424,017	475.5	196	36.0	6.3	.15	85.0
10. Brahmaputra	1,000	934,990	475.5	509	130.0	1,370.0	10.5	75.0
11. Irrawaddy		431,000	443.3	1,029	NA	700.0	NA	NA
12. Ganges	1,000	488,992	439.6	899	78.0	537.0	6.9	76.0
13. Mackenzie	2,035	1,766,380	403.7	229	39.0	65.0	1.7	
14. Ob	3,421	3,706,290	395.5	107	20.0	6.3	.3	50.0
15. Amur	2,700	1,843,044	349.9	190	10.9	13.6	1.1	20.0
16. St. Lawrence	1,560	1,010,100	322.9	320	51.0	5.0	.1	54.0
17. Indus	1,800	963,480	269.1	279	65.0	500.0	8.0	68.0
18. Zambezi	1,700	1,329,965	269.1	202	11.5	75.0	6.5	15.4
19. Volga	2,291	1,379,952	256.6	186	57.0	19.0	.3	77.0
20. Niger	2,600	1,502,200	224.3	149	9.0	60.0	6.7	10.0
21. Columbia	1,214	668,220	210.8	316	52.0	43.0	.8	34.0
22. Danube	1,777	816,990	197.4	242	75.0	84.0	1.1	60.0
23. Yukon	1,979	865,060	193.8	224	44.0	103.0	2.3	34.8
24. Fraser	850	219,632	112.4	512	NA	NA	NA	NA
25. San Francisco	1,987	652,680	107.7	165	NA	NA	NA	NA
26. Hwang-Ho (Yellow)	2,901	1,258,740	104.1	83	NA	2,150.0	NA	NA
27. Nile	4,157	2,849,000	80.7	28	5.8	37.0	6.4	10.0
28. Nelson	1,600	1,072,260	76.2	71	27.0	NA	NA	31.0
29. Murray-Darling	3,371	1,072,808	12.6	12	8.2	30.0	13.6	2.3

Note: t = metric ton. NA = not available.

Source: Columns B, C, and D from Ref. 14; Columns F, G, H, and I from Ref. 15. Column E is D/C. (Ed. note: For new information on solid transport, see also John D. Milliman and Robert H. Meade, 1983, World-wide delivery of river sediment to the oceans, *Journal of Geology* (91): 1–21.)

within it, and the river discharge is added to it. As mentioned previously, because of oceanic mixing, these values are higher than the 570 to 1250 years calculated.

The standard method of calculating residence time is to determine the stream runoff as a measure of flux (line F). This calculation (line G) illustrates the effect of the disproportionately large runoff into the Atlantic Ocean discussed previously and indicates that the introduction of elements by streams would quickly move them through the North Polar and Atlantic oceans relative to the Indian and Pacific oceans. For surface-only calculations this

disparity is even more striking, with both the Atlantic and North Polar showing less than 1000 years (line H). For water away from land in midocean it might be reasonable to calculate the residence time by the difference between evaporation and precipitation. The Atlantic loses significantly more water through evaporation than it gains through precipitation. Thus nonvolatile elements would be expected to be concentrated in the ocean and volatile elements quickly released to the atmosphere. One would expect this release of volatiles to be insignificant in the North Polar relative to water flow, minor in the Pacific, but major in the Indian and

Table 3-5. Oceanic Water Residence Times

	North Polar	Atlantic	Pacific	Indian	Total
A. Ocean volume (km^3 × 10^6)	8.85	350	695	295	1,349
B. Surface ocean volume (200 m) (km^3 × 10^6)	1.7	19.6	35.4	15.5	72.2
C. Ocean flow compensation (km^3/year)	3,000	−17,100	+28,000	−13,900	0
D. Total compensation time: A/C (year)	2,950	20,500	25,000	21,200	
E. Surface-only compensation time: B/C (year)	570	1,150	1,250	1,100	
F. Stream runoff to oceans (km^3/year)	2,600	19,400	21,100	5,600	39,700
G. Runoff residence time: A/F (year)	3,400	18,000	57,500	52,700	34,000
H. Surface runoff residence time: B/F (year)	650	1,000	2,900	2,800	1,800
I. Atmospheric cycling (precipitation minus evaporation) (km^3/year)	400	−36,500	15,900	−19,500	−39,700
J. Whole-ocean atmospheric-cycling residence time: A/I (year)	22,125	9,600	43,700	15,100	34,000
K. Atmospheric-cycling surface ocean residence time: B/I (year)	4,250	500	2,200	1,000	1,800

Source: Ref. 16.

Atlantic oceans. It is important to emphasize the predicted net movement of water from the Pacific and North Polar to the Atlantic and Indian oceans. However, this movement probably does not have a great diluting effect on the bordering seas, bays, and estuaries that appear to be the problem areas of pollution (17).

Atmosphere-Land Interface

The interchange of water between soil biota and atmosphere starts with precipitation (12, 13). The part that wets the aboveground vegetation and evaporates when the rain ceases is termed intercepted rain *(I)*; it is not available for transpiration by plants or for ground water recharge. The greater the canopy cover, the greater the amount of interception; the amount varies with tree species and with the amount of rainfall until capacity is reached. The rain that is not intercepted reaches the ground in three ways: throughfall *(Pt)*, canopy drop *(Pd)*, or stem flow *(Ps)*. Only the throughfall has the same properties as the rain above the canopy, and it can be distinguished from canopy drip only in the open. Thus the total amount of precipitation can be expressed as

$$P = I + Pt + Pd + Ps$$

Stem flow can be extremely effective, and a rainstorm may produce 500 L of water flowing down the stem and entering the soil at the base of a dominant beech tree whose branches and stems rise and have smooth surfaces—which facilitates such flow.

Finally, if the precipitation falls as snow, the effects are different. Interception is large but evaporation is low, and most of the snow either plops or drips to the ground, thus decreasing the true interception.

After reaching the ground, precipitation tends to infiltrate the soil as long as the input rate does not exceed the "infiltrability"; if it does, surface runoff is possible. The organic, upper soil layer in the forest provides a high storage capacity, and this minimizes runoff. The infiltrating water increases the amount of water in the soil pores and provides the reservoir from which plant roots extract water. The fate of the entering water depends on the soil's water retention ability and water conductivity. The total pore space of a soil consists of pores of various sizes, and the strength of the forces holding the water in the soil decreases with increasing pore size. If air and water are both present, air occupies the larger pores, and the system is called *unsaturated*. At least 10 percent of the total soil volume should be filled by air in the main root zone to provide sufficient air for the plant.

Because most of the pore space is of capillary dimension, capillary force causes a soil water potential (called soil water tension, capillary potential, or matrix potential). This potential is the force that causes the water to move. Such water movement is in the direction of negative hydraulic gradient and is proportional to such gradient by a coefficient called *permeability* or *soil water conductivity*. The coefficient is highly dependent on soil structure and the degree of water saturation of the soil.

Subsurface flow can occur in any direction be-

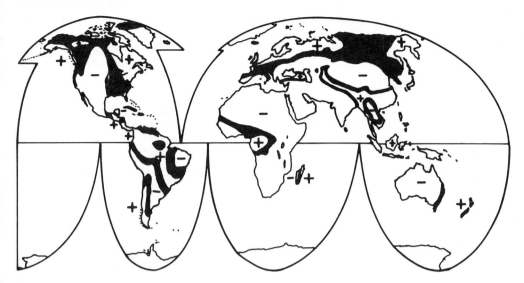

Figure 3-1. Regions of surplus (+) and deficit (−) of riverwater resources (±200 mm). (*Source:* Modified from (10).)

tween horizontal and vertical. The vertical aspect of such flow, termed *deep seepage,* is water that moves out of the root zone, is a primary soil-leaching agent, and replenishes the ground water. It is interesting to note that although the surface and subsurface flows have been considered as outputs, they can easily be inputs, either artificially (through irrigation) or naturally (downhill portion of slope).

The remaining outputs are evaporation and transpiration. Evaporation removes the water from the surface (using about 590 cal/cm^3 water), creating a hydraulic gradient that moves the moisture upward through the soil profile to the surface. Evaporation is a minor component under forest cover, partly because the low permeability of the organic layer limits upward movement despite steep vertical hydraulic gradients.

Transpiration can best be considered as a complete process that starts in the soil as soil-water potential and ends in the atmosphere as vapor pressure. Such potential deficiencies can amount to as much as hundreds of bars during sunlight hours. As long as the evaporative demand of the atmosphere can be met, the role of the plant is passive. It absorbs water in the root zone at a rate dominated by the energy available at transpiration surfaces, rather than by the plant's physiological requirements. When soil moisture is limited, plants begin to regulate the transpiration flux, generally wilting irreversibly when the soil-water potential is less than − 15 bars. The transpiration rate is a function not only of the matrix potential of the root

zone but also of the evaporation demand caused by weather conditions. Water that still exists in the soil at less than − 15 bars soil water potential—the level that supports plant transpiration—is termed *unavailable* (dead) *water.*

The atmosphere-land interface controls the water balance dynamics through the interactions of climate, soil, and vegetation discussed previously. A statistical dynamic formulation of water balance in terms of annual precipitation, potential evapotranspiration, and physical parameters of soil, vegetation, climate, and water table has recently been developed (18). Several tests indicate the applicability of the model for particular regions.

There are several generalized correlations of precipitation, evaporation, vegetation type, latitude, and runoff that guide predictions for particular basins (2, 3, 9). Maps of these variations have been prepared for UNESCO (10). The difference between precipitation and evapotranspiration is a measure of the amount of possible water runoff and thus is a measure of world regions with a surplus or deficit of river water resources. These areas are indicated in Figure 3-1. Regions with a significant surplus (annual precipitation is greater than annual evapotranspiration by more than 200 mm; indicated by +) are concentrated in Southeast Asia, the Amazon and Orinoco basins, the northeastern United States and Canada, northern Europe, northeast Asia, and the Congo Basin in Africa. Water deficit regions (precipitation is less than evapotranspiration by at least 200 mm; indicated by −) are shown to dominate Australia, southern

Asia, nearly all of Africa, the western half of North America, and large areas of South America. The shaded regions indicate areas of generalized water balance that could be sensitive to short-term climate effects.

NOTES

1. M. I. Lvovich, 1977, World water resources present and future, *Ambio,* 6 (1):13–21.

2. M. I. Lvovich, 1979, *World water resources and their future,* English translation, ed. R. L. Nace (Washington, D.C.: American Geophysical Union).

3. A. Baumgartner and E. Reichel, 1975, *The world water balance, mean annual global, continental, and maritime precipitation, evaporation, and runoff,* trans. Richard Lee (New York: Elsevier).

4. H. W. Menard and S. M. Smith, 1966, Hypsometry of ocean basin provinces, *Journal of Geophysical Research,* 71:4305–4325.

5. R. M. Garrels, and F. T. Mackenzie, 1971, *Evolution of sedimentary rocks* (New York: Norton).

6. M. I. Lvovich, 1973, The global water balance, *EOS,* 54 (1):28–42.

7. R. L. Nace 1967, Water resources: A global problem with local roots, *Environmental Science and Technology,* 1(7):550–560.

8. R. L. Nace, 1969, World water inventory and control, in *Water, Earth, and Man,* ed R. J. Chorley, (London: Methuen), 31–42.

9. USSR Committee for the International Hydrological Decade, U. I. Korzoun, ed. 1978, *World water balance and water resources of the earth* (Paris: UNESCO).

10. USSR Committee for the International Hydrological Decade, U. I. Korzoun, ed., 1978, *Atlas of world water balance* (Paris: UNESCO).

11. R. G. Barry, 1969, The world hydrological cycle, in *Water, earth, and man,* ed. R. J. Chorley (London: Methuen,) 11–29.

12. P. Benecke, 1976, Soil water relations and water exchange of forest ecosystems, in *Water and plant life,* ed. O. L. Lange, L. Kapper, and E. D. Schulze (Berlin: Springer-Verlag), 101–131.

13. R. M. Bloyd, Jr., 1974, *Summary appraisals of the nation's ground-water resources—Ohio region,* U.S. Geological Survey Professional Paper 813-A, (Washington, D.C.)

14. B. S. Browzin, 1972, Rivers, in *Encyclopedia britannica,* vol. 19 (: Britannica), 353–362.

15. M. Meybeck, 1976, Total mineral dissolved transport by world major rivers, *Hydrological Science Bulletin,* 21 (6):265–284.

16. D. H. Speidel and A. F. Agnew, 1979, The natural geochemistry of our environment, in *An overview of research in biogeochemistry and environmental health,* Committee Print 825, Committee on Science and Technology (Washington, D.C.: U.S. House of Representatives), 77–239.

17. K. T. Turekian, 1977, The fate of metals in the oceans, *Geochimica et Cosmochimica Acta,* 41:1139–1144.

18. P. S. Eagleson, 1978, Climate, Soil, and Vegetation, *Water Resources Research,* 14:705–776.

4

WATER SUPPLY:
THE HYDROLOGIC CYCLE
OFFICE OF TECHNOLOGY ASSESSMENT

The fundamental, unifying concept in the study and understanding of water is the hydrologic cycle (Figure 4-1). The cycle is the conceptual model that relates the interdependence and continuous movement of all forms of water through the vapor, liquid, and solid phases. It may be considered the central concept in hydrology.

The components of the hydrologic cycle are as follows:

- *Precipitation:* Water added to the surface of the earth from the atmosphere. It may be either liquid (e.g., rain and dew) or solid (e.g., snow, frost, and hail).
- *Evaporation:* The process by which a liquid is changed into a gas. In the context of the hydrologic cycle the most important form of evaporation is probably that which takes place from the seas and oceans. This is the main source of water on land areas.
- *Transpiration:* The process by which water vapor passes through a living plant and enters the atmosphere.
- *Infiltration:* The process whereby water soaks into or is absorbed by the surface soil layers.
- *Percolation:* The downward flow of water through soil and permeable rock formations to the water table.
- *Runoff:* The portion of precipitation that comprises the gravity movement of water in surface channels or depressions. It is a residual quantity, representing the excess of precipitation over evapotranspiration when allowance is made for storage on and beneath the ground surface.

All water involved in continuous cyclical movement according to the hydrologic cycle. Some of the water vapor in the atmosphere gives rise to precipitation through complex processes of condensation and freezing. Not all precipitation reaches the surface of the earth. Some evaporates while falling, and more importantly, some is intercepted by vegetation or artificial structures and is then returned to the atmosphere by subsequent evaporation.

The watershed, or river basin, is the fundamental geographic unit of hydrology. It is also the fundamental biophysical unit within which technologies to affect precipitation and runoff must be assessed. A watershed is a land area surrounded at its perimeter by highlands that cause precipitation falling within the watershed's bounds to flow generally toward its center to form rivers or streams. In 1970 the U.S. Water Resources Council divided the United States into geographic units based on the watershed, or river basin, for the collection and organization of hydrologic data (12) (Figure 4-2).

Water reaching the surface of the watershed follows one of three courses. First, it may remain on the surface as pools and surface moisture that eventually evaporates back into the atmosphere. Or it may be stored on the surface in the form of snow until air temperatures are high enough to allow melting and runoff. Storage as snow is a common occurrence during at least a portion of each year in much of the Western United States.

Second, precipitation reaching the ground may flow over the surface into depressions and channels to become surface runoff in the form of streams and lakes. It then moves by evaporation back into the atmosphere, by infiltration into the soil and toward the groundwater table, or by continued surface flow back into the seas.

Third, falling precipitation may infiltrate the sur-

This material is excerpted from Water supply and use in the western United States, in *Water-related technologies for substantial agriculture in U.S. arid and semiarid lands,* (1983), 48–66.

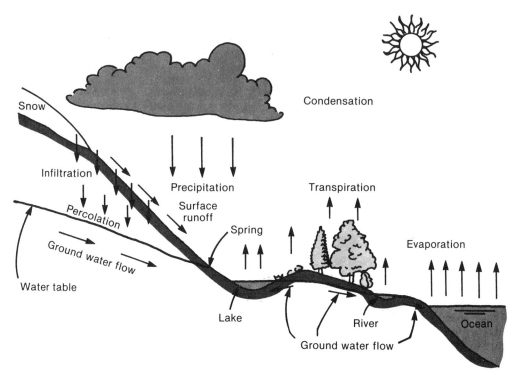

Figure 4-1. The hydrologic cycle. Water passes continuously through this cycle from evaporation from the oceans into the atmosphere through precipitation onto the continents and eventual runoff into the oceans. Human use of water may modify this cycle at virtually every point. The gross water budget of the conterminous United States is 4200 billion gallons per day (bgd) precipitation balanced by 2800 bgd evaporation and transpiration from surface water bodies, land surface, and vegetation; 300 bgd outflow to the oceans (of which 1230 is stream flow); and 1000 bgd consumptive use. *Source:* (4) and (11).

face and percolate to ground water. As ground water, it is stored for periods ranging from days to thousands of years. Ground water can be removed naturally by upward capillary movement to the soil surface and plant root zone, by ground water seepage, or by runout into surface streams, lakes, and oceans. Some of it is removed by pumping from wells, in which case it again arrives at the surface as artificial precipitation and follows one of the paths described above.

Generally acceptable estimates of the amounts of water passing annually through the various phases of the hydrologic cycle for the western United States have not been found in the literature. From estimates for the United States as a whole, however, more than 1500 million acre·ft of water are added to the western United States each year as precipitation, and the majority of this is consumed by evapotranspiration (12). Approximately 500 million acre·ft constitute the measured stream flow from the region (e.g., 5, 12), and 50 million acre·ft of water are added annually to the ground water reserves of the region.

Runoff is not uniformly distributed throughout the western United States. Stream flow to the Pacific Ocean, primarily from the Pacific Northwest region, is estimated to be over 335 million acre·ft annually, or nearly 70 percent of the total for the entire region. Almost all of the remaining surface runoff flows into the Mississippi River and ultimately into the Gulf of Mexico. In general, those areas with the lowest annual precipitation contribute runoff to rivers only during sporadic summer thunderstorms. The bulk of the runoff in the region originates from the melting mountain snowpack each spring and summer. Following snowmelt, runoff enters the river system of the region, where it is often stored in surface reservoirs until the period of peak demand in late summer.

THE COMPONENTS

The western United States has a wide range of hydrologic environments, both in terms of the absolute amount of water in the various hydrologic

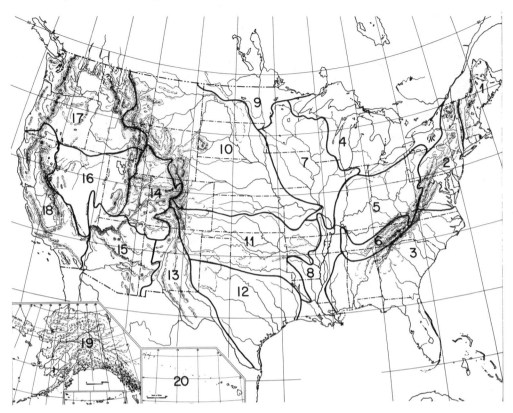

Figure 4-2. Water resources regions of the United States (major rivers and mountains are also indicated): The names of the regions are 1, New England; 2, Mid-Atlantic; 3, South Atlantic–Gulf; 4, Great Lakes; 5, Ohio; 6, Tennessee; 7, Upper Mississippi; 8, Lower Mississippi; 9, Souris-Red-Rainy; 10, Missouri; 11, Arkansas-White-Red; 12, Texas-Gulf; 13, Rio Grande; 14, Upper Colorado; 15, Lower Colorado; 16, Great Basin; 17, Pacific Northwest; 18, California; 19, Alaska; 20, Hawaii; and 21 (not illustrated), Caribbean. (*Source:* U.S. Water Resources Council and American Map Company. This diagram combines two figures from the original article.)

components and in terms of the interrelationships among the components.

Precipitation

The primary factor determining the amount of precipitation that falls over the 17 western states appears to be topography (Figures 4-2, 4-3). The four broad north-south zones are generally more uniform within themselves than are any two adjacent east-west zones. These general hydrologic zones are (1) the mountain ranges of the Pacific coast, consisting mainly of the Sierra and Cascade mountain ranges; (2) the interior basins; (3) the Rocky Mountains; and finally (4) the Great Plains, which extend from the eastern side of the Rocky Mountains to the western edge of the more humid portions of the continent, at approximately the 100th meridian, or the Missouri River.

Air masses that carry atmospheric moisture over

the region move generally onto the west coast of the continent and follow a west-to-east path. As these air masses cross the western portion of the United States, they are forced upward to cross each of the two major mountain chains in their path. The forced, or orographic, rise produces a band of increased precipitation associated with each of the major mountain chains. The subsequent descent on the downwind sides of these chains produces two belts of generally deficient rainfall.

Precipitation amounts in the region vary widely, depending largely on the geographic site with respect to these mountain chains and on the location of the major storm tracts. The snow/rain ratio is particularly important in understanding the role played by precipitation at a particular site. Maximum amounts of snowfall occur at the higher altitudes of the major mountain ranges and in the extreme northern section of the country, a result of the increasing length of the winter season with

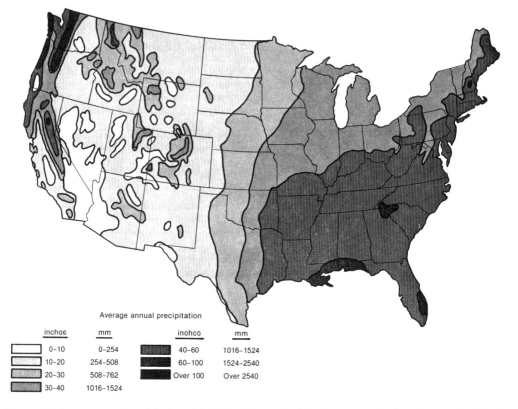

Average annual precipitation

inches	mm	inches	mm
0–10	0–254	40–60	1016–1524
10–20	254–508	60–100	1524–2540
20–30	508–762	Over 100	Over 2540
30–40	1016–1524		

Figure 4-3. Average annual precipitation of the United States. Precipitation patterns closely reflect a region's landforms, which are a primary factor in determining the amount of water available for use in any given area. (*Source:* Ref. 1.)

altitude and latitude. This snowfall represents the primary form of natural water storage for the region. It is stored during months of generally low water demand and released to surface runoff at a time coincident with peak demand. When western snowpacks melt in the spring and summer, they supply an estimated 70 to 100 percent (depending on location) of the total annual runoff for all the western river basins except for the Texas-Gulf region. Therefore they are more important to agriculture than an equivalent amount of rainfall received when demand is low or stored at high cost.

The greatest amount of precipitation in the western United States occurs in the Pacific Northwest, on the Olympic Peninsula, and on the west slope of the Cascade Mountains, where amounts total over 100 in. per year. At the opposite extreme, values of less than 5 in. per year are recorded in some of the southwestern deserts.

The annual regime of precipitation is highly variable from one part of the region to another. As much as half of the annual precipitation may fall

during the growing season in much of the eastern portion. On the Pacific coast the distribution is reversed, and virtually all of the total annual moisture falls during winter.

Evapotranspiration

Evaporation and transpiration are processes that return water to the atmosphere. These processes are controlled by the amount of energy available to convert liquid water to vapor and are limited also by the amount of available water. The term *evapotranspiration* is used to designate the loss of water from the soil by evaporation and from plants by transpiration.

Values of evapotranspiration are more difficult to evaluate than those of precipitation because in many areas of the West total evapotranspiration is limited only by the available water supply. Potential evapotranspiration is the amount of water that would be lost if precipitation were unlimited. Throughout the interior basins, the desert Southwest, and much of the southern portion of the Great Plains, actual evapotranspiration is a small fraction of potential evapotranspiration.

Actual evapotranspiration is determined in part by the seasonal distribution of precipitation and in part by air temperature regimes. If precipitation occurs largely during winter, as is the case in the mountain ranges of the western United States and along the Pacific coast, much of this precipitation runs off or infiltrates the soil. For most of the region, however, precipitation occurs during the summer, when evapotranspiration is at a maximum, and much of it is returned to the atmosphere without affecting other components of the hydrologic cycle.

The timing of precipitation and evapotranspiration is important to agriculture in the western United States because of its effect on available soil water and plant growth. Seasonal variations in soil water, as determined by the balance existing between precipitation and evapotranspiration for several selected stations in the region, are shown in Figure 4-4.

The average potential evapotranspiration in the western United States ranges from an estimated low of 15 to 20 in. in the high mountains of the Pacific Northwest and northern Rocky Mountains to a high of more than 60 in. in small isolated areas in the deserts of Arizona and southern California (Figure 4-5). It is less than 20 in. along the Canadian border and more than 60 in. in southern Texas. Although potential evapotranspiration and precipitation are independent climatic elements, potential evapotranspiration in arid regions is greater because of the higher daytime temperatures resulting from the absence of clouds and rain. High values in the Colorado and Gila deserts and in the lower Rio Grande Valley are examples. In the arid sections of the Columbia River Valley between Washington and Oregon, potential evapotranspiration is more than 30 in., whereas it is only about 20 in. at the same latitude in the eastern United States.

The variation of potential evapotranspiration through the year follows a uniform pattern in most of the region. It is negligible in the winter months as far south as the Gulf Coastal Plain. It rises to a maximum in July that ranges from 5 in. along the Canadian border to 7 in. on the Gulf coast. In some mountainous areas and along portions of the Pacific coast, it does not reach 5 in. in any month.

Infiltration and Percolation

Precipitation that falls on a surface and that is not immediately returned to the atmosphere by evaporation may infiltrate into the surface soil layers.

The amount of that which can infiltrate the surface layers is determined largely by the permeability of those layers (the ability to transmit water, which is governed by the size and the geometry of the spaces within the soil or rock layers) and the amount of water already present in those spaces. Infiltration rates are highest at the beginning of a rainstorm, gradually decreasing with time until some relatively constant value is reached. Some infiltrated water will be retained near the surface by capillary forces. Some will move by gravity flow either toward adjacent stream channels, where it will appear as runoff, or more commonly, downward by percolation to the water table, where it will enter into ground water storage.

All water that exists below the surface of the earth in interconnected openings (interstices) of soil or rock may be called *subsurface water*. That part of the subsurface water in interstices completely saturated with water is called *ground water*. The upper surface of the zone of ground water is known as the *water table*. Between the water table and the surface of the earth is the *zone of aeration*, where the interstices of the soil and rock may contain some varying amount of water, less than total saturation. The water table commonly rises and falls as the availability of water at the surface varies with time (e.g., as a result of climatic change) or as a result of ground water extraction practices.

Ground water is not uniformly distributed throughout the West. The major producing aquifers are deposits of unconsolidated sands, gravels, and clays located on preexisting outwash plains or in former lake beds and in the basalts of the Pacific Northwest. In general, the thickness of these aquifers ranges from tens of feet to several thousand feet. Both the amount of water they produce and the quality of that water are extremely variable, even from well to well within the same aquifer. The general locations of the more important ground water resource regions are shown in Chapter 7, Figures 7-8 and 7-9.

Surface Runoff

Surface runoff, as rivers or streams, generally occurs only after the requirements of evapotranspiration and soil and ground water recharge have been satisfied. Where the requirements of either or both processes are in excess of annual precipitation amounts, no runoff will take place. Water lost to evapotranspiration is completely lost to runoff. Water that infiltrates into the soil or percolates to ground water may ultimately appear as surface runoff at some point distant from that at which it

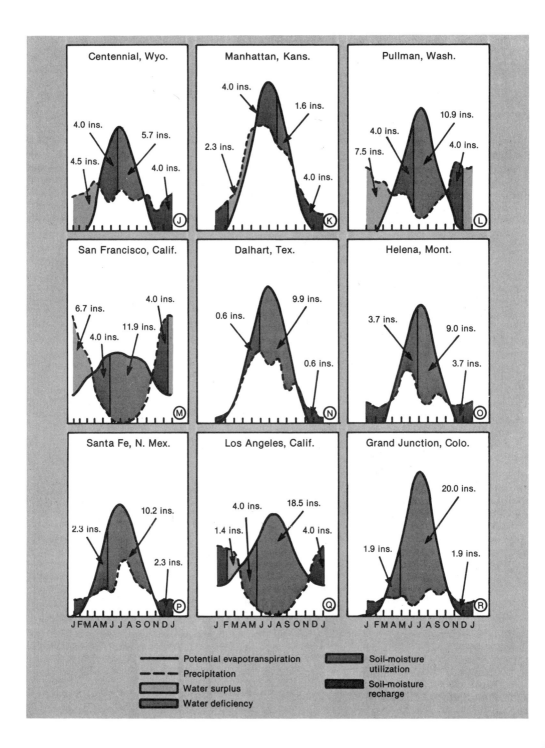

Centennial, Wyo.
4.0 ins.
5.7 ins.
4.5 ins.
4.0 ins.
J

Manhattan, Kans.
4.0 ins.
1.6 ins.
2.3 ins.
4.0 ins.
K

Pullman, Wash.
10.9 ins.
4.0 ins.
7.5 ins.
4.0 ins.
L

San Francisco, Calif.
6.7 ins.
4.0 ins.
4.0 ins.
11.9 ins.
M

Dalhart, Tex.
9.9 ins.
0.6 ins.
0.6 ins.
N

Helena, Mont.
3.7 ins.
9.0 ins.
3.7 ins.
O

Santa Fe, N. Mex.
10.2 ins.
2.3 ins.
2.3 ins.
P

Los Angeles, Calif.
4.0 ins.
18.5 ins.
1.4 ins.
4.0 ins.
Q

Grand Junction, Colo.
20.0 ins.
1.9 ins.
1.9 ins.
R

J F M A M J J A S O N D J   J F M A M J J A S O N D J   J F M A M J J A S O N D J

——— Potential evapotranspiration   Soil–moisture utilization
– – – Precipitation   Soil–moisture recharge
Water surplus
Water deficiency

42

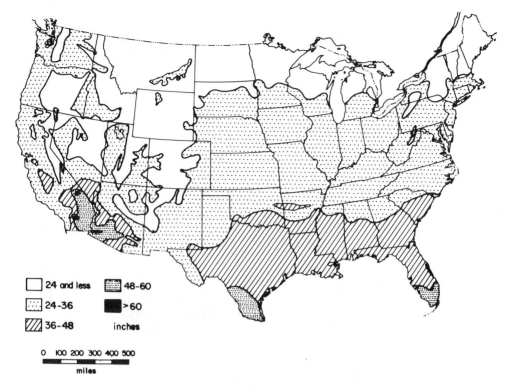

Figure 4-5. Potential evapotranspiration. Annual values (in inches) are indicated in a general pattern. Actual values will depend on the method of measurement used. (*Source:* After Plate 13, Ref. 3. This illustration has been substituted by the editors.)

fell as precipitation. This will be determined by the amount of transpiration losses, which depletes soil water, and by the ability of the rock formations at a given location to transmit water.

Surface runoff in the western United States is highly variable, both from one river basin to another and from one time of the year to another. In terms of total volume of annual discharge, the major river system of the region is the Columbia River, which has a mean annual flow in excess of 140 million acre·ft and represents nearly 36 percent of the total volume of surface water available for the entire region. The river system with the smallest annual discharge volume is that of the Rio Grande River, which has an estimated mean annual discharge between approximately 1.3 million acre·ft/year (1.2 million gal/day, mgd) (5) and 6.0 million acre·ft/year (5.4 mgd) (12).

All rivers of the western United States, except those flowing through the Texas Gulf region, have their headwaters in the mountain ranges of the region or in Canada (Figure 4-6). The period of peak runoff coincides with the period of spring snowmelt and generally occurs during May or June. There are two exceptions to this general pattern. First, rivers flowing into the Pacific Ocean from the west side of the Sierra and Cascade mountain ranges in Washington, Oregon, and northern California have a peak discharge in January or February. Second, the lower reaches of the Missouri and Snake rivers have a peak flow in March or April, reflecting the contribution of meltwater produced by the snow deposits of the plains.

The total amount of runoff contained in streams during the spring and summer months varies from over 90 percent of the annual total for some small

Figure 4-4. The relationship between precipitation and potential evapotranspiration. Monthly trends for selected stations in the western United States show the effects of precipitation and potential evapotranspiration on soil water conditions. For all stations precipitation exceeds evapotranspiration only during the winter months. During the summer months periods of soil water deficits occur and may last up to six months. (*Source:* Ref. 10)

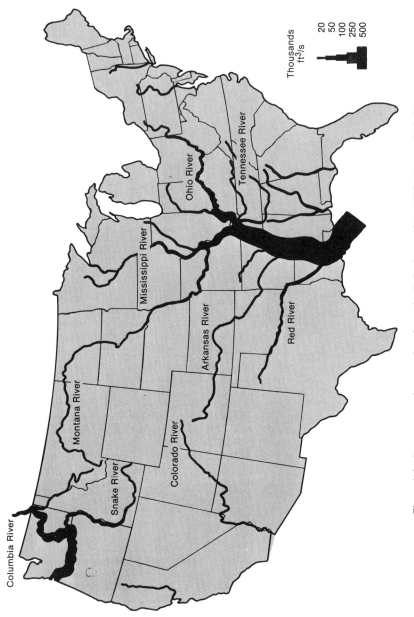

Figure 4-6. Average annual stream flow for major U.S. rivers, 1941–1970. (*Source:* Ref. 2.)

streams totally dependent on the mountain snow-pack to less than 15 percent for streams originating in the Cascade Range, where the contributions to flow are more uniformly balanced between winter rains and spring and summer snowmelt. Figure 4-7 shows the spatial pattern of the variations in areal contributions to surface runoff in the western United States. These values are the depth of runoff produced annually and underscore the importance of the mountainous portions of the region in determining water supply.

For the western United States as a whole, surface runoff estimates vary, depending on the data source (5, 7, 12). The range of estimates is between 515 and 550 million acre·ft/year (460 to 490 billion gal/day, bgd) for the amount of surface runoff that passes through the major river systems of the region.

VARIABILITY IN THE HYDROLOGIC CYCLE

Both human-caused and natural variations in the hydrologic cycle affect the timing and the volume of available water in the western United States. It is important to recognize that in the western United States very few areas remain where the hydrologic cycle operates naturally. Estimates of water availability in any particular component of the cycle must take into account human intervention at the specific site. The impacts of this intervention may vary from site to site. This is due partly to the particular nature of the human activity and partly to the natural hydrologic variability of the area. Thus it is important to understand the natural variability of western water resources as well as the variability when modified by humans.

Human Intervention

The primary approaches to accommodate natural variability of western rivers have been (1) construction of reservoirs to delay the surface runoff; (2) development of ground water resources; and (3) in limited cases importation of water from adjacent basins with greater natural supplies. It is estimated that in a natural state the runoff from the 17 western States would be approximately 590 million acre·ft/year (12). Human modification of the river systems of the region through the construction of storage reservoirs and water diversions for off-stream consumptive uses has reduced natural runoff by approximately 100 million acre·ft/year. Other components of the hydrologic cycle have also been affected by technological intervention. Human withdrawals from ground water, estimated to be

nearly 70 million acre·ft/year, affect the amount of recharge required to maintain the natural equilibrium (12).

Natural Variability

For any given watershed *wet* and *dry* years are defined with respect to the long-term average stream flow for that watershed. The definitions are based on the percentage of time that given flow volumes occur, as determined by a statistical analysis of the available stream flow record. For the *Second National Water Assessment* (12), a dry year has been defined in terms of the stream flow that would occur, as indicated by a statistical analysis of the data, 20 years out of every century, or 1 year out of 5. The volume of stream flow, as determined in this way, would be much less for a subregion that has a normally low volume of stream flow than for one where this volume was high. Where natural year-to-year variability of stream flow is low, little difference in the flow volume will exist between a dry year and a normal year. For those subregions with a high annual variability, the dry year may be a small fraction of the normal year flow volume.

It is generally recognized that the annual and seasonal variation in the flow of rivers in the western United States is significant, often varying by as much as ten times during a year or during two succeeding years. For example, Figure 4-8 reflects the variability of the Upper Colorado River, a pattern typical of western rivers. Because of such variability, the long-term average annual stream flow volume is not a particularly useful measure of the amount of water that will be available for any given year. Similarly, the monthly volume of flow fluctuates widely, with that occurring during the spring and summer months often representing as much as 90 percent of the total annual flow of many western rivers (Figure 4-9). Because of the extreme variability associated with both the annual and the monthly stream flow volumes, water management approaches that are based on a long-term average annual flow will generally be unrealistic for shorter time periods, such as a single year or month during a given year.

In a determination of the adequacy of existing reservoir storage facilities to meet water demand for agriculture during a series of dry years, it is more useful to know the year-to-year fluctuation of flow and the number of years that this may be expected to drop below an acceptable level than to know only the average flow for some period of years. In a determination of the extent to which a river will meet seasonal needs of irrigated agriculture, it is more useful to understand the nature of the seasonal variability of stream flow than to know

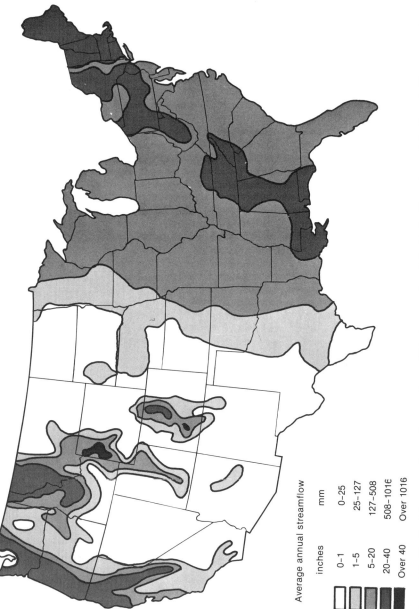

Average annual streamflow

inches	mm
0–1	0–25
1–5	25–127
5–20	127–508
20–40	508–1016
Over 40	Over 1016

Figure 4-7. The Spatial pattern of annual stream flows. With the exception of the Rocky Mountains and the Cascade-Sierra mountains, much of the western U.S. averages less than 1 in. of runoff or stream flow annually. (*Source*; Ref. 1.)

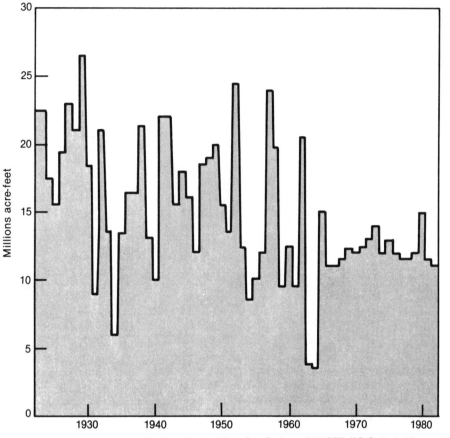

SOURCE: Office of Technology Assessment, compiled from National Water Data Exchange (NAWDEX), U.S. Geological Survey, 1983.

Figure 4-8. The annual variability of stream flow volume, Upper Colorado River, 1920–1980. The year-to-year variability is high. Where water is allocated on the basis of long-term mean flow volume, there will be insufficient water to meet that allocation during many years. The decreased variability beginning in the mid 1960s results from the construction of dams and reservoirs. (*Source:* Office of Technology Assessment.)

the annual flow volume. Most discussions of the adequacy of water supplies in the western United States have been developed in terms of annual mean values (e.g., 5).

Estimates of future water availability, including that for all types of agriculture, must be based on some estimates of climatic trends. Climatic fluctuations affect all components of food-producing ecosystems. Changes in food production can be caused by the effects of weather on pests, pathogens, weeds, and crop plants and by altering water supply and water use patterns. Western agriculture has developed during a particularly warm period in recent climatic history (9). Climatic records show that climate has varied in the past, however,

and significant fluctuations have occurred in recent history.

In addition to the natural variability of climate, there is growing speculation about human-induced climatic change. These include (1) the decreasing pH (increasing acidity) of rainfall, which may be caused by emissions from burning fossil fuel; (2) the gradual increases in the atmospheric fraction of carbon dioxide (CO_2) and other infrared absorbing gases, also largely a result of increased burning of fossil fuels; and (3) the associated changes in water quality, water quantity, and, specifically in the case of the infrared absorbing gases, air temperature increases.

Long-term agricultural planning and policy-

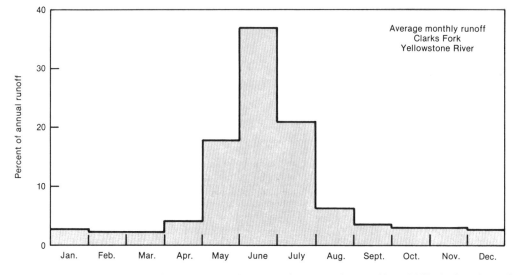

Figure 4-9. Average monthly runoff, Clarks Fork, Yellowstone River. Natural seasonable variability in the volume of flow of most western rivers is large. The spring and early summer snowmelt peak flow of some western rivers represents as much as 90 percent of the annual volume of flow of these rivers. (*Source:* Ref. 2.)

making must be undertaken with the knowledge that some climatic change is inevitable. The geographic extent of any changes in climate will be related to the frequency of the change. Changes on the order of a few years to a few decades will be more localized geographically than will those that persist for decades. To the extent that the ability to predict climatic trends is limited, so too is the ability to determine continuing availability of water for agriculture in the western United States. As stated in a National Research Council report (6), "Our knowledge of mechanisms of climate change is at least as fragmentary as our data." An improvement in the existing data base, as discussed in the next section, should be a first step toward improving the ability to factor climatic trends into agricultural planning.

Measurement

Water is in continuous movement through the hydrologic cycle. A variety of measurement techniques are required to monitor this movement. While all hydrologic processes take place over the surface area of a region, measurements of elements of the hydrologic cycle such as evapotranspiration or precipitation are made at discrete points within that region. In order to determine the volume of water involved in these transfer processes, one must combine the individual point measurements into a spatial pattern from which volume can be estimated.

Some of the problems inherent in all point measurements may be illustrated by those associated with determining the amount of rain that falls at a point. The uncertainties involved in even this apparently simple measurement are illustrated in Figure 4-10. In the development of average values representative of a particular place or time, the selection of the data to be included or excluded is critical.

Only surface runoff may be measured as an areal value, since all the surface runoff from a region must pass through a surface-gaging station. Thus for surface water the location selected for the placement of the gaging station is critical. In many cases the proximity of a gaging station to the point of use determines the usefulness of the data obtained.

In addition to the uncertainties of point measurements for estimating spatial volumes, there are uncertainties in developing time trends from estimates of selected time periods. The amount of water in each of the solid, liquid, and vapor phases changes naturally with time. For reduction of this continuous variation to terms meaningful for analysis, it is common to present data pertaining to elements of the hydrologic cycle as averages for selected time periods. Thus concepts such as *mean annual precipitation* or *mean monthly stream flow* have been introduced to simplify data manipulation. Ultimately, this simplifying process has produced concepts such as "the average precipitation for Arizona" or the "average runoff of the Upper Colorado River." In both cases a large amount of

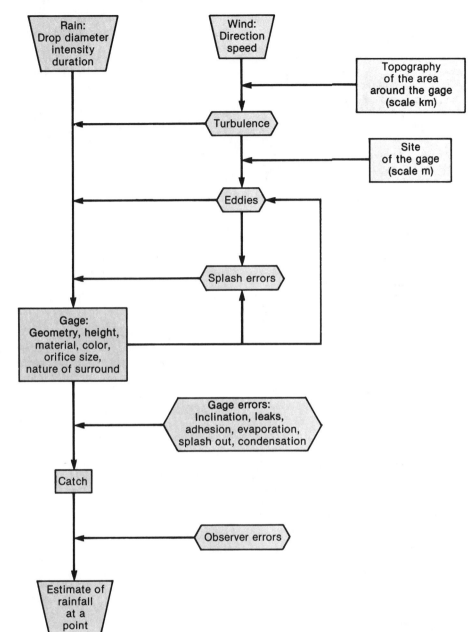

Figure 4-10. Potential errors in water measurements. Errors can occur in the measurement of any of the components of the hydrologic cycle and can affect the accuracy of the data. (*Source:* Ref. 13.)

spatial and temporal variation in the natural processes has been condensed in order to compare the environments of two or more hydrologic areas.

Also, as discussed previously, various components of the hydrologic cycle are modified by human intervention. For example, as water is stored in reservoirs or removed from the surface or subsurface and applied to some use such as irrigation, the fundamental natural relationships are altered. Virtually all the technologies discussed in this assessment are designed to modify to some degree the distribution of water within the natural hydrologic cycle. The degree to which the hydrologic cycle has been modified varies widely among the river basins of the western United States. This human-caused variability further complicates collecting, interpreting, and developing useful averages from existing data.

In the development of average values, short- and long-term syntheses are prepared. Short-term syntheses relate to daily, monthly, or annual fluctuations and are referred to as *climate* or as the *hydrologic regime* of a region. Climate is the average course or condition of the weather at a place over a period of years, as exhibited by air temperature, wind velocity, and precipitation. Taken together, these simple measurements of complex processes of water and energy transfer estimate the disposition of water among the various phases of the hydrologic cycle. These short-term syntheses are important in making decisions concerning water availability and use from one year to the next or from one growing season to the next.

Long-term syntheses involve the concept of climate change over decades, centuries, or longer. This type of synthesis uses the average values developed from short-term data collected over a few decades. Long-term change is identified as the climate slowly becomes wetter or drier, warmer or

cooler. An example of climate change that has been important for recent water planning involves the value for the average flow of the Colorado River used in the Colorado River Compact. Runoff in this river during the period used to determine an "average" flow for allocating the waters of the Colorado River was higher than the average annual flow that now exists. A change in the climate of that river basin has gradually decreased the flow of the river below the value used in the allocation of water between the upper and lower basin states.

Decisions on water availability and use in the western United States must reflect uncertainties associated with measurement. To some extent, all measurements of the elements of the hydrologic cycle are estimates. As concluded by another Office of Technology Assessment study, estimates of water volume or time trends from point estimates have varying degrees of reliability (8). The reliability of these estimates will be determined by (1) the ability of an instrument to accurately measure the processes involved; (2) the extent to which the measurement site is representative of the area in which it has been established, and (3) where point source data (e.g., precipitation measurements) are involved, the number of gages that are combined to develop the estimate. This reliability is also related to the length of record and the assumption of no climate change during the period of record.

NOTES

1. H. Anderson, M. Hoover, and K. Reinhart, 1976, *Forests and water; effects of forest management on floods, sedimentation and water supply,* USDA Forest Service General Technical Report PSW–18.

2. J. Bredehoeft, 1982, Physical limitations on water resources in the arid West (Paper presented at a conference on Impacts of Limited Water for Agriculture in the Arid West, Asilomar, Calif., September 28–October 1, Department of Land, Air, and Water Resources, University of California, Davis).

3. J. J. Geraghty, D. W. Miller, F. van der Leeden, and F. L. Troise, 1973. *Water atlas of the U.S.* (Syosset, N.Y.: Water Information Center).

4. H. Hengeveld and C. DeVocht, 1982, in *Urban Ecology,* 6:1–419.

5. C. Murray and E. Reeves, 1977, *Estimated use of water in the United States in 1975,* U.S. Geological Survey Circular 765.

6. National Research Council, National Academy of Sciences, 1975, *Understanding climatic change: A program for action.*

7. National Water Commission, 1973, *Water policies for the future: Final report to the Congress of the United States by the National Water Commission* (Syosset, N.Y.: Water Information Center).

8. Office of Technology Assessment, 1982, *Report on Use of models for water resources management, planning, and policy,* OTA–O–159.

9. R. Shaw, 1980, Climate change and the future of American agriculture, in *The future of American agriculture as a strategic resource* (Conservation Foundation).

10. R. Thornthwaite, 1948, An approach toward a rational classification of climate, *Geographical Review,* 28.

11. U.S. Geological Survey, 1984, *National water summary 1983—Hydrologic events and issues,* Water Supply Paper 2250.

12. U.S. Water Resources Council, 1978, *The nation's water resources 1975–2000,* vols. 1–4 (Washington, D.C.: U.S. Government Printing Office).

13. R. Ward, 1975, *Principles of hydrology* (New York: McGraw-Hill).

5

THE GREAT LAKES REDISCOVERED

ROBERT A. RAGOTZKIE

On the geological time scale the Great Lakes are very young, their beginnings dating from only about 20,000 years ago. Their present form is the culmination of a complicated series of events including several glacial advances and retreats and the subsequent tectonic uplift of the nothern part of the basin. Hough (1) gives the most complete account of these events and by radiocarbon dating traces the chronology to the present. The lakes have had various drainages at different stages, with a south drainage through the Chicago area occurring from time to time until as recently as 2000 to 3000 years ago. The early stage of Lake Superior, called Lake Duluth, drained southwestward through the St. Croix River up to about 9000 years ago. Lake Huron drained through North Bay as recently as 5000 to 6000 years ago.

Subsequent glacial retreat allowed the Great Lakes to take their final form, but with the three upper Great Lakes—Superior, Michigan, and Huron—joined to form Lake Nipissing. This lake had three outlets: at North Bay, Chicago, and Port Huron. Uplift of the northern part of the region led to the closing of the North Bay outlet. At about the same time erosive downcutting of the Port Huron outlet lowered the lake level below the Chicago outlet. Continued uplift in the Lake Superior region raised its water surface above that of Michigan-Huron and created the St. Marys Rapids, now bypassed by the Soo locks. The upper lakes now drain entirely through the Port Huron outlet to Erie and Ontario and thence via the St. Lawrence River to the Atlantic Ocean (see region 4, Figure 4-2).

The Great Lakes lie across the boundary between the Precambrian rocks of the Canadian shield and the adjoining Paleozoic formations. Lake Superior lies entirely within the shield, surrounded and underlain by the Keweenawan formation. Lake Huron lies along the boundary, with the north shore of Georgian Bay formed by the shield and the rest of the lake in Paleozoic formations. Lakes Michigan, Erie, and Ontario are entirely within the Paleozoic. The Niagaran Dolomite (Silurian) exerts dominant structural control, forming the west and north shore of Lake Michigan, the islands separating North Channel and Georgian Bay from Lake Huron, and the Niagara escarpment as this formation passes southward between Lakes Erie and Ontario (2). Due to their geological history and setting, the bottom topography of Lakes Superior and Huron and northern Lake Michigan is very rough and uneven, while Lakes Erie and Ontario and southern Lake Michigan resemble smooth troughs.

HYDROLOGY

The Great Lakes form a single great drainage system culminating in the St. Lawrence River. Their surfaces range from 600 ft above sea level for Superior to 245 ft for Ontario, and all but Erie extend below sea level in their deepest portions (see Figure 5-1 and Table 5-1).

Lake Superior, the largest and deepest of the Great Lakes, is, in terms of area, the largest freshwater lake in the world. Superior is also the purest of the Great Lakes, its water containing only 52 ppm total dissolved solids. Because of its relatively small drainage basin and large volume, its water has the longest residence time—about 500 years, assuming uniform mixing. Separated from Superior by the St. Marys River and Rapids, Michigan and

Dr. Ragotzkie is a limnologist and oceanographer. He is a professor in the Department of Meteorology and the director of the University of Wisconsin-Madison, Sea Grant Institute, Madison, Wisconsin 53705. In addition to his Great Lakes work, he has done research on the Georgia coast and on lakes in both the Arctic and Antarctic. This material is reprinted with light editing from *American Scientist*, 62 (4) (July–August 1974): 454–464. The epilogue "Great Lakes Research Ten Years Later" was added in 1984. Reprinted with permission of the author and *American Scientist*. Copyright © 1974 and 1984 by Sigma Xi, The Scientific Research Society.

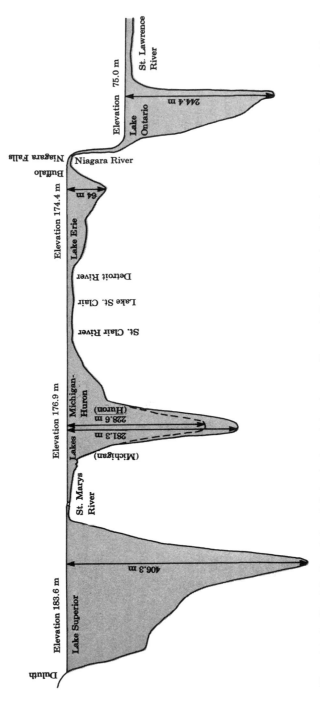

Figure 5-1. Profile of the Great Lakes. Relative depths and elevations are illustrated. Horizontal and vertical scales are distorted to convey a visual impression.

Table 5-1. Characteristics of the Great Lakes

Lake	Area Lake (km²)	Area Drainage Basin Land (km²)	Mean Depth (m)	Maximum Depth (m)	Elevation Above Sea Level (m)	Total Dissolved Solids— 1968 (ppm)	Mean Discharge (m³/s)
Superior	82,103	125,356	149.0	406.3	183.6	52	2076
Michigan	57,757	118,104	85.0	281.3	176.9	150	1558
Huron	59,570	131,313	59.4	228.6	176.9	118	5038
Erie	25,667	58,793	18.9	64.0	174.4	198	5545
Ontario	19,010	63,973	86.3	244.4	75.0	194	6624

Source: Physical data from U.S. Army Corps of Engineers, U.S. Lake Survey; chemical data from Ref. 37.

Huron are connected by the Straits of Mackinac. There is no difference in water level between them, and hydrologically, they can be considered a single lake. Lake Michigan, the only Great Lake wholly within the United States, resembles a long cul-de-sac and therefore flushes rather poorly, especially at its southern end.

Lake Huron drains into the St. Clair River, which widens into Lake St. Clair near Detroit. The Detroit River connects this lake with Lake Erie, the shallowest of the Great Lakes. Since Erie receives drainage from the upper three lakes as well as from its own watershed, it has a flushing time of only about three years. Thus it responds more quickly than the other lakes to changes in water quality.

Water from Lake Erie flows into the Niagara River and drops 225 ft in elevation, much of this at Niagara Falls, and thence into Lake Ontario. Lake Ontario has a longer flushing time, and though it receives water of relatively low quality from Lake Erie, it has a larger volume in which waste assimilation and mineralization can take place. The total water flow out of Lake Ontario into the St. Lawrence River averages about 6600 m³/s, or 208 km³/year which is slightly less than 1 percent of the total volume of the Great Lakes. The St. Lawrence River carries this flow about 350 mi to the Gulf of St. Lawrence and the Atlantic Ocean.

THE PLIGHT OF THE LAKES

During the late nineteenth century major settlement of the Great Lakes region took place. By 1890, widescale clearing of the forests occurred and agriculture had begun. During this same period a thriving commercial fishery developed, mainly for lake trout, whitefish, and lake herring. About 1900 this flourishing fishery began to experience declining stocks, especially of whitefish. For the first

time men began to realize that the resources of the lakes were not without limit. In search of reasons for the decline of the fish catch, several biological surveys and research programs were undertaken—in Lakes St. Clair and Erie by Reighard in 1894–1900 and in Lake Michigan by Ward (3). After a few years this research activity faded until the next crisis—the collapse of the Lake Erie cisco fishery in 1925, the first symptom of the deteriorating quality of the waters of that lake. This event triggered three major limnological surveys. Two of these, on Lake Erie, were sponsored by the U.S. Bureau of Fisheries, adjoining states, and the province of Ontario. A third survey on Lake Michigan in 1930–32 was carried out by the U.S. Bureau of Fisheries with support from Michigan and Wisconsin and from fishnet manufacturers. Following this effort, no extensive fieldwork was done on the Great Lakes by the Bureau of Fisheries for the next 15 years.

Meanwhile, with the opening of Ontario's new Welland Canal in 1932, the lamprey eel and the alewife, two marine species that were to have a profound effect on the entire Great Lakes system, were able to invade Lake Erie. By 1930 the lamprey eel had reached Lake Huron. Yet despite warnings by biologists, no research or control funds were forthcoming. By 1950 the lake trout fishery in Lake Huron had been destroyed, trout production in Lake Michigan was down by 95 percent, and the lamprey had reached Lake Superior.

In this climate of crisis the U.S. Fisheries Laboratory in Ann Arbor, Michigan, was reorganized and funded to take on a systematic fisheries research program including a major research effort on lamprey control. Electric fence barriers placed in streams used by the adult eels for spawning and selective larvicides applied to kill the larval eels before they returned to the lake were two of the methods developed. This vigorous international program succeeded in getting the lamprey under control and setting the stage for the rebuilding of the lake trout fishery of Lakes Michigan and Superior, an effort that is still going on.

Next came the alewife, a pelagic fish and a voracious plankton feeder with high biotic potential. In the absence of a predator lake trout population, the alewife population increased explosively and subsequently crashed, causing a severe nuisance as mass mortality in 1967 created windrows of dead fish around the entire shoreline of Lake Michigan. The successful adaptation of the alewife is believed to have seriously affected the commercially valuable perch and chub populations. Michigan and then the rest of the Great Lakes states subsequently introduced coho and other species of salmon in all the Great Lakes, primarily to develop a sport fishery but also take advantage of and perhaps control the alewife population. The creation of a sport fishery has succeeded, but whether control of the alewives has been achieved remains in doubt.

Thus the fisheries of the Great Lakes have been subjected to a whole series of man-made changes. Each crisis has been followed by a flurry of research activity, but despite amelioration of some of the problems, the Great Lakes commercial fishery has continued to decline in both quantity and quality. Largely in response to political factors management emphasis was shifted from a commercial to a sports fishery, but there is now a growing feeling that both commercial and recreational fishing can be accommodated within a single management scheme.

The deterioration of water quality that fishermen and biologists had recognized eventually began to be apparent to people of the cities and towns along the lakes. Although the Great Lakes states long enjoyed a virtually unlimited water supply for their coastal cities and industry and freely used the lakes as a convenient disposal system for domestic and industrial wastes, population growth and increased industrialization have brought inevitable conflict. What had once seemed to be a limitless water source and waste sink began to show signs of overload. The wastes of the Fox River Valley and the city of Green Bay so overloaded the waters of the lower part of Green Bay that the city was forced to obtain its drinking water by a 26-mi pipeline across the peninsula to Lake Michigan. In 1938 the city had to close its bathing beaches for sanitary reasons.

Chicago takes its drinking water from Lake Michigan, and to protect this supply, exports its wastes to the Mississippi River by means of the Chicago Drainage Canal, an artificial waterway that is flushed with Lake Michigan water. Similar water supply and waste disposal problems are being faced by many Great Lakes communities, and intercity and interstate legal conflicts have arisen. Illinois has filed suit against the city of Milwaukee for polluting Lake Michigan, which, ironically, is the drinking water supply for both Milwaukee and Chicago. Solutions to these problems are turning out to be more expensive than municipal governments can afford. Essential federal assistance is becoming available, but not as much or as quickly as local governments would like.

Increasing industrialization along the Great Lakes coasts has led to escalating water requirements, both for process water and for dilution of industrial wastes. The high costs to industry of pollution abatement will no doubt be passed on to the consumer, but the costs of nonabatement are even higher and will be borne by more people. One of the most controversial issues now affecting the Great Lakes is the siting of nuclear electric power plants.

As the largest source of cold water and the greatest heat sink in the interior United States, the Great Lakes, especially Lakes Michigan and Ontario, have been targeted for major electric power generation development. The effect on the lakes of the waste heat from these plants has aroused much scientific and emotional controversy and has stimulated extensive research efforts on the problem. However, it is clear that the need for electric power will produce immense public pressure to develop this relatively clean source of energy. The solution is unlikely to be either popular or cheap.

Another major factor that has spotlighted the plight of the Great Lakes is the recent upsurge of recreation on the lakes. Millions of people enjoy the boating, fishing, swimming, and aesthetic pleasures they afford. The annual rate of expansion in recreational use of the lakes exceeds the 10 to 12 percent annual rate estimated for all U.S. coasts. The successful introduction of the coho and other salmon and the revitalization of the lake trout fishery in Lakes Michigan and Superior have stimulated a sports fishery of national repute worth tens of millions of dollars in business for the coastal regions. But recreation and the economies that depend upon it require clean water.

Thus we see how three major uses of the Great Lakes—commercial fishing, water supply and waste disposal, and recreation—have converged to strain the limits of these inland seas to provide. Public concern about the problems outlined above has led to the rediscovery of the Great Lakes. People are seeking solutions for which there is an inadequate scientific base. With a few exceptions most of the research prior to 1950 was limited by funds and facilities to local areas near the coast, and thus it has not been representative of the Great Lakes as a whole.

A great deal of the government agency or government-sponsored university research on the Great Lakes has addressed specific pollution or fishery problem and sought short-term solutions. Support for research aimed at basic understanding of the physics, chemistry, and biology of the lakes has been all too scarce. The people of the United States and Canada now squarely face the question of whether the Great Lakes will become the victim of "the tragedy of the commons" (4) or whether through understanding and joint management the lakes can continue to serve the common good.

RESEARCH TODAY

A real understanding of the Great Lakes as a major and complex hydrologic system can come only from a holistic and interdisciplinary approach. The phytoplankton cell serves as the energy transducer for the entire biological community. At the same time this cell is subject to the steady and unsteady motions of the water in which it drifts and is dependent on a certain chemical environment, which in turn is a function of the surface geology and climate of the watershed and human activities. No component or phenomenon of the Great Lakes is independent. All are intertwined, some loosely and indirectly and some tightly and directly.

At the same time, in order to understand these interrelationships, one must understand the various components and phenomena in themselves. Thus strong disciplinary research is needed. To be effective, this research depends on the interaction of scientists engaged in the various disciplines. The biologist asks the physical limnologist or oceanographer for information about currents and thermal structure. The geologist, in tracing the history of the lake, looks to the biologist for insight on the significance of certain plankton assemblages and terrestrial pollen sources. The chemist needs to know the role of living organisms in the circulation of nutrients or the concentration of microcontaminants.

The most ambitious and comprehensive research program yet undertaken in the Great Lakes was the International Field Year for the Great Lakes (5, 6), an international effort combining government agencies and universities in a massive study of Lake Ontario during 1972–73. The primary emphasis was on water and heat budgets and circulation, but considerable chemical work was also undertaken.

In 1953 the Great Lakes Research Conferences were initiated by the University of Michigan's Great Lakes Research Institute. With the third conference in 1959 this meeting became an annual international event. The conferences resulted in the development of a unique international (United States and Canada) scientific community encompassing both academic and government organizations and, in recent years, industrial groups. In 1968 the International Association for Great Lakes Research was formed. Since then the association has sponsored the conferences held each year at a different Great Lakes research center in the United States or Canada.

The published proceedings of these conferences have become the primary source for literature on Great Lakes research. The Sixteenth Great Lakes Research Conference held in 1973 by Ohio State University at Sandusky, Ohio, was attended by over five-hundred scientists. During the three-day meeting 183 technical papers and 2 symposia were presented. In addition to the now voluminous *Proceedings,* an increasing number of journal articles and a flood of technical reports from academic groups and government agencies within the Great Lakes community have swelled the Great Lakes literature to great size. Of necessity, only a sampling of this source material is possible in this article. Though the aim was to select the most interesting and significant work, the personal knowledge and biases of the author obviously affected the choices.

THERMAL STRUCTURE

Situated in a continental temperate climate, the Great Lakes are subject to major seasonal changes in air temperature and radiation. Because of their large size and great depth, however, they have a relatively long thermal response time, and their annual temperature cycle lags behind that of the air and surrounding land. Therefore they exert a moderating influence on the climate of the coastal regions, much like that of the ocean.

All five lakes are dimictic—that is, they undergo vertical mixing at 4°C twice a year—and all five exhibit thermal stratification in both summer and winter. All but Superior warm sufficiently to develop a thermocline by mid to late June and remain stratified until surface cooling results in fall mixing in October or November. Lake Superior, being deeper, responds more slowly and does not usually stratify until late July—and then only near shore. The central region shows little stratification until well into August, and a sharp thermocline seldom develops there. Fall turnover occurs in early December, and thereafter the lake cools to 1° to 2°C.

The lower lakes, Erie and Ontario, being smaller, may freeze over 80 to 95 percent of their area, while Michigan and Huron seldom have an ice cover more than 5 or 10 mi from shore and may remain almost entirely open in some years. Lake Superior is more continental in location and, lying in the path of outbreaks of polar air from central Canada, is subject to much longer and colder winters. But due to Superior's great depth and the stormy weather of the region, ice forms late and seldom, if ever, covers the entire lake. The western arm sometimes freezes completely, however, and may retain its ice cover until the second week in June, as it did in 1972. Lake Superior is very close to being an arctic lake and certainly can be classified as subarctic.

The thermal structure of the lakes has been the subject of considerable interest and research in recent years. Rodgers (7, 8) has made extensive studies of the *thermal bar* in Lake Ontario. This term refers to a vertical division between the nearshore waters and the central lake by a vertical 4°C isothermal surface around the entire perimeter of the lake. This strange phenomenon develops as shore water warms above 4°C. Because the maximum density of fresh water occurs at 4°C, the central lake cannot stratify until it reaches 4°C, and therefore it continues to mix vertically. The thermal bar advances toward the center of the lake as heating progresses, and when the central portion reaches 4°C, the entire system apparently rotates 90°. Typical summer stratification results as the warmer water forms a surface layer over the entire lake and the 4°C water slides under it. The thermal bar has since been documented for Michigan and Superior.

The summertime temperature structure for Lake Superior has been reported by Smith (9), Ragotzkie (10, 11), and Ragotzkic and Bratnick (12). Data were obtained by recording thermographs at buoy stations, repeated bathythermograph sections off the Keweenaw Peninsula, a nine-depth termistor chain in deep water with year-round recording onshore, and by whole-lake airborne infrared radiometer surveys (see Figure 5-2). These observations showed that warmer water first occurs along the entire south shore, with an exceptionally strong horizontal temperature gradient along the Keweenaw Peninsula. This gradient suggested the presence of a relatively swift northeastward coastal current, which was later confirmed by bathythermograph cross sections and direct measurements by six recording current meters at two buoy stations (10, 13).

CIRCULATION

The Keweenaw Current (10) attains a speed of 1 knot (kn) or more and persists from late June until September at least. Comparisons of the direct velocity measurements by current meter and the velocities derived from dynamic height calculations based on the temperature structure suggest that this current is probably barotropic in June and early July and becomes more baroclinic as the temperature gradient increases later in the summer (9, 13). Continued investigations of the nature and dynamics of this current are being carried out by Yeske et al. (14) and Gilson et al. (15) by means of simultaneous bathythermograph sections, current meters, current drogues, photogrammetry of drift cards, and airborne infrared radiometry and imaging. These investigations have revealed similarities between this current and the Gulf Stream.

Coastal currents are now recognized as typical for all the Great Lakes. Observations on Lakes Huron and Ontario by Csanady (16, 17, 18), by Scott in Ontario (19, 20), and by various investigators in Lake Michigan, and the discovery of the Keweenaw Current in Lake Superior confirm this generalization. Extensive theoretical work done by Csanady (21, 22, 23) indicates that these currents are baroclinic in nature but behave as long edge waves with frequencies below the inertial period, about 16 h for the latitude of the Great Lakes.

Thought a great deal remains to be done on this problem, the nature and theoretical basis of coastal jetlike currents in the Great Lakes has been established. Coastal circulation has great significance for the distribution and dispersion of wastes introduced at the coast. Apparently, the wastes tend to become trapped in the coastal zone and do not mix readily with the central portion of the lake. Plant nutrients, plankton populations, and biological productivity all appear to be higher in the coastal waters than in the central waters, suggesting a "lake within a lake." Future research will no doubt shed more light on this and perhaps provide a more rational basis for waste disposal management in the lakes.

OSCILLATIONS

As large closed basins, the Great Lakes are subject to regular and large oscillations of their surfaces. These remarkable fluctuations of the water level, which may be likened to the sloshing back and forth of water in a bathtub, have piqued man's curiosity since the early explorers. Father André, the first Green Bay missionary, made extensive

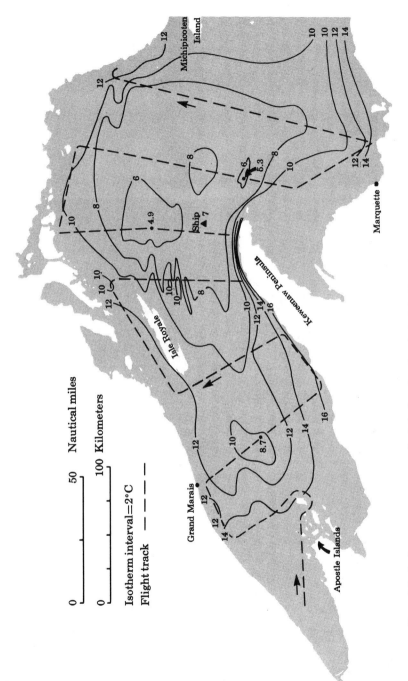

Figure 5-2. Surface temperature pattern for Lake Superior. Data is for July 29, 1964, from an airborne infrared radiometer.

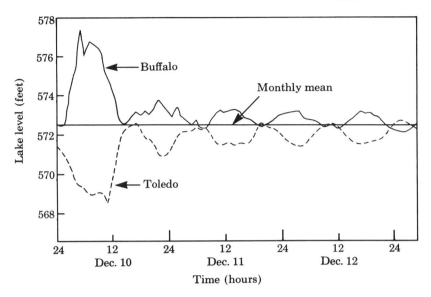

Figure 5-3. Water level records at Buffalo and Toledo, showing the wind-induced seiche on Lake Erie. (*Source:* After Ref. 50.)

observations of the "tides of the Bay des Puants [Green Bay]" and attempted to relate them to the passage of the moon and to the wind (24).

Platzman (25, 26) and Platzman and Rao (27) made the first dynamical studies of the wind seiche oscillations of Lakes Michigan and Erie. Mortimer (28) made a detailed study of long-period surface waves and tides on Lake Michigan. Using long-period water level records kept by the U.S. Army Corps of Engineers at various locations in the lake and in Green Bay, he found good agreement between the observed periods of the first six modes of the longitudinal sieche (about 9 h for the first mode) with the values calculated by F. Defant (29) and by himself using A. Defant's method (30). Mortimer also gives a thorough analysis of the transverse seiche and of the interaction between the oscillations of Green Bay and those of Lake Michigan as a whole, a relationship suspected by Father André. He also shows the existence of the lunar spring tide in Lake Michigan with a range of 0.1 to 0.2 ft in the open lake to 0.6 ft in Green Bay.

Internal oscillations also occur with great regularity on the deeper Great Lakes. In a study of 21 years of records of waterworks intake temperatures around Lake Michigan, Mortimer (31) has made a comprehensive and detailed analysis of these oscillations. Internal waves can produce rapid temperature changes along the coast, sometimes as extreme as a 15°C decrease in a few hours as the thermocline moves upward.

Aside from their scientific interest, surface seiches on the Great Lakes occasionally cause flooding and loss of life, especially in Lake Michigan. In Lake Erie, where the seiche may have an extreme range up to 13.5 ft, heavily loaded ships must time their entrances to harbors to coincide with high water in order to have sufficient water depth for safe navigation (see Figure 5-3). The horizontal motions associated with surface and internal seiches are important mechanisms for the exchange of water between semienclosed bays and the lakes proper, as shown by Bryson and Stearns (32) for South Bay in Lake Huron and Ragotzkie and Ahrnsbrak (33) for Chequamegon Bay in Lake Superior.

THE GREAT LAKES AS OCEANS

Past research has revealed that the Great Lakes have characteristics of both lakes and oceans. Their annual thermal cycle is similar to that of lakes, but their large thermal mass permits them to have some maritime climatic effects. Their currents are sometimes in geostrophic equilibrium, but they are not steady year round, as are the major ocean currents. The lake currents may reverse in a few days, or they may persist for some weeks. Their latitude range is too small to show the effects of a changing Coriolis parameter and vorticity, effects that are important in oceanic circulations, but the Great Lakes exhibit several unique phenomena. Coastal currents dominate the circulation, and the lakes exhibit strong and complex surface and internal oscillations, which are primarily wind-induced, though small lunar tidal effects have been documented for Lake Michigan.

EUTROPHICATION

Though the Great Lakes attained their present form only 2000 to 3000 years ago, the sediments of Lake Superior contain a historical record of about 9000 years (34). In the first 100 years of this record annual varvès reveal a very high rate of sedimentation, at about 1 cm/year followed by about 9000 years of record during which the sedimentation rate was quite constant at about 0.05 mm/year. In 1890 a sudden tenfold increase to 0.5 mm/year occurred, corresponding to human settlement and the clearing of the forests in the region. Thus man has already left his mark on the deepest and most oligotrophic of the Great Lakes.

Whether and how the waters of the Great Lakes have been changing due to human activity has been a subject of great interest recently. Beeton (35, 36) was the first to document these changes in terms of the chemistry and biology of the various lakes. Weiler and Chawla (37) carried the record of chemical changes to 1968 with a complete set of analyses for both major and minor ions for all five Great Lakes. The total dissolved solids of all the lakes except Superior began to increase in 1910, presumably as a result of increased human activity. Lakes Huron and Michigan show small increases, but the increase in Lakes Erie and Ontario has been dramatic, rising about 30 percent from 140 to 200 ppm (see Figure 5-4).

Beeton (35, 36) has correlated the chemical changes with changes in the plankton, bottom organisms, and fish populations, and demonstrates a trend of rapid eutrophication for Lakes Erie and Ontario and the beginning of this process for Michigan and Huron. [For a discussion of the concept of eutrophication, see Hutchinson (38).] In southern Lake Michigan total numbers of phytoplankton have increased over the past 33 years (39). Two species of zooplankton—*Bosmina longirostris* and *Diaptomus oregonensis*—have increased sharply in Lake Michigan in recent years. Both species have been shown to be sensitive indices of eutrophication in other lakes. Erie's plankton has also shifted toward a more eutrophic assemblage.

A comparison of the standing crop of bottom organisms for four Great Lakes with that of other eutrophic and oligotrophic lakes in North America (40) suggests that only Superior is oligotrophic, while Huron is mesotrophic, Michigan is moderately eutrophic, and Erie is eutrophic. The recent increase in eutrophy of Lake Erie is further confirmed by the replacement of mayfly nymphs by tubificids (worms) in the bottom and increased oxygen depletion in bottom waters (35).

FISHERIES

Fish populations have also changed in response to these chemical and biological changes. As noted earlier, in Lake Erie the cisco, a pelagic fish inhabiting oligotrophic and mesotrophic lakes, collapsed in 1925 and has never recovered; the blue pike suffered a similar demise in the late 1950s. These and several other species have been replaced by warm water fish such as the fresh water drum, carp, and perch. The total fish production of Lake Erie has remained relatively constant at about 50 million lb per year, but the species composition has changed toward lower quality.

The ubiquitous alewife, thriving in the absence of the predatory lake trout, unfortunately has no value as a food fish, and its use for fish meal and oil, barely economical to start with, has been plagued with severe odor and pollution problems. These problems, coupled with antiquated fishing regulations that limited trawling operations to a few months each year, led to insurmountable legal, political, and economic barriers. Thus this resource has not been effectively utilized by direct fishing.

Following the control of the lamprey in the 1960s a major effort to restore the salmonid fisheries was begun, and at the same time the decision was made to manage the Great Lakes primarily for sport fishing. Experimental plantings in Lake Michigan of the coho salmon by the state of Michigan succeeded, and soon the other states followed with a massive stocking program, first of coho and then of chinook salmon in all the Great Lakes; and recently Atlantic salmon have been introduced in Lake Superior. Because the coho and chinook salmon cannot reproduce in Great Lakes streams, this fishery is completely dependent on a continued stocking program.

A major lake trout–stocking program has also been undertaken in the upper lakes. In Lake Superior these fish have begun to reproduce naturally, and there is reason to hope that they will also do so in Michigan and Huron, as they did in the past. Reestablishment of naturally reproducing populations of lake trout in these lakes will probably not eliminate the need for continued stocking, but it will reduce the effort required.

With the successful introduction of the coho salmon, research on salmon migration, which had been underway for nearly 20 years in the laboratory and on Pacific coast rivers by University of Wis-

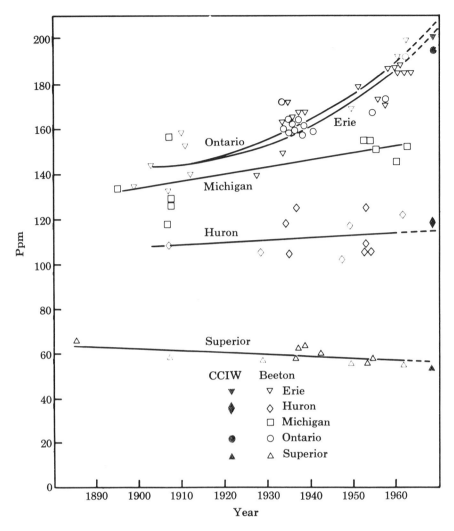

Figure 5-4. Total dissolved solids in the Great Lakes. Solid points are the averages of 12 or more determinations. (*Source:* Refs. 35 and 37 and The Canada Center for Inland Waters (CCIW).)

consin scientists under Arthur D. Hasler, returned to the Great Lakes. Hasler and his students and co-workers had shown that salmon return to their home streams by sensing the unique odor of the water from the stream where they hatched and grew to smolt stage (41, 42). Applied to the Great Lakes, this finding meant that to establish a "run" of salmon in a particular stream, the young fish had to be held in the home stream long enough to be "imprinted" with its odor before descending to the lake. A year and a half later they would return to that stream and attempt to swim back to the original holding pond (43, 44).

In a determination of the movements of the

salmon in the lake or ocean in its return to the home stream, a sonic tag or beeper is placed in the adult salmon's stomach, and the fish is released. The salmon is then tracked by boat by means of underwater directional hydrophones, and its movements are recorded in detail.

Hasler's group is now experimenting with artificial odors. If salmon can be imprinted with a known pure substance, this odor can be introduced in any selected stream and the adult salmon decoyed to that stream. First trials with this technique are giving excellent results, and it is now being used to a limited extent as a management tool in the Great Lakes salmon fishery. The potential implications of this technique for commercial salmon production on the Pacific coast are quite obvious. If a population of hatchery-raised salmon can be imprinted with a known artificial odor, these fish can then be brought directly to the cannery on the estuary or river, thus eliminating the need to capture the fish by conventional fishing methods.

The multiple-species salmon fishery and the re-established lake trout fishery, plus the apparently permanent presence of the alewife, the lamprey eel, and the smelt, have resulted in a truly chaotic situation in Great Lakes fisheries. Whether stability can be attained in this artificial and managed assemblage remains to be seen. One thing is certain: Continued management and predator control will be required from now on. Irreversible changes have precluded any possibility of a return to the natural endemic population of the past.

MICROCONTAMINANTS

Even more recently than the changes in the fish population, some of the Great Lakes have been subjected to pollution by what are known as *microcontaminants*. These include the heavy metals, such as mercury, copper, and lead, and man-made organic substances such as DDT and PCBs (polychlorinated biphenols). Although the concentrations of these substances are generally extremely low in the water, they are further concentrated by living organisms, with each successive trophic level tending to increase the concentration levels.

The heavy metals situation, though of continuing concern, does not appear to be as serious as was thought a few years ago. The high mercury levels found in the fish of Lake St. Clair have not been repeated downstream in Lakes Erie or Ontario. Mercury levels in fish are about the same for all the lakes and are thought to be a reflection of the natural geochemical sources for the lakes (45). For the other minor elements, including the heavy metals, Weiler and Chawla (37) found that the levels were generally below 10 μg/L in all the Great Lakes and probably reflect the composition of the rocks and soils in the drainage basin. Further, they suggest that sorption by oxides of manganese and iron and by other suspended material probably acts to remove these elements from solution and so maintain the observed low-equilibrium concentrations.

DDT and PCBs are more serious contaminants. Extensive agricultural use of DDT and related pesticides resulted in runoff to the lakes and its subsequent concentration by the organisms in the food chain. Fish-eating birds, at the top of the chain, contained the highest levels. Trout and salmon in Lake Michigan attained levels higher than 25 ppm, and as a result, an interstate embargo was placed on these fish for use as human food. A recent court decision banning the use of DDT has sharply reduced its input to the lakes, and already DDT levels in fish are beginning to decrease. DDT appears to be degraded in the bottom sediments much faster than originally thought (46, 47), and so the problem may be on the way to final solution.

PCBs have been recognized in the aquatic environment only in the past three years or so. This class of toxic substances is widely used industrially, and very small losses to the aquatic environment can result in high concentration in fish (48). Unlike DDT, PCBs are known to have harmful effects on humans. They can affect prenatal development of humans and are carcinogenic to primates (49). Industrial producers of this class of substances have taken the matter into their own hands and have instituted strict controls on their use and handling. It seems likely that the problem of PCBs in the Great Lakes will be solved by control of sources and self-policing within the industry.

Though other changes are necessarily occurring within the Great Lakes due to land use and human activities, these are the main trends at the present time. Whether the undesirable changes can be controlled or even reversed will depend on the continued progress of scientific research on the lakes and the will and resources of the local, state, and federal governments to bring about necessary waste disposal systems and land use reform.

LOOKING AHEAD

The clarion call for problem-solving research on the Great Lakes has been sounded many times. Unfortunately, it has been answered by narrow and short-term research projects. Research support by the multiplicity of federal and state agencies, with interest in or responsibility for the Great Lakes, has tended to be mission-oriented and limited to specific objectives. Because of the costs involved in providing facilities and logistics for Great Lakes research, only a few universities have undertaken major programs. International coordination has for the most part been left to individual scientists and, more recently, to the efforts of the International Association for Great Lakes Research, an entirely self-supporting association of scientists.

Several recent programs and events suggest that a change may be taking place, however. Canada has centralized its entire Great Lakes research and management program in the Canada Centre for Inland Waters at Burlington, Ontario. This center is well equipped with both laboratory and field facilities—including modern research vessels—and is staffed with top-quality scientists. On the United States side the National Sea Grant Program has enabled the Universities of Wisconsin and Michigan and the State University of New York to mount significant interdisciplinary research programs on the lakes. The International Field Year for the Great Lakes, which focused on Lake Ontario and represented the most comprehensive study ever made of a Great Lake, involved academic and government groups from both the United States and Canada.

The United States–Canadian agreement on water quality control and the U.S. Coastal Zone Act of 1973 are potentially major management initiatives, which could provide direction for more coherent research-management programs. The Great Lakes Basin Commission's Framework Study, aimed at providing a rational basis for comprehensive, long-range resource planning, is well underway.

Encouraging as these research programs and management developments are, there is not yet a stable long-term research program on the Great Lakes. Though major progress on the chemistry, geology, and physics of the lakes has been made, the level of basic research is still inadequate to build the understanding so necessary for successful and effective applied research programs.

Public concern about the deterioration of several of the Great Lakes and the threats to the others has reached the highest levels of government on both sides of the border. There is general agreement both in the management agencies and in the scientific community that the Great Lakes must be understood and managed as a total system. It is on this realization that we base our hopes for the future. Investment in research is essential for the preservation and continued use of these priceless resources.

NOTES

1. J. L. Hough, 1958, *Geology of the Great Lakes* (Urbana, Ill.: University of Illinois Press).

2. For a more complete account of the geology of the Lake Superior region, see J. S. Steinhart and T. J. Smith, eds., 1966, Section 1, *General geophysical and geological studies of the Lake Superior region: The earth beneath the continents,* Geophysical Monograph Series, no. 10, (Washington, D.C.: American Geophyics Union).

3. A. M. Beeton and D. C. Chandler, 1963, The St. Lawrence Great Lakes, in *Limnology in North America,* ed. D. G. Frey (Madison: University of Wisconsin Press), 535–558.

4. G. Hardin, 1968, The tragedy of the commons, *Science,* 162:1243–1248.

5. E. J. Aubert, 1972, International Field Year for the Great Lakes—United States viewpoint. *Proceedings of the 15th Conference on Great Lakes Research* (International Association of Great Lakes Research), 699–705.

6. J. P. Bruce, 1972, International Field year for the Great Lakes—Canadian viewpoint. *Proceedings of the 15th Conference on Great Lakes Research,* (International Association of Great Lakes Research), 706–709.

7. G. K. Rodgers, 1965, The thermal bar in the Laurentian Great Lakes. *Proceedings of the 8th Conference on Great Lakes Research* (University of Michigan Great Lakes Research Division), 358–363.

8. G. K. Rodgers, 1966, The thermal bar in Lake Ontario, spring 1965 and winter 1965–66. *Proceedings of the 9th Conference on Great Lakes Research* (University of Michigan Great Lakes Research Division), 369–374.

9. N. Smith, 1972, Summertime temperature and circulation patterns in Lake Superior (Ph. D. diss., University of Wisconsin-Madison).

10. R. A. Ragotzkie, 1966, *The Keweenaw Current: A regular feature of summer circulation of Lake Superior,* Technical Report no. 29, Nonr 1202(07) (Department of Meteorology, University of Wisconsin-Madison).

11. R. A. Ragotzkie, 1973, Temperature regime of Lake Superior at Silver Bay, 1971–1972 (Unpublished report, Marine Studies Center, University of Wisconsin-Madison).

12. R. A. Ragotzkie and M. Bratnick, 1966, Infrared temperature patterns on Lake Superior and inferred vertical motions, *Proceedings of the 8th Conference on Great Lakes Research* (University of Michigan Great Lakes Research Division), 349–357.

13. N. Smith and R. A. Ragotzkie, 1970, A comparison of computed and measured currents in Lake Superior, *Proceedings of the 13th Conference on Great Lakes Research,* (International Association of Great Lakes Research) 969–977.

14. L. A. Yeske, T. Green III, F. L. Scarpace, and R. E. Terrell, 1973, Measurements of currents in Lake Superior by photogrammetry (Paper presented at the 16th Conference on Great Lakes Research, International Association of Great Lakes Research, April 16–18, Huron, Ohio).

15. J. E. Gilson, T. Green III, and H. J. Niebauer, 1973, Short-term variations in the baroclinic flow of a coastal current (Paper presented at the 16th Conference of Great Lakes Research, International Association of Great Lakes Research, April 16–18, Huron, Ohio).

16. G. T. Csanady, 1970, *Coastal entrapment in Lake Huron,* presented at the 5th International Water Pollution Research Conference; (New York: Pergamon, 1971).

17. G. T. Csanady, 1972a, The coastal boundary layer in Lake Ontario. I. The spring regime, *Journal of Physical Oceanography,* 2(1):41–53.

18. G. T. Csanady, 1972b, The coastal boundary layer in Lake Ontario. II. The summer-fall regime, *Journal of Physical Oceanography,* 2(2): 168–176.

19. J. T. Scott and L. Lansing, 1967, Gradient circulation in eastern Lake Ontario, *Proceedings of the 10th Conference on Great Lakes Research* (University of Michigan Great Lakes Research Division), 322–336.

20. J. T. Scott and D. R. Landsberg, 1969, July currents near the south shore of Lake Ontario, *Proceedings of the 12th Conference on Great Lakes Research* (International Association of Great Lakes Research), 705–722.

21. G. T. Csanady, 1967, Large-scale motion in the Great Lakes, *Journal of Geophysical Research,* 72(16):4151–4162.

22. G. T. Csanady, 1971a, Baroclinic boundary currents and long-edge waves in basins with sloping shores, *Journal of Physical Oceanography,* 1(2):92–104.

23. G. T. Csanady, 1971b, On the equilibrium shape of the thermocline in a shore zone, *Journal of Physical Oceanography,* 1(4):263–270.

24. Father Louis André, 1676, Extrait d'une lettre du Père André escritte de la Baye des Puants, le 20 Avril 1676, in *The Jesuit relations and allied documents,* ed. R. G. Thwaites, vol. 60 (Cleveland: Burrow, 1900).

25. G. W. Platzman, 1958, A numerical computation of the surge of 26 June 1954 on Lake Michigan, *Geophysica,* 6:407–438.

26. G. W. Platzman, 1963, *The dynamical prediction of wind tides in Lake Erie,* American Meteorological Society, Meteorology Monograph, 4(26):1–44.

27. G. W. Platzman and D. B. Rao, 1964, The free oscillations of Lake Erie, *Studies on Oceanography,* ed. K. Yoshida (Seattle: University of Washington Press), 359–382.

28. C. H. Mortimer, 1965, Spectra of long-surface waves and tides in Lake Michigan and at Green Bay, Wisconsin. *Proceedings of the 8th Conference on Great Lakes Research* (University of Michigan Great Lakes Research Division), 304–325.

29. F. Defant, 1953, Theorie der Seiches des Michigansees und ihre Abwandlung durch Wirkung der Corioliskraft, *Arch. Met. Geophys. Bioklimatol.* (Vienna) A, 6:218–241.

30. A. Defant, 1961, *Physical Oceanography,* vol. 2 (London: Pergamon).

31. C. H. Mortimer, 1963, Frontiers in physical limnology with particular reference to long waves in rotating basins, *Proceedings of the 6th Conference on Great Lakes Research* (University of Michigan Great Lakes Research Division), 9–42.

32. R. A. Bryson and C. R. Stearns, 1959, A mechanism for the mixing of the waters of Lake Huron and South Bay, Manitoulin Island, *Limnology and Oceanography* 4(3):246–251.

33. R. A. Ragotzkie and W. F. Ahrnsbrak, 1961, Summer thermal structure and circulation of Chequamegon Bay, Lake Superior—A fluctuating system. *Proceedings of the 12th Conference on Great Lakes Research* (International Association of Great Lakes Research), 686–704.

34. L. J. Maher, 1973, Palynological studies in the western arm of Lake Superior (Unpublished report, Department of Geology and Geophysics, University of Wisconsin-Madison).

35. A. M. Beeton, 1965, Eutrophication of the St. Lawrence Great Lakes, *Limnology and Oceanography,* 10:240–254.

36. A. M. Beeton, 1966, Indices of Great Lakes eutrophication, *Proceedings of the 9th Conference on Great Lakes Research* (University of Michigan Great Lakes Research Division), 1–8.

37. R. R. Weiler and V. K. Chawla, 1969, Dissolved mineral quality of Great Lakes waters, *Proceedings of the 12th Conference on Great Lakes Research* (University of Michigan Great Lakes Research Division), 801–818.

38. G. E. Hutchinson, 1973, Eutrophication, *American Scientist,* 61:269–279.

39. K. E. Damann, 1960, Plankton studies of Lake Michigan. II. Thirty-three years of continuous plankton and coliform bacteria data collected from Lake Michigan at Chicago, Illinois, *Transactions of the American Microscopic Society,* 79:397–404.

40. W. P. Alley and C. F. Powers, 1970, Dry weight of the macrobenthos as an indicator of eutrophication of the Great Lakes. *Proceedings of the 13th Conference on Great Lakes Research* (International Association of Great Lakes Research), 595–600.

41. A. D. Hasler and W. J. Wisby, 1951, Discrimination of stream odors by fishes and its relation to parent stream behavior, *American Naturalist,* 85:(823):223–238.

42. A. D. Hasler, 1971, Orientation and fish migration, *Fish Physiology,* 6:429–510.

43. A. L. Jensen and R. N. Duncan, 1971, Homing of transplanted coho salmon, *Progress in Fish Culture,* 33(4):216–218.

44. D. M. Madison, A. Scholz, and A. D. Hasler, 1972, Behavioral evidence of "imprinting" to chemical cues in salmon, *American Zoology,* 12(4):643–644.

45. M. M. Thommes, H. F. Lucas, Jr., and D. N. Edgington, 1972, Mercury concentrations in fish taken from offshore areas of the Great Lakes, *Proceedings of the 15th Conference on Great Lakes Research* (International Association of Great Lakes Research), 192–197.

46. F. Matsumura, K. C. Petil, and G. M. Boush, 1971, DDT metabolized by microorganisms from Lake Michigan, *Nature,* 230:325–326.

47. R. C. O'Connor and D. E. Armstrong, 1973, Degradation of DDT in lake sediments (Paper presented to the

American Society of Agronomy, 1972, New York City; MS in preparation, Water Chemistry Lab, University of Wisconsin-Madison).

48. G. D. Veith and G. F. Lee, 1971, PCBs in fish from the Milwaukee region. *Proceedings of the 14th Conference on Great Lakes Research* (International Association of Great Lakes Research), pp. 157–169.

49. J. R. Allen and D. H. Normack, 1973, Polychlorinated biphenyl- and triphenyl-induced gastric mucosal hyperplasia in primates, *Science,* 179:498–499.

50. D. K. A. Gillies, 1960, Winds and water levels on Lake Erie, *Proceedings of the 3rd Conference on Great Lakes Research* (University of Michigan Great Lakes Research Division), 35–42.

EPILOGUE: GREAT LAKES RESEARCH TEN YEARS LATER

I have been encouraged to discuss how far research has progressed since the article "The Great Lakes Rediscovered" appeared in 1974. Since their rediscovery in the late 1960s, the Great Lakes have undergone unprecedented development and exploitation. Research on the lakes—which was once sporadic and in response to crises—has now been stabilized. The United States–Canadian Water Quality Agreements of 1972 and 1978 underscore the commitment of the two nations to protect and preserve this shared resource. The U.S. National Sea Grant College Program has expanded to include research programs in all eight Great Lakes states. Federal laboratories under the National Oceanic and Atmospheric Administration, the U.S. Fish and Wildlife Service, and the U.S. Environmental Protection Agency maintain important ongoing Great Lakes research programs. But pressures on the Great Lakes ecosystem continue to mount faster than has research for solving the problems these pressures cause.

Since 1974 the introduction of salmon to the Great Lakes has brought the alewife nuisance under control and created a billion dollar annual sport fishery. Research on this massive ecological experiment has resulted in highly successful computer models of the bioenergetics of predator-prey relationships. Yet 20 years of heavy stocking and intensive research have not succeeded in reestablishing naturally reproducing stocks of the originally native lake trout in Lake Michigan.

The water quality of the Great Lakes has attracted major public attention, and research on this subject has grown apace. An initial concern—waste heat from power plants—has turned out to be relatively benign, while the loading of organic microcontaminants and heavy metals has become a problem of monumental proportions. The DDT problem was essentially eliminated by banning its use, but PCBs continue to be a major concern. The contaminants problem is particularly insidious because recent research indicates that at least half and perhaps as much as 90 percent of PCBs and other compounds reach the lakes by atmospheric deposition rather than from potentially controllable point sources. New and possibly more dangerous substances like dioxin and benzofurans have also been found in parts of the lakes, where they are causing deformities in both fish and birds. The effects of controlling nutrient inputs, particularly phosphorus, on eutrophication of the Great Lakes have not yet been satisfactorily demonstrated and are now the subject of renewed scientific debate.

The hydrology of the lakes—once considered a matter of runoff, evaporation, and outflow—now includes ground water exchange. Along parts of the Wisconsin coast of Lake Michigan, the contribution of ground water ranges from 6 to 20 percent of surface runoff. The consequences of this finding in terms of water budget and the introduction of ground water contaminants have yet to be examined.

Despite the contaminants problem, improved water quality and the immensely popular sport fishery have further stimulated the recreational development of the coasts. Though economically productive, these activities have put great pressures on ecologically important wetlands and natural shorelines and have raised difficult social, economic, and legal questions.

Finally, the rediscovery of the Great Lakes has brought widespread recognition of the value of the water itself. The Great Lakes contain a fifth of all the surface fresh water in the world. In the United States alone 25 million people depend on the Great Lakes for drinking water, and industries throughout the region depend on the water for their activities. This vast reservoir of fresh water has now also been discovered by water-limited regions outside the Great Lakes basin. Great Plains farmers and residents of the fast-growing southwestern states are becoming desperately short of water and see the Great Lakes as a possible solution. This has created an entirely new issue—the interbasin transfer of Great Lakes water. The state, regional, federal, and international legal and political implications of this concept are only now beginning to be considered. This issue clearly will require major social, economic, and legal research to lay the necessary groundwork for the development of rational and politically acceptable policies.

While research on the Great Lakes has increased and stabilized during the past ten years, the need for greater understanding and the emergence of major new problems have outstripped our efforts to date. There is still no federal research program dedicated specifically to the Great Lakes; there is no ongoing support for Great Lakes research vessels; and, perhaps most important, the coordination of research between academic and government laboratories is still inadequate. The challenge for the future is to bring our research efforts up to the level required by the problems and needs of this unique and immensely valuable international resource we call the Great Lakes.

6

DESCRIPTION OF RUNOFF PROCESSES

THOMAS DUNNE

Rain water and melt water can follow one of several surface or subsurface paths down hillsides to stream channels. The migration of water along one of these paths is called a *runoff process*, and more than one such process usually operates at any given site. The relative magnitude of flows along each path depends on local climatic, soil, and geologic characteristics and particularly on vegetation cover, as is discussed later.

Runoff processes are the central issue in land-surface hydrology because they affect the volume, rate, and timing of stream flow and therefore the generation of floods and critically low dry weather flows, the prediction of which has been the subject of much effort. These efforts have responded mainly to design needs in water resource development and flood control, and their aims were limited. Consequently, it was sufficient to ignore the complex set of processes generating stream flow and to treat the problem of runoff prediction as a statistical problem in which measured output (stream flow) was related to measured input (rainfall or snowmelt) and various drainage basin characteristics such as area and slope. The resulting empirical relationship could be used for predicting future stream flow in the measured drainage basin or on unmeasured catchments. This method has yielded useful results that formed the basis of a great deal of successful engineering. However, the statistical approach has several limitations.

1. Most hydrologic records are short and unlikely to sample extreme events; the most important design events must often be estimated through extrapolation from an inadequate set of data.

2. Climatic fluctuations and human influences in the hydrologic cycle constantly alter the rainfall-runoff relationship in some regions, so that even long records do not provide a homogeneous set of data for statistical analysis.

3. The statistical approach provides little or no information about the physical characteristics or spatial distribution of runoff within a drainage basin. Recent interest in the scientific aspects of hydrology and in the application of hydrologic studies to a wider range of societal problems have focused attention on these runoff characteristics.

Runoff from hillsides entrains and transports into channels the sediment, chemical, organic debris, bacteria, viruses, and other materials that affect the characteristics of stream water. On its way to the channel runoff is associated with various degrees of soil wetness and pore pressure, which affect such important land characteristics as suitability for agriculture or waste disposal, trafficability, and stability against landsliding.

Human activity frequently alters the intensity of runoff mechanisms or introduces a process that did not occur in the area before disturbance. Thus changes in land use are often accompanied by increases in the size of floods on small streams, waterlogging of soils, and landsliding or high concentrations of sediment, chemicals, or biological pollutants in stream water. The disturbances spread and their effects accumulate while subject to a random array of meteorological events. Consequently, it is exceedingly difficult to predict the influences of human activity on the basis of em-

Dr. Dunne is a hydrologist and geomorphologist interested in field experimental and theoretical studies of hillslope erosion and the supply of sediment to river channels. He is a professor of geological sciences, University of Washington, Seattle, Washington 98195. This material was originally presented at the American Geophysical Union Meeting, San Francisco, December 1979, and subsequently published as Model of runoff processes and their significance, in *Scientific basis of water-resources management*, Geophysics Study Committee (Washington, D.C.: National Academy Press, 1982), 17–30. Pages 17–22 and 28–30 are reprinted here with the permission of the author.

piricism alone, because it is not possible to compile adequate data sets through repetition of controlled experiments. Various federal agencies have expended a great deal of effort on such experiments in small drainage basins, but the statistical limitations of the experimental designs and the inherent difficulty of interpreting spatially distributed processes on the basis of measurements at the basin outlet alone severely restrict the conclusions that can be drawn from such experiments. Thus after 30 years of such research there is still controversy and confusion about such issues as the influence of forest management on flood peaks, soil nutrient status, sediment production, and water quality.

Many questions concerning land and water management are likely to be of this type in the next few decades. For example, the problem of predicting non–point sources of phosphorus, nitrogen, or bacteria in streams and lakes or of the washoff of heavy metals and other pollutants in urban storm drainage requires that the amount and path of runoff and its hydraulic characteristics can be predicted. A statistical relationship that would predict peak runoff is not adequate. The sources and characteristics of the runoff and what it can entrain and transport to the stream need to be known. A model of these processes would need to account for spreading of the urbanized area in a drainage basin and the impact of strategies for modulating runoff rates.

Another example is the problem of predicting some effects of large-scale exploitation of tropical landscapes, such as the Amazon Basin. Concerns such as the potential for physical and chemical degradation, the alteration of biogeochemical cycles, and the stream transport of organic materials involve runoff. Yet only one small experimental investigation of runoff processes has been conducted in the 6.1-million-km^2 Amazon Basin (Northcliff et al., 1979) and only several in tropical forests in other parts of the world (Bonell and Gilmour, 1978; Leigh, 1978). We have virtually no basis for making predictions about the hydrologic consequences of disturbance; about the probable distribution of accelerated erosion; or of leaching, waterlogging, or any other impact.

In this and many other situations the consequences are likely to be predicted on the basis of simple statistical relationships that incorporate coefficients to predict volumes or peak rates of runoff, amounts of soil loss, or other requirements. These conceptually simple relationships are often applied to small portions of drainage basins and added together or otherwise manipulated by high-speed computers and presented as complex models. Close examination of such calculations frequently indicates that the application is fundamentally wrong.

The relationships are often developed with the wrong runoff process or erosion and transport mechanism in mind, or the parameters are unreasonable in light of field observations or physical possibilities. Despite the fact that such calculations may not describe what is happening in the field, they are often defended with such arguments as "The procedure can be calibrated against field measurements so that it yields correct answers," or "We have to have a simple uniform procedure that can be applied in all situations to meet the requirements of the law," or "We do not have the resources to train our prediction personnel in field observation or data collection."

If prediction efforts continue along these lines, they will lead to expensive failures and mismanagement of land and water resources. Errors are most likely to appear in the prediction of extreme events or disruptions of kinds that have not been well documented. There will also be a lack of intelligent approaches to problems other than prediction of some streamwater characteristic at the basin outlet. In many cases the manager or policymaker may be less interested in a precise calculation of (say) peak runoff than in anticipating the characteristics of certain zones of a landscape under various hydrologic scenarios or where to collect samples to monitor changes. Answers to this kind of question depend on a knowledge of the physics of the land phase of the hydrologic cycle.

During the past 20 years an alternative approach to hydrologic prediction has emerged through a combination of field experiments and mathematical models. These developments concentrate on the physics of runoff and of the erosion and transport for which it is responsible. They permit predictions not only of basin outflows but of the spatial distribution and characteristics of runoff. This chapter reviews the current status of field observation and physically based modeling capability.

DESCRIPTION OF RUNOFF PROCESSES

Horton Overland Flow

The paths that water can follow to a stream are indicated schematically in Figure 6-1. An important separation occurs at the soil surface, which can absorb water at a certain maximum rate known as the *infiltration capacity*. This rate is relatively high at the onset of rainfall but declines rapidly to an approximately constant value after about 0.5 to 2 h. If rainfall intensity at any time during a rainstorm exceeds the infiltration capacity of the soil, water accumulates on the surface, fills small depressions,

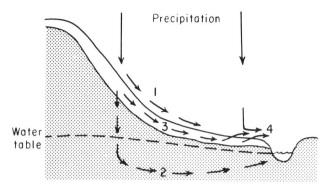

Figure 6-1. Paths of runoff from hillsides.

and eventually spills over to flow downslope as an irregular sheet. This runoff is known as *Horton overland flow* (path 1 in Figure 6-1) after Robert E. Horton (1933, 1940), who first developed a theory of the relationship between infiltration and runoff and their consequences for land and water management and who also conducted some of the earliest experiments on these processes. The occurrence of Horton overland flow depends mainly on the surface characteristics that control infiltration, the most important of which are vegetation and soil conditions. Dense vegetation cover protects the soil surface from the packing effect of raindrops, provides organic material, and promotes biological activity that forms open, stable soil aggregates. Coarse-textured soils can absorb rainfall at a higher rate than silty or clayey soils. Land use often involves the reduction of vegetation cover and the breakdown of soil aggregates, with a consequent lowering of infiltration capacity and an increase in the frequency and amount of Horton overland flow. In the extreme case of urban land use, the infiltration capacity of large fractions of the land surface is reduced to zero.

Horton overland flow tends to produce a major portion of the total runoff in arid and semiarid landscapes and in disturbed zones of humid landscapes, such as cultivated fields, paved areas, mine spoils, construction sites, and rural roads. These areas lack a dense vegetation cover and well-aggregated topsoil. In areas where Horton overland flow is the dominant producer of storm runoff, large areas of hillside commonly have a low infiltration capacity; but because infiltration may vary from place to place over a surface, the runoff is not distributed uniformly. An important problem in runoff prediction is the determination of whether overland flow will occur on a certain portion of the landscape in a particular rainstorm, as well as the amount that will be generated. Identification of runoff-producing zones is also important for recognizing areas that are susceptible to erosion and

the washing off of pollutants. Betson (1964) referred to the nonuniform generation of Horton overland flow as "partial-area runoff."

Horton overland flow typically travels over hillslope surfaces at velocities of 10 to 500 m/h, so that the whole of a 100-m-long hillside can contribute runoff to a stream channel in rainstorms of 0.2 to 10-h duration. This rapid influx to the channel is responsible for a rapid response of stream flow to rainfall and high peak rates of stream discharge relative to those from basins of similar size but subject to different runoff processes. Prediction of response time is another important goal of runoff modeling.

Subsurface Flow

In densely vegetated, humid regions the infiltration capacity is usually high enough to absorb all but the rarest, most intense storms. If the soil and underlying rock are deep and permeable, the infiltrated water percolates through the soil into the vadose zone and then follows a curving path through the phreatic zone to stream channels. Typical rates of this ground water flow (path 2 in Figure 6-1) are many orders of magnitude slower than velocities of overland flow, and the subsurface paths are generally long, so that ground water supplies dry weather stream flow and is a less important contributor to storm runoff.

In many soils or rocks, water percolating vertically encounters an impeding horizon and is diverted laterally. It migrates downslope through the soil and displaces into the stream channel water that is in storage and has been migrating slowly downslope during the preceding days or weeks. If this shallow subsurface flow (path 3 in Figure 6-1) is generated under a low-rainfall intensity onto a highly permeable soil, the water may travel downslope without saturating the soil. Flow is confined to the narrower, intergranular pores, and its velocity is usually of the order of 10^{-4} m/h or less.

Under intense rainstorms the soil often becomes saturated at some depth. Water is then able to migrate through the largest pores, rootholes, wormholes, and structural openings, and its velocity may increase up to about 0.2 m/h in highly permeable forest soils on steep slopes but to much lower values in most soils. Although it travels more slowly than Horton overland flow, some of this runoff arrives at the channel quickly enough to contribute to floods and is classified as *subsurface storm flow* (Whipkey, 1965). But small streams fed dominantly by this process respond an order of magnitude more slowly to rainfall than those with similar drainage area receiving Horton overland flow, and their peak rates of runoff tend to be more than an order of magnitude lower (Dunne, 1978). After a rainstorm shallow subsurface flow declines and may cease altogether in some soils. Hewlett (1961a, 1961b) showed, however, that in a mountainous region underlain by impervious rock, slow drainage of water from unsaturated soils may be responsible for dry weather flow also. Recent important work has been reported by Harr (1977), Anderson and Burt (1977, 1978), and Anderson and Kneale (1980).

Saturation Overland Flow

Vertical and horizontal percolation may cause the soil to become saturated throughout its depth on some parts of a hillside. The soil saturates upward from some restricting layer, and when saturation reaches the ground surface, rainfall cannot infiltrate but runs over the surface as *saturation overland flow*. Some of the water moving slowly through the topsoil may emerge and flow overland to the channel as *return flow* (Musgrave and Holtan, 1964). Thus *direct precipitation onto the saturated soil*, with or without return flow, generates saturation overland flow (path 4 in Figure 6-1), which is intimately related to subsurface runoff and soil conditions. The process occurs most frequently on gentle footslopes and hollows with shallow, wet soils (Dunne, 1970) but can also occur on other parts of a hillslope with thin or wet soils. During a storm the saturated zone may spread upslope and especially into topographic hollows and zones of shallow, wet, or less permeable soil. Slower fluctuations of the zone producing overland flow result from seasonal changes in wetness. Saturation overland flow usually moves more slowly (0.3 m/h to less than 100 m/h) than Horton overland flow because it travels through dense ground vegetation on low gradients. It generates stream flow with a lag time and peak discharge that are intermediate between those characteristic of Horton overland flow and of subsurface storm flow.

It is now generally agreed that in densely vegetated humid regions, most storm runoff is generated by a combination of shallow-subsurface storm flow and saturation overland flow, which in turn consists of various proportions of return flow and direct precipitation onto the saturated area. The relative contributions of subsurface storm flow and overland flow vary with soil and topographic conditions, and it is important that mathematical models be able to separate these contributions not only for prediction of the stream hydrograph but because the processes are associated with differences in soil drainage, entrainment of pollutants, and other facets of hillslope hydrology that are important in resource management. Where deep, highly permeable soils border narrow valley floors, almost all of the infiltrated water can travel beneath the soil surface to the stream. The stream hydrograph is dominated by subsurface storm flow, although the initial rapid response of stream flow to rainfall is often supplied by saturation overland flow from the saturated valley floor. Hewlett and Nutter (1970) and Harr (1977) have described conditions in forested mountains of the southeastern and western United States where subsurface flow is the dominant runoff mechanism. At the other end of the spectrum in humid landscapes are those regions that have long, gentle, concave footslopes covered with shallow, wet soils of low-to-moderate hydraulic conductivity. In these areas soils become saturated more frequently, and the area generating saturation overland flow may expand from less than 10 percent of the basin to more than 50 percent during a storm or a wet season. Figure 6-2 summarizes the conditions affecting the dominant runoff processes in a region.

The temporal and spatial variation of zones producing storm runoff by subsurface flow led early investigators to refer to their conceptual models of the process as the *variable-source-area concept* (Hewlett and Hibbert, 1967) and the *dynamic-watershed concept* (Tennessee Valley Authority, 1964). Most fieldwork on hillslope hydrology in humid regions is now focused on this concept, and a summary of progess up to about 1974 is provided in the book edited by Kirkby (1978).

Snowmelt Runoff

In cold-temperate and subarctic regions, snowmelt runoff generates floods, recharges ground water, and renders large areas of land temporarily uncultivable. When the economic and social impacts of snowmelt runoff are considered, as well as the sophistication of research into the physics of the melting process, it is almost inconceivable that so little effort has been spent on investigations of

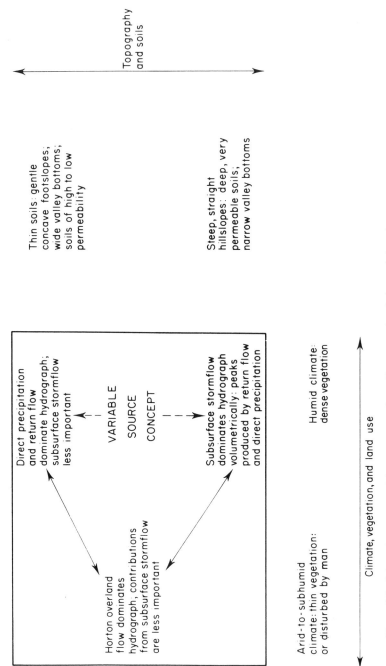

Figure 6-2. Schematic summary of the occurrence of runoff processes in relation to their major controls. (*Source:* After Dunne, 1978.)

runoff processes during snowmelt. Yet differences in the path of runoff can have much greater impacts on the timing and peak rates of runoff than do many of the melt parameters that have been studied in detail.

At a recent state-of-the-art conference on modeling of snowcover runoff (Colback and Ray, 1978), only 3 papers (out of 41) reported on field investigations of runoff processes. The difficulty and discomfort of making the necessary measurements probably contribute to this dearth, but the importance of the subject was attested to by several other papers that dealt with snowmelt runoff prediction without the assistance of field observations.

Major advances have been made recently in understanding the physics of snowpack metamorphism and the percolation of meltwater through the pack (Colbeck, 1971, 1978). After meltwater reaches the ground surface, it follows one or more of the paths outlined earlier, but several factors may alter the relative importance of the various runoff processes from the situation under rainfall at the same site. Snowmelt releases large quantities of water more slowly than in most important flood-producing rainstorms, so that the probability of the infiltration capacity being exceeded is low. Under some circumstances, however, a dense layer of ice may accumulate in the topsoil, rendering it almost impermeable. The equivalent of Horton overland flow occurs as saturated percolation through the lower layers of the snowpack. Horton (1938) was the first to recognize the importance of this process and to propose an approach to its computation and formal study. Colbeck (1974) presented a more refined theory, which was applied to a field investigation by Dunne et al. (1976). The opposite extreme of complete infiltration and the generation of subsurface stormflow was studied by Stephenson and Freeze (1974), and Dunne and Black (1971) documented an intermediate case of both surface and subsurface flow. The current state of knowledge on this subject was summarized by Wankiewicz (1978) and Price et al. (1978), but much remains to be done, particularly in areas of ephemeral soil freezing or saturation.

SUMMARY AND PROSPECT

Most hydrologic predictions are still based on the empirical-correlation approach in which measured variables such as flood peaks, mean annual floods, low-flow statistics, or characteristics of unit hydrographs are correlated with various geologic, geomorphic, and climatic variables, and the resulting empirical equation is used to estimate future hydrologic events or statistics in the measured basin or in ungaged watersheds. The approach is adequate for prediction of stream flow at a basin outlet during moderate-sized events and is still the preferred tool of many hydrologists, as indicated by the papers presented at a recent symposium of the International Association of Hydrological Sciences (1975). Limitations of the statistical predictions include the difficulty of estimating extreme events through extrapolation from short records and the fact that methods that treat the drainage basin as a lumped system fail to describe important characteristics of runoff processes that vary spatially. These spatial characteristics are required for rational prediction of other processes, such as the entrainment and transport of pollutants and impeded soil drainage, that may render fields impassable to farm equipment or reduce the suitability of a site for liquid-waste disposal.

Within the past two decades field experiments have enhanced understanding of the physics and spatially varying characteristics of runoff processes. The work has stimulated physically based mathematical modeling of the spatially distributed processes. These efforts have in turn provoked better field research by providing unified interpretations of data from disparate environments and by suggesting critical measurements that need to be made. This synergism bodes well for scientific hydrology and particularly for field research that seems likely to be conducted selectively in conjunction with physically based mathematical models of each process in order to verify that the model is an adequate description of the process and that the model incorporates correct values of important parameters. Closer cooperation between fieldwork and modeling efforts should encourage formulation of better field experiments and discourage much of the aimless collection of data that sometimes passes for field hydrology, especially at the watershed scale. A vigorous effort is necessary to improve the quality of hydrologic field studies. Publication trends in major hydrologic journals suggest that the most vigorous and sophisticated current developments in hydrology are due to the efforts of researchers concerned with physically based mathematical models. However, this expanding frontier will be hollow unless it is matched by equally sophisticated field experiments to discover unexpected hydrologic phenomena, to develop new concepts about familiar processes, and to guide the development of mathematical models based on sound physical insights into field conditions.

REFERENCES

Anderson, M. G., and T. P. Burt. 1977. Automatic monitoring of soil moisture conditions in a hillslope spur and hollow. *Journal of Hydrology*, 33:27–36.

———. 1978. Toward more detailed field montoring of variable source areas. *Water Resources Research*, 14:1123–1131.

Anderson, M. G., and P. E. Kneale. 1980. Topography and hillslope soil-water relationships in a catchment of low relief. *Journal of Hydrology*, 47:115–128.

Betson, R. P. 1964. What is watershed runoff? *Journal of Geophysical Research*, 69:1541–1551.

Bonell, M., and D. A. Gilmour. 1978. The development of overland flow in a tropical rainforest catchment. *Journal of Hydrology*, 39:365–382.

Colbeck, S. C. 1971. *One-dimensional theory of water flow through snow*. U. S. Army, Cold Regions Research and Engineering Laboratory Research Report, 296.

———. 1974. Water flow through snow overlying an impermeable boundary. *Water Resources Research*, 10:119–123.

———. 1978. The physical aspects of water flow through soil. *Advances in Hydroscience*, 11:165–206.

Colbeck, S. C., and R. M. Ray, eds. 1978. *Modeling of snow cover runoff*. U.S. Army Cold Regions Research and Engineering Laboratory. Hanover, N.H.

Dunne, T. 1970. *Runoff production in a humid area*. U.S. Department of Agriculture. ARS-41-160.

———. 1978. Field studies of hillslope flow processes. In *Hillslope hydrology*, ed. M. J. Kirkby, 227–293. New York: Wiley-Interscience.

Dunne, T., and R. D. Black. 1971. Runoff processes during snowmelt. *Water Resources Research*, 7:1160–1172.

Dunne, T., and W. E. Dietrich. 1980. Experimental investigation of Horton overland flow on tropical hillsides. *Zeit. Geomorphology Supplement*, 35.

Dunne, T., T. R. Moore, and C. H. Taylor. 1975. Recognition and prediction of runoff-producing zones in humid regions. *Hydrologic Science Bulletin*, 20:305-327.

Dunne, T., A. G. Price, and S. C. Colbeck. 1976. The generation of runoff from subarctic snowpacks. *Water Resources Research*, 12:677–685.

Freeze, R. A. 1974. Streamflow generation. *Reviews of Geophysics and Space Physics*, 12:627–647.

Harr, R. D. 1977. Water flux in soil and subsoil on a steep forested slope. *Journal of Hydrology*, 33:37–58.

Hewlett, J. D. 1961a. *Soil moisture as a source of baseflow from steep mountain watersheds*. U.S. Forest Service Southeast Forest Experiment Station Research Paper, 132.

———. 1961b. *Some ideas about storm runoff and baseflow*. U.S. Forest Service Southeast Forest Experiment Station Annual Report, 62–66.

Hewlett, J. D., and A. R. Hibbert. 1967. Factors affecting the response of small watershed to precipitation in humid areas. In *Forest hydrology*, ed. W. E. Sopper and H. W. Lull, 275–290.

Hewlett, J. D., and W. L. Nutter. 1970. The varying source area of streamflow from upland basins. In *Symposium on Interdisciplinary Aspects of Watershed Management*. American Society of Civil Engineers, 65–83

Horton, R. E. 1933. The role of infiltration in the hydrologic cycle. *EOS: Transactions of the American Geophysical Union*, 14:446–460.

———. 1938. *Phenomena of the contact zone between the ground surface and a layer of melting snow*. International Association Scientific Hydrology, Publication, 23:545–561.

———. 1940. An approach towards a physical interpretation of infiltration capacity. *Soil Science Society of American Proceedings*, 5:399–417.

International Association of Hydrological Sciences. 1975. *The hydrological characteristics of river basins*, IAHS Publication 117.

Kirkby, M. J., ed. 1978. *Hillslope hydrology*. New York: Wiley-Interscience.

Leigh, C. H. 1978. Slope hydrology and denudation in the Pasoh Forest Reserve. I. Surface wash: experimental techniques and some preliminary results. *Malaysia Nature Journal*, 30:179–197.

Musgrave, C. W., and H. N. Holtan. 1964. Infiltration. In *Handbook of applied hydrology*, V. T. Chow. New York: McGraw-Hill.

Northcliff, S., J. B. Thornes, and M. J. Waylen. 1979. Tropical forest systems: A hydrological approach. *Amazonizona*, 6:557–568.

Price, A. G., L. K. Hendrie, and T. Dunne. 1978. Controls on the production of snowmelt runoff. In *Modeling of snow cover runoff*, eds., S. C. Colbeck and M. Ray, 257–268. U.S. Army, Cold Regions Research and Engineering Laboratory Research Report.

Stephenson, G. R., and R. A. Freeze. 1974. Mathematical simulation of subsurface flow contributions to snowmelt runoff, Reynolds Creek, Idaho. *Water Resources Research*, 10, 284–294.

Tennessee Valley Authority. 1964. *Bradshaw Creek-Elk River: A pilot study in area-stream factor correlation*. Research Paper No. 4. Office of Tributary Area Development, Tennesse Valley Authority.

Wankiewica, A. 1978. A review of water movement in snow. In *Modeling of snow cover runoff*, eds. S. C. Colbeck and M. Ray, 222–252. U.S. Army, Cold Regions Research and Engineering Laboratory Research Report.

Whipkey, R. A. 1965. Subsurface stormflow from forested slopes. *International Association Scientific Hydrology Bulletin*, 10:74–85.

7

GROUND WATER

RALPH C. HEATH

The importance of ground water to mankind is difficult to overestimate. It is an important source of water in nearly all inhabited places on earth—and the only dependable source of water in most arid and semiarid regions. Much of the economic development of Third World countries, especially since World War II, has been made possible by the development of readily available ground water. But even in the more advanced countries ground water has been, and continues to be, an important factor in economic growth. For example, ground water withdrawals in the United States in 1980 amounted to about 124 km^3 (cubic kilometers) (29.7 mi^3 or 3.27×10^{13} gal). This represents about 40 percent of the fresh water used for all purposes except hydropower generation and electric power plant cooling.

Ground water is available in at least small amounts at nearly every point on the earth's surface, making it one of the most widely available of all natural resources. Consequently, it serves as the only, or the dominant, source of domestic water in all rural areas, as the largest source of water for irrigation and other purposes in arid and semiarid regions, and as an important source of water for urban and industrial purposes in humid areas. The importance of ground water is readily apparent from data on ground water use in the United States. The estimated use in 1980, by state, is shown in Figure 7-1 (see also Chapter 10).

The widespread use of ground water results not only from its general availability but also from economic and public health considerations. From the standpoint of economics ground water is commonly available at the point of need at relatively little cost and thus does not require the construction of distant reservoirs and long pipelines. It is usually of good quality, normally free of suspended sediment, and except in limited areas where it has been polluted, free of bacteria and other disease-causing organisms and thus not requiring extensive treatment and filtration prior to use. These characteristics are not only an economic benefit but also a benefit to health.

Although ground water is available nearly everywhere in the United States, the quantity available and the conditions controlling its occurrence and development differ from one part of the country to another.

GEOLOGY AND GROUND WATER

Ground water occurs in openings in the rocks that form the earth's crust. The volume of the openings and the other water-bearing characteristics of the rocks depend on the mineral composition and structure of the rocks. Therefore to understand the occurrence of ground water, one must first become familiar with the major groups of rocks in which ground water occurs.

Geologists divide all rocks exposed at the earth's surface into three major groups: (1) igneous, (2) sedimentary, and (3) metamorphic. *Igneous rocks* are rocks that have formed from a molten or partially molten state. Some types of igneous rocks, including granite, solidify at great depth below the land surface and are referred to as *intrusive igneous rocks*. Other igneous rocks form from lava or volcanic ash ejected onto the surface and are referred to as *extrusive igneous rocks*.

Sedimentary rocks are formed by the deposition of sediment by water, ice, or air. Most sedimentary rocks are unconsolidated (soillike) at the time of formation. If they are, in time, buried deeply enough and compressed, or if they undergo certain chemical changes, they may become consolidated.

Both igneous and sedimentary rocks may, over the course of geologic time, reach depths beneath the earth's crust at which they are subjected to

Mr. Heath is a hydrologist with the U.S. Geological Survey, Reston, Virginia. This material first appeared in *Ground-water regions of the United States* (U.S. Geological Survey Water-Supply Paper 2242, 1984), 1–19.

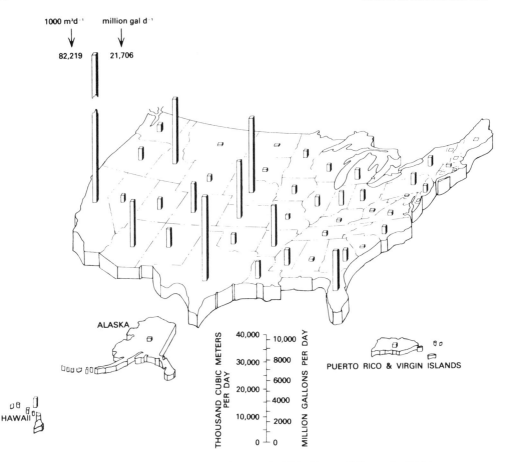

Figure 7-1. Ground water withdrawals, by state, in 1980. (*Source:* Solley et al., 1983.)

great heat and pressure. This may alter both their structural characteristics and their mineral composition to such an extent that they are changed into *metamorphic rocks*. Depending on their original mode of origin, they may be referred to, for example, as metavolcanic or metasedimentary rocks. Metamorphic rocks and intrusive igneous rocks are collectively termed *crystalline rocks* by some investigators.

The United States is underlain by many different types of rocks, including representatives of both the igneous and sedimentary groups and types of both groups that have been subjected to metamorphism. The nature of the water-bearing openings in these rocks depends to a large extent on the geologic age of the rocks as well as on the processes that formed the rocks. The youngest rocks are unconsolidated sedimentary deposits and extrusive igneous rocks. The openings in sedimentary deposits are pores between the mineral grains

(Figure 7-2). The openings in extrusive igneous rocks are, among other types, lava tubes, pores in ash deposits, and cooling fractures. Both of these geologically young types of rocks tend to have a larger volume of openings than do older rocks of the same types that have been subjected to consolidation and to the partial or complete filling of openings by the deposition of minerals and sediment.

Intrusive igneous rocks and metamorphic rocks are among the oldest water-bearing rocks. At the time of their formation they do not contain any appreciable openings. These rocks, as noted above, are formed at great depths below the surface. Thus at the time of formation they are under the pressures exerted by the weight of the overlying rocks. Over the course of geologic time most of these rocks have also been subjected to compressive forces acting more or less parallel to the land surface. As the rocks are gradually exposed by erosion of the

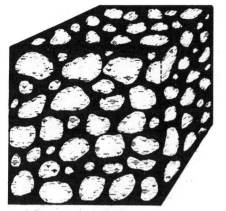

A. Pores in unconsolidated
sedimentary deposit

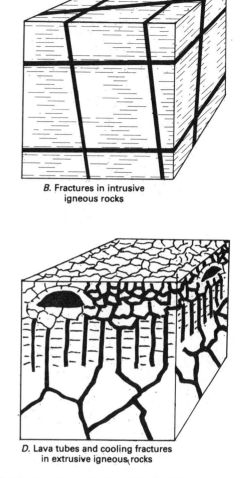

B. Fractures in intrusive
igneous rocks

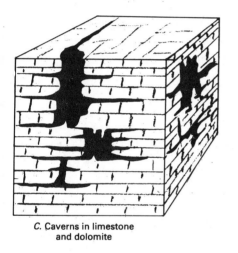

C. Caverns in limestone
and dolomite

D. Lava tubes and cooling fractures
in extrusive igneous rocks

Figure 7-2. Types of openings in selected water-bearing rocks. Block *A* is a few millimeters wide, block *B* is a meter or two wide, and blocks *C* and *D* are a few tens of meters wide.

overlying rocks, the vertical and lateral compressive forces on them are relieved, and they break along sets of horizontal and vertical fractures, which then serve as water-bearing openings (Figure 7-2). Similar fractures also form in sedimentary rocks that have been deeply buried, consolidated, and then exposed by erosion of the overlying rocks.

The earth's surface is underlain at most places by unconsolidated rocks. These may be relatively young sedimentary deposits composed of rock fragments of all kinds, volcanic ash, *alluvium, glacial drift,* sand dunes, or material derived from the breakdown (weathering) of the underlying igneous, metamorphic, or consolidated sedimentary rocks. This surficial layer of unconsolidated material, re-

gardless of its origin, is referred to by geologists as *regolith*. The regolith is underlain every place by consolidated rocks that, in different areas, are of igneous, metamorphic, or sedminentary origin. These consolidated rocks are referred to collectively as *bedrock*.

The thickness of the surficial layer of unconsolidated deposits differs widely from place to place. Under parts of the Atlantic and Gulf Coastal Plain, this layer consists of thousands of meters of gravel, sand, silt, clay, and limestone. In much of the south central part of the country, where it is composed of sandy and clayey material derived from weathering of consolidated sedimentary rocks, the surficial layer is generally only a few meters thick.

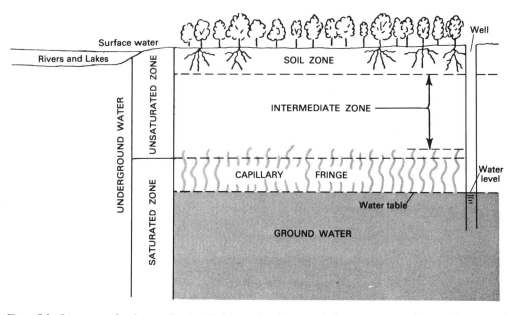

Figure 7-3. Occurrence of underground water. Underground water occurs in both an unsaturated zone and a saturated zone. The upper part of the saturated zone is occupied by water held in a capillary fringe by surface tension. The water table is the level in the saturated zone at which the water is under a pressure equal to atmospheric pressure. Its position is indicated by the water level in shallow wells.

BASIC GROUND WATER
CONCEPTS AND TERMINOLOGY

Water below the earth's surface is referred to as *underground water* (Figure 7-3). It occurs in two distinctly different zones. The uppermost zone extends from the land surface to depths ranging from less than a meter in parts of humid areas to a hundred meters or more in parts of some arid areas. Openings in this zone contain both water and air; as a consequence, the zone is referred to as the *unsaturated zone*. Below the unsaturated zone is a zone in which interconnected openings contain only water and which is referred to as the *saturated zone*. The *water table* is the level near the upper part of the saturated zone at which water occurs under a pressure equal to the atmospheric pressure. The position of the water table is indicated by the position of the water level in shallow wells. Water in the saturated zone above the water table occurs in a *capillary fringe* that is maintained by the strong surface tension of water. Water below the water table is referred to as *ground water*. It is this water that discharges through springs, seeps, and free-flowing or pumping wells.

Part of the precipitation that reaches the land surface in areas in which an unsaturated zone exists percolates downward across the unsaturated zone to the saturated zone (Figure 7-4). These areas are known as *recharge areas* because they are places where ground water recharge occurs. After reaching the saturated zone, water moves downward and laterally under the hydraulic gradients prevailing in the ground water system to *discharge areas*.

The capacity of rocks to transmit water is referred to as *hydraulic conductivity*. This term replaces the term *permeability,* which was in common use for many years. Saturated rock units whose hydraulic conductivity is large enough to supply water in a usable quantity to a well or a spring are referred to as *aquifers*. Less permeable intervening rock layers are referred to as *confining beds*. Because the hydraulic conductivity of aquifers is commonly several hundred to several thousand times that of confining beds, aquifers function as pipelines that transmit ground water from recharge areas to discharge areas (Figure 7-4). In ground water systems in which the aquifers and confining beds are essentially horizontal, confining beds impede the vertical movement of water. As a result, the most active lateral movement tends to occur in the shallowest aquifers.

The land surface in most areas is underlain by a relatively permeable layer, which in turn is underlain by less permeable materials. The surficial layer ranges in thickness, from one area to another, from

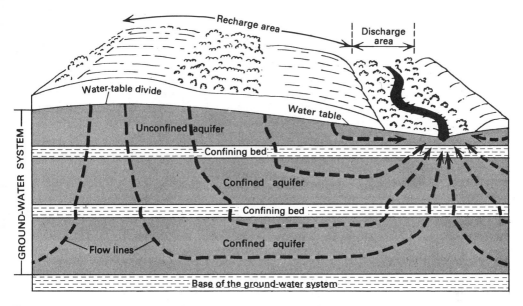

Figure 7-4. Movement of water through ground water systems. Water that enters a ground water system in recharge areas moves through the aquifers and confining beds comprising the system to discharge areas.

a few meters to a few hundred meters. Water that infiltrates the ground in these areas collects in a saturated zone above the underlying confining bed and thereby forms an *unconfined aquifer*—that is, an aquifer that is not full of water and in which the saturated zone ranges in thickness from one time to another. If water percolates through the confining bed and completely fills an underlying permeable bed, it forms a *confined aquifer*—that is, an aquifer that is full of water and that is overlain by a confining bed (Figure 7-4).

To facilitate comparison of the water-transmitting capacity of different types of rocks, hydraulic conductivity is expressed in terms of the volume of water that would be transmitted in a unit of time through a unit cross-sectional area of rock under a unit hydraulic gradient. The metric units commonly used to express hydraulic conductivity are cubic meters for volume, a square meter for area, and meter per meter for hydraulic gradient. The unit of time is commonly a day. In the inch-pound system of measurements the units are cubic feet (or gallons) per square foot under a hydraulic gradient of foot per foot. The water-transmitting capacity of an aquifer, in contrast to that of a unit volume of rock, is equal to the hydraulic conductivity times the aquifer thickness, when the thickness is expressed in either meters or feet. The resulting value is referred to as *transmissivity* (Figure 7-5).

The water in transit through ground water systems may also be viewed as water in storage.

Consequently, the storage properties of rocks are as important as their hydraulic conductivities. The volume of the openings in a rock is referred to as *porosity* and is generally expressed as a decimal fraction or as a percentage (Figure 7-6). When rock saturated with water is drained, water will remain in the smallest openings and as a film on the sides of the larger pores and other openings. Thus porosity can usefully be viewed as consisting of two parts: the part that will drain, which is referred to as *specific yield*, and the part that will be retained, which is referred to as *specific retention* (Figure 7-7).

WATER-BEARING ROCKS

All of the types of rocks mentioned in the preceding discussion of geology and ground water are water bearing and serve as sources of ground water in one area or another. Each group is composed of several different kinds of rocks, each kind having different water-bearing characteristics. It is, therefore, useful to identify and discuss briefly the kinds of rocks that serve as important sources of water or as barriers to ground water movement. Table 7-1 lists the rocks of most importance to ground water hydrology, either as sources of water (aquifers) or as layers that hamper the movement of water (confining beds).

Rocks that are most important as sources of large

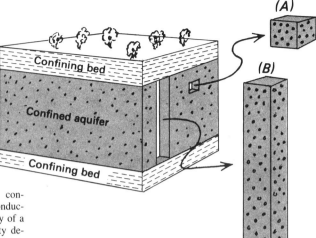

Figure 7-5. Difference between hydraulic conductivity and transmissivity. Hydraulic conductivity defines the water-transmitting capacity of a unit cube *(A)* of the aquifer. Transmissivity defines the water-transmitting capacity of a unit prism *(B)* of the aquifer.

amounts of ground water include (1) sand and gravel, (2) limestone and dolomite, (3) basalt, and (4) sandstone. The principal areas in which these rocks occur are shown on the map in figure 7-8. Other, relatively common rocks that serve as sources of small to moderate supplies are also listed in the table, but they are not shown on the map. Rocks that serve primarily as barriers to ground water movement include clay and shale.

Sand and gravel are the sources of most of the water pumped from wells in the United States. As shown on Figure 7-8, sand and gravel beds underlie much of the Atlantic and Gulf coastal plains, sev-

eral large areas in the north-central states, a large area east of the Rocky Mountains that extends from Texas to South Dakota, and numerous valleys in the west, including the Central Valley in California. Numerous glaciated valleys in the central and northern parts of the country are also underlain by highly productive deposits of sand and gravel. The importance of sand and gravel as a source of ground water is a result of both their widespread occurrence and their capacity to yield water to wells at large rates.

Limestone and dolomites are the sources of some of the largest well and spring yields in the United

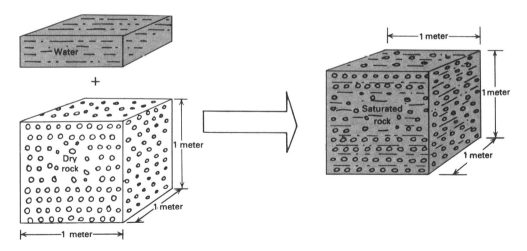

Figure 7-6. Porosity—the volume of a rock occupied by openings. Porosity determines the amount of water a dry sample of rock will absorb.

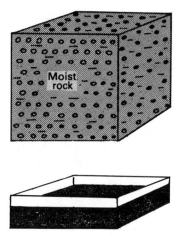

Figure 7-7. Specific yield and specific retention. Only a part of the water filling the pores of a rock will drain away under the influence of gravity. This part is referred to as specific yield. Water retained in small openings and as a film on the surface of the larger pores and openings is referred to as specific retention.

States. Of the 65 or more first-magnitude springs in the United States—springs having a maximum discharge of more than 2.83 cubic meters per second (m³/s) (100 ft³/s)—more than one-third are in limestone. Limestone and dolomite, collectively referred to as carbonate rocks, underlie large areas in the southeast and in the central part of the country, as seen in Figure 7-8. Their yields are large because they are more soluble in water than are other common aquifer-forming rocks. Openings present in them at the time of deposition, and openings that form later, are enlarged by circulating

water. One result of this is the formation of large and extensive cave systems.

Basalt and other volcanic rocks are among the most productive water-bearing materials. Basalt is a dark-colored volcanic rock that may spread in sheets or flows to form extensive plains, as has occurred in southern Idaho and eastern Washington and Oregon. The thickness of the flows ranges from a fraction of a meter to several tens of meters. Water-bearing openings in the basalt flows include *lava tubes, shrinkage cracks, joints,* and a fragmented and broken *(brecciated)* zone at the top of the flows. The interior of some flows is composed of basalt that cooled gradually, forming a dense rock in which most water-bearing openings are shrinkage cracks and joints. Hawaii is underlain by basalt laid down as a complex sequence of flows that formed dome-shaped masses around eruption centers.

Sandstone is the consolidated equivalent of sand and differs from it primarily by the presence of cementing material deposited between the grains or crystalline growth of the grains themselves. Any of several minerals, including calcium carbonate, silica, and iron oxide, may serve as "cement." Sandstone is most important as a source of ground water where the cementing minerals have been deposited only around the points of contact of the sand particles, resulting in the retention of appreciable intergranular porosity. Some geologists refer to such partially cemented rocks as *semiconsolidated* to differentiate them from other (consolidated) rocks in which all pores are filled with cementing minerals. Sandstone may be fractured along bedding planes and more or less perpendicular to the planes. Sandstone serves as an important source of ground water in a large area in the north-

Table 7-1. Rocks of Greatest Importance in Ground Water Hydrology

Sedimentary Rocks			Igneous Rocks	
Unconsolidated (Pores)	Consolidated (Pores, Fractures, and Solution Openings)	Metamorphic Rocks (Fractures)	Intrusive (Fractures)	Extrusive (Pores, Tubes, Rubble Zones, and Fractures)
GRAVEL[a]	Conglomerate[b]	Gneiss	Granite and other coarse-grained igneous rocks	BASALT and other fine-grained igneous rocks
SAND	SANDSTONE	Quartzite-schist		
Silt	Siltstone	Schist		
Clay[c]	*Shale*	Slate-schist		
Till	Tillite (rare)			
Marl	LIMESTONE-	Marble		
Coquina	DOLOMITE			

[a] Capitalized names indicate rocks that are major sources of large ground water supplies.

[b] Lowercase names indicate rocks of relatively wide extent that are sources of small to moderate ground water supplies.

[c] Italic names indicate rocks that function primarily as confining beds.

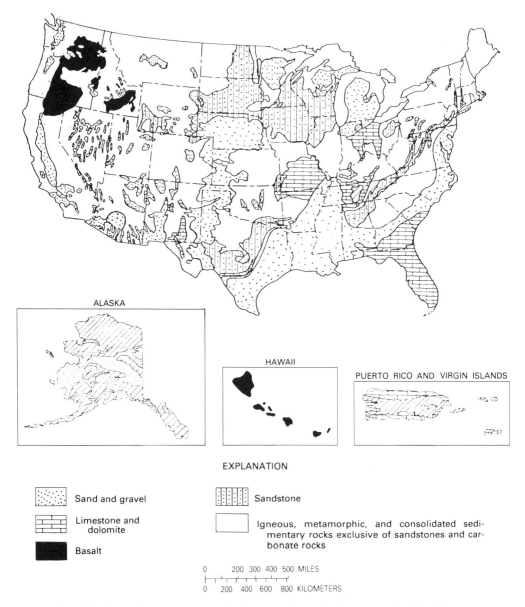

ALASKA

HAWAII

PUERTO RICO AND VIRGIN ISLANDS

EXPLANATION

Sand and gravel

Sandstone

Limestone and
    dolomite

Igneous, metamorphic, and consolidated sedimentary rocks exclusive of sandstones and carbonate rocks

Basalt

0   200 300 400 500 MILES

0   200   400   600   800 KILOMETERS

Figure 7-8. General occurrence of the principal types of water-bearing rocks in the United States.

central part of the country, in an area in northeastern Texas, in an area in west Texas (interbedded
with limestone), and in a relatively narrow zone
west of the Appalachian Mountains from Alabama
to Pennsylvania.

The fifth group of rocks shown on Figure 7-8
includes several different igneous, metamorphic,
and consolidated sedimentary rocks. Rocks in this
group include granite, gneiss, schist, quartzite,

slate, and interbedded shale and sandstone. It is
possible, from a ground water standpoint, to lump
together such a large and diverse group of rocks
because they contain water primarily in openings
developed along fractures and other breaks. Because all these rocks are relatively insoluble, there
is no appreciable enlargement of the fractures by
solution, as has occurred in the carbonate rocks.
The sandstones interbedded with shale, especially

in the central part of the country, may retain some intergranular porosity that makes them somewhat more productive than the other rocks in this group, but it is not feasible in the scope of this discussion to treat them separately from the remainder of the group. Because water is present in this group of rocks primarily along fractures, they are the least productive of the rocks covered in this discussion.

Most of the rocks discussed in this section are irregularly overlain by, and interbedded with, clay and shale, which serve as confining beds, or barriers to ground water movement. Clay has a large porosity, but because its pores are microscopic, the movement of water through it is extremely slow. Shale, like other consolidated rocks, is fractured; however, the fractures tend to be more closely spaced, and therefore the openings along them are smaller than those in carbonate rocks and sandstones.

CLASSIFICATION OF GROUND WATER REGIONS

To describe concisely ground water conditions in the United States, we must divide the country into regions in which these conditions are generally similar. Because the presence and availability of ground water depend primarily on geologic conditions, ground water regions also are areas in which the composition, arrangement, and structure of rock units are similar (Health, 1982).

To divide the country into ground water regions, we must develop a classification that identifies features of ground water systems that affect the occurrence and availability of ground water. The five features pertinent to such a classification are (1) the components of the system and their arrange-

ment, (2) the nature of the water-bearing openings of the dominent aquifer or aquifers with respect to whether they are of primary or secondary origin, (3) the mineral composition of the rock matrix of the dominant aquifers with respect to whether it is soluble or insoluble, (4) the water storage and transmission characteristics of the dominant aquifer or aquifers, and (5) the nature and location of recharge and discharge areas.

The first two of these features are primary criteria used in all delineations of ground water regions. The remaining three are secondary criteria that are useful in subdividing what might otherwise be large and unwieldy regions into areas that are more homogeneous and, therefore, more convenient for descriptive purposes. Each of the five features is listed in Table 7-2, together with explanatory information. The fact that most of the features are more or less interrelated is readily apparent from the comments in the column headed ''Significance of Feature.''

GROUND WATER REGIONS OF THE UNITED STATES

On the basis of the criteria listed in the preceding section, the United States, Puerto Rico, and the Virgin Islands are divided into 15 ground water regions. Table 7-3 contains a list of the regions and a checklist of the criteria that are believed to apply to each region. Because of the wide range in conditions within most of the regions, it is possible in a checklist such as that in Table 7-3 to indicate only the most prevalent conditions. The boundaries of all regions, except region 12, are shown in Figure 7-9. Region 12, which consists of those segments of the valleys of perennial streams

Table 7-2. Features of Ground Water Systems Useful in the Delineation of Ground Water Regions

Feature	Aspect	Range in Conditions	Significance of Feature
Component of the system	Unconfined aquifer	Thin, discontinuous, hydrologically insignificant Minor aquifer, serves primarily as a storage reservoir and recharge conduit for underlying aquifer The dominant aquifer	Affect response of the system to pumpage and other stresses. Affect recharge and discharge conditions. Determine susceptibility to pollution
	Confining beds	Not present, or hydrologically insignificant Thin, markedly discontinuous, or very leaky Thick, extensive, and impermeable Complexly interbedded with aquifers or productive zones	

Table 7-2. Features of Ground Water Systems Useful in the Delineation of Ground Water Regions (*Continued*)

Feature	Aspect	Range in Conditions	Significance of Feature
	Confined aquifers	Not present, or hydrologically insignificant Thin or not highly productive Multiple thin aquifers interbedded with nonproductive zones The dominant aquifer—thick and productive	
	Presence and arrangement of components	A single, unconfined aquifer Two interconnected aquifers of essentially equal hydrologic importance A three-unit system consisting of an unconfined aquifer, a confining bed, and a confined aquifer A complexly interbedded sequence of acquifers and confining beds	
Water-bearing openings of dominant aquifer	Primary openings	Pores in unconsolidated deposits Pores in semiconsolidated rocks Pores, tubes, and cooling fractures in volcanic (extrusive-igneous) rocks	Control water storage and transmission characteristics. Affect dispersion and dilution of wastes
	Secondary openings	Fractures and faults in crystalline and consolidated sedimentary rocks Solution-enlarged openings in limestones and other soluble rocks	
Composition of rock matrix of dominant aquifer	Insoluble Soluble	Essentially insoluble Both relatively insoluble and soluble constituents Relatively soluble	Affects water storage and transmission characteristics. Has major influence on water quality
Storage and transmission characteristics of dominant aquifer	Porosity	Large, as in well-sorted, unconsolidated deposits Moderate, as in poorly sorted unconsolidated deposits and semiconsolidated rocks Small, as in fractured crystalline and consolidated sedimentary rocks	Control response to pumpage and other stresses. Determine yield of wells. Affect long-term yield of system. Affect rate at which pollutants move
	Transmissivity	Large, as in cavernous limestones, some lava flows, and clean gravels Moderate, as in well sorted, coarse grained sands and semiconsolidated limestones Small, as in poorly sorted, fine-grained deposits and most fractured rocks Very small, as in confining beds	
Recharge and discharge conditions of dominant aquifer	Recharge	In upland areas between streams, particularly in humid regions Through channels of losing streams Largely or entirely by leakage across confining beds from adjacent aquifers	Affect response to stress and long-term yields. Determine susceptibility to pollution. Affect water quality
	Discharge	Through springs or by seepage to stream channels, estuaries, or the ocean By evaporation on floodplains and in basin ''sinks'' By seepage across confining beds into adjacent aquifers	

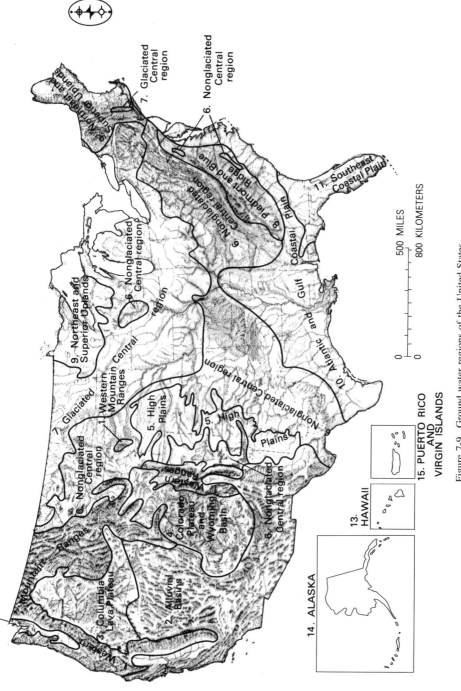

2. Alluvial Basin

1. Western Mountain Ranges

2. Alluvial Basins

3. Western Columbia Lava Plateau

4. Colorado Plateau and Wyoming Basin

5. High Plains

6. Nonglaciated Central region

7. Glaciated Central region

8. Piedmont and Blue Ridge

9. Northeast and Superior Uplands

10. Atlantic and Gulf Coastal Plain

11. Southeast Coastal Plain

13. HAWAII

14. ALASKA

15. PUERTO RICO AND VIRGIN ISLANDS

0   500 MILES

0   800 KILOMETERS

Figure 7-9. Ground water regions of the United States.

83

Table 7-3. Summary of the Principal Physical and Hydrologic Characte[r]

Region No.	Name	Unconfined Aquifer			Confining Beds			Confined Aquifers				Presence and Arrangeme[nt]		
		Hydrologically Insignificant	Minor Aquifer or Not Very Productive	Dominant Aquifer	Hydrologically Insignificant	Thin, Discontinuous, or Very Leaky	Interlayered with Aquifers	Hydrologically Insignificant	Not Highly Productive	Multiple Productive Aquifers	The Dominant Productive Aquifer	Single Unconfined Aquifer	Two Interconnected Aquifers	Unconfined Aquifer, Confining Bed, Confined Aquifer
1	Western mountain ranges		X		X				X				X	
2	Alluvial basins			X		X				X				
3	Columbia lava plateau		X			X				X				
4	Colorado Plateau and Wyoming Basin	X				X				X				
5	High plains			X	X			X				X		
6	Nonglaciated central region		X			X				X				
7	Glaciated central region		X			X				X				
8	Piedmont and Blue Ridge		X		X				X				X	
9	Northeast and superior uplands		X			X			X				X	
10	Atlantic and Gulf coastal plains		X			X				X				
11	Southeast coastal plain		X			X					X			X
12	Alluvial valleys			X	X			X					X	
13	Hawaii			X	X			X					X	
14	Alaska			X	X				X					
15	Puerto Rico and Virgin Islands			X	X					X				

he Ground Water Regions of the United States

	Characteristics of the Dominant Aquifers																		
Water-Bearing Openings				Composition			Storage and Transmission Properties							Recharge and Discharge Conditions					
Primary	Secondary			Degree of Solubility			Porosity			Transmissivity				Recharge			Discharge		
Pores in Semiconsolidated Rocks	Tubes and Cooling Cracks in Lava	Fractures and Faults	Solution-Enlarged Openings	Insoluble	Mixed Soluble and Insoluble	Soluble	Large (>0.2)	Moderate (0.01–0.2)	Small (<0.01)	Large (>2500 m^2/day)	Moderate (250–2500 m^2/day)	Small (25–250 m^2/day)	Very small (<25 m^2/day)	Uplands Between Streams	Losing Streams	Leakage Through Confining Beds	Springs and Surface Seepage	Evaporation and Basin Sinks	Into Other Aquifers
	X			X				X			X			X	X		X		
	X			X			X			X					X			X	X
X	X			X						X	X				X		X		
X	X			X				X			X			X		X	X		
	X			X			X			X					X		X		
X	X	X		X				X		X				X		X	X		X
X	X	X		X			X			X					X				X
		X		X				X			X		X	X			X		
	X			X				X			X	X		X			X		
		X		X			X			X				X		X	X		X
		X				X	X			X				X		X	X		
			X	X			X			X				X			X		
X				X			X			X				X	X		X		
	X			X			X			X				X			X		
		X				X	X			X				X		X	X		X

Table 7-4. Common Ranges on the Hydraulic Characteristics of Ground Water Regions of the United States

Region No.	Region	Geologic Situation	Common Ranges in Hydraulic Characteristics of the Dominant Aquifers							
			Transmissivity		Hydraulic Conductivity		Recharge Rate		Well Yield	
			(m²/day)	(ft²/day)	(m/day)	(ft/day)	(mm/year)	(in./year)	(m³/min)	(gal/min)
1	Western mountain ranges	Mountain with thin soils over fractured rocks, alternating with narrow alluvial and, in part, glaciated valleys	~100	5–5,000,000	0.0003–15	0.001–50	3–50	0.1–2	0.04–0.4	10–100
2	Alluvial basins	Thick[a] alluvial (locally glacial) deposits in basins and valleys bordered by mountains	20–20,000	2000–200,000	30–600	100–2000	0.03–30	0.001–1	0.4–20	100–5000
3	Columbia lava plateau	Thick sequence of lava flows interbedded with unconsolidated deposits and overlain by thin soils	2000–500,000	20,000–5,000,000	200–3000	500–10,000	5–300	0.2–10	0.4–80	100–20,000
4	Colorado Plateau and Wyoming Basin	Thin[a] soils over fractured sedimentary rocks	0.5–100	5–1000	0.003–2	0.01–5	0.3–50	0.01–2	0.04–2	10–1000
5	High plains	Thick alluvial deposits over fractured sedimentary rocks	1000–10,000	10,000–100,000	30–300	100–1000	5–80	0.2–3	0.4–10	100–3000
6	Nonglaciated central region	Thin regolith over fractured sedimentary rocks	300–10,000	3000–10,000	3–300	10–1000	5–500	0.2–20	0.4–20	100–5000
7	Glaciated central region	Thick glacial deposits over fractured sedimentary rocks	100–2000	1000–20,000	2–300	5–1000	5–300	0.2–10	0.2–2	50–500

#	Province / Region	Description								
8	Piedmont and Blue Ridge	Thick regolith over fractured crystalline and metamorphosed sedimentary rocks	9–200	100–2000	0.001–1	0.003–3	30–300	1–10	0.2–2	50–500
9	Northeast and superior uplands	Thick glacial deposits over fractured crystalline rocks	50–500	500–5000	2–30	5–100	30–300	1–10	0.1–1	20–200
10	Atlantic and Gulf coastal plains	Complexly interbedded sands, silts, and clays	1500–10,000	5000–100,000	3–100	10–400	50–500	2–20	0.4–20	100–5000
11	Southeast coastal plain	Thick layers of sand and clay over semiconsolidated carbonate rocks	1000–100,000	10,000–1,000,000	30–3000	100–10,000	30–500	1–20	4–80	1000–20,000
12	Alluvial valleys	Thick sand and gravel deposits beneath floodplains and terraces of streams	200–50,000	2000–500,000	30–2000	100–5000	50–500	2–20	0.4–20	100–5000
13	Hawaiian Islands	Lava flows segmented by dikes, interbedded with ash deposits, and partly overlain by alluvium	10,000–100,000	100,000–1,000,000	200–3000	500–10,000	30–1000	1–40	0.4–20	100–5000
14	Alaska	Glacial and alluvial deposits in part perennially frozen and overlying crystalline, metamorphic, and sedimentary rocks	100–10,000	1000–100,000	30–600	100–2000	3–300	0.1–10	0.04–4	10–1000
15	Puerto Rico and Virgin Islands	Alluvium and limestones overlying and bordering fractured igneous rocks	100–10,000	1000–100,000	3–300	10–1000	3–300	0.1–10	0.04–10	10–3000

Note: All values rounded to one significant figure.

[a] An average thickness of about 5 was used as the breaking point between thick and thin.

Figure 7-10. Alluvial valleys ground water region of the United States.

ALASKA

HAWAII

PUERTO RICO AND
VIRGIN ISLANDS

that are underlain by sand and gravel thick enough to be hydrologically significant (thicknesses generally more than about 8 m), is shown in Figure 7-10.

The nature and extent of the dominant aquifers and their relations to other units of the ground water system are the primary criteria used in delineating the regions. Consequently, the boundaries of the regions generally coincide with major geologic boundaries and at most places do not coincide with drainage divides. Although this lack of coincidence emphasizes that the physical characteristics of ground water systems and stream systems are controlled by different factors, it does not mean that the two systems are not related. Ground water systems and stream systems are intimately related.

Ranges of values for selected hydrologic characteristics of each region are listed in Table 7-4. The range of values within regions indicates the wide range in conditions from one part of a region to another. The differences in values from one region to another indicate the wide range in ground water conditions from one part of the country to another.

This article continues with detailed discussions of each region—the editors.

REFERENCES

Heath, Ralph C. 1982. Classification of ground water systems of the United States. *Ground water,* 20(4) (July–August:393–401).
Solley, Wayne B., Edith B. Chase, and William B. Mann IV. 1983. *Estimated use of water in the United States in 1980.* U.S. Geological Survey Circular 1001.
U.S. Geological Survey. *Summary Appraisals of the nation's ground water resources.* U.S. Geological Survey Professional Paper 813. The chapters are as follows:

Chapter	Water Resources Region	Authors
A	Ohio	R. M. Bloyd, Jr.
B	Upper Mississippi	R. M. Bloyd, Jr.
C	Upper Colorado	Don Price and Ted Arnow
D	Rio Grande	S. W. West and W. L. Broadhurst
E	California	H. E. Thomas and D. A. Pheonix
F	Texas-Gulf	E. T. Baker, Jr., and J. R. Wall
G	Great Basin	T. E. Eakin, Don Price, and S. R. Harrill
H	Arkansas, White, Red	M. S. Bedinger and R. T. Sniegocki
I	Mid-Atlantic	Allen Sinnott and E. M. Cushing
J	Great Lakes	W. G. Weist, Jr.
K	Souris-Red-Rainy	H. O. Reeder
L	Tennessee	Ann Zurawski
M	Hawaii	K. J. Takasaki
N	Lower Mississippi	J. E. Terry, R. L. Hosman, and C. T. Bryant
O	South Atlantic-Gulf	D. J. Cederstrom, E. H. Boswell, and G. R. Tarver
P	Alaska	Chester Zenone and G. S. Anderson
Q	Missouri Basin	O. J. Taylor
R	Lower Colorado	E. S. Davidson
S	Pacific-Northwest	B. L. Foxworthy
T	New England	Allen Sinnott
U	Caribbean	F. Gomez-Gomez and J. E. Heisel

PART III

WATER USE

The major uses of water can be classified as *instream use* and *offstream use*. Instream use is further classified as *flow use* or *on-site use*. Flow uses are dependent on the existence of freely running water and include the maintenance of fish populations, hydroelectric power generation, waste dilution and removal, sediment control, and ecosystem controls such as saltwater advancement into estuaries. On-site uses occur when the water in a stream, lake, or reservoir is used directly. Examples are recreation, navigation, and sources of supply for offstream use. Figure III-1 illustrates how demand for instream water varies during the year for some uses.

Uses compete. Allocation of water resources is not always controlled, as discussed in the introduction to Part II. However, in many instances legal compacts (interstate and international) have been established to control water distribution and use. Even this type of arrangement does not eliminate conflicts. An example is the recent decision of the Delaware River Basin Commission, a four-state compact. During the summer of 1985 the amount of water that New York City could withdraw from its two wholly owned reservoirs on branches of the upper part of the Delaware River was limited, in part to ensure enough rapidly moving water to avoid fish-kills of trout and other species dependent on moving water. Another reason was to limit the salt water front in the estuary and to maintain that front below the intake valves for the Philadelphia municipal water supply.

Some nomadic people live on as little as 5 liters (L) of water per day. Generally, 40 to 50 L/day is required for personal and domestic hygiene, and still greater amounts are needed for

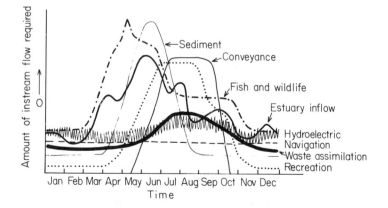

Figure III-1. Instream flow uses of water. In practically all water regions, the fish and wildlife use is the dominant instream flow use. Note the variations during the year except for navigation, where a minimum draft is constant. The amount needed for hydroelectricity increases with the demand for electricity during the summer. (*Source:* U.S. Water Resources Council, 1978, *Second national water assessment.*)

rural industry and watering of animals. Thus rural villagers need 100 L of water per person per day. With the increasing demands of large-scale organized irrigation and development of industry, water use can approach 500 L per person per day. Water, like energy, is a natural resource that increases in use with increasing development.

At the present stage of worldwide development, the major offstream uses are for agriculture (mostly irrigation), industry (mostly for steam generation in thermoelectric power plants), and domestic uses either through individual supplies or through municipal systems. Chapter 8, ''Water Demand'' by the United Nations Environment Programme, and Chapter 9, ''Fresh Water Supplies and Competing Uses'' by Sandra Postel, give a picture of the international scene. Irrigation accounts for about 70 percent of the total use, with industrial use accounting for another 20 percent. Great variations occur between countries. Comparing Figure 8-1 with Table 9-4, we see that the distribution diagram does not indicate the total amount of water used and could give the impression that India and Poland, for example, can be directly compared in water use. Postel's table indicates that India has a daily use of 1058 billion L, the third largest. The high proportion of water use for agriculture in India is thus much more important when one is calculating worldwide water uses than is the high industrial use of water in Poland.

Chapter 10, ''Estimated Use of Water in the United States in 1980'' by Wayne B. Solley, Edith B. Chase, and William B. Mann IV of the U.S. Geological Survey (USGS), summarizes water use in the United States. Irrigation in the West and industrial use in the East constituted the major demands for water in 1980. There is uncertainty in the numbers because not all uses in all river basins are well known. Indeed, the USGS figures are only one set of several different estimates. For example, the U.S. Water Resources Council published the *Second National Water Assessment* in 1978; its water use diagram for the United States is presented in Figure III-2. Also, the Soil Conservation Service of the U.S. Department of Agriculture, state agencies, and the various river basin commissions derive their own sets of numbers for various uses. So a key word in the title of Chapter 10 is *estimated*.

The remaining chapters in Part III deal with specific uses. Chapter 11, ''The Future of Irrigation'' by Kenneth D. Frederick, examines the effect on irrigated agriculture of past trends in irrigated acreage, availability and emerging demands for water, water institutions, rising energy costs and depth of water on water costs, emerging technologies, and environmental problems associated with irrigation. He feels that physical availability of water is not a problem.

''Don't drink the water!'' We have all heard this warning as we travel to areas where the availability of potable water is limited. In Chapter 12, ''Historical Review of Drinking Water'', Charles C. Johnson establishes three major periods for concepts about drinking water, separated by perceptions of diseases and their transmittal. However, knowing why you are getting sick and being able to do something about it are two different things.

Chapter 13, ''Maintaining Drinking Water Quality'' by Joseph A. Cotruvo and Ervin Bellack, describes the role of federal and state governments in the provision of acceptable drinking water. One of the key diagrams in this book is Figure 13-1, a diagram from a paper by Dr. Cotruvo that the editors added to this chapter. Almost 80 percent of the U.S. population is supplied with water by only 5 percent of the total number of water delivery systems. At the other end, two-thirds of the systems supply only 2 percent of the population. The level of management needed to supervise these two extremes is understandably quite different.

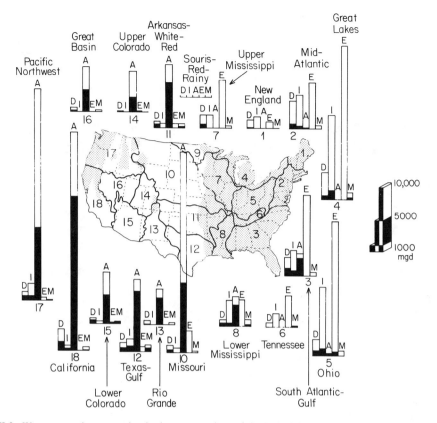

Figure III-2. Water use and consumption in the water regions of the United States. Heavy lines outline the regions. Shading indicates those areas where annual precipitation is greater than annual evapotranspiration. The water use in each region is given by the length of the bars (million gallons per day) for domestic (D), industrial (I), agricultural (A), offstream energy production (E), and mineral industry (M) uses. The shading within each bar indicates the amount of water consumed. (*Source:* Information from Water Resources Council, 1978. Diagram first appeared in David H. Speidel, 1982, Earth resources, in *Physical geology*, A. Ludman and N. Coch, Chap. 22. New York: McGraw-Hill. Copyright © by D. H. Speidel.)

One way to maintain drinking water quality is to purify it prior to use. A schematic diagram illustrating a conventional filtration supply system is given in Figure III-3. Raw water is mixed with chemicals to coagulate fine-grained particles and precipitate other material. Then the water moves through a series of sediment traps and a sand filter (with or without charcoal) to remove the remaining solid materials. Chlorine or some other biocidal agent and fluorine can be added to the water at this time, if needed. Storage and delivery complete the water supply diagram. Usually, only the very large systems follow all of the steps shown here. In Pennsylvania, for instance, there are more than 2500 water supply systems that are unfiltered and unregulated.

Domestic uses of water go beyond drinking water, though. For example, a recent drought emergency in New Jersey limited water use to 50 gal per person per day. Table III-1 illustrates a guide for how such a limit can be met. Note that washwater can be used to supply the water for toilets or, perhaps, the water for the lawn—a use not covered under the 50-gal maximum.

Water use in the production of energy is the subject of the next two chapters. Chapter 14,

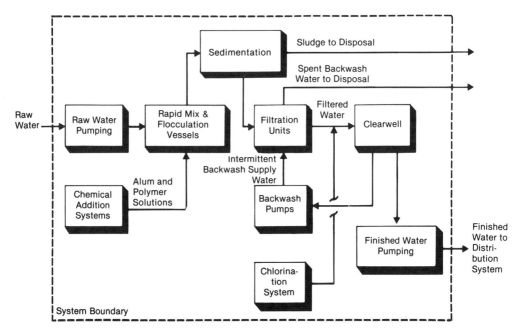

Figure III-3. Conventional filtration system for drinking water treatment. Raw water is pumped to a rapid-mix tank in which chemical (e.g., alum and polymer) solutions are added to enhance flocculation. A flocculation vessel allows sufficient time for the suspended solids to aggregate into larger particles, or flocs, which are removed by sedimentation and filtration. The waste sludge of solids and water is removed from the sedimentation basins by discharge to a municipal sewer or hauled to a landfill for disposal. Clarified water flows onto the filtration units, which are steel or concrete vessels containing granular materials such as graded sands, anthracite, and garnet. Solids are strained from the water as it passes through the filters. When the filters become clogged with accumulated solids, a backwash stream of filtered water passes through the units in reverse flow, cleaning the filter beds. This backwash is also sent to a sewer. Filtered water is disinfected with chlorine, stored, and then pumped to the water supply distribution system. (*Source:* G. DeWolf, P. Murin, J. Jarvis, and M. Kelly, 1984, *The cost digest: Cost summaries of selected environmental control technologies.* U.S. Environmental Protection Agency, EPA–600/8–84–010.) Washington, D.C., EPA.

''Water and Energy'' by John Harte and Mohamed El-Gasseir, proposes that the geographic and temporal variability of water supply constrains the choice of energy source and its level of use. These authors feel, for example, that the conversion of coal to synthetic fuel can be severely limited by water availability. In contrast, Kenneth Frederick, the author of Chapter 11, feels that water supplies appear to be adequate to meet all needs for developing domestic energy resources. He sees the impact of water on energy development as similar to the impact of water on western agriculture; that is, water systems and laws are appropriate only for conditions of abundance.

Another energy-producing use of water—generation of electricity through the use of moving water, or hydroelectric power—is described in Chapter 15, ''Water for Electricity'' by Charles M. Payne. Hydroelectric power is the fourth largest source of electricity in the United States, behind the three fossil fuels but ahead of nuclear power and all other minor supplies.

As previously discussed, the bulk of water used by industry is for the production of electricity. This use is generally not a problem because there is little consumptive loss and there is usually only the disposal of hot water to worry about. In other industries water is used as a coolant, a lubricant, and a component. Three industries—chemicals, primary metals, and pa-

Table III-1. How to Survive on 50-gal of Water per Person per Day

Use	What It Uses	Water Budget	Percent Used
Toilet	5–7 gal	15 gal for 3 flushes	30
Dishwashing	20 gal per load for automatic; 5–10 gal if by hand	10 gal	20
Shower	3 gal/min	9 gal	18
Tub	25–35 gal	Don't use	
Hygiene	3 gal/min to run tap	9 gal	18
Clothes washing	30–40 gal per load	5 gal (wash once per week)	10
Other	2 gal/drinking, cooking, watering, other	2 gal	4

Source: N.J. Department of Environmental Protection, 1985.

Note: To cut consumption, flush toilets only when necessary, take short showers, fill bathtubs less than halfway, wash dishes by hand, keep bottled water in the refrigerator for drinking, shave in a partially filled basin, place flow restrictors on taps, and fix leaky pipes.

per—account for 70 percent of the industrial water use, as shown in Table III-2. When the petroleum and coal industries and the food industry are added to the three largest users, the water use of these industrial groups is more that 85 percent of the total industrial water use. As you might imagine, disposal of this industrial waste water is something of a problem. (The specific problems generated by industrial waste water are discussed later in Chapter 21).

But not all domestic and industrial waste water must be disposed of. Some of it can be reused. Chapter 16, "Waste Water Reuse" by the Office of Technology Assessment, considers what is involved in preparing such water for reuse. Often public attitudes toward the application of recycled water is the limit to its use rather than technological or economic issues.

Table III-2. Industrial Water Use in the United States in 1978

Industry (SIC code)	Water Intake (Billion gal)	Percent of Total
Chemicals (28)	4,363.7	31.9
Primary metals (33)	3,414.7	25.0
Paper (26)	1,985.8	14.5
Petroleum and coal (29)	1,180.6	8.6
Food (20)	830.1	6.1
Transportation (37)	261.2	1.9
Stone, clay, and glass (32)	257.3	1.9
Nonelectrical machinery (35)	237.5	1.7
Rubber, miscellaneous plastics (30)	232.3	1.7
Lumber and wood products (24)	226.7	1.7
Textiles (22)	187.2	1.4
Fabricated metals (34)	160.8	1.2
Electrical equipment (36)	149.5	1.1
Other (21, 25, 31, 38, 39)	145.9	1.1
Total	13,679.0	

Source: U.S. Bureau of the Census, 1981, *Water use in manufacturing.*

8

WATER DEMAND

UNITED NATIONS ENVIRONMENT PROGRAMME

GENERAL PATTERN OF DEMAND

Recent estimates placed the total water use in 1980 in the order of 2600 to 3000 km^3/year; this is projected to reach 3750 km^3 in 1985—about 8 to 10 percent of the average runoff in all continental river basins. Accurate data for withdrawals are not available, but rough estimates for the three major uses in terms of volume of water taken from the source are presented in Table 8-1: Irrigation accounted for 73 percent, industry accounted for 21 percent, and the remainder went to domestic, livestock, recreational, and other uses. The proportion withdrawn and returned to its supply source, or the consumptive use, is large in the case of agriculture, where as much as 80 to 90 percent may be lost by evaporation and transpiration, and is usually small for domestic and industrial uses.

The differing national patterns of water withdrawals are illustrated by the data for 16 countries. (Figure 8-1). Some, such as Czechoslovakia, the German Democratic Republic, the Federal Republic of Germany, and Poland, made their greatest demands in the industrial sector. Others, such as India and Mexico, withdrew most heavily for irrigation. In all but a few the domestic-sector withdrawal was low.

The Organization for Economic Cooperation and Development (OECD) reported in 1979 that "in certain countries, while domestic withdrawals increased markedly between 1965 and 1975, growth now seems to have slowed due to reduced rates of urban expansion and small growth in the use of household appliances. Furthermore, there is evidence that the increasing cost of freshwater is beginning to affect industrial abstraction, except for cooling, and is leading to increased recycling sometimes in association with pollution control".

Table 8-1. Increase of Water Withdrawal over the World (km^3/year)

Water Used	1970	1975	Estimated Percentage Used Consumptively
Domestic water supply	120	150	1–15
Industry	510	630	0–10
Agriculture	1900	2100	10–80
Total	2530	2880	

Source: Ref. 1.

Figures for 1975 presented to the United Nations Water Conference by 14 governments from developed countries and 7 from countries with centrally planned economies indicated that all but 2 of them estimated that in the period 1981–2000 their annual percentage increase in all water use would be less than during 1970–1980 (2, p. 486). Sweden estimated in that connection that its annual water use would decline by 4 percent during the 1980s, and only the German Democratic Republic and the Federal Republic of Germany predicted increases in their annual rate of change.

IRRIGATION

Growth in the use of water for irrigation was linked to two factors. First and most important, the area served by regular water supplies was expanded by numerous large projects and by individual pump installations drawing upon ground water. In all regions the area of land reported to be served by irrigation increased by at least 15 percent during 1968–1978 (Table 8-2; compare with Table 9-4).

This material is excerpted from *The world environment, 1972–1982*, ed. Martin W. Holdgate, Mohammed Kassas, and Gilbert F. White, with assistance from David Spurgeon (Dublin: Tycooly International), 131–138. Reprinted by permission of United Nations Environment Programme. Copyright © 1982 by United Nations Environment Programme.

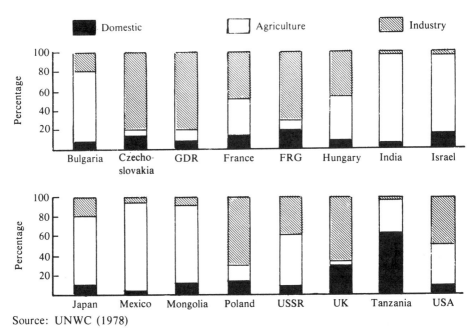

Source: UNWC (1978)

Figure 8-1. Distribution of withdrawals among major categories of water uses; selected countries, 1965 (*Source:* Ref. 2.)

The largest absolute gains were in Asia, principally in China and India, and the largest percentage gain in the Soviet Union (62.7 percent).

The stability of food production was enhanced thereby to the extent that land supplied by a water delivery system was actually cultivated. There is often a considerable time interval between the construction of a major water storage or delivery scheme and the completion of the distribution system, the preparation of the soil for cultivation, and the provision of suitable drainage. Thus surveys made

of irrigation projects in Africa south of the Sahara in 1965 and 1975 showed that the irrigation potential of the projects increased from 2.1 million to 3.04 million ha but that the actual cultivated acreage was less, and that in the event of full development the systems could reach 4.3 million ha (2, p. 536). In some developing countries the costs and difficulties of land development impeded or even reversed the expansion of irrigation. In developed countries in Europe and North America, much of the enlargement of irrigated land was through the

Table 8-2. Irrigated Area in Major Regions (1000 ha)

Region	1961–1965	1968	1978	Percent Change 1968–1978
World	149,474	162,634	200,913	23.5
Africa	5,870	6,772	7,831	15.6
North and Central America	18,606	20,366	23,543	15.6
South America	4,862	5,403	6,663	23.3
Asia	100,363	108,184	130,950	21.0
Europe	8,957	10,266	13,670	33.2
Oceania	1,198	1,443	1,656	14.8
Soviet Union	9,618	10,200	16,600	62.7

Source: Ref. 4.

installation of spray and similar pumping rigs for supplemental supply to lands already cultivated.

The second factor affecting the expansion of irrigation was the extent to which techniques of application and land management allowed economies in the amount of water required and the avoidance of environmental hazards. The development of deep wells and application systems that did not require extensive distribution channel networks reduced costs and in many cases also became more efficient in the sense of requiring less water for a given crop output. However, the pump and spray systems made it possible to deplete ground water at a rapid rate. Where the rates exceeded the natural recharge, the aquifers were depleted. Such "mining" of the supply occurred in areas like the high plains of the western United States. The gains in crop production were therefore temporary.

The lifetime of other irrigation projects was jeopardized by waterlogging and salinization resulting from inadequate design or operation and maintenance. On a large scale there was recognition, spurred on by the Food and Agriculture Organization of the United Nations (FAO), of the necessity to provide suitable drainage facilities, and this was expressed in the project reviews of the World Bank and the United Nations Development Programme (UNDP). There was substantial rehabilitation of waterlogged land in the Indus Valley. The intensity of use of irrigated land was estimated to have risen between 1965 and 1975 from 77 to 89 percent in Latin America, from 80 to 95 percent in the Near East, and from 119 to 129 percent in Asia (2, p. 912). Sophisticated techniques such as drip irrigation and measures to reduce water losses by canal lining, scheduling of water delivery in relation to soil moisture measurements, and other improved farm management systems spread slowly during the decade (2, p. 91). In the Federal Republic of Germany new spray techniques made it possible to irrigate in 1976, with only 210 million m^3 of water, the same area of land that in 1958 required 1230 million m^3.

INDUSTRY AND ENERGY

Industrial water withdrawals, as already noted for OECD countries, were on the increase, but with a marked trend in efficiency of water use. Water withdrawn per unit of manufacturing output decreased in many places, and greater attention was given to reuse within the plant. Pulp and paper plants began to employ processes that discharged little or no water and drew fresh supplies only to make up for evaporation and product losses (2, p. 2456). In Japan the total industrial water withdrawal moved from about 50 million m^3 daily in 1965 to 120 million m^3 in 1974, but by the mid 1970s two-thirds of the water was recycled, in contrast to one-third at the beginning of the period (2, p. 2440).

A large part of the additional use in many areas was for cooling thermal power installations. Air cooling was a substitute adopted in a number of plants. Where, however, major new power plants were in prospect, the demand for water was massive. In Ghana, for example, withdrawals were expected to grow 500-fold when the country supplemented its Volta River hydroelectric plant with a thermal installation (2, p. 543). There were also substantial new hydropower developments, and in several countries there was debate over their environmental impact, especially on rivers prized for their scenery and sport fisheries.

DOMESTIC

The domestic water supply picture varied. Whereas the proportion of the urban population in developing countries with access to safe water supply rose from 67 percent in 1970 to 77 percent in 1975 and then declined slightly to 75 percent in 1980 (as shown in Table 8-3), the proportion of rural people served by safe water supply increased from 14 percent in 1970 to 29 percent in 1980. During that period the number of countries reporting

Table 8-3. Estimated Service Coverage for Drinking Water Supply in Developing Countries, 1970–1980

	1970		1975		1980	
	Population Served (Millions)	Percentage of Total Population	Population Served (Millions)	Percentage of Total Population	Population Served (Millions)	Percentage of Total Population
Urban	316	67	450	77	526	75
Rural	182	14	313	22	469	29
Total	498	29	763	38	995	43

Source: Ref. 5.

Note: Figures do not include the People's Republic of China.

Table 8-4. Water Supply Coverage, Urban and Rural, by Region, for Countries Reporting in Both 1970 and 1980

Region	Number of Countries	1970			1980			Change in Percentage Covered
		Total Population (Millions)	Water Coverage (Millions)	Percentage of Total Population	Total Population (Millions)	Water Coverage (Millions)	Percentage of Total Population	
Africa (ECA members)								
Urban	29	62.8	51.5	82	96.2	78.9	82	0
Rural	23	187.8	40.2	21	239.6	64.6	27	+6
Latin America (ECLA members)								
Urban	18	153.1	115.6	76	212.6	157.8	74	−2
Rural	15	110.6	25.2	24	129.1	27.8	22	−2
Western Asia (ECWA members)								
Urban	9	13.9	13.3	96	22.5	19.8	88	−8
Rural	7	18.0	6.1	34	18.4	6.2	34	0
Asia and the Pacific (ESCAP members)								
Urban	14	220.5	130.2	59	300.3	209.5	70	+11
Rural	12	737.3	77.6	11	917.3	298.6	32	+21

Source: Refs. 5 and 6.

Note: The European (ECE members) region countries qualifying for technical assistance under UNDP procedures are not included, as only one country reported in both years, and the figures listed for it for 1970 are not consistent.

changed, and enumeration methods varied. The data therefore are of uneven quality and, when aggregated as shown in Table 8-4, may be misleading. A less rough picture of changes may be gained from comparing countries reporting in both 1970 and 1980. Table 8-3 excludes the developing-country members of the Economic Commission for Europe (ECE). Among the major regions only the Economic and Social Commission for Asia and the Pacific (ESCAP) members—the most numerous in population—reported significant gains in the proportion of both urban and rural populations covered by safe water supply. The Economic Commission for Africa (ECA) members reported an extension of coverage for rural dwellers but no increase in urban coverage. In Latin America and western Asia the proportions remained the same or declined (see Table 8-4).

Although the estimates are rough (and are not based on a uniform definition of what constitutes safe water and reasonable access) a few aspects of domestic supply became apparent as the decade drew to a close. Massive improvements were made in the availability of supply. The number of rural dwellers served increased by 157 percent. The urban dwellers served expanded by 66 percent. However, considering the total population in need of service, the urban gains were modest; and while the rural proportion doubled, it still left more than two-thirds without safe service. In only one major region was the rate of improvement in excess of the rate of population growth. A continuation of the 1970–1980 trends would leave the total population only slightly better off. A disturbing aspect of the situation not revealed by the statistics (and, indeed, not precisely documented) was the probably large number of rural improvements that had fallen into disrepair. In addition, many urban systems, such as those in Nepal and Pakistan, provided only intermittent service (ESCAP, 1980).

The waste water situation was even less heartening. While a high proportion of the developed urban populations had adequate services, the proportion of developing-country urban population served by sewers, latrines, or other sanitary measures for excreta disposal declined during the decade from 71 to 53 percent (Table 8-5).

In rural areas the numbers served were 11 percent in 1970 and little better in 1980. Regional data are so incomplete that comparisons between 1970 and 1980 are not warranted. In 1980, however, it appeared that the proportion of urban populations covered was Africa, 56 percent; Latin America, 54 percent; western Asia, 70 percent; and Asia and the Pacific, 50 percent. For the rural populations the estimates were Africa, 15 percent; Latin America, 23 percent; western Asia, 20 percent; and Asia and the Pacific, 10 percent. Urban sanitation efforts clearly had not kept up with population growth, and rural improvement had barely kept pace.

INLAND NAVIGATION

Inland navigation on international rivers has been acceptable for decades, because of the evident mutual interests of riparian countries and because of its nonconsumptive use of water. However, with the increasing demand for water for other purposes, this situation is gradually changing, and great effort and cost are involved in harmonizing the different interests.

The decade witnessed increasing awareness of the environmental impact of inland navigation and developed new techniques to prevent possible pollution. The measures taken included the provision of facilities to retain sewage on board boats, for later discharge ashore. Care was taken to minimize pollution from oil spillage and leaks, bilge water tank refuse, and water liquids. In many cases these were being reclaimed and recycled. The disposal of dredged material from channel beds in the course of routine annual maintenance was another major environmental concern. Research was undertaken in a number of countries, especially the United

Table 8-5. Estimated Service Coverage for Sanitation in Developing Countries, 1970–1980

	1970		1975		1980	
	Population Served (Millions)	Percentage of Total Population	Population Served (Millions)	Percentage of Total Population	Population Served (Millions)	Percentage of Total Population
Urban	337	71	437	75	372	53
Rural	134	11	209	15	213	13
Total	471	27	646	33	585	25

Source: Ref. 5.

Note: Figures do not include the People's Republic of China.

States, to evaluate its environmental impact and develop improved dredging and disposal techniques.

RECREATION

The use of inland waters for recreation, and in particular, bathing and fishing, gained popularity during the past ten years. In most industrial countries this brought demands for improvement in the quality of inland waters, to make more of them "fishable" and "swimmable". The formulation of minimum health or environmental quality standards enabled sites to be more easily protected for bathers and facilitated measures to control pollution caused by indiscriminate waste discharge from factories and conurbations. Many governments also supported action to preserve or improve fishing. Because the presence of fish in inland waters is one of the better indices of the quality of the aquatic environment, the concerns of advocates of water-based recreation and of water quality management frequently converged. The importance of inland wetlands for wildlife conservation is widely recognized.

NOTES

1. U.S.S.R. Committee for the International Hydrological Decade 1974, 1978, *World water balance and water resources of the earth,* English translation (Paris: UNESCO).

2. United Nations Water Conference, 1978, *Proceedings of the United Nations Water Conference, Mar dal Plata, Argentina, 1977,* 4 vols. (Oxford: Pergamon).

3. Organization for Economic Cooperation and Development, 1979, *The state of the environment in OECD member countries,* (Paris: OECD).

4. Food and Agricultural Organization, 1980, *1979 production yearbook* (Rome: FAO).

5. United Nations, 1980, *International drinking water supply and sanitation decade; present situation and prospects,*. A/35/367 (New York: United Nations).

6. World Health Organization, 1973, *World health statistics,* Report 26, 11 (Geneva: WHO).

7. Economic and Social Commission for Asia and the Pacific, 1980, *Document E/ESCAP/71,* (New York: United Nations).

9

FRESH WATER SUPPLIES AND COMPETING USES

SANDRA POSTEL

Numbers alone fail to tell water's true story. Enough rain and snow falls over the continents each year to fill Lake Huron 30 times, to magnify the flow of the Amazon River 16-fold, or to cover the earth's total land area to a depth of 83 cm. Yet lack of water to grow crops periodically threatens millions with famine. Water tables in southern India, northern China, the Valley of Mexico, and the U.S. Southwest are falling precipitously, causing wells to go dry. Rivers that once ran year-round now fade with the end of the rainy season. Inland lakes and seas are shrinking.

Unlike coal, oil, wood, and most other vital resources, water is usually needed in vast quantities that are too unwieldy to be traded internationally. Rarely is it transported more than several hundred kilometers from its source. Thus while fresh water everywhere is linked to a vast global cycle, its viability and adequacy as a resource is determined by the amount available locally or regionally and by the way it is used and managed.

Each year the sun's energy lifts some 500,000 km^3 of water from the earth's surface—86 percent from the oceans and 14 percent from land. (One cubic kilometer equals one billion cubic meters or one trillion liters; in standard United States usage the equivalent is about 264 billion gal.) An equal amount falls back to earth as rain, sleet, or snow but fortunately not in the same proportions. Some 110,300 km^3 falls over land (excluding Greenland and Antarctica), whereas only 71,500 is evaporated from it. Thus this solar-powered cycle annually distills and transfers 38,800 km^3 of water from the oceans to the continents. To complete the natural cycle, the water then makes its way back to the sea as runoff.[1]

By virtue of this cyclic flow between the sea, air, and land, fresh water is a renewable resource. Under the planet's existing climatic conditions approximately the same volume is made available each year. Today's supply is the same as when civilizations first dawned in the fertile river valleys of the Ganges, the Tigris-Euphrates, and the Nile. Viewed globally, fresh water is still undeniably abundant: For each human inhabitant there is now an annual renewable supply of 8300 m^3, which is enough to fill a 6-m^2 room 38 times, and several times the amount needed to sustain a moderate standard of living.[2]

Natural variations in climate and the vagaries of weather easily cast shadows over this picture of plenty, however, for water is not always available when and where it is most needed. Nearly two-thirds of each year's runoff flows rapidly away in floods, often bringing more destruction than benefit. The other third is stable and is thus a reliable source of water for drinking or irrigating crops year-round. Water that infiltrates and flows underground provides the base flow of rivers and streams and accounts for most of the stable supply. The controlled release of water from lakes and reservoirs adds a bit more, bringing the total stable supply to about 14,000 km^3 or 3000 m^3 per person—the present practical limit of the renewable fresh water supply.

Asia and Africa are the continents facing the greatest water stress. Supplies for each Asian today are less than half the global average, and the

Sandra Postel is a senior researcher with Worldwatch Institute. She is author of several Worldwatch papers, and coauthor of the Institute's annual *State of the world* report. She studied geology and political science at Wittenberg University and resource economics and policy at Duke University. This material is excerpted from Managing freshwater supplies, in *State of the world 1985* (New York: Norton), Chapter 3, 42–50. Reprinted by permission of the author and Worldwatch Institute, Washington, D.C. Copyright © 1985 by Worldwatch Institute.

continent's runoff is the least stable of all the major land masses (see Table 9-1). Lofty mountain ranges and a monsoon climate make rainfall and runoff highly variable. China's Huang He, or Yellow River, has had at least one major change of course every century of the 2500 years of recorded Chinese history. In India 90 percent of the precipitation falls between the months of June and September, and most of the runoff flows in the Ganges and Brahmaputra basins in the north. Failure of the 1979 monsoon led to one of the worst droughts of recent record and reduced India's production of foodgrains by 16 percent. In Africa the Zaire River (formerly the Congo)—second in volume only to the Amazon—accounts for about 30 percent of the continent's renewable supplies but flows largely through sparsely populated rain forest. Two-thirds of the African nations have at least a third less annual runoff than the global average. Drought conditions that persistently plague the continent's dry regions have in recent years threatened over twenty nations with famine.[3]

North and South America and the Soviet Union all appear to have abundant water resources for their populations, though again great geographic disparities exist. South America appears the most richly endowed continent, yet 60 percent of its runoff flows in the channel of the Amazon, remote from most people and a hard source to tap. North and Central America together have a per capita water supply twice the global average, but natural supplies are limited in broad areas of the West, particularly in the southwestern United States and northern Mexico. The Soviet Union's three largest rivers—the Yenisei, the Lena, and the Ob'—all flow north through Siberia to the Arctic Seas, far from the major population centers. Finally, Europe joins Asia as a continent with a substantially greater share of the world's people than of its fresh water. The continent's per capita runoff is only half the global average, and supplies are especially short in southern and eastern Europe. Fortunately, for much of the continent a generally temperate climate and a large number of smaller rivers with fairly steady flows allow a comparatively large share of the runoff to be tapped.

A detailed breakdown of supplies by country confirms water's unequal distribution (see Table 9-2). Per capita runoff ranges from over 100,000 m^3 in Canada to less than 1000 in Egypt. Yet even these national figures hide important disparities. On a per capita basis Canada is the most water-wealthy nation in the world, but two-thirds of its river flow is northward, while 80 percent of its people live within 200 km of the Canadian-U.S. border. Similarly, Indonesia appears to be a relatively water-rich nation, yet over 60 percent of the population live on the island of Java, which has less than 10 percent of the country's runoff. Especially for the water-poor nations of Europe, Africa, and Asia, water flowing in from neighboring countries can be a vital addition to the runoff originating within their own borders. (The runoff estimates in Table 9-2 are consistent with a global water balance and thus include only runoff originating within each particular country.) Inflow accounts for roughly 70 percent of Czechoslovakia's water supplies, for example, roughly half of East and West Germany's, and 90 percent of Bulgaria's. Egypt, one of the most water-short nations in the world, is almost entirely dependent on the water of the Nile that enters the country from Sudan.[4]

Given existing climatic conditions and current population projections, the per capita global water supply at the end of the century will have declined by 24 percent, while the stable, reliable component

Table 9-1. Distribution of Renewable Fresh water Supplies, by Continent

Region	Average Annual Runoff (km^3)[a]	Share of Global Runoff	Share of Global Population (%)	Share of Runoff That Is Stable
Africa	4,225	11	11	45
Asia	9,865	26	58	30
Europe	2,129	5	10	43
North America[a]	5,960	15	8	40
South America	10,380	27	6	38
Oceania	1,965	5	1	25
Soviet Union	4,350	11	6	30
World	38,874	100	100	36[b]

Sources: Adapted from M. I. L'vovich, 1979, *World water resources and their future,* translation ed. Raymond L. Nace (Washington, D.C.: American Geophysical Union); population figures are mid-1983 estimates from Population Reference Bureau, 1983, *1983 World population data sheet* (Washington, D.C.).

[a] Includes Central America, with runoff of 545 km^3.   [b] Average.

Table 9-2. Average Annual per Capita Runoff
in Selected Countries, 1983,
with Projections for 2000

Country	1983 (1000 m³)	2000 (1000 m³)	Change (%)
Canada	110.0	95.1	−14
Norway	91.7	91.7	0
Brazil	43.2	30.2	−30
Venezuela	42.3	26.8	−37
Sweden	23.4	24.3	+4
Australia	21.8	18.5	−15
Soviet Union	16.0	14.1	−12
United States	10.0	8.8	−12
Indonesia	9.7	7.6	−22
Mexico	4.4	2.9	−34
France	4.3	4.1	−5
Japan	3.3	3.1	−6
Nigeria	3.1	1.8	−42
China	2.8	2.3	−18
India	2.1	1.6	−24
Kenya	2.0	1.0	−50
South Africa and Swaziland	1.9	1.2	−37
Poland	1.5	1.4	−7
West Germany	1.4	1.4	0
Bangladesh	1.3	0.9	−31
Egypt	0.09	0.06	−33
World	8.3	6.3	−24

Sources: M. I. L'vovich, 1979, *World water resources and their future*, translation ed. Raymond L. Nace (Washington D.C.: American Geophysical Union); population figures are mid-1983 estimates from Population Reference Bureau, 1983, *1983 World population data sheet* (Washington D.C.).

Note: Estimates are for runoff originating within each specific country and do not include inflow from other countries.

ity—as a factor of production in agriculture, industry, or household activities. Yet water in rivers, lakes, streams, and estuaries also is home to countless fish and plants, acts as a diluting and purifying solvent, and offers a source of aesthetic enjoyment and richness that adds immeasurably to the quality of life. No society can draw on all its available supplies and hope to maintain the benefits water freely offers when left undisturbed. The need to protect these natural functions is thus a critical backdrop to considering society's pattern of water use.

Although the practice of irrigation dates back several thousand years to early Egyptian and Babylonian societies, and although water has been tapped to supply homes and small industries for centuries, for most of humanity's history water use expanded at a moderate pace. Over this century, however, demands have soared with rapid industrialization and the need to feed an expanding world population. According to estimates prepared by Soviet scientists in the early seventies for the U.N. International Hydrological Decade (1965–1974), which are among the most comprehensive historical data available, world water use in 1900 was 400 billion m³, or 242 m³ per person. By 1940 global use had doubled, while population had increased about 40 percent (see Figure 9-1). A rapid rise in water demand then began at midcentury: By 1970 annual per capita withdrawals had climbed to over 700 m³, 60 percent higher than in 1950. Both agricultural and industrial water use increased twice as much during these 20 years as they had over the entire first half of the century.[6]

Today, humanity's annual water withdrawals equal about a tenth of the total renewable supply and about a quarter of the stable supply—that which is

of that water will have dropped from 3000 to 2280 m³ per person. Population continues to grow fastest in some of the most water-short regions. Per capita supplies in Kenya and Nigeria, for example, will diminish by 50 and 42 percent, respectively. Supplies per person in Bangladesh and Egypt will diminish by a third, and in India by a fourth. Moreover, if projected climatic shifts from the rising concentration of atmospheric carbon dioxide materialize, water supplies may diminish in some areas already chronically water-short, including major grain-producing regions of northern China and the United States.[5]

COMPETING USES

When analysts speak of the "demand" for water, they typically refer to water's use as a commod-

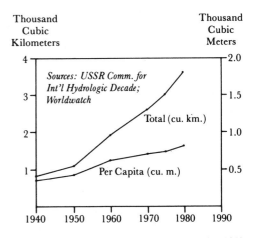

Figure 9-1. World water use, total and per capita, 1940–1980.

typically available throughout a year. Agriculture claims the lion's share of world water use, accounting for about 70 percent of total withdrawals. As fertile land became more scarce, irrigation enabled farmers to get higher yields from existing fields, essentially substituting water for new cropland. With a controllable, year-round source of water, farmers also found it profitable to invest in fertilizer and to plant higher-yielding crop varieties. Yields of rain-fed rice, for example, typically increase by 50 percent if the effects of flood and drought can be eliminated, by 130 percent if controlled irrigation and drainage and some fertilizer are introduced, and by 280 percent or more if advanced irrigation techniques, generous amounts of fertilizer, pest control, and high-yielding seeds are used.[7]

Roughly a third of today's harvest comes from the 17 percent of the world's cropland that is irrigated. Irrigation thus greatly helps meet the challenge of feeding an ever-growing population. Since 1950 the irrigated area worldwide has increased from 94 million to 261 million ha. During the sixties irrigation water was brought to an additional 6 million ha each year; since 1970 an additional 5.2 million ha have been added annually (see Table 9-3). At today's average rates of water use (some 11,000–12,000 m^3 per irrigated hectare per year), and assuming irrigation continues to expand at a slightly diminishing rate, an additional 820 km^3 of water will be needed for irrigation each year by the turn of the century—a 25 to 30 percent increase over existing levels.[8]

Besides demanding a large share of any region's available supplies, irrigation results in a large volume being "consumed"—removed from the local water supply through evaporation and transpiration. Crops must consume some water in order to grow, but typically much more water is transported and applied to fields than the crops require. Often less than half the water withdrawn for irrigation returns to a nearby stream or aquifer, where it can be used again. In the United States, for example, 55 percent of agricultural withdrawals are consumed, which in turn accounts for 81 percent of all the water consumed annually nationwide.[9]

Industry is the second major water-using sector of society, accounting for about a quarter of water use worldwide. Producing energy from nuclear and fossil-fueled power plants is by far the largest single industrial water use. Water is the source of steam that drives the turbogenerators, and vast quantities are used to cool power plant condensers. Unlike in agriculture, however, only a small fraction of this water is consumed. Most existing power plants have "once-through" cooling systems that return water to its source immediately after it passes through the plant. United States power plants, for example, consume only 2 percent of their withdrawals. Thus especially when plants are situated next to large lakes or rivers, the volume of cooling water withdrawn is usually of less concern than the discharge of heated water back to the source. If lake or stream temperatures get too high, oxygen levels may drop, threatening fish and other aquatic life.[10]

Excluding energy production, two-thirds of the remaining industrial withdrawals go to just five industries: primary metals, chemical products, petroleum refining, pulp and paper manufacturing, and food processing. In countries with an established industrial base and water pollution laws in effect, withdrawals for these industries are not likely to increase. Most pollution control techniques involve recycling and reusing water, thus reducing an industry's demand for new supplies. Industrial use has declined, or is expected to decline soon, in countries such as Finland, Sweden, and the United States. In contrast, Portugal, the

Table 9-3. Growth in Irrigated Area, by Continent, 1950–1982

Region	Total Irrigated Area, 1982 (Million ha)	Growth in Irrigated Area (%)		
		1950–1960	1960–1970	1970–1980[a]
Africa	12	25	80	33
Asia[b]	177	52	32	34
Europe[c]	28	50	67	40
North America	34	42	71	17
South America	8	67	20	33
Oceania	2	0	100	0
World	261	49	41	32

Source: W. R. Rangeley, 1983, Irrigation—Current trends and a future perspective (Washington, D.C.: World Bank Seminar).

[a]Percentage increase between 1970 and 1982 prorated to 1970–1980 to maintain comparison by decade.   [b]Includes the Asian portion of the Soviet Union.   [c]Includes the European portion of the Soviet Union.

Soviet Union, Turkey, and several of the Eastern bloc nations are projecting a doubling of their industrial withdrawals over the century's last quarter. Increases of no more than 50 percent are expected in Czechoslovakia, France, and East and West Germany.[11]

Industry typically accounts for less than 10 percent of total withdrawals in most Third World countries, compared with 60 to 80 percent in most industrial nations (see Table 9-4). Much of the developing world is just embarking on the industrialization path taken by other countries four decades ago. Water demands for power production, manufacturing, mining, and materials processing are thus poised for a rapid increase if industries adopt the water-intensive technologies that those of the industrial world did. Industrial water use in Latin America, for example, is projected to jump 350 percent during the century's last quarter, compared with nearly 180 percent for municipal uses and 70 percent for irrigation (see Figure 9-2). Among the targets set for the United Nations Second Development Decade is an 8 percent average annual rate of industrial growth for the Third World. Though this may prove too ambitious a goal, given the debt burden many of these countries face, the developing world's industrial water use could easily double by the end of the century.[12]

Water used by households—for drinking and cooking, bathing, washing clothes, and other activities—varies greatly with both income levels and the way in which water is supplied. In urban households with piped water available at the touch

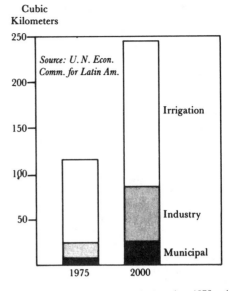

Figure 9-2. Water demands in Latin America, 1975, with projections for 2000.

of a tap, daily use typically ranges between 100 and 350 L per person. Households with water-intensive appliances, such as dishwashers and washing machines, and those where water is used to irrigate large lawns and gardens can use over 1000 L per person daily. In many developing countries, where water is supplied through a public

Table 9-4. Estimated Water Use in Selected Countries, Total, per Capita, and by Sector, 1980

Country	Daily Water Withdrawals		Share Withdrawn by Major Sectors (%)		
	Total (Billion L)	Per Capita (1000 L)	Agriculture	Industrial	Municipal[a]
United States	1683	7.2	34	57	9
Canada	120	4.8	7	84	9
Soviet Union	967	3.6	64	30	6
Japan[b]	306	2.6	29	61	10
Mexico[b]	149	2.0	88	7	5
India[b]	1058	1.5	92	2	6
United Kingdom	78	1.4	1	85	14
Poland	46	1.3	21	62	17
China	1260	1.2	87	7	6
Indonesia[b]	115	0.7	86	3	11

Sources: U.S. data, U.S. Geological Survey; Canadian data, Harold D. Foster and W. R. Derick Sewell, 1981, *Water: The emerging crisis in Canada* (Toronto: Lorimer); Soviet, U.K., Polish data, U.N. Economic Commission for Europe; Japanese, Indian, Indonesian data, *Global 2000 report;* Mexican data, U.N. Economic Commission on Latin America; Chinese data, Vaclav Smil, *The bad earth.*

[a] Along with residential use, figures may include commercial and public uses, such as watering parks and golf courses.   [b] 1975 figures for Mexico; 1977 for India, Indonesia, and Japan.

hydrant, daily use ranges between 20 and 70 L per person. Areas such as Kenya, where women may walk several kilometers to draw water for their families, can record use close to the biological minimum—2 to 5 L per person daily.[13]

Residential and other municipal uses of water account for less than a tenth of water withdrawals in many nations, and only about 7 percent of total withdrawals worldwide. In industrial countries where population growth is slow and most households are already adequately supplied with water, growth in domestic demand is slowing and probably will continue to do so. In parts of Europe that are still converting from community wells to individual piped-water systems—including Czechoslovakia, Poland, Portugal, Romania, and Turkey—demand for drinking water is expected to double over the next two decades. The largest increase will probably occur in the Third World, where fresh water supplies are not yet universally available. The World Health Organization estimates that as of 1980 only 75 percent of the developing world's urban dwellers and 29 percent of its rural population were served with drinking water. The United Nations has set a goal of providing safe water to all by 1990, which, although unlikely to be met, will contribute to a probable doubling of Third World domestic water demands by the end of the century.[14]

Even given these large increases in water withdrawals for irrigation, industrial, and domestic needs, total use worldwide by the year 2000 is still likely to be less than half the stable renewable supply. Yet projections by leading hydrologists show that meeting demands in North Africa and the Middle East will require virtually all the usable fresh water supplies in these regions. Use in southern and eastern Europe, as well as central and southern Asia, will also be uncomfortably close to the volume of supplies these regions can safely and reliably tap.[15] Moreover, even if supplies appear more than adequate, no region is immune from the consequences of mismanagement and abuse that are already arising and that are bound to worsen as competing demands escalate.

NOTES

1. Frits van der Leeden, 1975, *Water resources of the world* (Port Washington, N.Y.: Water Information Center).

2. An annual supply of 1000 m³ per person is typically given as necessary for a decent standard of living. See Carl Widstrand, ed., 1980, *Water conflicts and research priorities* (Oxford: Pergamon).

3. Vaclav Smil, 1984, *The bad earth: Environmental degradation in China* (Armonk, N.Y.: Sharpe); Malin Falkenmark and Gunnar Lindh, 1976, *Water for a starving world* (Boulder, Colo.: Westview Press); Gary S. Posz et al., 1980, Water resource development in India, American Embassy, New Delhi, June; foodgrain reduction from B. B. Sundaresan, 1984, Water: A vital resource for the developing world, in *Water and sanitation: Economic and sociological perspectives,* ed. Peter G. Bourne (Orlando, Fla.: Academic Press); Zaire River flow from van der Leeden, *Water resources of the world;* famine threat in Africa discussed in Worldwatch Institute, 1985, *State of the world 1985.*

4. Inquiry on Federal Water Policy, 1984, *Water is a mainstream issue: Participation paper,* (Ottawa: Canadian Minister of Supply and Services); Mardjono Notodihardjo, 1983, Indonesia's water resources, in *Water for human consumption,* ed. W. Hall C. Maxwell, Proceedings of the Fourth World Congress of the International Water Resources Association (Dublin: Tycoly International); U.N. Economic Commission for Europe (UNECE), 1981, *Long-term perspectives for water use and supply in the ECE region* (New York: United Nations)

5. William W. Kellogg and Robert Schware, 1982, Society, science and climate change, *Foreign Affairs* (Summer).

6. USSR Committee for the International Hydrological Decade, 1974, *World water balance and water resources of the earth* (Paris: UNESCO).

7. Rice yields with different degrees of water control from Asit K. Biswas, 1983, Major water problems facing the world, *International Journal of Water Resources Development* (April).

8. Estimate of global irrigated area and its contribution to production from W. R. Rangeley, 1983, *Irrigation—Current trends and a future perspective,* (Washington, D.C.: World Bank Seminar). Irrigation demands based on FAO estimate for the 1974 World Food Conference that gross water demand for harvested crop irrigation averages 11,400 m³/ha. Estimate takes into account that rice requires twice as much water as wheat and other dry cereals. This figure, applied to an average expansion of irrigation of 4 million ha annually, led to estimate of an additional 820 billion m³ annually for irrigation by 2000.

9. Wayne B. Solley et al., *Estimated use of water in the United States in 1980* (see Chapter 10 of this book).

10. See Chapter 14 of this book and Norman L. Dalsted and John W. Green, 1984, Water requirements for coal-fired power plants, *Natural Resources Journal,* (January); percent consumed from Solley et al., Chapter 10 of this book.

11. Major water-using industries from United Nations, 1977, *Resources and needs: Assessment of the world water*

situation, prepared for the U.N. Water Conference, Mar del Plata, Argentina, March 1977; trends in European countries from UNECE, *Long-term perspectives,* Swedish Preparatory Committee for the U.N. Water Conference, 1977, *Water in Sweden* (Stockholm: Ministry of Agriculture).

12. Development decade goals cited in Biswas, Major water problems.

13. United Nations, *Resources and needs;* Solley et al., *Estimated use of Water.*

14. UNECE, *Long-term perspectives;* World Health Organization, 1981, *Drinking water and sanitation,* 1981–1990 (Geneva).

15. Widstrand, *Water conflicts.*

10

ESTIMATED USE OF WATER IN THE UNITED STATES IN 1980

WAYNE B. SOLLEY, EDITH B. CHASE, and WILLIAM B. MANN IV

Water use in this chapter is considered as offstream use and instream use. The difference between these two types of use is explained next.

INSTREAM USE

Instream use is a water use not dependent on withdrawal or diversion from ground or surface water sources, and it usually is classified as flow uses or on-site uses. Examples of flow uses, which depend on water running freely in a channel, are hydroelectric power generation, fresh water sweetening of saline estuaries, maintenance of minimum stream flow to support fish propagation, and the disposition and dilution of wastewater. On-site uses may occur when water is used directly in a water course, lake, reservoir, or other body of water—an example is evaporation from a lake or reservoir associated with hydroelectric power generation.

Quantitative estimates for most instream uses are difficult to make. However, because such uses reflect the level of competition with offstream uses and affect the quantity and quality of water resources for all uses, effective water resources management requires that methods and procedures be devised to enable instream uses to be quantitatively determined.

The only instream use discussed in this report is hydroelectric power generation. Unlike other instream uses, the water used for hydroelectric power generation is a measurable quantity, because the water is passed through the plant and can be documented. Consumptive use in hydroelectric power generation generally is negligible and is not discussed.

OFFSTREAM USE

Offstream use is a water use that depends on water being withdrawn or diverted from a ground or surface water source. In a determination of the amount of water used, three factors are involved:

1. *Withdrawals:* The amount of water withdrawn or diverted from a ground or surface water source.
2. *Delivery/release:* The amount of water delivered at the point of use and the amount released after use. The difference between these volumes will in some instances be the consumptive use, the amount of water that is no longer available for subsequent use.
3. *Return flow:* The amount of water that reaches a ground or surface water source after release from the point of use and becomes available for further use.

Figure 10-1 explains how these factors are aggregated for three industries (X,Y,Z) in the same SIC class.

In this chapter withdrawal data and estimates of consumptive use are given for four categories of offstream use: public supply (water delivered to domestic, commercial, and industrial users), rural use (self-supplied domestic and livestock use), irrigation use, and self-supplied industrial use. Data on delivery/releases and return flows were not adequate for detailed discussions; however, generalized return flow information is presented graphically for each category of use.

It should be noted that each category of use has characteristically different effects on the usability

The authors are hydrologists with the United States Geological Survey. This material is excerpted from U.S. Geological Survey Circular 1001 (1983), pp. 6, 7, 8, 12, 25, 28, 32, 46, 47, 48, 56; Figures 5, 6, 9b, 12, 13; Tables 13, 15, 17, 22.

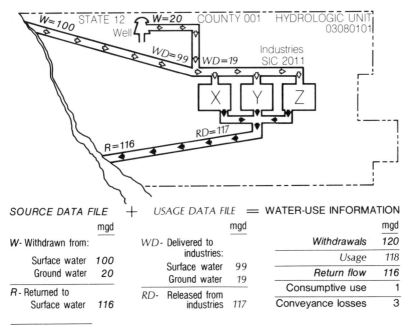

SOURCE DATA FILE	+	USAGE DATA FILE	=	WATER-USE INFORMATION	
mgd		mgd			mgd

SOURCE DATA FILE		USAGE DATA FILE		WATER-USE INFORMATION	
W- Withdrawn from:		WD- Delivered to industries:		Withdrawals	120
Surface water	100	Surface water	99	Usage	118
Ground water	20	Ground water	19	Return flow	116
R- Returned to		RD- Released from industries	117	Consumptive use	1
Surface water	116			Conveyance losses	3

Withdrawals (W), Usage (WD), Return flow (R), Consumptive use (WD–RD),
Conveyance losses [(W–WD)+(RD–R)]

Figure 10-1. Hypothetical aggregated water use information for industries X, Y, and Z derived from source and usage data files.

or reuse potential of return flows. Reuse potential is a measure of the quality and quantity of water available for subsequent use. For example, irrigation return flow may be contaminated by pesticides and fertilizers, and because of the high consumptive use in irrigation, the mineral content of the return flow often is increased substantially. Consequently, irrigation return flow frequently has little reuse potential. This reuse potential is a significant contrast to that of water discharged from thermoelectric plants, where the principal change in the water is an increase in temperature, and the return flow has maximum potential for further use. The National Water-Use Information Program is now documenting return flows, and future reports will contain such data. Future plans also include obtaining information on water quality changes associated with the various uses of water.

SUMMARY OF OFFSTREAM AND INSTREAM USES

The estimated withdrawal of 450 billion gallons per day (bgd) for all offstream uses (public supply, rural, irrigation, and self-supplied industrial use)

in 1980 was about 8 percent greater than the withdrawals estimated for 1975. Ground water withdrawals accounted for 89 bgd, a 7 percent increase over 1975; of this amount 88 bgd was fresh water. Surface water withdrawals accounted for 360 bgd, a 9 percent increase from 1975, of which 71 bgd was saline water. Reclaimed sewage amounted to 0.5 bgd in 1980, an 11 percent decrease from 1975.

Fresh water consumptive use in 1980 was estimated at 100 bgd, a 7 percent increase from 1975. The percentages of water consumed by the various use categories were nearly the same as in 1970 and 1975. Irrigation water accounted for the largest amount of water consumed, 83 bgd. In addition, conveyance losses associated with irrigation were estimated at 24 bgd. Geographically, 80 percent of the consumptive use was in the western states, a decrease of 4 percent since 1975 and 6 percent since 1970, whereas the 20 percent consumed in the eastern states reflects an increase of 6 percent since 1970.

Several tables and illustrations are included in this section to summarize the vast amount of data given in this chapter. The percentages of water withdrawn and consumed by the four offstream water use categories are shown in Figure 10-2. A comparison of withdrawals from ground and surface water sources for both states and water resources regions is shown in Figure 10-3.

The per capita withdrawals and consumptive use for the United States and for the eastern and western water resources regions are given in Table 10-1. The total offstream water use (withdrawals, conveyance losses, and consumptive use) is given by water resources regions in Table 10-2. A summary of withdrawals for the offstream water use categories is given by water resources regions in Table 10-3. Total offstream withdrawals by source and disposition are shown in Figure 10-4.

Public Supply

Public supply refers to water withdrawn by public and private water suppliers and delivered to a variety of users for domestic or household use, public use, industrial use, and commercial use.

Public suppliers served about 186 million people in 1980, about 81 percent of the total population, a slight increase in percentage since 1975. Domestic use includes such activities as drinking, food preparation, bathing, washing clothes and dishes, flushing toilets, and watering lawns and gardens. Public use includes water for firefighting, street washing, and municipal parks and swimming pools. Many industrial and commercial establishments use public supplies, especially where the volume of water required is small and the quality of water must be high. However, some industries that require large amounts of water also use public supply for principal or auxiliary water. Among commercial users are hotels, restaurants, laundry services, office facilities, and institutions, both civilian and military. Data on population served by public supply and public supply withdrawals and deliveries usually are reliable because local government agencies generally maintain relatively complete files.

Total water withdrawn for public supply in 1980

Table 10-1. Per Capita Water Withdrawals and Consumptive Use;
Eastern and Western Water Resources Regions and United States, 1980

	Conterminous United States Water Resources Regions		United States (50 States, District of Columbia, Puerto Rico, and Virgin Islands)
	Eastern (9 Regions = 31 States)[a]	Western (9 Regions = 17 States)[a]	
Population, in millions			
Total	155.7	69.1	229.6
Served by public supplies	123.5	58.1	186.1
Self-supplied (rural)	32.2	11.0	43.5
Per capita water use (gal/day)			
Offstream use			
Total withdrawals[b]	1,600	2,900	2,000
Public supplies			
All uses[c]	160	230	180
Domestic and public uses and losses[c]	100	150	120
Rural domestic use[d]	73	98	79
Irrigation[b]	82	2,000	660
Self-supplied industrial[b]	1,300	660	1,100
Consumptive fresh water use[b]	120	1,200	450
Instream use			
Hydroelectric power[b]	8,900	27,000	14,000
Total offstream and instream use[b]	10,000	30,000	16,000

Note: All per capita data calculated from unrounded figures and rounded to two significant figures.

[a] Approximate boundaries.
[b] Based on total population.
[c] Based on population served by public supplies.
[d] Based on rural population.

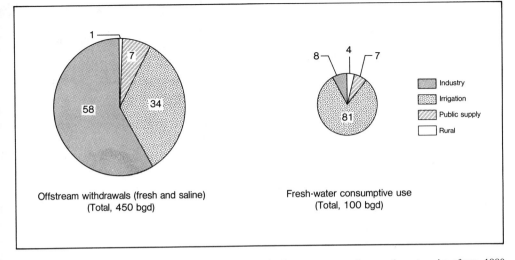

Offstream withdrawals (fresh and saline)
(Total, 450 bgd)

Fresh-water consumptive use
(Total, 100 bgd)

Industry
Irrigation
Public supply
Rural

Figure 10-2. Percentage of total offstream withdrawals and fresh water consumptive use, by categories of use, 1980.

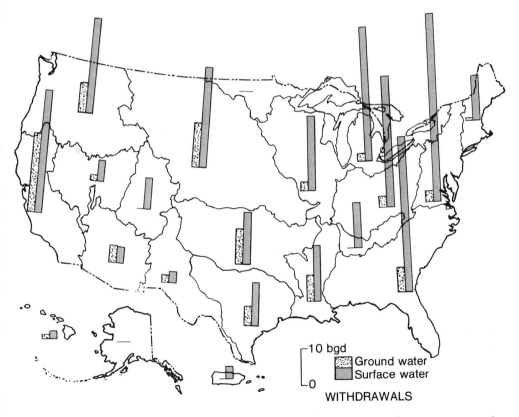

10 bgd

Ground water
Surface water

WITHDRAWALS

Figure 10-3. Withdrawals for offstream use from ground water and surface water sources, by water resources regions, 1980.

Table 10-2. Total Offstream Water Use, by Regions (mgd, Except as Noted), 1980

Water Resources Region	Population (thousands)	Per Capita Use, Fresh Water (gal/day)	Ground Water Fresh	Ground Water Saline	Ground Water Total	Surface Water Fresh	Surface Water Saline	Surface Water Total	Reclaimed Sewage	Total, Excluding Reclaimed Sewage Fresh	Total, Excluding Reclaimed Sewage Saline	Total, Excluding Reclaimed Sewage Total	Conveyance Losses	Consumptive Use, Fresh Water
New England	11,941	450	650	1.0	650	4,800	7,500	12,000	0	5,400	7,500	13,000	0.5	360
Mid-Atlantic	38,881	630	2,400	12	2,400	22,000	28,000	50,000	160	24,000	28,000	52,000	1.7	1,700
South Atlantic–Gulf	29,449	1,100	6,600	44	6,600	27,000	15,000	42,000	0	34,000	15,000	49,000	38	5,100
Great Lakes	21,489	1,700	1,600	420	2,000	36,000	0	36,000	30	37,000	420	38,000	0	1,300
Ohio	21,461	1,800	2,500	24	2,500	35,000	0	35,000	0	38,000	24	38,000	0.1	1,700
Tennessee	3,677	3,200	260	0	260	12,000	0	12,000	0	12,000	0	12,000	0.2	370
Upper Mississippi	21,083	1100	2,600	15	2,600	20,000	0	20,000	0	23,000	15	23,000	0	1,500
Lower Mississippi	6,874	3,000	6,700	19	6,700	14,000	390	15,000	0	21,000	410	21,000	960	7,100
Souris-Red-Rainy	796	280	110	0	110	110	0	110	0.2	220	0	220	5.9	130
Missouri Basin	9,761	4,000	12,000	26	12,000	27,000	0	27,000	1.7	39,000	26	39,000	6,000	16,000
Arkansas-White-Red	7,900	3,000	9,400	95	9,500	14,000	2.0	14,000	15	24,000	97	24,000	360	9,600
Texas-Gulf	12,524	820	5,100	0	5,100	5,200	6,600	12,000	55	10,000	6,600	17,000	140	6,500
Rio Grande	1,775	2,700	1,900	0.9	1,900	2,800	0	2,800	0	4,700	0.9	4,700	290	2,400
Upper Colorado	548	16,000	140	3.5	150	8,400	0.7	8,400	0.1	8,500	4.2	8,500	830	2,300
Lower Colorado	3,241	2,700	4,500	0.2	4,500	4,200	0	4,200	18	8,700	0.2	8,700	950	4,900
Great Basin	1,782	4,200	1,600	13	1,600	5,800	55	5,900	4.8	7,400	68	7,500	1,000	3,900
Pacific Northwest	7,870	4,400	8,200	0	8,200	26,000	42	26,000	17	34,000	42	34,000	6,800	12,000
California	23,671	1,900	21,000	250	21,000	23,000	9,800	33,000	160	44,000	10,000	54,000	5,800	25,000
Alaska	403	550	49	0	49	170	0	170	0	220	0	220	0	35
Hawaii	965	1,400	800	0	800	510	1,200	1,700	10	1,300	1,200	2,500	300	680
Caribbean	3,500	230	320	5.0	320	500	2,500	3,000	0	820	2,500	3,300	30	310
Total	229,592	1,600	88,000	930	89,000	290,000	71,000	360,000	470	380,000	72,000	450,000	24,000	100,000

Note: Water use data generally are rounded to two significant figures; figures may not add to totals because of independent rounding.

114

Table 10-3. Summary of Water Withdrawals for Offstream Water Use Categories, by Regions (mgd, Except as Noted), 1980

Water Resources Region	Public Supply		Rural Use			Irrigation			Self-Supplied Industrial Thermoelectric Power		Other Uses		Total, Excluding Reclaimed Sewage	
	Population Served (Thousands)	Withdrawals (mgd)	Domestic Use	Livestock Use	Domestic and Livestock	Irrigated Land (Thousand Acres)	Thousand acre·ft/year	mgd	Fresh	Saline	Fresh	Saline	Fresh	Saline
New England	10,000	1,500	130	9.2	140	79	59	53	2,300	7,400	1,500	78	5,400	7,500
Mid-Atlantic	34,100	5,400	430	110	550	230	280	250	15,000	25,000	3,400	2,100	24,000	28,000
South Atlantic–Gulf	21,400	3,800	720	240	960	3,400	4,300	3,800	19,000	15,000	5,900	330	34,000	15,000
Great Lakes	21,500	3,900	270	84	350	450	380	340	27,000	0	5,700	420	37,000	420
Ohio	15,300	2200	310	150	470	84	170	150	30,000	0	5,000	24	38,000	24
Tennessee	2,680	410	61	41	100	14	7.6	6.8	9,300	0	2,000	0	12,000	0
Upper Mississippi	12,600	1,900	300	270	570	820	420	380	16,000	0	3,300	15	23,000	15
Lower Mississippi	5,330	920	94	42	140	2,900	8,700	7,700	7,700	180	4,300	230	21,000	410
Souris-Red-Rainy	494	57	23	14	37	120	72	64	54	0	9.3	0	220	0
Missouri Basin	8,090	1,400	230	390	630	14,000	32,000	28,000	8,200	0	680	26	39,000	26
Arkansas-White-Red	6,090	1,600	160	240	390	7,000	12,000	11,000	10,000	0	840	97	24,000	97
Texas-Gulf	10,100	3,000	120	190	310	5,200	6,200	5,500	980	5,500	520	1,100	10,000	6,600
Rio Grande	1,370	320	33	32	65	1,400	4,800	4,300	17	0	16	0.9	4,700	0.9
Upper Colorado	357	120	58	94	150	1,300	8,400	7,500	140	0.7	590	3.5	8,500	4.2
Lower Colorado	2,910	720	37	17	54	1,400	8,500	7,600	90	0	250	0.2	8,700	0.2
Great Basin	1,570	810	36	46	82	1,900	6,600	5,900	130	5.2	500	63	7400	68
Pacific Northwest	5,320	1,300	270	55	320	7,700	33,000	29,000	29	0	3,700	42	34,000	42
California	22,300	4,100	140	86	220	10,000	42,000	38,000	2,000	9,200	480	820	44,000	10,000
Alaska	286	53	11	0.2	11	0	0	0	30	0	130	0	220	0
Hawaii	965	200	3.9	5.5	9.4	140	1,000	910	140	1,200	45	7.0	1,300	1,200
Caribbean	3,260	350	8.1	30	38	76	350	310	3.0	1,500	120	930	820	2,500
Total	186,000	34,000	3,400	2,200	5,600	58,000	170,000	150,000	150,000	65,000	39,000	6,300	380,000	72,000

Note: Water use data generally are rounded to two significant figures; population data are rounded to three significant figures. Figures may not add to totals because of independent rounding.

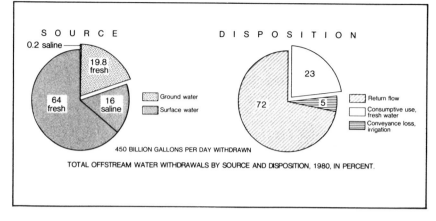

Figure 10-4. Total offstream water withdrawals, by source and disposition, 1980, in percentages.

was estimated as 34 bgd, or an average of 183 gal/day for each individual served (see Table 10-3.) This amount represents a 15 percent increase from 1975 when 29 bgd of water was withdrawn for public supply or a per capita use of 168 gal/day. Part of this increase is due to the fact that nearly 2 bgd of water erroneously identified in previous reports as self-supplied industrial withdrawals is now included in the public supply category. Another factor in the increase in this category is a 6 percent increase from 1975 in population served by public supplies along with higher per capita use. Combined daily average for domestic and public uses accounted for almost two-thirds of the public supply withdrawals and was estimated at 22 bgd, or an average of 120 gal/day for each individual served, compared with a per capita use of 117 gal/day in 1975. Included in the 22 bgd is water lost in the distribution system. Industrial and commercial users received the other third of the public supply withdrawals, about the same distribution as in 1975.

Water consumed by public supply users increased 6 percent to 7.1 bgd in 1980 and accounted for about 21 percent of the public supply withdrawals, approximately the same proportion as in 1965, 1970, and 1975. The larger cities were supplied principally by surface water sources, which furnished about two-thirds of the public-supplied water.

California, New York, and Texas, the three most populated states, withdrew the most water for public supplies and accounted for about 30 percent of the nation's total withdrawal by public suppliers. Per capita domestic use from public supplies averaged 100 gal/day for the eastern states and 150 gal/day for the western states (see Table 10-1). The two most populated water resources regions,

California and Mid-Atlantic, withdrew the most water for public supplies and accounted for about 28 percent of the total withdrawal by public suppliers.

The range in public supply fresh water withdrawals by water resources regions is shown in Table 10-3. The source of and disposition of withdrawals for public supply are shown in Figure 10-5.

Rural Use

Water for rural use includes self-supplied domestic use, drinking water for livestock, and other uses such as dairy sanitation, evaporation from stock-watering ponds, and cleaning and waste disposal. The number of people served by self-supplied systems was determined by subtracting the total number of people served by public supply systems from the total population, as derived from the U.S. Bureau of Census (1982) advance population data for 1980. The difference between these totals showed that 44 million people were served by their own water supply systems in 1980, compared with 41 million people in 1975. Rural self-supplied systems rarely are metered, and few "hard" data exist. Therefore water for rural use can only be estimated.

The quantity of fresh water withdrawn for rural domestic and livestock use in 1980 was 5.6 bgd, a 14 percent increase from 1975. Rural domestic withdrawals were 3.4 bgd, a 23 percent increase from 1975. This large increase is the result of the increased population being served by self-supplied systems and an increase in the per capita use, which was about 79 gal/day compared with about 68 gal/day in 1975. The increase in per capita use reflects the application of more realistic estimating

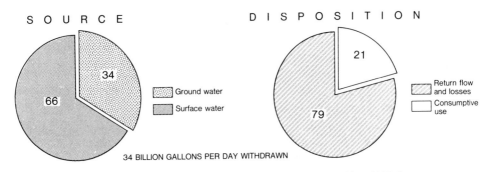

Figure 10-5. Water withdrawn for public supply, by source and disposition, 1980, in percentages.

techniques, which also indicate that previous estimates were probably too low. The quantity of water used by livestock increased slightly from 2.1 bgd in 1975 to nearly 2.2 bgd in 1980.

The consumptive use of fresh water for rural domestic use and livestock use in 1980 was about 2.0 bgd and 1.9 bgd, or 57 and 88 percent of withdrawals, respectively. Total consumptive use was 69 percent of total rural withdrawals. Only about 5 percent of the rural domestic water was surface water, but some 45 percent of the water used for livestock was surface water.

Rural domestic and livestock water use is fairly evenly distributed among the states, with Texas and Florida the major users, accounting for 7 percent and 6 percent, respectively. The South Atlantic–Gulf water resources region withdrew the most water for total rural use, and it also experienced the largest volume increase in rural domestic withdrawals. The Missouri Basin region withdrew the most water for rural livestock use and accounted for about 18 percent of the total withdrawals for livestock use. Rural water use data by water re-

sources regions are given in Table 10-3. The source of and disposition of withdrawals for rural use are shown in Figure 10-6.

Irrigation

Irrigation of crops developed along with the settlement of the arid West, because most years farmers needed to irrigate to raise any crops. In the humid eastern states irrigation has been used to supplement natural rainfall in order to increase the number of plantings per year and yield of crops per acre and to reduce the risk of crop failures during drought periods. Irrigation also is used to maintain recreational lands such as parks and golf courses. Estimates of withdrawals for irrigation vary greatly. In some instances they are based on subjective amounts of water required to raise an acre of a given crop. In other instances accurate records of water application rates are available. Reliable estimates of water withdrawn for irrigation can be made if the number of acres irrigated and the water application rates are known. It usually is difficult

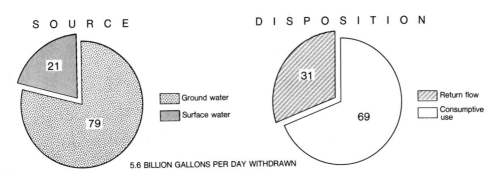

Figure 10-6. Water withdrawn for rural domestic and livestock use, by source and disposition, 1980, in percentages.

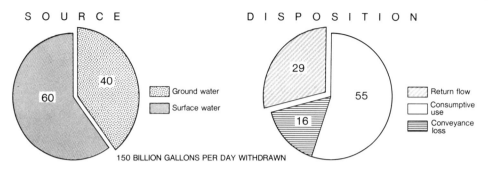

Figure 10-7. Water withdrawn for irrigation, by source and disposition, 1980, in percentages.

to obtain reliable estimates for consumptive use and for conveyance loss. Thus some of the estimates of consumptive use and conveyance loss may be only rough approximations of actual conditions. Nevertheless, it is likely that better estimates were made of water used per acre in 1980 than in 1975, and in particular, the values given for conveyance loss for 1980 are more realistic because of progressively better records being kept by the water users.

The quantity of water withdrawn for irrigation in 1980 was estimated at about 170 million acre·ft, or 150 bgd (see Table 10-3). The water was used on approximately 58 million acres of farmland. This represents an increase in both water use and irrigated acreage of about 7 percent from the 1975 estimate. Where irrigation is used primarily to supplement natural rainfall, it is to be expected that there normally will be large differences in irrigation withdrawals from year to year.

The consumptive use of irrigation water was estimated to be 93 million acre·ft, or 83 bgd in 1980. This was 55 percent of the irrigation water withdrawn and accounted for about 81 percent of the total consumptive use by the nation. Conveyance loss was estimated at about 26 million acre·ft (24 bgd), or 16 percent of 1980 irrigation withdrawals. Consumptive use and conveyance losses in 1980 were slightly higher than in 1975 but were essentially in the same proportion to irrigation water withdrawn as they were in 1975.

Surface water was the source of about 60 percent of the irrigation water (the same as 1975), and except for a small fraction of 1 percent that was reclaimed sewage, groundwater furnished the remainder.

The nine western water resources regions (regions 10–18 of Figure III-2), led by the California region, accounted for 91 percent of the total water withdrawn for irrigation in 1980, compared with

93 percent in 1975. In the eastern regions most of the water used for irrigation was in the South Atlantic–Gulf and Lower Mississippi regions, which together withdrew over 3 bgd more water in 1980 than in 1975. The state of California was by far the largest user of irrigation water, withdrawing about 37 bgd, 25 percent of the national total, which is more than the next two largest users, Idaho and Colorado, combined. Nebraska and Georgia showed the largest increase in number of acres irrigated from 1975 to 1980. The source of and disposition of withdrawals for irrigation use are shown in Figure 10-7.

Self-Supplied Industrial

Self-supplied industrial water use is categorized in this chapter as thermoelectric power (electric utility) and "other" self-supplied water-using industries (see Table 10-3). "Other" self-supplied water-using industries include, but are not limited to, steel, chemical and allied products, paper and allied products, mining, and petroleum refining. Thermoelectric power plants can be powered by fossil fuel, geothermal, or nuclear energy and account for the largest quantity of water withdrawn for offstream use. Because of the magnitude of water required for thermoelectric power generation, the estimates of use are discussed here as part of the total self-supplied industrial use and in more detail in the next section. Self-supplied industrial water systems often are metered, and estimates of water withdrawn and consumed generally are reliable. It is likely that better estimates were made in 1980 than in 1975 because more comprehensive inventories were obtained and more accurate and complete records were available from the users.

More water continues to be withdrawn for industrial use than for any other category. In 1980 the amount of self-supplied industrial water with-

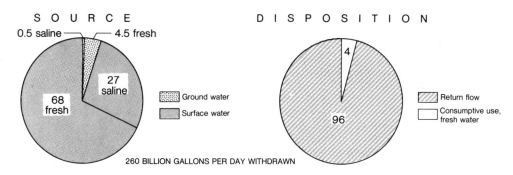

Figure 10-8. Water withdrawn for all self-supplied industrial use, by source and disposition, 1980, in percentages.

drawn was estimated at 260 bgd, of which about 72 bgd was saline (see Table 10-3). This is an increase of 8 percent from the 1975 estimate. Of the 260 bgd, about 210 bgd, or 83 percent of all industrial withdrawals, was withdrawn by thermoelectric power plants. Withdrawals for thermoelectric power plants showed a 9 percent increase from 1975, and withdrawals for other industrial uses (about 45 bgd) remained about the same as in 1975. Saline water constituted about 28 percent of the total self-supplied industrial withdrawals, approximately the same proportion as in 1965, 1970, and 1975. Public supply systems delivered about 2 bgd for thermoelectric power generation and about 10 bgd for other industrial and commercial uses. The withdrawal estimates for thermoelectric power plants include the water from public supplies; however, public supplies are not included in the estimate for total self-supplied industrial use but are summarized in the public supply category.

Consumptive use of fresh water by thermoelectric plants was about 2 percent and for other self-supplied industrial uses about 13 percent, giving a combined consumptive use of about 4 percent for all types of self-supplied industries. Saline water consumed by thermoelectric plants also was about 2 percent of the saline withdrawals and about 15 percent for other industrial uses. These consumptive-use figures are higher than in previous years and indicate an increased reuse of water.

The relative proportion of source of supply has remained constant since 1965—ground water still supplied nearly 5 percent, surface water about 95 percent, and reclaimed sewage only a fraction of 1 percent.

The Mid-Atlantic water resources region withdrew slightly more water for industrial use in 1980 than in 1975 and withdrew the most saline water and total water (fresh and saline). The Ohio region withdrew about 6 percent more water for industrial

use in 1980 than in 1975 and accounted for the most fresh water withdrawals. Withdrawals in the state of Illinois for self-supplied industrial use increased 50 percent from 1975 to 1980, based on a more complete inventory of industrial users, making Illinois the second largest user of self-supplied industrial water behind Florida.

The range in self-supplied industrial water withdrawals by water resources region is tabulated in Table 10-3. The source and disposition of withdrawals for self-supplied industrial use are shown in Figure 10-8.

Self-Supplied Industrial: Thermoelectric Power

Thermoelectric power generation is categorized as a self-supplied industrial water use. However, because of the magnitude of water required, separate estimates were made of the source, use, and disposition of water for the thermoelectric power industry. These estimates usually are reliable inasmuch as relatively complete data files are maintained by federal and state agencies.

Thermoelectric power plants furnish practically all their own water; less than 1 percent is obtained from public supplies. In 1980 water withdrawn by thermoelectric power plants was about 210 bgd, an increase of about 9 percent from the 1975 estimate and an increase of about 26 percent from the 1970 estimate. The thermoelectric power industry continues to withdraw the largest quantity of water for offstream use, more than 1.4 times the water withdrawn for irrigation, the next largest water use category.

About 99 percent of the total water withdrawn by thermoelectric plants was used for condenser and reactor cooling of generators. Plants vary widely as to the techniques used in disposal of the cooling water after it has passed through the condensers. Where water is expensive or scarce, cooling towers or ponds (Federal Power Commission, 1969) are employed so that the same water can be used repeatedly in the condensers. Prevention of thermal pollution of the receiving water body is another factor that has caused some plants to resort to water-cooling devices. The quantity of water consumed by steam plants will increase as reuse of water becomes more prevalent. About 2 percent of the water withdrawn in 1980 was consumed, compared with 1 percent in 1975 and only one-half of 1 percent in 1970. Surface water constituted 98 percent of total thermoelectric withdrawals in 1980, and 30 percent was saline, compared with 33 percent in 1975 and 28 percent in 1970.

Public supply systems delivered about 2 bgd for thermoelectric power generation. In previous reports in this series a major part of this water was erroneously identified as self-supplied industrial withdrawals.

The amount of water withdrawn in the Mid-Atlantic water resources region by thermoelectric power plants in 1980 was approximately the same as in 1975 and accounted for the most saline water and total water (fresh and saline) withdrawals. The Ohio region withdrew about 13 percent more water for thermoelectric power plants in 1980 than in 1975 and accounted for the most fresh water withdrawals. Withdrawals in the state of Illinois for thermoelectric power generation increased 54 percent from 1975 to 1980, based on a more complete inventory of industrial users, making Illinois the second largest user of water for thermoelectric power after Florida. Arkansas withdrawals increased from about 2 bgd to nearly 10 bgd from 1975 to 1980 as the result of nuclear power plants coming on-line between 1975 and 1980.

Thermoelectric power water use data by water resources regions are given in Table 10-3. The source of and disposition of withdrawals for ther-

moelectric power generation are shown in Figure 10-9.

Hydroelectric Power

Estimated quantities of water used for hydroelectric power generation may differ because of the manner in which individual estimates are made of the amount of water passed through the plants. Where the water is passed through the plant only one time, good estimates of water use can be obtained. However, where hydroelectric plants have pumped-storage facilities and recycle the same water through the plant a number of times, it is difficult to obtain net water use. The magnitude of the effect of pumped storage on water estimates for hydroelectric power generation is not known, but as pumped storage becomes more prevalent, it becomes an important factor in making accurate water use estimates.

Water used for hydroelectric power generation showed an increasing trend from 1950 to 1975. The trend leveled off from 1975 to 1980, as the amount of water used remained approximately the same, an estimated 3300 bgd or 3700 million acre·ft, which is 2.75 times the average annual runoff in the conterminous United States (Langbein et al., 1949). From 1975 to 1980 there was about a 17 percent increase in developed hydroelectric capacity. But during this period utility hydroelectric production decreased 8 percent, from 301 billion kilowatt hours (kWh) to 277 billion kWh, primarily because less stream flow was available in 1980 than in 1975.

Although a very small quantity of water is evaporated (consumptive use) in the generation of hydroelectric power, some depletion of the available water supply will occur as a result of evaporation from reservoirs associated with hydroelectric power generation, repeated reuse of water within a pumped-storage power plant, and the cumulative use that

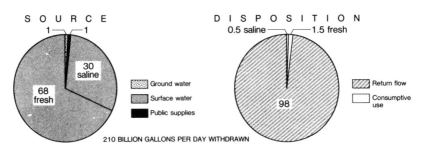

Figure 10-9. Water withdrawn for the generation of thermoelectric power, by source and disposition, 1980, in percentages.

Figure 10-11. Water used for hydroelectric power generation in the United States, by source and disposition, 1980, in percentages.

now occurs in successive plants downstream (3300 bgd used compared with a total surface water supply of 1200 bgd). This can be confirmed by the fact that evaporation from reservoirs and regulated lakes, which has been classified as an instream use, contributes to the reduction of available water for all uses. For example, the annual evaporation from the principal reservoirs and regulated lakes in the western United States is estimated to be 11 bgd (Meyers, 1962, Table 6). This amount is equivalent to about 11 percent of consumption by all offstream uses in 1980.

The Pacific Northwest water resources region was by far the largest user of water for hydroelectric power generation in 1980, accounting for almost one-half the total water used for hydroelectric power in the nation. Washington and Oregon used more water for hydroelectric power generation than the combined total of all water used for hydroelectric power in the eastern states.

The range in hydroelectric power water use by water resources regions is shown in Figure 10-10. The source of and disposition of water used for hydroelectric power are shown in Figure 10-11.

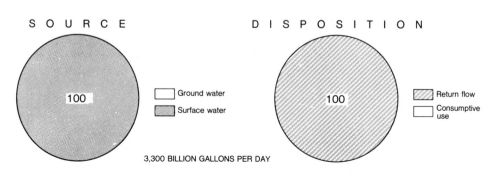

Figure 10-10. Hydroelectric power water use, by water resources regions, 1980.

TRENDS IN WATER USE, 1950–1980

Water use for public supply, rural needs, irrigation, industry, and hydroelectric power generation has increased steadily from 1950 to 1980. This trend is shown graphically in Figures 10-12 and 10-13. Data in Table 10-4, which is a summary of estimated water use—offstream withdrawals, source of withdrawals, consumptive use, and instream use (hydroelectric power)—at 5-year intervals for the period 1950–1980, also confirm this trend. Table 10-4 also shows the percentage increase or decrease for the various categories of water use and sources of supply for the periods 1970–1975 and 1975–1980.

Trends established over the period 1950–1975 did not change significantly during the 1975–1980 period. For most categories of use the general slackening in the rate of increase that was observed from 1970 to 1975 is again detectable for the 1975–1980 period. There are two exceptions to this trend: Public supply and rural withdrawals increased 15 and 14 percent, respectively, compared with corresponding increases of 8 and 10 percent from

1970 to 1975. Part of the increase for public supply is due to the fact that nearly 2 bgd of water previously identified as self-supplied industrial withdrawals was actually public-supplied water, and it is now identified in the public supply category. The increase in rural withdrawals resulted from an increase in the population being served by self-supplied systems and an increase in per capita use. This per capita use increase reflects the application of more realistic estimating techniques, which indicate that previous estimates were probably too low.

Irrigation water use declined from 1955 to 1960, when there was a decrease in the amount of surface water used, but irrigation water use has continued to increase since 1960. The amount of surface water used for irrigation increased 7.1 percent from 1975 to 1980—nearly double the 3.7 percent increase from 1970 to 1975. In contrast, the amount of ground water used for irrigation has increased steadily since 1950; however, the increase from 1975 to 1980 was only 5 percent compared with 27 percent from 1970 to 1975. The average amount of water required per acre for irrigation in 1980

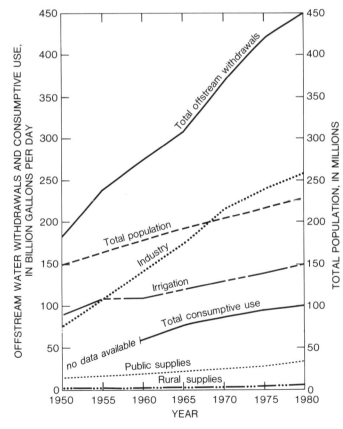

Figure 10-12. Trends in withdrawals, consumptive use, and population, 1950–1980.

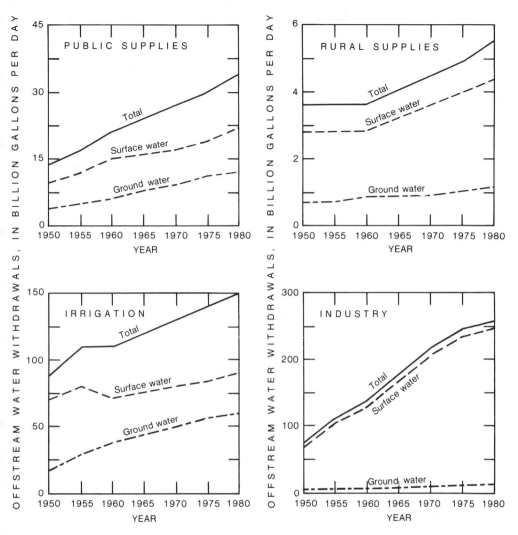

Figure 10-13. Trends in water withdrawals for public supplies, rural supplies, irrigation, and self-supplied industry, 1950–1980.

(2.9 acre·ft/acre) was the same as in 1975. Although the acreage irrigated in 1980 was about 7 percent greater than in 1975, it was less than the 9 percent increase that took place from 1970 to 1975 and the 13 percent increase that took place from 1960 to 1965 and from 1965 to 1970.

More water continues to be withdrawn for industrial use than for any other category, even though the rate of increase in water withdrawals for thermoelectric power continued to decline—a 33 percent increase from 1965 to 1970, an 18 percent increase from 1970 to 1975, and a 9 percent increase from 1975 to 1980. Withdrawals for other

industrial uses remained about the same in 1970, 1975, and 1980.

Water used for hydroelectric power generation had been increasing steadily from 1950 to 1975, but in 1980 hydroelectric power water use was approximately the same as in 1975, compared with a 21 percent increase between 1970 and 1975.

A shift in the source of total withdrawals also is shown by Table 10-4, which indicates that the withdrawal of fresh surface water increased by 10 percent between 1975 and 1980, compared with a 5 percent increase between 1970 and 1975. Fresh ground water and saline surface water, which showed

Table 10-4. Summary of Estimated Water Use in the United States (bgd) at 5-Year Intervals, 1950–1980

	Estimated Water Use (bgd)							Percentage Increase (+) or Decrease (−)	
	1950[a]	1955[a]	1960[b]	1965[b]	1970[c]	1975[d]	1980[d]	1970–1975	1975–1980
Population; in millions	150.7	164.0	179.3	193.8	205.9	216.4[e]	229.6	+5	+6
Offstream use									
Total withdrawals	180[e]	240	270	310	370	420	450	+12	+8
Public supply	14	17	21	24	27	29	34	+8	+15
Rural domestic and livestock	3.6	3.6	3.6	4.0	4.5	4.9	5.6	+10	+14
Irrigation	89[e]	110	110	120	130	140	150	+11	+7
Self-supplied industrial Thermoelectric power use	40	72	100	130	170	200	210	+18	+9
Other industrial uses	37	39	38	46	47	45	45	−6	+1
Source of withdrawals									
Ground water									
Fresh	34	47	50	60	68	82	88	+22	+7
Saline	([f])	0.6	0.4	0.5	1	1	0.9	−6	−5
Surface water									
Fresh	140[e]	180	190	210	250	260	290	+5	+10
Saline	10	18	31	43	53	69	71	+31	+2
Reclaimed sewage	([f])	0.2	0.6[e]	0.7	0.5	0.5	0.5	+2	−11
Consumptive use	([f])	([f])	61	77	87[g]	96[g]	100[g]	+10	+7
Instream use									
Hydroelectric power	1100	1500	2000	2300	2800	3300	3300	+21	−2

Source: Data for 1950–1975 adapted from MacKichan (1951, 1957), MacKichan and Kammerer (1961), Murray (1968), and Murray and Reeves (1972, 1977). The data generally are rounded to two significant figures: however, the percentage changes are calculated from unrounded numbers.

[a] 48 States and District of Columbia.
[b] 50 States and District of Columbia.
[c] 50 States, District of Columbia, and Puerto Rico.
[d] 50 States, District of Columbia, Puerto Rico, and Virgin Islands.
[e] Corrected from published report.
[f] Data not available.
[g] Fresh water only.

substantial increases from 1970 to 1975 (22 and 31 percent, respectively) only increased 7 and 2 percent, respectively, from 1975 to 1980. The slowdown in the rate of increase in total withdrawals, 8 percent increase between 1975 and 1980, more closely follows the rate of increase in total population of 6 percent during the same period. This is in contrast to the rate of increase in total withdrawals during the period 1970–1975, which was more than double the rate of population growth. The rate of increase in consumptive use of fresh water has steadily decreased from 13 percent for the period 1965–1970 to 7 percent for the period 1975–1980. The changes shown in Table 10-4 and Figures 10-

12 and 10-13 can be attributed to several important factors:

1. Demands on the ground water system influence the pumping lift, flow rate, or quality of the water supply. Each of these factors also influences the cost of water and makes users, especially irrigators, more selective and efficient with their use of ground water.
2. The price of water influences the volume used, encourages efficient use, and may determine when the use of reclaimed water and increased reuse are viable alternatives.
3. Availability of water in a particular year, es-

pecially stream flow, strongly affects the quantity of water used for irrigation and hydroelectric power development.

Although 1980 estimates of water use were higher than the 1975 estimates for all offstream categories, trends established during the periods 1970–1975 and 1975–1980 indicate a general slackening in the rate of total withdrawals in comparison to the period 1965–1970. Even with the slackening of the rates of water withdrawal and consumptive use, major attention must be given to water management problems, because in addition to the need for an adequate water supply, water quality conditions must be suitable if supply and demand are to be in balance. The degree to which the different uses of water degrade the supply vary widely and affect the potential reuse of the return flows.

Projections of future water use are beyond the scope of this chapter, although the trends established over the past 30 years provide some basis for estimating future water demands. Many other agencies and commissions have made projections of national water use to the year 2000. Notable examples are studies by the Senate Select Committee on National Water Resources (U.S. Congress, 1961), Resources for the Future, Inc. (Wollman and Bonem, 1971), the National Water Commission (1973), and the U.S. Water Resources Council (1968 and 1978). Summaries of these national projections and projections for individual states to the year 2000 are included in a report prepared by the Congressional Research Service (Viessman and DeMoncoda, 1980). The projections vary greatly based on availability of reliable data and different assumptions of future population growth, economic conditions, environmental regulations, and energy resources development. Regardless of which projection proves correct, major attention must be given to water management problems to ensure that maximum benefits will be obtained from use of the nation's water resources.

REFERENCES CITED

Federal Power Commission. 1969. *Problems in disposal of waste heat from steam-electric plants: Staff study.* Bureau of Power.

Langbein, W. B., et al. 1949. *Annual runoff in the United States.* U.S. Geological Survey Circular 52.

MacKichan, K. A. 1951. *Estimated water use in the United States, 1950.* U.S. Geological Survey Circular 115.

———. 1957. *Estimated water use in the United States, 1955.* U.S. Geological Survey Circular 398.

MacKichan, K. A., and J. C. Kammerer. 1961. *Estimated use of water in the United States, 1960.* U.S. Geological Survey Circular 456.

Mann, W. B., IV, J. E. Moore, and E. B. Chase. 1982. *A national water-use information program.* U.S. Geological Survey Open-File Report 82–862.

Meyers, J. S. 1962. *Evaporation from the 17 western states,* with a section on Evaporation rates, by T. J. Nordenson. U.S. Geological Survey Professional Paper 272–D.

Murray, C. R. 1968. *Estimated use of water in the United States, 1965.* U.S. Geological Survey Circular 556.

Murray, C. R., and E. B. Reeves. 1972. *Estimated use of water in the United States in 1970.* U.S. Geological Survey Circular 676.

———. 1977. *Estimated use of water in the United States in 1975.* U.S. Geological Survey Circular 765.

National Water Commission. 1973. *Water policies for the future.* Washington, D.C.: U.S. Government Printing Office.

U.S. Bureau of the Census. 1982. *Census of the population, characteristics of the population, number of inhabitants— 1980.* Published separately by states, Puerto Rico, and outlying areas. PC 80–1–A1 to A57a, and A57b.

U.S. Congress. Senate. 1961. *Report of the Select Committee on national water resources.* 87th Cong., 1st sess. S. Rep. 29.

U.S. Geological Survey. 1980. Hydrologic unit map of the United States (2 sheets, east and west). Scale 1:250,000 (1 inch = about 4 miles).

U.S. Water Resources Council. 1968. *The nation's water resources.* Washington, D.C.: U.S. Government Printing Office.

———. 1978. *The nation's water resources, 1975–2000.* Vols. 1–4. Washington, D.C.: U.S. Government Printing Office.

Viessman, Warren, Jr., and Christine DeMoncoda. 1980. *State and national water use trends to the year 2000.* Report prepared for the U.S. Senate Committee on Environment and Public Works. 96th Cong., 2nd sess., Committee Print 96–12.

Wollman, N., and G. E. Bonem. 1971. *The outlook for water quality, quantity, and national growth: Resources for the future.* Baltimore: Johns Hopkins University Press.

GENERAL REFERENCES

Davis, G. H., and L. A. Wood. 1974. *Water demands for expanding energy development*. U.S. Geological Survey Circular 703.

Giusti, E. V., and E. L. Meyer. 1977. *Water consumption by nuclear powerplants and some hydrologic implications*. U.S. Geological Survey Circular 745.

James, I. C., II, J. C. Kammerer, and C. R. Murray. 1977. *How much water in a 12-ounce can? A perspective on water-use information*. U.S. Geological Survey Annual Report, fiscal year 1976, pp. 17–27.

Missouri River Basin Commission. 1981. *Assessment of water use information needs and application*. Omaha.

Walling, F. B. 1977. *Water and industry in the United States*. U.S. Geological Survey leaflet.

11

THE FUTURE OF IRRIGATION

KENNETH D. FREDERICK

Irrigation of about 25 million additional acres in the West was an important factor in raising U.S. crop production by 70 percent from 1950 to 1977 without any net increase in total harvested acreage in the United States. Two broad trends have dominated the rate and location of the growth of western irrigated acreage for the past 25 years, and both are likely to continue in the future. First, the rate of growth of irrigated acreage, which has declined steadily over the last three decades, will continue to decline. Second, the locus of the growth, which has moved from south to north, will remain in the north. From 1945 to 1954, 79 percent of the growth of irrigated acreage was in a southern belt extending from Texas and Oklahoma to California. In subsequent decades this region contributed 31 percent and then only 1 percent of the overall growth of western irrigation. Analysis of the water supply and demand conditions and the adverse impacts of rising energy prices suggest that irrigated acreage has probably started or will soon start to decline in this southern belt. In contrast, the central and northern High Plains contribution to the expansion of western irrigation rose from about 10 percent from 1945 to 1954 to over 90 percent from 1964 to 1974. Since Nebraska is one of the few western states for which significant further expansion of irrigated acreage appears likely, this percentage is likely to rise even further.

WATER SUPPLIES

Total surface water withdrawals for irrigation have fluctuated around a level trend since the mid 1950s, and the combined impact of several factors suggests

this situation may continue for several more decades. Total water requirements exceed average year stream flows within the West's principal irrigated areas, and there is little water available for expansion in most of the other areas. Where there is water for expansion, the increases will go primarily to nonagricultural users who can afford to pay higher water costs.

The comparative stability of surface water use for irrigation in the face of increasing water scarcity reflects in part the insulation of most surface water costs from both market considerations and rising energy costs. Undoubtedly, there will be transfers of water from agricultural to other uses, particularly within the water-scarce areas, as farmers are presented with increasing opportunities to sell their water rights. However, since large percentage increases in other water uses can be accommodated with small changes in irrigation water use, at least over the next several decades, the impacts of such transfers on irrigation will be gradual and relatively minor from the perspective of the entire West. Moreover, transfers of water rights from irrigation to other uses in the water-scarce areas are likely to be offset in part by modest increases in irrigation within the regions where surface water is still available for appropriation. Federal irrigation projects, which are encountering increasing resistance to their high investment costs and the high opportunity costs of diverting more scarce water to irrigation, will not make any further significant contribution to the expansion of western irrigation.

For several decades the growth of western irrigation has been based on ground water withdrawals, which rose threefold between 1950 and 1975. Currently, ground water withdrawals, which ac-

Dr. Frederick is a senior fellow at Resources for the Future, where he has been director of the Renewable Resources Division since 1977. He has a Ph.D. in economics from M.I.T. and has been a member of the economics faculty at the California Institute of Technology and an economic advisor in Brazil for the U.S. Agency for International Development. This material is excerpted from Summary and conclusions, in *Water for western agriculture* (Baltimore: Johns Hopkins University Press, 1982), Chapter 7, 210–217 and 232–236. Reprinted by permission of Resources for the Future, Inc. Copyright © 1982 by Resources for the Future, Inc.

count for about 39 percent of all western irrigation water, result in the mining of more than 22 million acre·ft/year from western aquifers. Even though ground water stocks are still large in relation to current use, and mining is not a threat to exhaust physically the water stored in any of the water resource regions or subregions in the foreseeable future, the combination of the overdrafts and rising energy costs threaten the economic supply of water for many irrigators. In some regions, especially the High Plains, the combination of increased pumping depths, lower well yields, and higher energy costs already has started to curtail pumping and irrigated acreage. In the absence of sharp increases in real crop prices, future expansion likely will be limited to areas with relatively low pumping depths or access to energy supplies under relatively favorable terms. But combining one or both of these conditions with favorable growing conditions will become increasingly rare.

The water available for irrigation is not likely to be augmented significantly in the next several decades through development of unconventional sources of supply. Although some forms of weather modification, particularly winter cloud seeding, appear technologically and economically promising, institutional obstacles are apt to inhibit its widespread adoption. Cost factors associated in part with their high energy intensity suggest that water importation and desalinization of sea water will not be profitable for use in irrigation. And a combination of cost and technical and institutional uncertainties make icebergs an unlikely source of western water for the foreseeable future.

Institutional Factors

Western water law and management institutions were developed when water was plentiful in relation to demand. Important objectives of early water law and policy were to give investors clear and unambiguous title to water rights, to encourage the development of water resources, and to minimize uncertainty and conflict among users.

Today the West faces a situation where water is scarce in relation to demand and the costs of developing new supplies are high in relation to its value in irrigation. Urban, industrial, recreational, and wildlife needs for fresh water are becoming increasingly competitive with irrigation. Yet there has not been a corresponding adjustment in the laws and institutions that control and manage the resource. All too often the laws and institutions governing water use limit rather than facilitate the transfer of water to higher-valued uses and stifle rather than encourage conservation measures. Such deficiencies are common with most state water institutions, but they are especially severe in the case of federal projects, which provide irrigators enormous subsidies but little to no opportunity or incentive to benefit from conservation.

Ground water users do not have the security of long-term access to low-cost water enjoyed by the owners of senior surface water rights. Ground water supplies often are depletable resources threatened by the addition of new wells, and the costs of pumping are closely linked to energy costs. Individually, pumpers have improved the efficiency of their water use; collectively, they have sought government help in limiting depletion and curbing water cost increases. Many states have enacted or are considering legislation to limit pumping in order to extend the life of the aquifers. In terms of achieving a long-term efficient use of the resource, one problem with the current groundwater situation is that farmers' costs do not include the loss to neighboring farmers and future users of depleting an aquifer. Theoretically, taxes on pumping could internalize these costs, but practically, it would be difficult to approximate the ideal level for such a tax, and any tax would be strongly resisted by pumpers.

Economic Factors

In general, ground water irrigators pay the most for their water and are the most susceptible to further cost increases stemming from both rising energy costs and declining water levels. Ground water costs vary widely depending on the pumping depth and the type and cost of the fuel. For a typical farmer pumping from 200 ft with electricity, water at the wellhead costs about $24 per acre-foot; distributing this water through a center-pivot system adds another $20 to $30 per acre-foot to irrigation costs (assuming 1980 energy prices and deflating all costs to 1977 constant dollars). A doubling of electricity prices would add another $32 to the costs of irrigation if no adjustments were made to the higher energy costs. Such cost levels would make it difficult to irrigate profitably grains, cotton, and many other crops in the absence of significant increases in crop prices. Farmers confronted with even higher pumping depths and declining aquifers will be even harder-pressed to compete. Unless crop price increases compensate for rising water costs, many farmers will be forced to reduce or terminate their irrigation or radically alter their irrigation techniques over the next several decades. An increase in corn prices of roughly 25 percent would be required to offset the cost increases resulting from a doubling of electricity prices.[1]

Surface water costs also vary widely depending on the distance and height the water must be transported to arrive at the farm, the availability of subsidies, and the need for on-farm pumping to get it to the field. Surface water costs, however, tend to be considerably lower than ground water costs and not as subject to change. Farmers with senior rights to neighboring surface waters that can be distributed through gravity have the lowest costs and are the least susceptible to future changes. For these fortunate farmers water is virtually a free resource, and it is probably treated as such unless the farmer has an opportunity to sell or put to alternative uses any water saving. Another fortunate group of irrigators comprises the farmers receiving water from federal water projects. This highly subsidized water is used on nearly 20 percent of the West's irrigated acreage.

The opportunities for expanding irrigation with low-cost surface water are virtually nonexistent within the West's principal irrigated areas and very limited in other areas of the West. For the past several decades ground water has been the basis of changes in western irrigation, and it is likely to remain so for the foreseeable future. The direction of future change will depend in large part on the ability of these farmers to adapt to rising water and energy costs.

Total water costs, of course, depend both on the unit cost of water and on the quantity of water applied. Farmers in arid zones who depend on irrigation to supply virtually all the crop water requirements are more affected by a rise in water costs than are those who rely on irrigation as a supplement to precipitation during the growing season. Most irrigators, however, have a wide range of opportunities for responding to high energy and water costs short of abandoning irrigation. Measures such as improvement of pump efficiency, tailwater reuse systems, and irrigation scheduling already are profitable under a variety of conditions. Future innovations undoubtedly will provide further opportunities for increasing the yields per unit of irrigation water and reducing the cost of water. Nevertheless, these innovations are not likely to alter the adverse impacts that rising energy costs and water scarcity are having on the profitability of irrigated relative to dryland farming, since dryland farmers can also be expected to benefit from future technological developments. On the other hand, if real crop prices rise as some analysts predict, irrigators with their higher-than-average yields will tend to benefit more than dryland farmers.

Environmental Factors

A variety of environmental problems are associated with irrigation, but the only ones likely to have any significant effect on the role of irrigation are low stream flows, ground water depletion, and salinity. Current water requirements, defined to include instream uses, already exceed average stream flows in the most favorable areas for irrigation. If irrigation water use were curtailed to the levels that will "ensure maintenance of stream flow for optimum fish and wildlife habitat and other environmental values," irrigated acreage would have to be reduced by 22 percent in comparison to the moderate export run for 1985 without environmental restrictions, according to analysis undertaken by the U.S. Department of Agriculture as part of the Agricultural Resource Assessment Systems.[2] And, of course, in those areas where the aquifers are being mined, withdrawals eventually must be reduced.

An estimated 25 to 35 percent of the West's irrigated lands have salinity problems. But the lands where salinity is likely to significantly curtail production comprise a much smaller percentage. The productivity of several million acres, primarily in the Lower Colorado River Basin and California's San Joaquin Valley, are threatened by high salt levels. The annual damages to agricultural plus municipal and industrial users of the salt-laden waters already probably exceed $100 million, and damages will rise unless preventive measures are taken. Improved basinwide and on-farm water management have great potential for reducing salinity levels and mitigating the damages from high salt levels. However, institutional obstacles to adopting improved basinwide management schemes and the high costs of some of the structural measures that might achieve the same result suggest an increasing number of farmers will be confronted with serious salt problems. Nevertheless, although the impacts will be serious within the affected areas, the overall impact on the productive potential of western irrigation is not likely to exceed 2 to 3 percent over the next several decades.

Another environmental problem that could affect the growth of irrigation over the next several decades is ground water pollution from the infiltration of agricultural chemicals. Agricultural chemicals, especially nitrogen, are readily leached into the ground water of the Nebraska Sandhills, which is the most likely area to experience a significant expansion of irrigation. Good on-farm water management can keep nitrate levels within tolerable levels, but failure to adopt such practices could lead to state intervention to enforce better management practices, possibly limiting the growth of irrigation in the area.

PROJECTIONS OF IRRIGATED ACREAGE

Two separate estimates of irrigated acreage are presented in the [*Second National Water*] *Assess-*

ment—the National Futures (NF) estimate, which was developed by group consensus of representatives from the federal agencies involved in the *Assessment,* and the State and Regional Futures (SRF) developed by the committees established by the *Water Resources Council* (WRC) in each region. These committees were composed of local river basin, U.S. Geological Survey, and Soil Conservation Service officials and others.

Both the 1975 base-year levels and the 1985 and 2000 irrigation projections of the two groups differ significantly. Not only are the SRF base-year estimates higher than the NF estimates (46.3 versus 40.5 million acres), the rates of expansion of the SRF projections also are much higher. The NF projects a 7 percent increase in irrigated acreage by 1985 to 43.4 million acres and a further increase of only 1.2 million acres by 2000. In comparison, the SRF projects a 16 percent rise to 53.6 million acres in 1985 and an additional 14 percent rise to 61.0 million acres by 2000. By the turn of the century the SRF projections show 37 percent more acres irrigated than do the NF projections. The *Assessment* does not provide justification for either set of irrigation projections.

Over the next several decades changes in total irrigated acreage will depend in large part on what happens to agricultural prices. Indeed, the relation between crop price levels and the expansion of irrigation will be much stronger than in the past. Past expansion was primarily a function of technological developments, water projects, and a learning process as farmers followed what their more progressive neighbors had proved successful. Future expansion will be influenced by very different cost and resource conditions that make past trends of limited use for making projections and the profitability of irrigated relative to dryland farming in many areas dependent on higher crop prices.

Although irrigation in many areas of the West is constrained by physical and institutional limits on developing new water supplies, there are additional lands with access to water for irrigation. Within these areas the important constraints on irrigation are economic. Significant increases in product prices would offset the negative impacts of high energy prices and increasing pumping depths; the development of new irrigated lands would be stimulated, and the decline in irrigated farming in areas with significant ground water mining would be slowed.

Table 11-1 summarizes what should be viewed as very rough projections of irrigated acreage by farm production region during the decade from 2000 to 2010 under alternative assumptions of no change and a 25 percent increase in real crop

Table 11-1. Projections of Irrigated Acreage By Farm Production Region Under Alternative Crop Price Scenarios (Millions of Acres)

		Projections for 2000 to 2010	
		No Change	25 Percent
	1977 NRI	in Real	Rise in Real
Region	Estimates	Crop Prices	Crop Prices
Northern Plains	10.7	15.0	16.5
Southern Plains	9.0	7.0	8.0
Mountain	17.2	16.5	17.5
Pacific	13.3	15.0	16.0
17 western states	50.2	53.5	58.0

prices.[3] These projections are impressionistic estimates drawing on analysis of recent trends, water supplies, and institutional, economic, and environmental factors.

Even in the absence of any significant change in crop prices, some net expansion of western irrigation seems likely. Most of the expansion will be in the Northern Plains states and, more specifically, within the Nebraska Sandhills. This area has considerable potential for expansion, and on the basis of recent investments in wells and center pivots, irrigation investment in the area is profitable at current price levels. Modest increases in irrigation within the Dakotas is also likely, although irrigation will remain relatively unimportant in these states. The increases within the Northern Plains will be partly offset by some reduction in irrigation in the Kansas and Nebraska High Plains.

The impacts of high energy costs and declining ground water tables will force a significant decline in irrigation by the turn of the century within the High Plains of Texas and Oklahoma. Increases in irrigated acreage within the eastern areas of these states will be modest in comparison to the declines in the western areas. Thus with no change in crop prices, a 2-million-acre decline in irrigation is projected for the Southern Plains.

The mountain region will experience little net change in irrigated acreage, but within this large, heterogeneous area comprising eight states, significant changes in the location of irrigation are likely. The southern areas and eastern Colorado, where current levels of irrigation are dependent on nonrenewable water sources, will experience some decline in irrigated acreage. In the absence of high crop prices, these declines will be offset only in part by some modest increases in irrigation within the rest of the mountain region.

In the Pacific region a modest overall expansion

of irrigation is likely even with no increase in crop prices. A 1974 California water plan by the Department of Water Resources projected its state's irrigation would reach 9.5 to 10.6 million acres by 1990 and 9.8 to 12.1 million by 2010.[4] These levels compare to the National Resources Inventory (NRI)'s estimate of 8.9 million acres in 1977 and the state's estimate of 9.3 to 9.6 million as of 1980.[5] The increases implied by the higher projections of the plan are improbable in view of the other demands being made on the state's water, the high cost of developing new supplies, ground water mining in some regions stemming from current water use patterns, the prospective loss of nearly 1 million acre·ft/year of Colorado River water once the Central Arizona Project is completed, and the salinity problems threatening a million acres in the San Joaquin Valley. Indeed, the 1974 water plan did not even consider water as a constraint on expanding irrigation. Correcting for this oversight makes the high projections of the 1974 plan appear much too optimistic. Overall, California's irrigated acreage is not likely to rise significantly above 10 million acres by 2000. While the Pacific Northwest is the area least affected by water shortages, the region's water is becoming increasingly valuable for use in hydropower production and preserving the wildlife and amenities of the area. Furthermore, very inexpensive electricity is no longer available for expanding pumping. Consequently, additions to irrigated acreage in the Northwest over the next several decades are not likely to push the totals for Washington and Oregon much beyond 5 million acres without significant crop price increases.

A 25 percent increase in real farm prices might add another 4.5 million irrigated acres in the West, more than doubling the growth of irrigated acreage over the next two to three decades. Yet even assuming such a substantial increase in crop prices, irrigation is projected to rise only 15.5 percent over two to three decades. In comparison, western irrigation doubled over the previous 25 years, rising 21 percent from 1967 to 1977. The long-term elasticity of irrigation with respect to crop prices implied in the projections of Table 11-1 is only about 0.3 (that is, a 25 percent increase in crop prices increases irrigated acreage by about 8 percent.) This relative insensitivity to real price levels is primarily a reflection of two factors—the fact that no foreseeable crop price levels will make irrigation competitive with most municipal and industrial water uses and the importance of institutional and resource factors in determining the level of irrigation. Few areas have the water to support a major expansion of irrigation under any realistic projections of crop prices. Ground water is often the only water available for expansion, but many western aquifers are already being mined at significant rates. Major expansion of surface water irrigation outside of the areas that have other significant handicaps to high-productivity farming requires institutional changes that will provide irrigators with incentives to conserve water. Conservation would enable the available supplies to be spread over more acres. In the absence of such changes ground water will remain the primary water source for new irrigation. While high crop prices will encourage greater pumping, the impacts of increased pumping on total irrigated acreage may be relatively short-lived in view of the non-renewable nature of much of the ground water.

QUALITATIVE CHANGES IN WESTERN IRRIGATION

The nature as well as the rate of growth of irrigation will be very different from these characteristics of irrigation in the past when irrigated farming was stimulated by the availability of inexpensive water and energy. Water withdrawals will be reduced as it becomes more profitable both to make water-saving investments and to reduce the water delivered to plant even if it means some reduction in crop yields. While total irrigated acreage may not peak until the first decade of the next century, total withdrawals for irrigation probably will peak much sooner, perhaps within the next decade. Improved yields to water inputs and shifts to higher-value crops will enable the value of production from irrigated farms to rise even after the quantities of land and water in irrigation have peaked and started to decline. Thus both water withdrawals and acreage are likely to become increasingly poor indicators of changes in the contribution of irrigated output to national agricultural production.

As is evident from a recent report of an interagency task force group that assessed the potential for improving the efficiency of irrigation water use and management in the United States, most water conservation opportunities actually imply a substitution between water and other inputs such as capital, labor, and improved water and agronomic management. There are, of course, limits on the substitutions that are profitable, and these limits depend importantly on the cost of water. The task force group concluded that in addition to "ongoing programs, public and private investments of up to five billion dollars should be made over the next three decades to implement needed water conservation measures."[6] These investments could "result in decreasing gross annual diversions by 15 to

20 million acre-feet and making two to five million acre-feet of water available for new uses."[7]

The conclusions of the task force study draw on the results of Soil Conservation Service (SCS) field estimates of the measures required to achieve a "reasonable level of irrigation water management" in the 17 western states. The major means of reducing conveyance losses is lining canals or piping to reduce seepage losses. Further reductions in conveyance losses could be made through consolidation, realignment, or enlargement of canals and control structures. In total, an estimated 3.1 million acre·ft of water per year could be saved through off-farm investments of $6.2 billion.

On-farm water losses can be reduced through lining or piping of field distribution systems, land leveling, water control or -measuring structures, automation, tailwater recovery systems, and changing to sprinkler or drip irrigation. The SCS survey concluded that "about one-fourth of the irrigated land in surface systems requires land leveling, over one-third could use tailwater recovery systems to reduce flows, and about 60 percent should use more effective ways to schedule and apply water to meet the needs of the crops and reduce deep percolation and excessive return flow."[8] In total, an estimated investment of $8.4 billion plus an annual cost of $142 million for irrigation water management could reduce on-farm water losses by 4.5 million acre·ft/year.

In addition to the reduction in irrecoverable or incidental water losses, other major changes in water management are implied by these infrastructure investments. For example, the proposed measures would result in a 38.6-million-acre·ft reduction in water withdrawals, a 35.3-million-acre·ft reduction in return flows, a 4.3-million-acre·ft increase in crop water consumption (which helps increase farm output by $500 million per year), and a $109-million reduction in energy requirements.[9]

Future federal water projects likely will focus more on improving water use efficiency and preserving the productivity of existing irrigation rather than continuing the past concentration on developing new supplies. Numerous opportunities for reducing on-farm and off-farm water losses through structural means are noted in the task force study cited earlier. Even if the economics of such projects are not more favorable, they have political advantages relative to supply-oriented projects. Well-organized and powerful environmental interests can be expected to oppose efforts to develop additional western stream flows for use in agriculture, but there is no established opposition to conservation or salinity control projects other than those concerned with the budgetary considerations of any major government project.

Nonstructural measures may prove to be even more important than the structural alternatives for improving the returns to water in irrigation. Improved irrigation scheduling, higher-value or less water-using crops, tillage practices designed to conserve soil moisture, and seed varieties offering higher returns to water will become increasingly attractive to farmers confronted with higher water costs. Furthermore, the technologies available for responding to high water costs can be expected to expand significantly in the coming decades. For example, California's Department of Water Resources believes there is considerable potential for reducing evaporation from cropped areas and increasing the portion of transpiration directed to commercially desirable features of the plant. Three requests for proposals were issued by the department in the fall of 1980 for preparing (1) a report on cropping pattern changes to conserve water, (2) a study of the relationships between evaporation from soil and water surfaces and plant transpiration to identify means of reducing evaporation from irrigated lands and to evaluate the potential for water conservation savings by such means, and (3) a report on the potential for saving water through the use of improved cultivars.[10]

The overall potential for increasing irrigated production through a combination of structural and nonstructural measures without increases in total water use is great. The extent to which this potential is realized depends in large part on the institutions and policies affecting water use. Before we turn to the overall policy implications of the analysis, however, the environmental issues are reviewed, since alternative policies should be evaluated for their environmental as well as for their production implications.

Environmental Implications of Anticipated Qualitative and Quantitative Changes in Western Irrigation

The changing character of the growth of irrigation is having some beneficial environmental effects. The environmental problems of erosion, sedimentation, and water quality degradation that are associated with irrigation generally diminish as the application efficiency of water increases. Thus the same investments and irrigation practices that are becoming attractive to irrigators as water costs rise also help reduce the environmental damages caused by overirrigation.

Adoption of improved water management practices, however, largely has been limited to farms with expensive water. Although water costs are generally rising, a large segment of surface water users remain insulated from the impacts of increasing water scarcity. Little improvement in the environmental impacts of their farming practices is likely to occur as long as these farmers pay little for their water and nothing for the environmental damages resulting from its use.

Understanding the environmental implications of irrigated agriculture requires looking beyond the impacts associated with the irrigation process itself. Irrigation is both a land-conserving technology and a means of expanding the land that can be used for cropping in arid and semiarid regions. Since irrigated yields tend to be much higher than the overall average, any change in the number of irrigated acres affects the total number of cropped and pastured acres required to produce a given output. Consequently, an increase in irrigation reduces the pressures on other agricultural land resources, an outcome of increasing importance when the marginal lands that would be brought into production in the absence of irrigation are apt to be particularly susceptible to erosion. The anticipated reduction in the growth of irrigation suggests that future growth of agricultural output will place greater pressures on the land base.

Several studies suggest the environmental costs of substantially increasing the cropland base would be high. A study of the potential for and costs of converting woodland and pastureland to cropland in the Mississippi Delta region estimates that conversion would increase soil erosion by an average of 8.38 to 14.06 tons per acre per year depending on the number of acres and the soil groups converted and the crop rotations selected.[11] A similar study of the State of Iowa indicates that converting up to 2.5 million acres to cropland in Iowa would require using larger portions of highly erodable lands with steep slopes.[12] To some extent, these are environmental costs that have been avoided by the growth of western irrigation; assessing policies that will affect the future role of irrigation should take these impacts into account. For example, restricting water withdrawals to ensure stream flow for optimum fish and wildlife habitat (and thereby decreasing irrigation water use by 22 percent) undoubtedly would hasten the conversion of highly erodable lands to crops. These conversions in turn would have adverse impacts on water quality in the affected areas.

CONCLUSIONS AND POLICY IMPLICATIONS

The West is not running out of water. It is running out of low-cost water, however, and no set of policies and programs will alter the fact that water is becoming increasingly valuable in the West.

As a resource becomes scarcer and more costly, development tends to move in directions that conserve on the use of that resource. These changes take two forms. For a given activity it becomes increasingly profitable to substitute other inputs for the increasingly scarce factors of production. In addition, the mix of activities tends to shift toward those that produce higher values per unit of the scarce resource. The West is undergoing such a transition in response to the increasing value of its water. But this transition is proving unnecessarily costly and disruptive to the region's overall development because the laws, institutions, and policies governing the allocation and use of western water often preserve an illusion of cheap water for some users, which increases the scarcity imposed on others.

Most of the legal and institutional arrangements that influence western water use evolved during and are most appropriate for conditions of relative water abundance. An important objective of the 1902 Reclamation Act was to stimulate settlement of the arid West, and the subsidies built into federal irrigation projects undoubtedly encouraged western rural settlement during the first five or six decades of this century. State water laws helped attract investment to the West by providing assurances of continued access to water, and these laws provided settlers some protection from unscrupulous developers by restricting the transfer of water rights. These water policies and programs that fostered western development in the first half of this century are having negative effects on the region under the water scarcity conditions prevalent in the 1980s. By insulating some water users from the increasing value of the resource, other users and potential users find it more difficult and costly to fulfill their needs. Administration of federal water projects as well as state water laws and policies not only allow an inefficient use of western water but also often ensure inefficiency by reducing or eliminating the incentives and opportunities for transferring water to higher value uses.

Farmers have never had to pay for water rights, only for the cost of getting water from its source to the farm. Even these costs often have been subsidized. Charging farmers for only the costs of delivering water to their fields is an economically sound pricing policy under conditions of abundant water. When there is sufficient water in a stream to satisfy all users and no environmental costs are associated with a given use, there are no additional costs to withdrawing surface water. Likewise, as long as ground water pumping does not exceed natural recharge or affect neighboring users, the farmer's pumping costs reflect the social costs of ground water use. But these conditions no longer prevail in most of the West. Commonly, water requirements exceed stream flows, and ground water use exceeds recharge. Consequently, the water itself has a value that should be reflected in user's costs if water is to be used efficiently.

Improving the efficiency of western water use does not require forcing the owners of water rights to pay for what has been legally given to them. Indeed, any attempt to abrogate these rights would be futile and potentially damaging to the region since it would threaten the entire legal and institutional structure that has brought order to the allocation of western water. A more acceptable alternative for providing incentives to conserve water in areas where it is scarce is to allow and facilitate the sale of water that is not used by the owners of the water rights. Then even if a farmer does not have to pay for the water, there would be an opportunity cost to putting the water to any particular use. Water laws vary among states, but in many western states the principal changes in existing water laws required to implement water markets would be to classify water sales as a beneficial use and to eliminate provisions making water rights appurtenant to a given piece of land. Of course, farmers must have an opportunity as well as a right to transfer water if a more efficient allocation is to be achieved. Establishing effective water markets will require the creation or transformation of institutions to provide for the transfers of money and water.

Since irrigation is a relatively low-value user, a more market-oriented allocation system is likely to transfer water from irrigation to municipal, industrial, and other uses. However, these negative impacts on irrigation might be more than offset by the added incentive that would be provided to increase the returns to the water used in irrigation. Moreover, since irrigation is such a predominant user of western waters, a small improvement in the efficiency of irrigation might save enough water to satisfy nonagricultural water needs without significantly reducing irrigated acreage. Where irrigated acreage is reduced, market forces would focus the impacts on the lower-value uses such as forage and pasture. A reduction in the environmental damages from irrigation practices would be a further benefit of policies designed to make water costs more nearly reflect the scarcity value of the resource.

If the transition from water abundance to scarcity allows for an efficient use of the resources over time, irrigation will contribute to agricultural production and growth for many more decades. Future potential lies primarily in increasing the returns to water, not in the development of new water supplies. Realization of this potential requires providing incentives to conserve water and encourage the development of more effective ways for farmers to respond to higher water and energy costs and salinity levels. The social returns to agricultural research have been very high in the past, and the benefits from research to develop improved irrigation practices, new crops and seed varieties requiring less water, improved understanding of the transpiration of plants, and a host of other promising areas for increasing the options for irrigated farming are likely to be very high in the future. These research efforts, which represent the real hope for the long-term contribution of western irrigation to agricultural production, should be encouraged. Nevertheless, innovations will be adopted in a timely way only if the institutions provide the correct incentives regarding use of the region's scarce resources.

Western irrigation is not likely to make the contribution to agricultural growth that it did when water was abundant and cheap even if the relevant research is encouraged and institutional reforms are made. The inevitable adjustments to declining ground water supplies will bring hardship to some areas. Nevertheless, the socially most expensive response would be to provide subsidies to either enable farmers to pump to greater depths or to import water. An area where agriculture depends on declining ground water supplies inevitably will become a higher-cost irrigated producer. Spreading the cost increases among the general public clearly helps farmers in the affected region. But it also ensures higher overall production costs due to inefficient use of society's resources. The serious problems will emerge if we attempt to keep water inexpensive when it is not. Such efforts will ensure its inefficient use and push the social costs of irrigated production to levels well above those of the dryland alternatives.[13]

The underlying resource conditions as well as the national interests in the use of western water have changed dramatically within the past several decades; there is now need for a corresponding adjustment in the laws and institutions that control and manage the resource. The potential beneficiaries of such reform surely outnumber the potential losers. Irrigators are not helped by existing policies that prevent them from selling water rights. To the contrary, most farmers would welcome the opportunity to sell their water for many times its value in agriculture. Nor are current policies justified as necessary protection of the nation's capacity to produce food and fiber. To the contrary, these policies may limit the long-term role of irrigation by discouraging farmers from adopting water conservation measures and by hastening the mining of nonrenewable supplies. Existing water laws and institutions, however, have proved useful to some groups as a vehicle for at least delaying developments that they consider undesirable. For instance, water rights issues have been used to stall energy projects by groups concerned primarily with impacts on the environment and life-styles. The concerns of such groups may be legitimate and, indeed, may warrant major changes in the design or even cancellation of a project. But there are environmental laws and regulations that enable the

airing of such concerns, and these should be strengthened if society believes they provide inadequate protection. Such concerns do not justify retention of outmoded and inefficient water institutions.

NOTES

1. For a farmer producing 105 bushels per acre, receiving $2.70 per bushel, pumping 2.5 acre·ft/acre a height of 200 ft, and using a center-pivot system, a 28.5 percent increase in corn prices would be required to compensate for a doubling of electricity prices.

2. Paul Fuglestad, Robert Niehaus, and Paul Rosenberry, 1978, *Agricultural resource assessment system: Alternative future analysis,* vol. II (Washington, D.C.: U.S. Department of Agriculture), 68–89.

3. A range of 2000 to 2010 for the projection period is used to indicate the very rough nature of the projections as well as to reflect a suspicion that total irrigated acreage in the West will peak during this decade in the absence of major institutional changes, providing the owners of surface water rights with strong incentives to adopt water conservation practices. In 1977 constant dollars, the average prices received by farmers in the United States from 1975 to 1980 was $2.21 per bushel of corn, $3.01 per bushel of wheat, $3.69 per hundred weight of sorghum, and $0.56 per pound of cotton lint. Inflating these prices by 36 percent would convert them to 1980 prices. These price levels are based on 1980 preliminary prices from U.S. Department of Agriculture, 1981, *Agricultural outlook* (Washington, D.C.), 27; and 1975–1979 prices from U.S. Department of Agriculture, 1980, *Agricultural prices annual summary 1979* (Washington, D.C.).

4. State of California, The Resources Agency, Department of Water Resources, 1974, *The California water plan: Outlook in 1974,* Summary Report, Department of Water Resources Bulletin no. 160–74 (Sacramento), 34–35.

5. From agricultural commissioner reports for each county and the Resources Agency's land use surveys, Warren Cole estimates there were 9.3 million acres with full irrigation and 0.3 million with low water applications as of 1980. Personal communication, February 28, 1980, with Warren Cole, supervising engineer of the California Department of Water Resources.

6. U.S. Department of Interior, U.S. Department of Agriculture, Environmental Protection Agency, 1979, *Irrigation water use and management: An interagency task force report* (Washington, D.C.: U.S. Government Printing Office), ix.

7. Ibid.

8. Ibid., 84–85.

9. Ibid., and Tables 14, 15, and 18. California's Department of Water Resources has serious reservations about whether the water savings identified in the task force report actually refer to water that would otherwise be unusable. Gerald Meral, deputy director of the Department of Water Resources, suggests that the analysis for California did not differentiate savings subject to reuse from those that would otherwise be lost to the system. It is conservation of the latter type that can be viewed as an alternative to developing new supplies through construction of reservoirs or interbasin transfer facilities. (Meral's reservations were expressed in a December 4, 1980, letter to the author.)

10. Department of Water Resources, The Resources Agency, State of California, requests for proposals nos. 80 AG–CONS–2, 5, and 8.

11. Robert N. Shulstad, Ralph D. May, and Billy E. Herrington, Jr., 1979, Cropland conversion study for the Mississippi Delta region (Report prepared for Resources for the Future, April 30, University of Arkansas, Fayetteville), 139–142.

12. Orley M. Amos, Jr., 1979, Supply of potential cropland in Iowa (Ph.D. diss., Iowa State University, preliminary draft), 149.

13. Even now society may be paying more for farm output produced on its irrigated lands. Certainly, irrigated production would not be as great as it is if farmers paid the full cost of getting water to their farms. Moreover, as noted earlier, results of the national agricultural model developed at Iowa State University suggest that our nation's agricultural output could be produced at lower cost with fewer irrigated acres.

12

HISTORICAL REVIEW OF DRINKING WATER

CHARLES C. JOHNSON, JR.

Next to the air we breathe, the water we drink is the most important part of our human existence. It has been said that a person can live 5 minutes without air, 5 days without water, and 5 weeks without food. History has repeatedly warned us that each of these substances must be safe for human consumption if they are to serve their most beneficial purpose.

My role is to present a brief historical review of drinking water. This review will embrace three periods that seem to set out significant changes in approaches to producing acceptable drinking water. I have characterized these periods as follows:

1. *Ancient:* That period from the earliest recorded knowledge on water quality to the years just preceding the advent of the germ theory of disease.
2. *Progressive:* That period following the establishment of the germ theory of disease (approximately 1880) through the control of bacterial and other acute waterborne diseases.
3. *Contradictive:* The period that indicated concern for water contamination and lifetime or chronic health effects.

THE ANCIENT PERIOD

The ancient period covered a time when most population groups depended upon individual initiatives for the quality of the water consumed. The one prominent exception to this was the water supply for the city of Rome, which had its beginning about 313 B.C. From the earliest of time humans have exhibited a concern for the quality of water they drink. As early as 2000 B.C. a quotation in Sanskrit said, "It is good to keep water in copper vessels, to expose it to sunlight, and filter through charcoal." Later, in his writings on public hygiene, Hippocrates (460–354 B.C.) directed attention principally to the importance of water in the maintenance of health, but he also stated that rain water should be boiled and strained or it would have a bad smell and cause hoarseness. We are told that Cyrus, the great king of Persia, when going to war, took boiled water in silver flagons loaded on carts drawn by mules.

From earliest time water has been treated by one of the processes we now take for granted—sedimentation, coagulation, filtration, disinfection. We still use the same general principles to treat water today. These precautions probably were taken more for aesthetic considerations than for health purposes, yet they do illustrate an early concern for water quality.

Historical records indicate that standards for water quality, except for occasional references to aesthetics, were absent up to and including most of the nineteenth century. Yet some advances were made in the processes of community water treatment. The first municipal filtration works was built in Paisley, Scotland, about 1832. The historical basis for requiring water treatment emanates from the mandating of this treatment technique. A law was passed in 1882 in London that henceforth all water should be filtered. This occurred approximately ten years before Pasteur and others demonstrated the germ theory of disease.

Charles C. Johnson, Jr., is president of the environmental engineering firm of C. C. Johnson and Associates, Inc. He is assistant surgeon general (retired) of the Public Health Service and past chairman (1975–1981) of the U.S. Environmental Protection Agency's National Drinking Water Advisory Council. This material is taken from *Drinking water and human health* (Chicago: American Medical Association, 1984), 5–11. Reprinted with the permission of the author and American Medical Association. Copyright © 1984 by the American Medical Association.

Apparently, ancient people deduced by observation, and in the absence of scientific proof, that certain waters promoted good health, while others produced infection. And though they knew little or nothing about the causes of disease, they must have, in some instances, recognized the health-giving properties of pure, wholesome water. Unfortunately, the record suggests that a century of observations on deaths caused by waterborne diseases was necessary to clarify these facts.

Not until the cholera epidemics of the 1800s was a clear-cut relationship established between water and disease. We are familiar with the classic study of Dr. John Snow that linked the Broad Street well in London to the transmission of cholera. It is important to emphasize again that in Dr. Snow's time the germ theory of disease had not yet been established.

We may reiterate several important points. The concern for the quality of drinking water has existed since ancient time. Early movements toward improving drinking water quality through treatment were not predicated on a scientifically based, cause-and-effect relationship. And the main problems associated with drinking water quality were directly related to the concentration of population and to the waste disposal practices of the time.

THE PROGRESSIVE PERIOD

The progressive period, which began about 1880, was characterized by rapid improvement in and wide acceptance of water treatment technology, control of waterborne bacterial diseases, and passage of national legislation and promulgation of standards designed to assure safe drinking water.

In the United States, cholera was not a problem after the mid 1800s, the waterborne disease of particular concern being typhoid fever. Research proved the efficacy of the slow sand filter in reducing the death rate from this disease. Much of this is attributed to the Lawrence Experiment Station established by the Massachusetts Board of Health in the late 1800s. The use of slow sand filtration by the Lawrence Experiment Station demonstrated a 79 percent reduction in the death rate from typhoid fever and a reduction in the death rate from all causes of 10 percent.

The popularity of the slow sand filter was soon surpassed by the essentially American innovation of combining coagulation and rapid sand filtration. Research carried out at the Louisville (Kentucky) Water Company at the end of the eighteenth and beginning of the nineteenth centuries led to widespread use of rapid sand filtration. The Louisville experiments showed that even the most turbid of waters could be treated successfully to eliminate the turbidity and color and remove about 99 percent of the bacteria present. These conditions were considered to be a standard by which the quality of treated water should be judged.

Then followed the most important advance in the history of water treatment practice. The introduction of chlorination occurred about 1908. Chlorination provided an inexpensive, reproducible method of ensuring the bacteriological quality of water. Nothing in the field of water purification has come into use as rapidly or as widely as chlorination. Chlorination was introduced about the time that adequate methods of bacteriological examination of water had developed, thus permitting an objective evaluation of the efficiency of treatment. It was easily demonstrated that it was possible to remove most of the bacteria in raw water, to 0.1 percent of the preceding concentration, a tenfold reduction over that achieved by filtration alone. Since that time continued improvement in chlorination and other water treatment processes has allowed the establishment of the more rigid standard of one coliform per 100 mL of treated water. This sequence of events also established *the principle of attainability in the setting of water quality standards*.

With the successful use of the basic water treatment processes of sedimentation, coagulation, filtration, and disinfection, the essential features of water treatment techniques were known by 1914. Since that time many engineering refinements have been made, but there have been no changes in basic concepts. Thus the stage was set for the development of drinking water standards. It is important to note that the availability of treatment technology has always influenced the development of standards for drinking water.

There were approximately three thousand community water supply systems in the United States at the turn of the century. Health and sanitary conditions were still in need of improvement. The community water supply systems were contributing to major outbreaks of waterborne disease, since pumped and piped water when contaminated provided a highly efficient vehicle for the delivery of pathogenic bacteria. Fortunately, Congress recognized that water supplies serving the public must meet minimum quality standards. The first standards of the U.S. Public Health Service were promulgated in 1914. A maximum contaminant level (MCL) of 2 coliforms per 100 mL of water was established. With this *the concept of maximum, permissible, safe limit was introduced*. Furthermore, the importance of sanitary surveys to identify undesirable watershed conditions was stressed. Since that time the national drinking water regulations have been amended five times (1925, 1942, 1946, 1962, 1974) up to and including the passage of the

Safe Drinking Water Act in 1974. These modifications resulted in addition of new standards, strengthening of existing standards, and, in 1974, broadening of their application from supplies serving interstate carriers to all community water supplies.

One set of deliberations of the advisory committee for the 1962 revisions to the drinking water regulations is worthy of comment. For the first time a regulation was included for a substance for which there was no clear-cut data on a cause-and-effect relationship between level of the contaminant and the health of the consumer. The 1962 revisions limited the concentration of radioactivity in water. The effects on large population groups of chronic exposure to low levels of radioactivity were noted as "not yet well defined."

In 1962 scientific opinion held that potentially harmful radioactivity above background levels should be limited in our drinking water supplies. The MCL could not be set with absolute certainty. The limits were an effort to derive an initial standard on the best information then available. It was recognized that the new standard might have to be adjusted upward or downward as better data became available. The inclusion of this standard emphasized the true measure of public health practices in producing safe water. The standard said "we do not know how bad it is for us, but we do know it is not good for us." The decision was made to limit radioactivity in drinking water based on the best information then available.

In retrospect, the success of this national effort is testament to the many persons in the water supply industry and the several levels of government that joined together in the effort that marks the progressive period. With rare exceptions epidemics traceable to waterborne diseases are no longer a part of our way of life. People expect to travel anywhere in the United States and drink water without fear of getting sick. In other times and even today in other countries, the accomplishment of that task would be considered an idealistic dream. Yet the nation's waterworks industry, under the unifying controls of federal and state health regulations, made that dream a reality in this country. Three ingredients were paramount to this success: an understanding of the problem, a willingness to overcome the problem, and recognition of the need to protect the public's health.

THE CONTRADICTIVE PERIOD

The contradictive period was fashioned from the industrial progress that accompanied World War II, but its impact on drinking water was not recognized until the 1960s. There is mounting evidence that our water supply sources are being subjected to contamination by substances of unknown significance and public health concern.

Today over seven hundred organic chemical contaminants have been found in drinking water supplied by public water systems. Many are probably of no consequence, but others pose a potential health risk to consumers. Even when the contaminant is known to be harmful to the health of persons under some conditions of exposure, the significance of its presence in trace quantities in drinking water is unknown. We do know that given the option we would not knowingly add these contaminants to our water. They are not known to serve any useful purpose when taken into the human body in this manner. Yet we are reluctant to take the steps necessary to bring this problem under control.

In earlier days a concept prevailed that the public's requirements for a domestic water supply in both quantity and quality were guaranteed top priority in the hierarchy of uses of our water resources. Today they are captive to the need to produce energy, reverse the balance of payments, and dispose of the waste products of our progress.

At one time there was a regulation, and it was accepted practice, that water supplies be obtained from a protected source, and that every effort should be made to prevent or control pollution of water supply sources. Yet the sanitary survey has been all but abandoned, and our watersheds have become unprotected victims of community and industrial development and of unregulated waste disposal.

Many communities have had to abandon their water supply sources owing to the discovery of contamination by synthetic organic chemicals for which we have been unable to establish acceptable limits and for which some say there is no current health problem. Yet these episodes have failed to spur the broad initiatives required to provide a rational determination of the need for taking such drastic steps.

We decry the practice of water reuse. Yet every community water treatment plant that exists below the discharge of a wastewater treatment plant is in a mode of unplanned reuse. Evidence abounds that major ground water aquifers are being contaminated by hazardous and toxic wastes. Yet we have not found it necessary to implement the corrective programs required to adequately control this threat to water supply sources for half the population of this country.

For a brief period, with the promulgation of an MCL for radioactivity in drinking water, the principle of preventive public health emerged. Now if we cannot count the patients in the hospital or the corpses in the street, we are reluctant to support comprehensive regulation of synthetic organic contaminants in our drinking water.

Many water treatment plants are experiencing difficulty in producing aesthetically satisfactory water and in removing or reducing the levels of these contaminants that come from raw water sources polluted with industrial, municipal, and agricultural wastes. We are learning that water quality is dynamic. The products of population growth and industrial development are reflected in a deteriorating quality of our drinking water in many areas across the country.

Advanced water treatment techniques are being used in a number of water treatment plants in Europe to combat the presence of organic contaminants in their raw water supplies. It is estimated that more than thirty treatment plants in Western Europe are now using granular activated carbon (GAC) on a routine basis for removal or reduction of these chemical contaminants. While GAC is still a subject of research, a consensus seems to be emerging in Europe that the use of GAC is an essential tool for modern water treatment practice. This is a very different attitude than the one that prevails in this country on control of synthetic organics in drinking water.

A recent attempt to mandate the use of GAC treatment technology in American practice was successfully defeated. An effort is underway to amend the Safe Drinking Water Act in a manner that would make it more difficult to write regulations intended to protect the public's health. Some sources indicate that as far as trace levels of synthetic organic contaminants in water are concerned, no known harm is being done. Let us wait until we know what the health effects are. More research will tell us if and when we need to regulate these synthetic organic contaminants.

Should attitudes like these prevail in the national debate that is now underway, I see a bleak future for the continued progress of drinking water protection in this country. The water industry is not likely to voluntarily implement the changes required to improve the chemical quality of our drinking water. Our research efforts are unlikely to provide the degree of certainty required by the skeptics for support of new regulations. The funds required to protect and clean up our sources of water, already in short supply, are likely to face further reductions at the federal level. Efforts to generate and promulgate national regulations will continue to be impeded by demands for cost-benefit analyses, which we should know, in this instance, are impossible to equitably assess.

Our only defense must be a public attitude that says, "An ounce of prevention is worth a pound of cure." Only then will we be ready to support the drinking water standards and construct the water treatment systems that are necessary to provide acceptable protection to the public health. Is it not time that we eliminate contradictions in our rhetoric about the safety of our water supplies and assume a preventive approach with respect to drinking water quality?

SUMMARY

In this historical review we have traveled almost four thousand years in time. A concern by the public for the quality of its drinking water has been and is ever present. We have seen that when we ignore the manner in which the population and industry dispose of their waste, our water supply sources are placed in grave peril, and often disease and death are the result. If proper motivation is given, the technology and standards required to ensure an acceptable quality of water under most circumstances can be made available. In the past when a crisis was evident, we were able to marshal our collective scientific, technical, and political strength to resolve the problem at hand. It is unfortunate that history does not provide us with the crystal ball that tells us what we should do before we pay the price for not having done it.

We do not have to wait for this to happen. I do not think we will ever know with any degree of certainty what harm our contaminated water is causing us. I do not believe this lack of knowledge provides anyone with a license to produce contaminated drinking water that is potentially harmful to the public's health. In my opinion we must move ahead on the basis of the best evidence that is available to produce the best water that we can. We do know that these modern contaminants are not good for us; we just do not know how bad they are for us.

Now is the time for us to require the best treatment that technology can provide to protect the public from the myriad of toxic chemicals we have discharged to our environment and that ultimately find their way into our water supplies. I believe now is the time for us to adopt a nondegradation policy with respect to the quality of the water we drink. I believe now is the time for the American public, its legislators, and its regulators to say we are not going to play Russian roulette with our health and the future posterity of this nation.

AINING
Dri.. KING WATER QUALITY

JOSEPH A. COTRUVO and ERVIN BELLACK

Drinking water is everyone's concern, and each of us has a role in ensuring that the water we drink is not only wholesome but also aesthetically acceptable. The federal and state governments have legal responsibilities for the provision of high-quality drinking water to the public. In recent years this responsibility has been extended to individual suppliers of water, whether the water utility is publicly or privately owned. The consumer has an obligation as well. If his drinking water comes from an individual well, he has almost all of the responsibility for maintaining the quality of the water. Even if his drinking water comes from a public supply, he should take an active interest not only in the quality of water being delivered but also in the problems the water utility may face in providing that water.

Drinking water quality and safety is affected by contamination that can occur at the source, during treatment, or during transit from the water treatment plant to the consumer's tap (Table 13-1). Contamination of some source rivers by biological and chemical products of human and natural origin is well known. Although treatment removes most potentially hazardous substances, some chemical by-products of chlorination are often added. Passage of water through defective or inappropriate pipes can add bacteria or other microorganisms from growths or infiltration. Metals such as lead and copper can leach from pipes in contact with corrosive water.

The aim of the Safe Drinking Water Act is to

Table 13-1. Sources of Contamination

Source	Contaminants
Raw water	Industrial wastes
	Natural organic matter
	Agricultural runoff
	Surface runoff
	Municipal wastes
	Natural minerals
Treatment	Disinfection by-products
	Intentional additives
	Unintentional additives
Distribution	Corrosion products
	Disinfection by-products
	Unintentional additives
	Cross-connection contaminants
	Microbiological growths

ensure that the water that reaches the consumer's tap is free from harmful contaminants. Until recently, the federal government's role in maintaining drinking water quality was limited to preventing the spread of contagious disease across state lines. While the U.S. Public Health Service issued relatively comprehensive drinking water standards, most recently in 1962, the only authority for those standards was the Interstate Quarantine Regulations; for the most part the standards were only guidelines. Only standards for microbiological quality

Dr. Cotruvo is director of the Criteria and Standards Division, Office of Drinking Water, U.S. Environmental Protection Agency. He was manager for Research and Development, ChemSampCo Division, Albany International Corporation. He has served on several advisory groups for the World Health Organization, North Atlantic Treaty Organization, and the Organization for Economic Cooperation and Development. Dr. Ervin Bellack is retired from his position of chemist, Office of Drinking Water, EPA, after over forty years of government service. This material is taken from *Drinking water and human health* (Chicago: American Medical Association, 1984), 12–21. Reprinted by permission of the authors and American Medical Association. Copyright © 1984 by American Medical Association.

were legally enforceable—and then only in the case of public water supplies that served interstate carriers.

THE SAFE DRINKING WATER ACT

In 1974 the Safe Drinking Water Act was passed "to assure that the public is provided with safe drinking water." The passage of the act greatly expanded the government's role. The act applies to all public water systems, and such systems are defined as those that regularly serve 25 or more persons or consist of at least 15 service connections. The Environmental Protection Agency (EPA) has categorized public water systems into community water systems, those that serve resident populations at least 60 days out of the year, and noncommunity water systems, those that serve primarily transient populations.

There are about 60,000 community water systems, serving about 2 million persons, and perhaps 160,000 noncommunity water systems in the United States. The majority of community water systems are very small, but larger systems serve the vast majority of the population, as shown in Figure 13–1. The act provides the authority to regulate

substances in drinking water that "may have any adverse effect on the health of persons," and also provides that authority to regulate any substance in drinking water that may adversely affect its odor or appearance or that may otherwise adversely affect the public welfare. The Federal Food and Drug Administration retains the responsibility for ensuring the safety of bottled drinking water by developing applicable regulations after EPA produces National Drinking Water Regulations.

The Safe Drinking Water Act has the following principal provisions affecting the federal government's role in maintaining drinking water quality: (1) a mandate for the development of national drinking water regulations, both primary (health-related) and secondary (aesthetics); (2) a provision for assigning primary enforcement responsibility (primacy) to the states; (3) a mandate for the development of regulations to protect underground sources of drinking water; (4) a mandate for informing the public of the existence of violations of the provisions of primary regulations.

National Primary Drinking Water Regulations

The act defines a primary drinking water regulation as one specifying contaminants that may have any

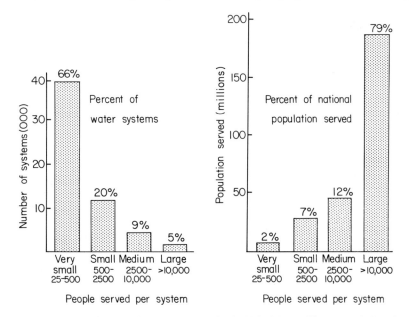

Figure 13-1. Number and size of community water systems in the United States. The vast majority of systems serve between 25 and 250 persons but account for only 2 percent of the total population. The 2700 large systems (serving more that 10,000 persons each) serve over 80 percent of the population, even though they are only 5 percent of the total number of systems. (*Source:* J. Cotruvo, 1981, EPA policies to protect the health of consumers of drinking water in the United States, in *Water supply and health,* ed. H. van Lelyveld and B. C. J. Zoeteman. New York: Elsevier. This illustration has been added by the editors and used with permission of the author and publisher.)

adverse effect on the health of persons, and for each contaminant it either states a maximum contaminant level (MCL) or, if it is not economically or technologically feasible to ascertain the level of such contaminant, specifies treatment known to reduce the level of the contaminant sufficiently to protect health. Primary regulations also can contain criteria and procedures ensuring a supply of drinking water that dependably complies with the MCLs, including quality control and testing procedures.

Public Notification

One feature of primary regulations that should be of particular interest to physicians is the requirement for public notification when an MCL is exceeded or when some other provision of the regulation is violated. Notices regarding MCL violations are expected to explain to consumers the health significance of the violation in nontechnical language and in terms that inform but do not unnecessarily alarm the readers.

The intent of Congress in instituting this requirement was to see that the public was educated "as to the extent to which public water systems serving them are performing inadequately in light of the objectives and requirements" of the Safe Drinking Water Act. Congress believed that such public education was necessary in order to develop public awareness of the problems facing public water systems, to encourage a willingness to support greater expenditures to assist in solving those problems, and to advise the public of potential or actual health hazards.

Variances and Exemptions

Primary regulations also contain provisions for variances and exemptions. Variances are waivers that can be granted if, for certain reasons, the MCL cannot be attained despite application of the best generally available technology or treatment technique. Exemptions are temporary waivers permitting a system to achieve compliance where noncompliance is due to compelling factors such as costs. Before a variance or exemption can be granted, there must be a finding that the grant will not result in an unreasonable risk to health. For each contaminant EPA has developed guidance that can assist a state in making the determination that the health risk from noncompliance is "unreasonable."

Regulations Development

The act establishes a sequence for development of primary drinking water regulations as follows:

1. National Interim Primary Drinking Water Regulations, to be developed rapidly, based on the U.S. Public Health Service Drinking Water Standards of 1962 as modified by the EPA Advisory Committee on the Revision and Application of Drinking Water Standards.
2. A review and report by the National Academy of Sciences, providing health effects and recommended MCLs (RMCLs).
3. A list of contaminants with recommended maximum contaminant levels—levels at which no known or anticipated adverse health effects would occur with an adequate margin of safety.
4. Revised Primary Drinking Water Regulations, with MCLs as close to RMCLs as feasible or with required treatment that will reduce the level of the contaminant to a level as close to the RMCL as is feasible.

National Interim Primary Drinking Regulations were first issued in 1975, amended to include radionuclides in 1976, amended to include trihalomethanes in 1979, and further amended in 1980 to include requirements for sodium monitoring and to include the initiation of actions to deal with corrosive water. The interim regulations, in effect since June 1977, now contain MCLs for 10 inorganic substances, 10 organic substances, radionuclides, coliform bacteria, and turbidity (Table 13-2). The regulations also contain monitoring requirements and specify analytical methods. Fifty out of 57 states and territories have qualified for primary enforcement responsibility, qualification depending in part on the adoption of state regulations that are at least as stringent as the national regulations. Where a state has not qualified for primacy, EPA is enforcing the regulations through its regional offices.

National Academy of Sciences

The National Academy of Sciences' Safe Drinking Water Committee completed its initial review of the health effects of drinking water contaminants in 1977. Under the title *Drinking Water and Health,* that review included an assessment of the adequacy of existing limits to protect human health and examined the health effects of some unregulated contaminants. The academy did not make recommendations regarding MCLs since such recommendations required regulatory decisions including other factors beyond the realm of science. The academy has since produced three additional reports updating the earlier report and addressing additional areas of interest. Volume 2 of *Drinking Water and Health* addresses the disinfection of drinking water and the chemistry of disinfectants in water, and it

Table 13-2. National Interim Primary and Secondary Drinking Water Regulations (1984)

Primary Constitutents	Secondary Constituents	Maximum Allowable Concentration by U.S. EPA[a]
A. Inorganic		
1. Arsenic		0.050
2. Barium		1.000
3. Cadmium		0.010
4. Chromium		0.050
5. Fluoride		1.4–2.4
6. Lead		0.050
7. Mercury		0.002
8. Nitrate (as N)		10.000
9. Selenium		0.010
10. Silver		0.05
	1. Chloride	250.000
	2. Copper	1.000
	3. Iron	0.300
	4. Manganese	0.050
	5. Sulfate	250.000
	6. Zinc	5.000
B. Radionuclides		
1. Radium 226 and 228 (combined)		5.0 pCi/L
2. Gross alpha particle activity		15.0 pCi/L
3. Gross beta particle activity		4 mrem/year
C. Organics		
1. Endrin		0.0002
2. Lindane		0.0040
3. Methoxylchor		0.1000
4. 2.4-D		0.1000
5. 2.4-5-TP Silvex		0.0100
6. Total trihalomethanes[b]		0.1000
7. Toxaphene		0.0050
D. Biological		
Coliform bacteria		1/100 mL (mean)
E. Physical characteristics		
Turbidity		1–5tu
	1. Color	15 color units
	2. Corosivity	Noncorrosive
	3. Foaming agents	0.5
	4. Odor	3 (threshold odor number)
	5. pH	6.5 to 8.5 units
	6. Total dissolved solids (T.D.S.)	500.00

Source: Primary constituents: U.S. Environmental Protection Agency, 1982, Maximum contaminant levels (subpart B of part 141, National interim primary drinking water regulations): U.S. Code of Federal Regulations, 1982, Title 40, Parts 100 to 149, revised as of July 1, pp. 315–318. *Secondary constituents:* U.S. Environmental Protection Agency, 1982, Secondary maximum contaminant levels (section 143.3 of part 143, National secondary drinking water regulations): U.S. Code of Federal Regulations, 1982, Title 40, Parts 100 to 149, revised as of July 1, p. 374.

Note: This table has been reorganized by the editors in order to group classes of pollutants, that is, inorganics and so on.

[a] All data in mg/L unless otherwise stated (pCi/L = picocurie/L; mrem = millirems; tu = turbidity).

[b] Includes the sum of the concentrations of bromodichloromethane, dibromochloromethane, tribromomethane (bromoform), and trichloromethanes (chloroform).

contains an evaluation of activated carbon for drinking water treatment. Volume 3 discusses epidemiological studies on cancer frequency in relation to the presence of certain organic constituents of drinking water, the problems of risk estimation, and the contribution of drinking water to mineral nutrition in humans, and it includes toxicological information supplementing volume 1. Volume 4, published in 1982, contains supplemental toxicological information, and it addresses the chemical and biological quality of water in distribution systems and the health implications of distribution system deficiencies.

Revised Primary Drinking Water Regulations

The EPA is currently in the process of developing revised primary regulations as required by the Safe Drinking Water Act. Development of the revised regulations will be accomplished in four phases: (1) development of regulations for volatile synthetic organic chemicals most commonly detected in ground water; (2) development of regulations for synthetic organic chemicals (including pesticides), inorganic chemicals, and microbiological factors—this will involve a comprehensive reassessment of the current requirements in the interim regulations along with consideration of other contaminants detected in drinking water; (3) development of regulations for radionuclides—The current requirements in the interim regulations will be assessed along with the evaluation of other radionuclides; (4) development of regulations for disinfection by-products—the current requirements for trihalomethanes will be revised on the basis of implementation experience and results of current research projects; other disinfection by-products will also be addressed.

National Secondary Drinking Water Regulations

Most of the drinking water constituents listed in the Public Health Service (PHS) Drinking Water Standards with nonmandatory or recommended limits are now included in secondary drinking water regulations. These constituents, such as iron, chlorides, and color, do not usually have a direct effect on health, but they may affect water quality to such a point that consumers will seek other, less healthful drinking water sources. The federal role in dealing with non-health-related drinking water quality remains largely one of providing guidance and recommendations. Secondary regulations were issued in 1979 (Table 13-2). These regulations may be adopted by the states, but the Safe Drinking Water Act provides no enforcement authority.

Protection of Underground Sources of Drinking Water

The Safe Drinking Water Act's provisions for protecting the quality of drinking water from underground sources are largely limited to the authorization of a program for controlling the underground injection of fluids that may impact on water sources. While national regulations for the control of underground injection are issued, enforcement is primarily a state function. States where underground injection control programs are necessary may authorize underground injections by means of permits. Where it has been determined that an underground aquifer is the sole or principal source of drinking water for an area, no federal funds may be used for any project that according to EPA may contaminate the aquifer through a recharge zone and create a significant hazard to public health.

FEDERAL GOVERNMENT FUNCTIONS UNDER OTHER AUTHORITIES

The Safe Drinking Water Act (SDWA) specifically limits control of source water quality to those contaminants for which there is no applicable removal treatment. Thus criteria for raw water source quality remains essentially within the purview of the Federal Water Pollution Act and the amendments contained in the Clean Water Act. The Clean Water Act is a comprehensive act intended to restore and maintain the chemical, physical, and biological integrity of the nation's waters. It addresses such diverse topics as the discharge of pollutants into navigable waters; quality of water used for the protection and propagation of fish, shellfish, and wildlife; the quality of water for recreation; and waste treatment and the technology necessary to eliminate pollutant discharge into oceans. The act also provides financial assistance for the construction of publicly owned water treatment works. Under this authority EPA has established criteria for 129 contaminants applicable to raw water sources. These criteria are not drinking water standards and are not federally enforceable. They are to be used by the states to control the quality of ambient waters that may find application as drinking water sources.

Drinking Water Additives

The EPA has responsibility for both direct and indirect additives to drinking water. This includes responsibility for impurities in treatment chemicals and in substances contributed to drinking water by contact, such as paints and coatings. In the case of water treatment chemicals, EPA has contracted with the National Academy of Sciences for the production of a Water Treatment Chemical Codex, a compendium that specifies the limits for impurities in these chemicals that might adversely affect the quality of drinking water. In the case of coagulants and other indirect additives to drinking water, EPA had provided informal toxicological opinions on the suitability of these products. The EPA now is in the process of developing a program, possibly with industry participation, for managing the quality of additives to drinking water.

Disease Outbreaks

The federal Center for Disease Control (CDC) includes among its functions the control of waterborne disease outbreaks, whether caused by microorganisms or chemicals. In case of an outbreak the CDC might investigate the epidemiological and etiological aspects of the disease, while the EPA might investigate the water supply deficiencies that were contributing factors. The CDC publishes an annual summary of waterborne-disease outbreaks, and the EPA contributes information on the water supply engineering aspects of the outbreaks.

Pesticides

Under the authority contained in the Federal Insecticide, Fungicide and Rodenticide Act (FIFRA) and additional authorities transferred from Food and Drug Administration (FDA), the EPA has control over the contribution of certain pesticides to drinking water quality. This includes control over disinfectants by means of a registration process and restrictions on the application of those pesticides that might impact on drinking water sources.

Ground Water

The protection of ground water sources has become a major issue in recent years; a number of federal laws in addition to the Safe Drinking Water Act have been enacted to preserve the quality of these sources. They include the Solid Waste Disposal Act, the National Environmental Policy Act, the Resource Conservation and Recovery Act, the Toxic Substances Control Act, the Clean Water Act, and the Comprehensive Environmental Response Compensation and Liability Act, better known as Superfund.

NONREGULATORY ACTIVITIES

While the Safe Drinking Water Act applies only to public water systems, and national regulations cannot cover every conceivable contaminant in drink-

Table 13-3. Health Advisories

Carbofuran	Uranium
Chlordane	Carbon tetrachloride
1,1-Dichloroethylene	cis-1,2-Dichloroethylene
Dichloromethane	1,2-Dichloroethane
Formaldehyde	Ethylene glycol
n-Hexane	Fuel oil #2/kerosene
1,1,1-Trichloroethane	Methyl ethyl ketone
trans-1,2-Dichloroethylene	Tetrachloroethylene
Trichloroethylene	Xylenes
Benzene	p-Dioxane
PCBs	Toluene

ing water, the federal government's role in maintaining drinking water quality is not limited to the authority granted by that and other acts. Where contamination has been detected, whether in individual wells or because of the presence of unregulated contaminants, EPA's Office of Drinking Water will provide information regarding possible health effects and advisory recommendations. Health advisories have been issued for 22 substances (Table 13-3). Technical assistance includes recommended treatment techniques, evaluation of analytical methods, and assistance in interpreting analytical results.

Over the years the EPA has conducted special studies to determine the nature and extent of contamination of drinking water supplies. These studies include the National Organic Reconnaissance Survey, the Community Water Supply Survey, and the Ground Water Survey, which principally are concerned with the contamination of water supplies by volatile organic chemicals. Recently, a study was conducted on the quality of drinking water available in rural areas from municipal supplies and from individual wells. The study showed that rural drinking water supplies frequently are inferior to urban supplies in quality and microbiological quality.

THE PHYSICIAN'S ROLE

It is incumbent on a practicing physician to be aware of the role of drinking water contamination as it contributes to human disease. While public water systems regularly are monitored for certain contaminants, individual supplies, usually wells, may be tested occasionally for coliform bacteria. The supplies are seldom tested for other contamination, unless contamination is apparent because of unusual taste, odor, or color. A knowledge of the health effects of potential contaminants can assist the physician who is considering the possibility that an illness may be due to drinking water or who wishes to eliminate drinking water as the source of the cause of illness.

Specific areas where a physician can have a role include public notification of MCL violations and the granting of variances and exemptions by states from primary drinking water regulations. As noted earlier, public notification is intended to inform, but not alarm, consumers as to the health implications of the violation. The physician may be able to assist in the preparation of the notice or to interpret the meaning of the notice for the patient and citizen. Variances and exemptions are contingent on the absence of an unreasonable risk to health. While any violation of an MCL implies the existence of a risk to health, determination of an "unreasonable" risk requires judgment on the part of someone with a knowledge of the health effects of a contaminant and of the margin of safety associated with the contaminant.

Since drinking water frequently contributes significantly to the human intake of a number of chemicals, the physician should be aware of the composition of water in relation to the patient's condition. For example, sodium content of drinking water in public systems is reported to health authorities so that physicians can prescribe alternative water sources for hypertensives and others who must restrict sodium intake. No such reporting requirement is applicable to individual well supplies.

Even with expanded federal and state roles in maintaining drinking water quality, some adverse health effects, including outbreaks of disease, still occur. Over 63,000 cases of waterborne disease were reported in the past decade in the United States, about half of which were of unknown etiological origin. Between 1972 and 1980 there were 38 reported waterborne outbreaks of giardiasis alone with 20,000 cases. Giardiasis is caused by a protozoan that can be removed from drinking water

sources by the simple expedient of filtration. The physician should be aware of the health risk to patients who drink water from unfiltered surface sources. Other prominent waterborne diseases include hepatitis, gastroenteritis, and shigellosis.

Hemodialysis

While tap water is sometimes used in hemodialysis, there are limitations to the quality of tap water that could make this hazardous to the patient. The Association for the Advancement of Medical Instrumentation (AAMI) has issued standards for the water to be used in hemodialysis. Fluoride, chlorine, and nitrate are among substances commonly present in potable drinking water that may be of concern during dialysis. Recent changes in quality of a given water supply, such as the substitution of chloramine for chlorine, require that physicians be alert to the health significance of these changes.

CONCLUSION

There is ample opportunity for public input at every stage of the regulatory process. The medical community can further the cause of health by participating in the development and implementation of rules issued by government, during the process of developing regulations, and by providing expertise and assisting communities as they deal with drinking water quality.

While the federal government has a major role in providing direction for maintaining drinking water quality, the states have a role in implementing regulations, and public water systems have the responsibility of providing safe drinking water, there must be participation by the community and by individual citizens. Each has a responsibility, because drinking water is everyone's concern.

14

WATER AND ENERGY

JOHN HARTE and MOHAMED EL-GASSEIR

Providing energy for human use consumes water, and providing water consumes energy. Our objective is to assess the constraints that limited and unpredictable supplies of fresh water in the United States may place on energy development.

Energy technologies use water resources in numerous ways. For example, the cooling of electric generating plants or coal gasification and liquefaction plants may consume fresh water. Coal and oil shale conversion processes require water as a chemical feedstock. Coal mining and land reclamation subsequent to surface mining require water. Solar bioconversion plantations are likely to require irrigation water. Hydroelectric power consumes water in the sense that artificial lakes enhance evaporation losses. In fact, nearly every imaginable energy system demands water. Because of the limited fresh water supply in many regions of the United States, and because of the unpredictable nature of precipitation, it is important to understand the fresh water requirements for each of the many energy technologies available to society during the next several decades. Water consumption requirements place serious constraints on the future level of development of many of this country's energy options.

During the coming decades the United States will have to find energy sources that can replace natural gas and petroleum. Because many end uses, especially transportation and home heating, rely today on these two fuels, it appears that only three paths are available. One option is to adapt such end uses to electricity. The second is to replace dwindling gas and petroleum supplies with syn-

thetic gaseous and liquid fuels. A third path, which could ease the demand on gaseous and liquid fuels for space heating and cooling, is to expand active and passive use of the sun. Part of our concern here will be with the comparative impacts of these three paths on water consumption.

FRESH WATER SUPPLY AND DEMAND

To estimate the consequences of the water requirements for energy production, a distinction must be made between water withdrawal and water consumption. Water withdrawn is water taken from a water supply but not necessarily consumed. Water consumed is water rendered unavailable for specified further uses. The water consumption of a given activity depends on the ways water is used in the activity and the ways it is needed by downstream users, including the spatial distributions and time schedules of all such uses (1).

Thus heavily polluted water that is discharged from a coal gasification plant is consumed water for many competing uses, although not, perhaps, for mine floor wetting. Water evaporated from a wet cooling tower or an artificial lake or from surface-mined land under reclamation is consumed water from the viewpoint of other users in the region because the evaporated water cannot be expected to fall as rain on the same region. Also, water used as a source of hydrogen for synthetic fuel production is consumed water—notwithstanding the fact that this water is regenerated when the fuel is eventually burned.

Dr. Harte is professor of physics in the Energy and Resources Group, University of California, Berkeley. His research interests include water resources, exotoxicology, inadvertent climate modifications, and acid deposition in the western United States. Mr. El-Gassier was a graduate student in the Energy and Resources Program, University of California, Berkeley. This material is taken from Energy II: Use, conservation and supply, ed. Phillip H. Abelson and Allen L. Hammond (Washington, D.C.: American Association for the Advancement of Science, 1983). This article, originally titled Energy and water, is based on a study done in connection with the National Academy of Sciences/Nuclear Regulatory Commission Committee on Nuclear and Alternate Energy System report. Lightly edited and reprinted with permission. Copyright © 1978 by AAAS.

Technical, economic, and policy considerations in the development of new energy sources can change the balance between water withdrawal and consumption. In this chapter we assume that the rate of consumptive use of water varies from the minimum rate believed to be achievable under strict conservation and purification efforts to the maximum rate where water-conserving practices and adequate water treatment are not even attempted. This maximum consumptive rate, in some instances, would simply be the withdrawal rate.

We are concerned here primarily with fresh water consumption requirements of alternative energy systems. Withdrawal, while less worrisome than consumption, is nonetheless an important environmental problem for several reasons. First, the rate at which water is withdrawn provides a rough measure of the rate at which aquatic habitat is temporarily destroyed and aquatic organisms are killed or injured. Organisms, for example, can be killed by entrainment in cooling condensers (2). Second, the larger the withdrawal, the greater is the need for a storage reservoir for operation in times of low flow. Because of the great range and intensity of environmental hazards associated with the damming of rivers to create reservoirs (3, 4), including, though by no means limited to, large consumptive losses caused by excessive evaporation and bottom seepage, the size of withdrawal requirements should not be overlooked in assessing possible future energy sources.

WATER CONSUMPTION REQUIREMENTS OF ENERGY ALTERNATIVES

In this section we estimate the water requirements for a variety of energy technologies that are candidates for major expansion in the United States. From these estimates judgments are given about the relative impacts on water resources of some technologies that are competitive in the sense that they could provide similar benefits to society. We adopt as an energy reference the quantity 10^{18} J. Note that the commonly used unit of energy called the quad (10^{15} British thermal units, Btu) is approximately equal to 1.05×10^{18} J. Some energy quantities, pertinent to the following discussions, are listed in Table 14-1.

Coal and Oil Shale: Mining, Reclamation, and Conversion to Synthetic Fuels

Coal and oil shale are the major fossil fuel resources of the United States and potentially form the base for a large and long-lasting energy supply.

Table 14-1. Some Useful Energy Quantities for the United States in 1975

Energy Category	Energy (10^{18} J/year)
Total energy consumption	72.7
Liquid fuels consumption	34.5
Natural gas consumption	21.3
Coal consumption	14.1
Steam-generated electricity output	6.1
Energy yield from 1 km^2 average western surface-mined coal	0.1 to 0.2
Annual average sunlight on 1 km^2	0.0056

Source: Data adapted from Ref. 13.

Most of the explored coal and oil shale deposits are found in five water resource regions (see Figure 14-1); about 50 percent of the total recoverable coal reserves and 30 percent of the surface-mineable reserves are in the Ohio and Upper Mississippi regions, which are also close to major demand centers. The remaining coal resources (which happen to be more attractive commercially) and the principal oil shale reserves are found in sparsely populated, arid or semiarid areas of the West, with the oil shales confined to a far smaller region than the coal.

Estimates of water consumption for various shale and coal conversion pathways are given in Table 14-2 (5). Clearly, the major part of the water consumption occurs at the conversion stage itself. Two other major categories of water consumption are reclamation of surface-mined land and coal transport via slurry pipelines. Unless the water used in a slurry pipeline is adequately treated and returned to its source, it must be considered to be consumed water in its region of origin.

Among the entries in Table 14-2, one with an especially large range of uncertainty is that for land reclamation in the West. This uncertainty is mostly due to a lack of understanding of environmental factors such as soil-binding properties and the conditions under which detrital and soil microorganism–based nutrient cycles can be reestablished in dry, disrupted terrain (6). The uncertainties include not only the unknown requirements for annual irrigation but also the unknown number of years for which irrigation would be necessary for reestablishing a viable ecosystem. Successful revegetation (though not necessarily restoration to original conditions) is likely to be necessary in order to

Figure 14-1. Water resources regions of the United States and major coal and oil shale deposits. Recently found eastern oil shale deposits of uncertain commercial value are not shown. (*Source:* Refs. 23, 24, 32.)

reduce problems of erosion, mine drainage (with subsequent deterioration of downstream water quality), and possibly flooding. The lower value given in Table 14-2 in our judgment has a low probability of leading to genuine revegetation (6, 7). We have not attempted to include in our estimates the additional water consumption resulting from secondary impacts of erosion, drainage, or flooding, should land reclamation be unsuccessful.

Table 14-2 shows the water consumption for converting shale to be smaller than the consumption for syncrude production from coal. However, the listed ranges of water consumption mean very different things in the two cases, and the actual situation could turn out to be more complicated than these numbers indicate. The effects of coal mining have long been recognized. Mine drainage, soil erosion, and alteration of runoff characteristics are among the important ones. These effects are also expected from shale mining. However, mining of oil shale results in a volume of processed shale that is about 1.2 times greater than the raw shale. The resulting difficulty of storing the wastes in the excavated areas has led to proposals to use natural canyons as storage space for spent shale. Such action would lead not only to permanent loss of many canyon lands but also to the destruction of natural habitats, many of which are homes for a number of rare and endangered species (7, 8), and

to an alteration of the hydrologic regime of the region. Furthermore, the stability of the spent shale when subjected to precipitation and snowmelt is questionable (9).

These, and also economic, considerations have directed attention toward *in situ* technology for extracting oil from shale. One water-related problem with *in situ* processes is particularly worrisome. The significant shale deposits of the Piceance Basin in Colorado are in themselves an integral part of the mechanism by which ground water quality and flow are naturally maintained (10). A disruption of this system could affect the flow and quality of the White River and ultimately the Green and Colorado rivers by causing the release of artesian, saline ground water into fresh water systems. Our listed range of uncertainty in Table 14-2 does not cover the case of aquifer disruption leading to alteration of the White, Green, and Colorado rivers; it includes only the more narrow range of water requirements associated with the range of technological options.

Problems affecting water availability and quality in the Upper and Lower Colorado regions are already serious. Overallocation of water, low-flow conditions, salinity, and erosion are well recognized (6, 8–10). An oil shale industry, whether based on surface or deep mining, aboveground or *in situ* retorting, poses the risk of serious ecological

Table 14-2. Water Consumption for the Production of Synthetic Fuels from Coal and Oil Shale in the United States

	Category of Use						
	Mining[a]	Reclamation[b]	Transport by Slurry Pipelines[c]	Conversion[d]	Associated Urban[e]	Total with Slurry Pipelines	Total Without Slurry Pipelines
LOW-BTU GAS							
Eastern coal							
Surface-mined	0.0028 to 0.0035	0.0 to 0.030	0.045 to 0.057	0.083 to 0.58	0.018	0.15 to 0.69	0.10 to 0.63
Deep-mined	0.0062 to 0.0078	0.0	0.045 to 0.057	0.083 to 0.58	0.018	0.15 to 0.66	0.11 to 0.61
Western coal							
Surface-mined	0.0028 to 0.0070	0.0028 to 0.14	0.045 to 0.11	0.083 to 0.58	0.018	0.15 to 0.86	0.11 to 0.74
Deep-mined	0.0062 to 0.010	0.0	0.045 to 0.11	0.083 to 0.58	0.018	0.15 to 0.72	0.11 to 0.61
HIGH-BTU GAS							
Eastern coal							
Surface-mined	0.0035 to 0.0042	0.0 to 0.036	0.057 to 0.069	0.083 to 0.58	0.049	0.19 to 0.74	0.14 to 0.67
Deep-mined	0.0078 to 0.0095	0.0	0.057 to 0.069	0.083 to 0.58	0.049	0.20 to 0.71	0.14 to 0.64
Western coal							
Surface-mined	0.0035 to 0.0085	0.0036 to 0.17	0.057 to 0.14	0.083 to 0.58	0.049	0.20 to 0.95	0.14 to 0.81
Deep-mined	0.0078 to 0.012	0.0	0.057 to 0.14	0.083 to 0.58	0.049	0.20 to 0.7	0.14 to 0.64

SYNCRUDE

Eastern coal							
Surface-mined	0.0031 to 0.057	0.0 to 0.048	0.051 to 0.093	0.11 to 0.74	0.029	0.19 to 0.92	0.14 to 0.82
Deep-mined	0.0070 to 0.013	0.0	0.051 to 0.093	0.11 to 0.74	0.029	0.20 to 0.88	0.15 to 0.78
Western coal							
Surface-mined	0.0031 to 0.011	0.0032 to 0.23	0.051 to 0.19	0.11 to 0.74	0.029	0.20 to 1.2	0.14 to 1.0
Deep-mined	0.0070 to 0.017	0.0	0.051 to 0.19	0.11 to 0.74	0.029	0.20 to 0.98	0.14 to 0.79
OIL FROM SHALE							
Surface technology							
Surface-mined	0.0040 to 0.0056	0.033 to 0.053	NA	0.030 to 0.044	0.0069 to 0.0092	NA	0.074 to 0.11
Deep-mined	0.0041 to 0.0056	0.032 to 0.056	NA	0.030 to 0.044	0.0082 to 0.011	NA	0.074 to 0.12
In situ technology							
Modified *in situ*	0.0019 to 0.0026	0.014 to 0.030	NA	0.027 to 0.047	0.0087 to 0.010	NA	0.052 to 0.090
True *in situ*	NA	0.0 to 0.0077	NA	0.0 to 0.044	0.0088 to 0.010	NA	0.009 to 0.062

Source: The data and the references were derived from Ref. 7. All calculations are based on coal energy content of 28, 22, and 14 million J/kg of bituminous, subbituminous, and lignite coals (19) and on conversion efficiencies of 67 to 85, 55 to 67, and 41 to 75 percent for low- and high-Btu gasification and liquefaction, respectively (20–22).

Note: Data are expressed as cubic kilometers per 10^{18} joules of synthetic fuel product.

[a]In the East surface and deep mining consume 2.3 and 5.2 m³/10^{12} J coal mined. In the West consumption is 2.3 to 4.7 and 5.2 to 6.8 m³/10^{12} J mined, respectively (23). [b]In the East land disturbance is 22 to 65 m²/10^{12} J coal mined (20), and annual water consumption is 0 to 0.015 m³/m² over a one- to two-year period (7). In the West the corresponding figures are 3.9 to 31 m²/10^{12} J coal mined (20) and 0.30 to 0.61 m³/m² over two to five years (7). The shale estimates include consumption for revegetation as well as processed shale disposal (24). [c]Slurry pipelines consume 38 and 37 to 76 m³/10^{12} J coal mined in the East and the West, respectively (25). [d]For coal conversion, see Refs. 25 and 26; for shale extraction, see Ref. 24. [e]For coal conversion, see Ref. 21; for shale, see Ref. 24.

impacts in its competition for water. The geographic confinement of commercially attractive and explored oil shales to this region is in contrast to coal, which is found in significant quantities across a spectrum of meteorological, topographical, hydrological, and ecological conditions. In choosing between coal and oil shale, the greater flexibility of coal-mining sites and uncertainties about aquifer disruption from oil shale activities must be considered along with the numbers in Table 14-2.

Cooling Requirements for Steam Electric Plants

The fresh water required for the major ways of cooling steam electric power plants is listed in Table 14-3. The listed range of requirements reflects variation in regional evaporation rates, differences in the temperature to which the cooling water is heated, and some uncertainties arising from the complex mechanisms by which open water dissipates heat. The thermal efficiency of the electric generating system is assumed to be 38 percent, typical of a modern coal-burning plant.

Table 14-3 shows that the use of wet cooling towers is not necessarily preferable to once-through cooling. Wet tower cooling reduces the withdrawal requirements, while once-through cooling reduces consumption requirements, provided that additional water storage is not needed to meet withdrawal needs of a once-through system. In areas where water is scarce and river flow is variable, the large withdrawal needs of a once-through system may not be met without providing for additional storage. If storage must be added with a once-through

Table 14-3. Water Requirements for Electric Power Plant Cooling

Cooling Mode	Withdrawal	Consumption
Once-through (no storage)	28.0 to 40.0	0.2 to 0.4
Once-through (storage[a])	28.5 to 41.5	0.5 to 1.5
Wet cooling tower[b]	0.6 to 0.8	0.4 to 0.6

Note: Data are expressed as cubic kilometers of water per 10^{18} J of electric output. It is assumed that the thermal efficiency is 38 percent and that 17 percent of the waste heat is dissipated directly to the atmosphere in the form of hot stack gases (11).

[a]Reservoir capacity is assumed to meet backup storage requirements of 1000 MWe-sized plants for 90 days; lake surface evaporative loss is assumed to be in the range 0.75 to 1.5 m/year. For further assumptions, see Ref. (11). [b]Wet tower consumption is the sum of evaporative loss plus drift; withdrawal is equal to consumption plus blowdown.

cooling system, then wet cooling is preferable. In this circumstance wet tower cooling not only reduces water consumption but also avoids problems of thermal pollution in aquatic habitats, as well as the many ecological hazards associated with damming free-flowing rivers (4). In circumstances where additional storage is not required but water consumption is a problem (for example, western lakes), the once-through method may be preferable (11).

Lest is appear that dry cooling is an unqualified blessing because of savings in water, we note that a coal-burning, dry-cooled, electric generating plant is likely to have a thermal efficiency of about 1.5 percentage points lower than a plant with a once-through or wet tower cooling (12). Thus more fuel will be required for a given electric output, and extra water will be consumed for mining and land reclamation. Consider, for example, two electric power plants producing the same electric output from western surface-mined coal. Assume that one operates at 38 percent efficiency and employs wet tower cooling, while the other operates at 36.5 percent efficiency and is dry-cooled. From Tables 14-2 and 14-3 it can be calculated that the dry-cooled plant indeed leads to less total water consumption than the water-cooled plant. The dry-cooled plant would consume an additional 0.0005 to 0.0095 $km^3/10^{18}$ J electric at the mine site, whereas the wet tower–cooled system would consume an additional 0.4 to 0.6 $km^3/10^{18}$ J electric at the power plant. However, one must consider that the additional water used at the mine may be environmentally more critical in terms of a possible shortage in local water supply.

Coal and Uranium for Electricity

Coal and uranium are often viewed as alternative sources of energy for future electric generation. Although these are not the only candidates for meeting future demand for electricity, it is nevertheless interesting to look at the coal-nuclear issue from the perspective of water resources.

The cooling required for a light-water reactor (LWR) is considerably greater than that for a modern coal- or petroleum-fueled plant producing the same electric power. Consider, for example, an LWR with a typical thermal conversion efficiency of 33 percent and a fossil fuel plant operating at 38 percent efficiency (12). For the same power output this difference in efficiency results in the release of about 24 percent more waste heat by the LWR. Because a nuclear plant releases all but from 0 to 5 percent of its waste heat through its cooling condensers, whereas a coal-burning plant typically releases 15 to 20 percent of its waste heat directly into the atmosphere with flue gas (12), the LWR actually requires about 39 to 50 percent more cooling water than does the fossil plant. Together, these differences cause an additional consumptive loss of water by the LWR of 0.16 to 0.30 $km^3/10^{18}$ J electric as compared with the fossil fuel plant, if both employ wet tower cooling. Moreover, with once-through riverwater cooling, the need for storage reservoirs for the LWR will be greatly increased because of the 39 to 50 percent increased water withdrawal requirement.

The future water needs for uranium mining, even on a per-unit-energy basis, are difficult to predict because of uncertainty over available reserves of high-grade uranium ore. Today uranium fuel can be obtained from ores containing 50 times the energy content of coal per unit weight (13). As long as such rich sources of uranium fuel are available, the water required for uranium mining and reclamation will be considerably less than they are for surface mining of coal. But as these rich supplies dwindle, nuclear reactors will require the use of low grade ores. One possible ore, the Chattanooga shale, has an energy content per unit weight roughly twice that of coal (13). Should such ores be mined, their geographic location and depth would be decisive factors in comparing impacts of water requirements of coal and uranium mining.

One worrisome possibility is that the last remaining rich supplies of uranium ores might happen to lie either in areas of special ecological value or in regions of especially scarce water. Economic pressure to exploit these supplies might be difficult to resist. In contrast, coal, being more widespread geographically, would then offer a wider choice of mining sites. These issues and the actual water impacts of uranium mining will become clearer when the amount and distribution of uranium fuel reserves become better known.

Taking into account the entire fuel cycle, we may question how coal and nuclear electric generation today compare with respect to water consumption. On the one hand, coal stripping and reclamation require an additional 0.004 to 0.09 $km^3/10^{18}$ J electric of water compared with uranium mining. On the other hand, nuclear plants that are wet tower–cooled required 0.16 to 0.30 $km^3/10^{18}$ J electric more water than a coal-fired plant. Thus the nuclear plants are more fresh water-intensive. With dry cooling or sea water cooling, the situation is, of course, reversed.

Future efficiencies of power plants are quite uncertain. Pollution control equipment on fossil fuel plants could reduce their efficiency, but fluidized-bed combustion could eventually provide ways for controlling emissions at efficiencies higher than today's coal-burning plants (14). The breeder reactor is likely to have a higher efficiency than the LWR. And finally, cogeneration of process steam and electricity, combined-cycle, fossil fuel plants, and development of uses for waste heat will make the water bookkeeping more complicated than presented here.

The Solar Options

The various solar energy technologies differ greatly with respect to water consumption. Because solar radiation is most intense and most predictable in parts of the United States where runoff is lowest and least predictable, water impacts of solar energy technologies must be thoroughly examined.

Several solar options for electricity generation are attractive on this score because the only water they would consume would be that used during the manufacturing of materials and the installation and maintenance of operating facilities (15). Wind energy is an example, because wind-generated electricity requires no cooling water. Certain methods of photovoltaic conversion provide other examples. Among the solar thermal conversion systems that have been suggested, either open-cycle Brayton generation (gas turbine) or Rankine cycle conversion with dry cooling towers would require minimal amounts of water. Although thermal generation of electricity by solar energy is likely to be less than 20 percent efficient, the steam cycle should operate at about the same efficiency as a fossil fuel plant, and therefore water consumption for wet tower or once-through cooling will be approximately same as for coal-fired plants (16).

Bioconversion is a possible means of producing gaseous and liquid fuels. One of the most efficient crops for energy plantations is sugar beets, which could have an annual yield of 10^{18} J on about 8000 km^2. On this basis, approximately 8.5 percent of the land area of the conterminous 48 states would be required to meet all current U.S. energy needs, provided that this land were sufficiently irrigated and fertilized and had high insolation and warm temperatures. Irrigation requirements alone for such a crop are estimated to be 10 $km^3/10^{18}$ J of biomass (17), about half of which would be consumed. The water consumed in meeting the current U.S. annual energy demand of 80×10^{18} J by bioconversion would exceed *all* current water consumption in the United States by almost a factor of 3. If such bioconversion plantations were located in the Southwest, as would be favored by factors such as climate and land availability, their annual water withdrawal requirements would exceed the mean annual runoff of all rivers in the conterminous United States west of the Mississippi. Evidently, such plantations could be maintained only by a massive system of water imports. Bioconversion schemes using artificial ponds for fresh water algal culture would result in comparable water consumption on a per-unit-energy basis unless evaporation-preventing protective covers were used.

On a smaller scale, however, biconversion systems designed to process agricultural or feedlot wastes or designed in tandem with sewage treatment facilities could actually have a net beneficial effect on water resources and could make small but useful contributions to U.S. energy needs.

Solar rooftop panels and passive systems for domestic and commercial heating appear quite favorable from the viewpoint of water conservation. Indiscriminate cutting down of trees in the vicinity of houses could lead to greater household water consumption for maintenance of lawns and low shrubs, but coordinated efforts of landscape architects and solar engineers should avoid such problems.

How Should Coal Be Used to Heat Homes?

Coal can be used to heat homes directly or by conversion to synthetic fuels or electricity. Direct heating is not environmentally acceptable. Establishment of a major synthetic-fuel industry is likely to require massive amounts of natural resources, and it is therefore imperative that a careful assessment of the consequences of such an industry be made. An assessment procedure that avoids some pitfalls of cost-benefit analysis is to estimate and compare the environmental impacts and the consumption of natural resources that will accompany the provision of a given measure of a particular end use via alternative technologies. Here we compare the amounts of water consumed for two home-heating methods, one using electricity produced from coal and the other using synthetic high-Btu gas produced from coal.

Electricity production is less efficient than synfuel production. However, there is also a considerable gap between the number of joules of electricity and of gas required for space heating at the point of use. This is illustrated in the first two columns of Table 14-4, which show the energy requirements, at the point of use, for electrically heated and gas-heated model unit houses in two locations, as developed by the Federal Energy Administration (FEA) (18). The difference between the number of joules of electricity and gas is attributed to the lower system efficiency of gas-heated homes. First, at the point of conversion to heat, the gas furnaces of today are less efficient than electric heaters, the difference being about 20 percent. Second, and more important, is the higher heat loss rates in gas-heated homes today, arising

from duct and ventilation losses. Electricity also allows for individual zonal or room thermostat settings, in contrast to most gas-heated homes. It is quite difficult to predict improvements in the efficiency of coal conversion plants or electric generating plants, and also of home-heating systems. In principle, one can build homes so that human warmth and electric lighting suffice for space heating. Concern over indoor air pollution may influence progress toward this ideal by gas-heated homes.

We show our results for three cases in Table 14-4. In case A, which is our worst case from the viewpoint of water consumption, cooling is carried out by the once-through method with storage; minimal water conservation and treatment is assumed in the production of synfuels (see Table 14-2); and home insulation and heating appliances are typical of those in use today. In case B cooling is carried out by the once-through method without storage; water consumption in synfuel production is assumed to be midway between the worst and best cases (see Table 14-2); and home insulation and home-heating appliances are taken from FEA estimates (18) of improved 1990 homes. In case C dry cooling is employed; maximal water treatment and conservation is assumed in the production of synfuels; home insulation is superior to case B (18); and home heating is carried out with heat pumps, with one-half of the waste heat from the gas-fired heat pump captured and used in the home.

In case C, which minimizes water consumption, both coal and water needs for home heating are sufficiently low that resource considerations would probably not be an important factor in deciding between the electric and the synfuels path. Where they could be an important factor, in either of the first two cases, the electric path appears to be superior to the synfuels path. From the perspective of water consumption the use of active or passive solar space heating would be preferable to either of these coal paths.

CONCLUSIONS

We have examined constraints of fresh water on the expansion rate of particular energy options and have answered specific questions that were posed in terms of rather narrow sets of choices among alternative technological means to common objectives. From technology comparisons and scenario analyses, the availability of fresh water is clearly a paramount factor to be considered in setting energy policy. Our conclusions are based solely on the factor of water consumption; numerous other factors, including land use, air and water pollution,

Table 14-4. Water and Energy Consumption for Home Heating by Synthetic Gas and Electricity Derived from Coal

Region	End Use Energy Consumption (10^9 J/house/year)[a]		Coal Consumption (10^9 J/house/year)		Water Consumption (m^3/house/year)			
					Gas[d]		Electricity[e]	
	Gas	Electricity	Gas[b]	Electricity[c]	Surface Mining	Deep Mining	Surface Mining	Deep Mining
				CASE A				
East	220	79	390	230	150 to 160	152	88 to 100	89 to 100
West	120	52	210	150	83 to 100	83	58 to 77	58 to 63
				CASE B				
East	160	72	280	210	66 to 72	67	25 to 40	26 to 36
West	86	47	150	140	36 to 50	36	16 to 34	16 to 21
				CASE C				
East	60	26	97	78	8.8 to 11	9.1	0.77 to 6.2	0.99 to 4.9
West	60	26	97	78	9.0 to 18	9.1 to 9.2	0.64 to 11	0.68 to 3.4

Note: Case A denotes little or no conservation of energy or water. Case C represents the other extreme, while case B is intermediate.

[a]These are estimates of the energy to be delivered to a single-family, one-story detached house for the purpose of space heating. Cases A and B are based on synthesized (model) demand (18). In case A the demand reflects 1970 conditions. Case B is based on projected reductions of, respectively, 28 and 9 percent in gas and electricity consumption per home (relative to 1970). In case C the house is designed according to NEMA standards (single thermostat) with net heating requirements amounting to 52×10^9 J/year and 60×10^9 J/year for the electric- and gas-heated home, respectively (18). The homes are equipped with a gas or an electric heat pump of equal coefficients of performance (COP = 2). The gas heat pump has a mechanical efficiency of 33 percent, but half of the heat not converted is recovered. In cases A and B East denotes a Michigan house dependent on eastern coal, while West refers to a New Mexico location fueled with western coal. East and West in case C denote only the source of coal. [b]Based on regional distribution and pipeline transport (1600 km average) losses of 0.7 and 7 percent, respectively (27). In cases A and B efficiency of conversion (to high-Btu gas) is assumed to be 61 percent. In C the efficiency is 67 percent (7). [c]A transmission loss of 8.6 percent is assumed (250 km) (27). Power plant thermal efficiency is 38 percent of cases A and B and 36.5 percent for case C (dry tower cooling) (12). [d]Slurry pipelines are not included. In case A conversion water consumption is 0.58 $m^3/10^9$ J of gas (at the plant). In case C it is 0.083 $m^3/10^9$ J. Case B assumes the mean of these two values. Other assumptions are the same as in Table 14-2 (high-Btu gasification). [e]Cooling by once-through in cases A and B (A uses storage, B does not) and by dry tower in case C. Water consumption estimates include mining and reclamation (see Table 14-2), cooling (1.0, 0.3, and 0.0097 $m^3/10^9$ J electric for cases A, B, and C, respectively) [Table 14-3 and (12)], coal cleaning (0.012 to 0.062 $m^3/10^9$ J electric) in the East and none in the West (28), and air pollution control (0 to 0.10 $m^3/10^9$ J electric) (12) (all joules electric refer to the power plant).

economics, and occupational hazards, must be included in any overall planning effort.

Our analysis suggests several conclusions. One is that a coal conversion industry in the United States supplying as much as 8×10^{18} J/year of synthetic fuels will be constrained by a scarcity of fresh water. An annual production of 8×10^{18} J of synthetic fuels is not even enough to replace the present consumption of natural gaseous and liquid fuels in only those end uses for which direct burning of coal is inappropriate (for example, transportation and home heating). This deficiency, coupled with the low likelihood that bioconversion can meet these present needs in an environmentally acceptable fashion, suggests the importance of directing greater R&D effort toward ultimate end use modification that would permit the use of electricity in place of natural gaseous and liquid fuels. It also

emphasizes the acute need for more stringent energy conservation in transportation and home heating.

A second finding is that production of steam-generated electricity as a substitute for natural gaseous and liquid fuels would cause conflicts in the use of fresh water unless dry cooling were extensively used. Technologies for electricity production that do not depend on water, such as wind and photovoltaics, as well as solar active or passive home heating look especially desirable in this light.

Combining these two observations we conclude that limited availability of fresh water is likely to be a severely constraining factor in future energy development. Even if no overall growth in energy consumption were to take place in this country, the need for substitutes for natural gaseous and liquid fuels could pose staggering problems for water resource management and for natural ecosystems that depend on relatively free-flowing fresh water. Overall growth in U.S. energy consumption would, of course, exacerbate these problems.

The degree of dependence of energy development on fresh water hinges on a number of unknown factors: the extent to which water conservation practices, including water pollution treatment, are carried out in coal conversion plants and mining operations; the economic feasibility of dry cooling or cooling with agricultural waste water; the economic feasibility of desalination; the results of further research on ground water and its management as a renewable resource rather than as a commodity to be mined and lost; the results of further experience with land reclamation, especially in areas hard to reclaim such as the northern Great Plains; and the feasibility of piping sea water inland for use in cooling power plants. The consequences to society of use of fresh water for energy will depend also on what the future demand will be in competing sectors of the water economy such as agriculture, municipal use, and industry. Moreover, decisions on acceptable limits of water use for energy will require greater understanding of rivers, lakes, and estuaries and greater knowledge of climatic variability.

Resolving these uncertainties will not be easy. Information on biological and climatic constraints is likely to be especially elusive. Yet planning must proceed, even in the face of uncertainty. Water constraints on energy development are sufficiently great to warrant far more attention. Two broad and urgent needs are identified. First is the need to develop adequate criteria for acceptable water consumption based on considerations of ecosystem balance, human well-being, nonuniform distribution of water, and the vicissitudes of its abundance under a capricious climate. Second is the need to set energy policy and water management on a course compatible with the criteria that are chosen. That course is certain to be characterized by a vital and enormous role for energy and water conservation (29).

NOTES

1. For an overview of water use taxonomy, see F. W. Sinden, 1976, In *Boundaries of analysis: An inquiry into the Tocks Island Dam controversy,* ed. H. A. Feiveson, F. W. Sinden, and R. H. Socolow. Cambridge, Mass.: Ballinger, 163–215.

2. J. F. Storr, in *Thermal ecology* (U.S. Atomic Energy Commission, U.S. Government Printing Office, Washington, D.C., 1973), pp. 291–295.

3. A. Jassby, in (4). pp. A1–A26.

4. J. Harte, ed., *Energy and the fate of ecosystems,* report of the Ecosystem Impacts Resource Group of the Risk/Impact Panel of the Committee on Nuclear and Alternative Energy Systems (National Academy of Sciences-National Research Council, Washington, D.C., in press).

5. A detailed discussion of the environmental impacts associated with the water requirements for coal conversion, including effects on aquatic life which could result from untreated effluent water, are discussed in (7).

6. R. Curry, in *Practices and problems of land reclamation in western North America,* M. K. Wali, ed. (Univ. of North Dakota Press, Grand Forks, 1976).

7. M. El-Gasseir, in (4), pp. EI1–EIX.52.

8. A. McDonald, *Shale oil, an environmental critique* (CSPI Oil Ser. 3, Center for Science in the Public Interest, Washington, D.C., 1974).

9. F. M. Pfeffer, *Pollutional problems and research needs for an oil shale industry* (U.S. Environmental Protection Agency, EPA-660/2-74-067, Washington, D.C., June 1974).

10. G. D. Weaver, *Oil shale technology,* Hearings before the Subcommittee on Energy, U.S. House of Representatives, 93rd Congress, 2nd Session, 1974; J. B. Weeks, G. H. Leavesly, F. A. Welder, G. J. Saulnier, Jr., *U.S. Geological Survey Professional Paper 908* (1974).

11. L. King, in (4), pp. K1–K74.

12. T. H. Pigford et al., *Fuel cycles for electrical power generation,* parts I and II, *Towards comprehensive standards: The electric power case* (Teknekron, Inc., Berkeley, Calif., 1974).

13. P. Ehrlich, A. Ehrlich, J. Holdren, *Ecoscience: Population, resources, environment* (Freeman, San Francisco, 1977).

14. J. M. Beer, *The fluidised combustion of coal,* Sixteenth Symposium (International) on Combustion (Combustion Institute, Pittsburgh, Pa., 1976), p. 439.

15. A rough estimate of water consumed in the provision of materials for a solar central-receiver system can be worked out from data provided by Davidson and Grether (16). Most of the water consumption will result from the cement requirements for the heliostat. Averaged over an assumed 30-year plant lifetime, this water consumption will be about $0.002 \text{ km}^2/10^{18}$ J of electric output. We thank B. Wolfson for information on this subject.

16. M. Davidson and D. Grether, *The central receiver power plant: an environmental, ecological, and socioeconomic analysis* (LBL-6329, UC-62, Lawrence Berkeley, Laboratory, Berkeley, Calif. 1977).

17. C. Calef, *Environment* 18, 17 (1976).

18. For cases A and B, see Federal Energy Administration, *Project independence—Residential and commercial energy use patterns 1970–1990* (Government Printing Office, Washington, D.C., November 1974), vol. 1, pp. 78–81. For case C see Energy Utilization Systems, Inc., *Energy consumption and life-cycle costs of space conditioning systems,* sponsored by Electric Environmental Equipment Section (National Electrical Manufacturers Association, New York, 1976).

19. V. E. Swanson, J. H. Medlin, J. R. Hatch, S. L. Coleman, G. H. Wood, S. D. Woodruff, R. T. Hilderbrand, *Collection, chemical analysis and evaluation of coal samples in 1975* (U.S. Geological Survey Open-file Report 76-468, 1976).

20. E. M. Dickson et al., *Impacts of synthetic liquid fuel development-Automotive market,* Environmental Protection Agency, Interagency Energy-Environment Research and Development Program Report (EPA-600/7-76-0046, Government Printing Office, Washington, D.C., July 1976), vol. 2.

21. U.S. Department of the Interior, Energy Research and Development Administration, *Synthetic fuel commercialization program,* Draft Environmental Statement (ERDA-1547, Government Printing Office, Washington, D.C., December 1975), vol. 4.

22. U.S. Department of the Interior, Bureau of Mines Process Evaluation Group, Morgantown, W. Va. Prepared for the Energy Research and Development Administration, ERDA 76-47, 76-48, 76-49, 76-52, 76-57, 76-58, 76-59, March, 1976 (National Technical Information Service, U.S. Department of Commerce, Springfield, Va., 1976).

23. Committee on Interior and Insular Affairs, U.S. Senate, *Factors affecting coal substitution for other fossil fuels in electric power production and industrial uses* (The National Fuels and Energy Policy Study, Serial No. 94-17, Government Printing Office, Washington, D.C., 1975).

24. Federal Energy Administration, *Project independence blueprint final task force report,* Project Independence-Interagency Task Force on Oil Shale (Government Printing Office, Washington, D.C., November 1974).

25. Western States Water Resources Council, *Western states water requirements for energy development to 1990* (Western States Water Resources Council, Salt Lake City, Utah, November 1974).

26. W. H. Smith and J. B. Stall, *Coal and water resources for coal conversion in Illinois,* State Water Survey State Geological Survey Cooperative Report 4, 1975; Water Resources Council, *Project independence—Water requirements, availabilities, constraints, and recommended federal actions,* Federal Energy Administration Project Independence Blueprint Final Task Force Report (Government Printing Office, Washington, D.C., November 1974), p. 22.

27. Colorado Energy Research Institute, *Net energy analysis: An energy balance study of fossil fuel resources* (Denver, Colorado, April 1976).

28. C. R. Murray and E. B. Reeves, *U.S. Geological Survey Circular 765* (1977).

29. We thank the Energy Research and Development Administration for their support. We are also grateful for the numerous discussions we have held with many members of the National Academy of Sciences' Nuclear and Alternative Energy Systems Study. L. King, a participant in that study, has especially contributed to our thinking. P. Benenson, P. Fox, J. Holdren, L. King, D. Levy, and especially H. Malde, suggested numerous improvements in both substance and style in an earlier draft of this article.

15

WATER FOR ELECTRICITY

CHARLES M. PAYNE

Water. Electricity. The creation of electric power through the transformation of energy present in falling or moving water is not a new idea. At the turn of the century a large proportion of the electricity produced in the United States was hydroelectric power.

After a period of decline in relative importance in the nation's overall electric energy supply mix, hydroelectric power is once again under active consideration as a potential partial solution to the national energy supply problem.

Basically, the dynamics of hydroelectric power are simple and well-understood concepts. Water is diverted through a conduit to turbine blades. The blades spin the turbine shaft, which in turn spins a generator shaft. A spinning, copper wire–wound generator armature creates the magnetic field recognized as electricity.

Advantages of hydroelectric power are many, not the least of which is the nonconsumptive use of a naturally occurring renewable resource. Water is diverted from a stream or other water body, used to generate electricity, then discharged back into the stream. Other public benefits may also be present.

Flood control, fishery enhancement, streamflow augmentation and regulation, and recreational opportunities are usually associated with hydroelectric projects. Dams are also constructed to provide water supplies for crop irrigation and for municipal, industrial, and manufacturing purposes. Hydroelectric generating plants can be installed at many of these water supply dams without interfering with the purposes for which the dams were constructed.

The key element in any hydroelectric project is a dam that may impound a reservoir or a diversion structure with no impoundment. The latter is referred to as *run-of-the-river*. Public benefits may

be incidental to the production of electricity or they may be purposefully designed into the project.

PROS AND CONS

More specifically, hydroelectric power offers many advantages from the electric power production standpoint.

- The equipment has a long operating life, generally a minimum of 50 years. Operation and maintenance costs are low. Hydroelectric power is readily amenable to remotely controlled operation. It can be brought on-line within minutes; no lengthy warm-up or start-up time is required.
- Hydropower is combustion-free; thus it is free from air pollution problems. Operating efficiency is high: 85 percent or more. Once installed, a hydro project is virtually inflationproof. It is not subject to escalating fuel costs. And if it is remotely controlled, there are no direct labor costs.

There are major drawbacks, as well as advantages, to hydroelectric power.

- It is capital-intensive at the front end, resulting in negative cash flow in the early years of a project; thus the profit break-even point may be extended for several years. It is site-specific; therefore standardization of equipment is difficult. Because of variations in terrain, construction techniques and the electric plant to be installed usually will have to be customized.
- If the project involves a new dam, expensive fish passage facilities may have to be provided. Variations in stream flow may create problems in

Mr. Payne is a supplemental energy resources specialist with the Rural Electrification Administration. This material is taken from Harnessing water for electricity—One more time, in *Using our resources, 1983 yearbook of agriculture,* ed. Jack Hayes, (Washington, D.C.: U.S. Department of Agriculture), 354–360.

assigning dependable generating-capacity values to a project. Minimum stream flows may have to be released at the dam to prevent dewatering immediately below the dam; this may result in loss of power generation.

- Hydroelectric power is heavily regulated. There are at least 17 federal laws that impinge on hydropower development. The major laws involved are the Federal Power Act, the National Environmental Policy Act, and the Fish and Wildlife Coordination Act.

These drawbacks will cause rejection of some sites for development. However, careful planning and design can usually offset most disadvantages. In addition, mitigation measures can sometimes be provided to make development of a particular site acceptable. Many hydroelectric plants constructed during the 1920s and earlier are still operating as environmentally and economically sound ventures.

HISTORY OF HYDROPOWER

The earliest known hydroelectric generating plants were placed in operation in 1882 at Niagara Falls in New York and St. Anthony Falls in Minnesota. The first commercial application of hydroelectricity was street and business lighting for the city of Minneapolis. The first central-station hydroelectric plant appears to have been a 25-kW plant on the Fox River at Appleton, Wisconsin.

Early in this century hydroelectricity played a major role in the nation's burgeoning industrial and economic base. By the midthirties, however, discovery of vast oil, natural gas, and coal fields had produced huge stocks of cheap fossil fuels.

Consequently, very large fossil fuel-fired central generating stations were constructed to take advantage of the economies of scale afforded by these low-priced fuels. Hydroelectric power started to decline as a percentage of the nation's total energy supply—to about 30 percent by 1935, 20 percent by 1965, and about 12 percent at the current time.

It is estimated that by the early seventies about three-quarters of all hydroelectric generating plants of 5000 kW or less had been abandoned, dismantled, or simply not repaired or replaced as they became inoperable.

RURAL ELECTRIFICATION

Plentiful, low-priced fuel, coupled with large central-station generating plants, resulted in virtually complete electrification of the nation's larger cities and urban industrial centers by 1935. But this was not the case for rural areas.

Only about 11 percent of the nation's farms had access to central-station electricity in 1935. At that time it appeared it would never be economically feasible to provide electricity to the sparsely settled and remote areas of the country. There was no profit to be made in constructing electric distribution lines many miles long to serve very few customers.

Legislation enacted in 1936 made it possible for the federal government to make low-interest, long-term loans to groups of farmers banded together into rural electric cooperatives for the purpose of helping them to electrify rural America. These loans, along with financial and technical advice, were (and still are) provided through the Rural Electrification Administration, an agency within the U.S. Department of Agriculture. That the program has been successful is attested by the fact that today almost all farms and rural areas in the United States have access to electricity.

The rural electric cooperative system has become a vital part of the nationwide electrical supply system. The amount of electricity consumed represents less than 10 percent of national consumption; however, the cooperatives serve about 70 percent of the land area of the United States.

Continued success of the rural electrification program will depend on ability of the cooperatives to modernize and maintain their existing systems and to expand to accommodate a growing rural population. This is particularly true in view of the latest census data, which indicates large population shifts from urban areas to the countryside.

Use of hydroelectric power by the cooperatives parallels that of the industry at large. That is, during the early fifties, rural electric cooperatives had 43 hydroelectric power plants in operation. By 1980 only 18 of these plants were still in service.

CHANGING EVENTS

Events of the past ten or so years promise to reverse this downward trend in hydroelectric plant development. The national electrical load growth pattern has dropped dramatically from 6 to 7 percent annually in the early seventies to less than 2 percent in the early eighties. Sharply escalating construction costs and interest rates have caused constricted cash flows for the electric industry all across the country.

Load growth uncertainty and the long lead times and high costs of large generating plants have prompted the industry to take a new, hard look at new power plant planning. Smaller generating plants located closer to load growth areas lessen many of the risks associated with large-scale generating plants. Hydroelectric power plants, especially small-scale ones, fit well into this concept.

The single most important event affecting the national energy supply during the 1970–1980 time frame was probably the oil embargo.

The Arab oil embargo of 1973, and subsequent skyrocketing of petroleum prices, sparked nationwide interest in a search for domestic energy supplies sufficient to foster energy independence for the United States. Oil prices were shocking enough, but long-term availability became an even more serious concern as political confrontations among the oil-producing Mideast nations and their neighbors escalated into armed conflicts.

RENEWED INTEREST

In part, the search for energy independence has focused on the prospect for development of renewable energy resources—water, wind, solar, tidal, geothermal, biomass. Interest in these resources is self-evident, for they are naturally occurring resources and thus can be assumed to be, collectively, a virtually unlimited domestically available source of energy.

Many of these renewable resources are currently in the process of research and development. Some have advanced to the stage of demonstration projects. For the most part, however, full-scale commercialization—and hence long-term availability—of these energy sources lies many years in the future.

The one exception is hydroelectric power. Hydropower is available now. It is an industry with proven reliability over more than seventy years. The technology behind the industry is well established and widely known. No research and development effort or demonstration projects are required, except possibly to improve upon the already high operating efficiencies of the generating equipment.

The magnitude of renewed interest in hydroelectric power can be illustrated by information available from the Federal Energy Regulatory Commission (FERC), the agency responsible for licensing nonfederal hydroelectric projects. Applications for preliminary permits (prelicensing feasibility study authorizations) numbered 36 in fiscal year 1976 and 70 in fiscal year 1977. During fiscal years 1981 and 1982 these filings increased to 1859 and 944, respectively.

The great majority of the hydroelectric power plants under study are proposed for installation at already existing dams. A survey conducted by the Army Corps of Engineers in 1979 indicated there are about fifty thousand existing dams scattered through the country. These dams serve a multitude of purposes—such as flood control, navigation channels, irrigation, municipal water supply, and industrial and manufacturing process water supply.

Many existing dam sites are amenable to hydroelectric power plant installation without impinging upon the purposes for which the dams were created.

CO-OPS SEEK HYDRO SITES

Renewed interest in hydropower on the part of rural electric cooperatives mirrors that of the electric industry at large. During fiscal years 1976 and 1977 two applications for preliminary permits seeking authorization to study hydro sites were filed at FERC. In fiscal years 1981 and 1982, 70 such applications were filed.

Many of these sites have since proved not to be economically feasible. Some were lost to other applicants competing for particular sites. Nevertheless, at the current time 23 preliminary permits are in effect, and applications are pending for another 9 sites.

Issuance of a preliminary permit by FERC establishes and reserves to the permittee certain priorities for a subsequent license application. Issuance of a license authorizes construction of a hydro project.

Over the past two years, 8 licenses (or amendments to existing licenses, or exemptions from FERC licensing) have been issued to rural electric cooperatives, and 15 applications for licenses are presently pending.

Development of the hydroelectric power potential in the United States certainly cannot be said to be a panacea for the nation's energy supply problems, nor will it solve all the energy problems of rural areas.

SHORTFALLS PREDICTED

Some industry experts have recently begun to predict national electric generating capacity shortfalls for the nineties. It appears likely that continued development of all forms of generation will be required to meet the nation's future electric energy needs. Nuclear energy, coal, oil, natural gas, synthetic fuels, renewable resources—all will be needed.

And just as certainly as hydropower played a major role in the past development of electric energy in this country, it can and should play a significant part in the future national energy supply mix. To ignore this hydroelectric power potential would be to ignore a naturally occurring, renewable, nonconsumptive energy resource.

16

WASTE WATER REUSE

OFFICE OF TECHNOLOGY ASSESSMENT

Waste water reuse, defined as the use of land to renovate sewage effluent from municipal or industrial sources, is receiving increased attention as a possible way to augment irrigation water supplies and to reduce the water pollution that might otherwise occur if such waste water were released directly to waterways. Those who advocate waste water reuse consider that waste water and the nutrients it contains are resources rather than refuse. Waste water provides water and nutrients to plants. In return, biological and chemical processes that occur in the soil, microorganisms, and plants are thought to cleanse the waste water. According to supporters of this practice, renovated, safe water may then percolate downward to recharge the ground water reservoir or be discharged directly to surface water (Figure 16-1).

The idea of using land to treat waste water is not new. Parker (1) notes that "treatment and disposal of sewage by land extensive schemes of irrigation is the oldest form of the modern methods of purification." These methods dominated U.S. municipal sewage treatment systems until the early twentieth century but gradually diminished in importance as metropolitan areas expanded, large expanses of land adjacent to urban areas became limited, and concerns grew about possible public health and water quality effects. Sewage treatments that used less land were developed as well.

Since the early 1970s interest in land application technologies has revived. In part, this interest reflects a greater public awareness of the costs of treatment to meet health standards and of potential pollution and degradation caused by the discharge of partially treated wastes into waterways. It also reflects concerns about the growing scarcity of unallocated surface water supplies and rates of ground water depletion, especially in the West. Finally, federal actions—for example, passage of

the Water Pollution Control Act of 1972 (Public Law 92–500)—have stimulated interest in reuse technologies in balance treatment costs.

WASTE WATER TREATMENT METHODS

Waste water contains two major categories of contaminants: biological and chemical. Biological contaminants include bacterial or viral pathogens and intestinal parasites. Chemical contaminants include substances such as nitrates, sodium, heavy metals (e.g., cadmium, lead, and zinc), oil, grease, and pesticides.

Levels of treatment generally recognized for waste water are primary, secondary, and tertiary. This classification is based on the removal of suspended solids and on the reduction of biochemical oxygen demand (BOD)—that is, the oxygen needed to meet metabolic needs of aerobic microorganisms in water containing organic matter. Primary treatment consists of mechanical and physical removal of suspended solids. This process is estimated to remove approximately 35 percent of the BOD and 60 percent of the suspended solids in raw sewage water (2).

Secondary treatment introduces biological processes—for example, activated sludge or trickling filters—to remove much of the remaining suspended solids and organic matter. Secondary treatment will remove from 80 to 95 percent of the BOD and suspended solids (2).

CONVENTIONAL SECONDARY WASTE WATER TREATMENT (3)

Waste water treatment processes that achieve effluent levels of 30 mg/L or less of 5-day biochem-

This material is excerpted from Selected technologies affecting water and land management, in *Water-related technologies for substantial agriculture in U.S. arid and semiarid lands* (Washington, D.C.: Office of Technology Assessment, 1983), Chapter 11, 304–309.

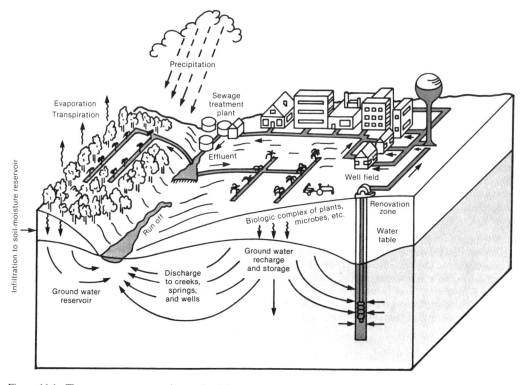

Figure 16-1. The waste water renovation cycle. After treatment at a sewage treatment facility, effluent is applied to the land. Biological and chemical processes that occur in the soil and plants provide further treatment for the waste water. Renovated water then percolates to the ground water or is discharged directly to the surface water. (*Source:* Ref. 2.)

ical oxygen demand, BOD_5, and 30 mg/L or less of suspended solids, SS, are referred to as *conventional secondary treatment*. Those systems that achieve effluent levels of 10 mg/L or less of BOD_5 and 10 mg/L or less of suspended solids are referred to as *advanced waste water treatment*. Both types of treatment systems can use a number of combinations of unit processes to achieve these effluent levels. Advanced waste treatment plants use the same process operations as conventional secondary treatment plants, with additional processing steps to achieve greater removal of pollutants. Individual treatment plants of either kind can differ in details of component equipment configurations and specifications because of differences in influent water characteristics, treatment objectives, and other site-specific considerations.

The typical conventional secondary treatment system considered in this section contains the following major process modules:

• Preliminary treatment.
• Influent pumping.
• Primary clarification.

• Activated sludge secondary treatment.
• Secondary clarification.
• Effluent disinfection by chlorination.
• Sludge treatment.

A typical advanced waste water treatment system contains, in addition to the above, the following process modules:

• Primary chemical addition (prior to primary clarification).
• Secondary chemical addition (prior to secondary clarification).
• Granular media filtration of secondary clarifier effluent.

An additional process module, granular activated carbon treatment, could be used after granular media filtration but is not considered here.

Configuration of a typical system for conventional secondary treatment is shown conceptually in Figure 16-2.

Influent enters a preliminary treatment module, where debris and large suspended solids such as

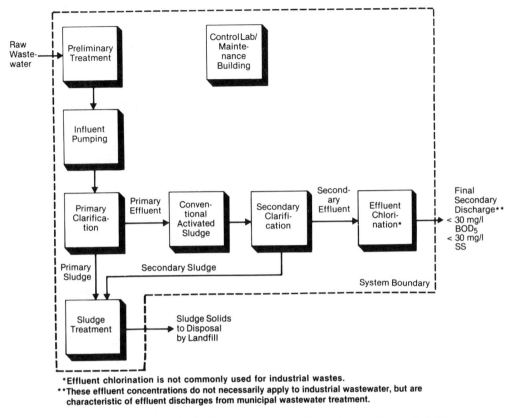

*Effluent chlorination is not commonly used for industrial wastes.
**These effluent concentrations do not necessarily apply to industrial wastewater, but are
  characteristic of effluent discharges from municipal wastewater treatment.

Figure 16-2. Conventional secondary treatment system for waste water (*Source:* Ref. 3.)

grit are removed. Sometimes, the flow in preliminary treatment is equalized in a large holding basin to dampen the effect of fluctuations in influent flow rates and waste loadings on downstream process modules. Flow equalization enhances the downstream removal of contaminants by providing a more uniform waste stream.

Effluent from preliminary treatment flows to the primary clarifiers. The clarifiers provide a relatively long detention time so that a large portion of the suspended solids can settle out. Chemical coagulants and coagulant aids can be used to enhance the removal of solids. Conventional systems sometimes use chemicals; advanced treatment systems nearly always use chemicals in this step.

Clarifiers can be either rectangular or circular and fabricated of either concrete or steel. Sludge (settled suspended solids) is removed from the bottom of the clarifier vessel and is pumped to sludge treatment. Clarifier effluent flows to the activated sludge aeration tanks for further treatment.

Conventional activated sludge treatment is a continuous-flow biological process. A suspension

of aerobic microorganisms is mixed into the waste water; the mixture of microorganisms and waste water, called mixed liquor, is agitated by air bubbles rising from diffuser pipes in the bottom of the aeration vessel or by mechanical surface aerators. The microoganisms oxidize soluble and colloidal organic compounds to carbon dioxide and water. The mixture flows from the aeration vessel to secondary clarifiers for separation of solids. These clarifiers are similar to the primary clarifiers discussed previously.

Secondary clarifiers remove some of the suspended solids from the activated sludge aeration vessel effluent. A portion of the solids settled out in the secondary clarifiers is returned to the aeration tank inlet as recycle sludge to seed biological activity in the incoming waste water. Excess sludge resulting from microorganism growth is routed to the sludge treatment processes for disposal.

In the clarifiers chemical addition can be used to enhance settling. In conventional secondary treatment the clarified secondary effluent may be disinfected prior to discharge. In advanced waste water treatment the secondary effluent passes through granular media filters, which further reduce suspended solids and BOD_5 to the required advanced waste water treatment levels ($BOD_5 \leq 10$ mg/L, $SS \leq 10$ mg/L for municipal waste water).

For some high-strength industrial wastes some of the secondary clarifier effluent is recycled to the activated sludge aeration vessels in order to dilute high levels of BOD_5. Lower BOD_5 levels in the aeration vessel may be necessary to ensure the required removal efficiency. Another approach with high-strength wastes is to provide a longer detention time in the aeration vessel than for low-strength wastes.

Sludge treatment is used to reduce the volume of sludge from both primary and secondary clarifiers and to render the sludge more acceptable for final disposal. A number of sludge treatment options may be used. One common method is thickening, digestion, dewatering, and final disposal by landfill.

Secondary clarifier sludge, which contains about 95 percent water, is commonly concentrated in a gravity thickener. From this process the sludge is transferred to a digester, which chemically and physically alters the sludge solids to facilitate ultimate disposal.

Sludge digestion can be either aerobic or anaerobic. Anaerobic digestion, which is most commonly employed, converts sludge into methane, carbon dioxide, and a residual organic material. The digestion takes place in the first of two tanks in series. The second tank provides for settling of solids and separation of supernatant liquid, which is routed to a previous process step. Combustible gas is collected from both stages and used as heater fuel in the treatment plant. Sludge is dewatered to increase the solids content prior to final disposal.

Dewatering can be accomplished by sand bed drying, vacuum filtration, or centrifugation, depending on the physical properties of the sludge. Landfill, incineration, land spreading, and other methods are used for final dewatered sludge solids disposal.

Conventional secondary or advanced waste water treatment using the activated sludge process can be applied to both domestic waste water and biodegradable industrial waste water. It is not uncommon for municipal and industrial waste waters to be combined for treatment. In these cases the industrial waste cannot contain toxic materials that would render the biological treatment process inoperative or refractory materials that would result in effluent standards being exceeded. Also, the industrial waste might require special provisions for oil and grease separation as part of preliminary treatment.

Advanced waste water treatment achieves higher-quality effluent than can be achieved by conventional secondary treatment. If the nonbiodegradable organic portion of the waste is large enough to cause problems in receiving water bodies, granular activated carbon treatment might be required to reduce effluent organic concentrations.

ADVANCED WASTE WATER TREATMENT

Tertiary, or advanced waste water, treatment is used after primary and secondary treatments to reduce BOD further, remove suspended solids, lower nutrient concentrations, and improve the effectiveness and reliability of disinfection. Land application of waste water is considered one method of tertiary treatment, along with others such as chemical coagulation, clarification, filtration, activated carbon treatment, and reverse osmosis.

Advanced waste water treatment by land application can be achieved in a variety of ways, depending on the goal of the treatment, the composition of waste water, and characteristics of the waste site. The three most commonly used methods for land application are slow-rate irrigation, overland flow, and rapid infiltration-percolation. Table 16-1 compares the three methods by use objectives.

Slow-rate irrigation (Figure 16-3a) is probably the method used most often and with most potential for agricultural use. In this process waste water is applied to the soil surface by a fixed or moving sprinkler system or by surface irrigation. Water application rates are generally low and are largely determined by climate, soil, and the water and nutrient needs of the crops. Treatment proceeds as vegetation and soil microorganisms act to remove and alter waste water as it percolates through the soil.

Overland flow reuse systems (Figure 16-3b) also rely on a vegetative cover to effect waste treatment but differ from slow-rate methods because crop production is usually not a major objective. In this process waste water is applied over the upper parts of vegetated terraces and allowed to flow in a thin sheet down the relatively impermeable surface to runoff collection ditches. Only small amounts of waste water infiltrate into the soil or percolate to the ground water. Renovated water that is collected may be reused or discharged directly to surface water. Treatment occurs by physical, chemical, and biological means.

Rapid infiltration-percolation, the third method, uses less land area to effect treatment than do the other two methods. The main objective in this system is ground water recharge (Figure 16-3c). Waste water is applied to highly permeable soils in basins by flooding or sprinkling. The basins may or may not be vegetated. Renovation occurs through

Table 16-1. Comparison of Irrigation, Overland Flow, and Infiltration-Percolation of Municipal Waste Water

Objective	Type of Approach		
	Irrigation	Overland Flow	Infiltration-Percolation
Use as a treatment process with a recovery of renovated water[a]	0–70% recovery	50–80% recovery	Up to 97% recovery
Use for treatment beyond secondary			
1. For BOD[b] and suspended solids removal	98 + %	92 + %	85–99%
2. For N removal	85 + %[c]	70–90%	0–50%
3. For P removal	80–99%	40–80%	60–95%
Use to grow crops for sale	Excellent	Fair	Poor
Use as direct recycle to the land	Complete	Partial	Complete
Use to recharge ground water	0–70%	0–10%	Up to 97%
Use in cold climates	Fair[d]	—[d]	Excellent

Source: Ref. 2.
[a]Percentage of applied water recovered depends on recovery technique and the climate.
[b]BOD—biochemical oxygen demand.
[c]Dependent on crop uptake.
[d]Conflicting data—woods irrigation acceptable; cropland irrigation marginal.

natural processes in the soil as water percolates downward.

The degree of success for waste water treatment varies with the type of method used to apply wastes. Table 16-2 shows expected treatment performance for these three processes.

ASSESSMENT

Municipal waste water is now used for irrigation. However, information on the number of communities that use land treatment for sewage renovation and the number of individuals who use waste water to irrigate agricultural crops is inexact, because funding for these systems may be through the federal government, industry, or private sources. With federal funding estimates are that approximately one thousand municipalities use land application to treat wastes (4). Unpublished data from the U.S. Geological Survey (USGS) indicate that in 1980, 0.5 billion gal/day (bgd) of effluent was used for irrigation compared to 290 bgd of fresh surface water and 88 bgd of fresh ground water.

Some communities use municipal waste water for irrigation for parks, golf courses, and greenbelts, and federal and state guidelines have been developed for its application (e.g., 1, 2). In California, for example, irrigation return flows of relatively good quality are applied to wetland areas

Table 16-2. Expected Quality of Treated Water from Land Treatment (mg/L)

Constituent	Slow Rate[a]		Rapid Infiltration[b]		Overland Flow[c]	
	Average	Maximum	Average	Maximum	Average	Maximum
BOD	<2	<5	2	<5	10	<15
Suspended solids	<1	<5	2	<5	10	<20
Ammonia nitrogen as N	<0.5	<2	0.5	<2	0.8	<2
Total nitrogen as N	3	<8	10	<20	3	<5
Total phosphorus as P	<0.1	<0.3	1	<5	4	<6

Source: Ref. 2.
[a]Percolation of primary or secondary effluent through 1.5 m (5 ft) of soil.
[b]Percolation of primary or secondary effluent through 4.5 m (15 ft) of soil.
[c]Runoff of municipal wastewater over about 45 m (150 ft) of slope.

166   WATER USE

A. Slow-rate irrigation

Wastewater is applied to the soil surface and allowed to percolate downward. Treatment proceeds as soil, vegetation, and soil micro-organisms remove nutrients and suspended solid material.

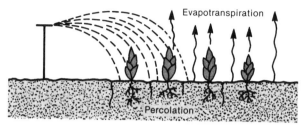

B. Overland flow reuse

Wastewater is applied to a sloping surface and allowed to flow over the soil surface to runoff collection ditches. Treatment is a result of physical, chemical, and biological processes.

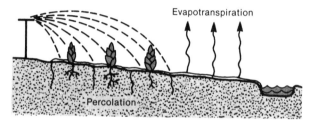

C. Rapid-infiltration percolation

Wastewater is applied by flooding or sprinkling to highly permeable soils in basins. As the wastewater percolates into the soil, renovation occurs.

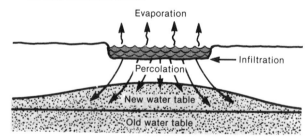

Figure 16-3. The major types of wate water treatment. (*Source:* Ref. 2.)

to maintain natural vegetation. These areas are then used for cattle grazing in summer and hunting in the fall. Chemical wastes from potato-processing plants have also been used for irrigation. In spite of these signs of acceptance, widespread adoption of reuse systems for most agricultural crops is hindered by numerous biological, social, economic, and legal questions and a lack of long-term research on the subject. Table 16-3 presents a summary of issue areas that require resolution.

Three examples illustrate the complexity of these issues. First, regarding the question of land productivity with reclaimed water, data indicate that the yields of crops irrigated with waste water usually increase or remain the same (2, 5), but crops seem to vary in their tolerance to waste water application (Table 16-4). For example, research in Hawaii on sugarcane tested the dilution of waste water required for optimal sugar yield (6). Five treatments for the two-year cane cycle were tested:

Table 16-3. Summary: Issues Surrounding Water Reuse for Irrigation

Resource issues	2. Social factors
1. Effluent quality	Public attitudes toward application
Nutrient content	Public attitudes by consumers of products
Heavy metal content	Attitudes of nearby residents
Pathogen content	3. Economic considerations
2. Soil productivity	Water pricing
Salt buildup	Transportation costs
Toxicity buildup	Subsidies for those who use water
Viral contamination	Facilities for water storage
Physical degradation	Value in alternate uses
3. Crop production	Type of material contained in water
Fertilizer and water requirements	*Institutional issues*
Crop growth and yields	1. Water treatment facilities
Crop uptake of nutrients	Adequacy and reliability of treatment prior to
Crop uptake of toxics and pathogens	application
4. Animal health	Adequacy of storage facilities during periods
Animal uptake of nutrients	of nonapplication
Animal transmission of pathogens to human	2. Monitoring
consumers	Need for monitoring air, effluent, ground water,
5. Ground water quality	crop, and soil quality
Path of water to water table	3. Legal issues
Quality of water reaching ground water	Ownership and sale of water
6. Air quality (with sprinkler irrigation)	Water rights
Health effects for workers and nearby residents	Liability for damages
Odor considerations	Responsibility for monitoring
Social and economic issues	Guidelines for water reuse (e.g., crops to be
1. Human health effects	grown, amount of water to be applied)
Contact with effluent by farmworkers	Effect on downstream users (third parties), if
Contact with plant and animal products by	water previously was part of return flows
consumers	

Source: Ref. 9.

(1) conventional irrigation water, (2) 12.5 percent effluent diluted with irrigation water, (3) 25 percent sewage water, (4) 50 percent effluent diluted with ditch water, and (5) effluent the first year and irrigation water the second year. Scientists found that sugar yields for waste water concentrations up to 25 percent, or for waste water the first year and irrigation water the second year, were equal to those from conventional irrigation supplies. When waste water concentrations increased to 50 percent, however, sugar yields and juice quality declined significantly. The researchers concluded that chlorinated, secondarily treated sewage effluent with nitrogen concentrations could be used in furrow irrigation for the two-year crop cycle of sugarcane if waste water were diluted with fresh water so that the concentration of effluent was 25 percent or less. They cautioned, however, that effluent quality must be constantly monitored for nitrogen content, pesticides, heavy metals, and pathogenic viruses. Such substances in the soils or waterbodies could prove difficult, if not impossible, to eliminate. In addition, fieldworkers were warned to practice careful sanitation and personal hygiene to protect against infection.

A second illustration provides a sample of the economic questions that surround application of waste water for irrigation. Although effluent was generally recognized as a valuable resource for water and nutrients, few farmers actually measured the fertilizer value of the water. Similarly, those in local governments responsible for the operation of reuse systems acknowledged the economic value of the effluent but had not established procedures to charge landowners or farm operators for the value that they received. Instead, they took the

Table 16-4. Crop Yields at Various Levels of
Application of Waste Water, Pennsylvania
State University

Crop	Waste Water Application Rates (in./week)[a]		
	0[b]	1.0	2.0
	(BUSHELS PER ACRE)		
Wheat	48	45	54
Corn	73	103	105
Oats	82	113	88
	(TONS PER ACRE)		
Alfalfa	2.2	3.7	5.1
Red clover	2.4	4.9	4.6
Corn stover	3.6	7.3	8.5
Corn silage	4.3	6.4	6.0
Reed canarygrass	1.4	—	5.0

Source: Ref. 2.
[a]Metric units in original document have been converted to English units.
[b]Control areas received commercial fertilizer ranging from 10 tons/acre of 0–20–20 for oats to 40 tons/acre of 10–10–10 for corn.

view that landowners performed a service in disposal of municipal waste effluent.

Third, with regard to social concerns, public reaction may be an obstacle to waste water reuse. A survey of selected California communities indicated that respondents favored water treatment options that protected public health, enhanced the environment, and conserved scarce water (7). However, use of reclaimed water for ground water recharge and drinking supplies was perceived as a threat to human health. Effluent used for industrial purposes or for irrigation of animal feed and fiber crops was considered to be an acceptable practice (8). The inconsistency arises when contaminants from effluent reach the ground water secondarily, as the result of its use in industry or irrigation.

Waste water reuse has been adopted by relatively few communities and farmers in the United States. It can potentially supplement irrigation water supplies and reduce reliance on added fertilizer but generates many questions on long-term impacts for soil and water quality and, ultimately, public health. Wider application of reuse systems will require careful planning and monitoring by municipalities and irrigators. The costs and danger of handling waste water will have to be balanced with economic benefits of its reuse. In addition, when it is applied on a massive scale, questions are raised about the impacts of this water shift on other aspects of the hydrologic cycle, especially stream flow, if the treated water has previously been part of return flows. Moreover, legal considerations for downstream users may be complex. Much research has been done on this topic, but additional long-term research on the effects of these systems on crops, soils, ground water, and human and animal consumers is needed.

NOTES

1. C. D. Parker, 1982, Land extensive processes for wastewater treatment and disposal—A perspective, *Water Science and Technology,* 14:393–406,

2. W. E. Sopper, 1979, Surface application of sewage effluent, in *Planning the uses and management of land,* ed. Marvin T. Beatty et. al. (Madison, Wis.: American Society of Agronomy), 633–663.

3. This section is added by the editors and taken from G. DeWolf, P. Murin, J. Jarvis, and M. Kelly, 1984, *The cost digest: Cost summaries of selected environmental control technologies,* EPA–600/8–84–010 (Environmental Protection Agency), 27–30.

4. Personal communication (1982) with R. Thomas, land treatment specialist with the U.S. Environmental Protection Agency.

5. L. A. Christensen, 1982, *Irrigating with municipal effluent: A socioeconomic study of community experience,* U.S. Department of Agriculture, Economic Research Service, Natural Resource Economics Division, ERS–672.

6. L. S. Lau, P. C. Ekern, P. C. S. Loh, R. H. F. Young, G. L. Dugan, R. S. Fujioka, and K. T. S. How, 1980, *Recycling of sewage effluent by sugarcane irrigation: A dilution study, October 1976 to October 1978, Phase II–A,* University of Hawaii, Water Resources Research Center, Technical Report No. 130.

7. W. H. Bruvold, and J. Crook, 1980, *Public evaluation of wastewater reuse options,* U.S. Department of the Interior, Office of Water Research and Technology, 1980.

8. W. H. Bruvold, B. H. Olson, and M. Rigby, 1981, "Public Policy for the Use of Reclaimed Water," *Environmental Management* 5:95–107.

9. William H. Bruvold, 1982, Agricultural use of reclaimed water (unpublished paper prepared for the National Science Foundation, January 1982, Office of Technology Assessment).

P A R T I V

PROBLEMS AND HAZARDS

Several chapters in Part III, "Water Use," also included significant discussion of problems associated with water use. For example, Chapter 11 discussed the problem of increased salinity of water in certain irrigated areas. The chapters in this part focus specifically on problems and hazards of water and water use. Several recent overviews of these water-related problems are specific to the United States; see, for example, *National Water Summary 1983* (U.S. Geological Survey Water Supply Paper 2250) and *The Nation's Water Resources, 1975–2000* (U.S. Water Resources Council, 1978). A worldwide view of some of these same problems is given in Chapter 17, "Environmental Problems Affecting Water Supply." These problems can be classified as water quantity (availability), water quality, related land use and hydrologic hazards, and institutional and management issues. Supply of water clearly exceeds demand, but increases of demand with increasing population growth, irrigated agriculture, and industrial growth may place stresses on this supply-demand relationship. Local problems can be severe as deforestation affects stream runoff, precipitation patterns change, and pollution limits the suitability of the water.

WATER QUANTITY PROBLEMS

Water quantity problems vary regionally. Figure IV-1 indicates the consumptive use of water (i.e., water not returned to the source) and the average renewable water supply available for the various water resource regions of the United States. Note the Lower Colorado region. Consumptive use can exceed renewable supply only at the cost of a net depletion of ground water. Accounts must balance. Thus the shaded patterns indicate those areas where surface water availability is a major problem.

Reservoir development has been extensive in the shaded areas. In most instances the heavy use of water in the areas was initiated because the reservoirs guaranteed water availability. These problems were discussed briefly in Chapter 4.

Problems of ground water availability are indicated in Figure IV-2, where the shaded areas show water level declines of at least 40 ft or more since pumping began. By itself, however, this figure does not indicate that ground water is not available. It does indicate that this resource is being artificially depleted by humans faster than it can be recharged by nature.

Another problem of water availability is drought, a period of time when water availability is constrained because of extended period of below-normal rainfall. It is a natural condition that can have many social problems associated with it. Chapter 18, "Drought Impacts on People" by Howard F. Matthai, indicates that reactions change from concern to panic until the drought

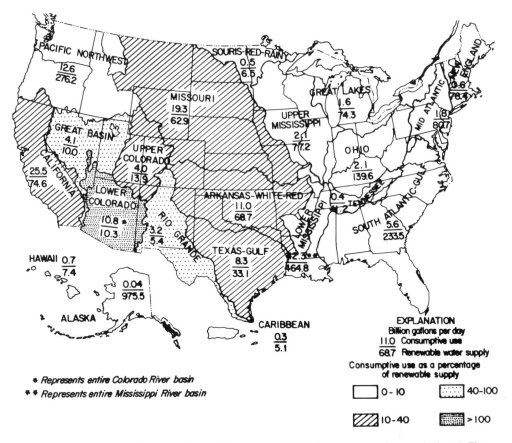

Figure IV-1. Average consumptive use and renewable water supply. Water resources regions are indicated. The consumptive use as a percentage of renewable supply is indicated for the water resources regions. (*Source:* U.S. Geological Survey. 1984. *National water summary 1983,* U.S. Geological Survey Water Supply Paper 2250.)

is ended by sufficient rain—and then apathy returns. Given the high capital costs of constructing water projects, it seems likely that severe conservation measures will remain the overwhelming choice for dealing with the lack of sufficient water during droughts.

WATER QUALITY PROBLEMS

The quality of water determines the usability. Figure IV-3 illustrates such variations for domestic, irrigation, and industrial uses. Water for domestic use should have a good smell and taste, should be free from contaminants that are harmful to health, and should not damage plumbing or appliances. Thus domestic water should be low in hardness, sediments, and the compounds indicated. The requirements for industrial water vary with the industry. Industrial water, however, should not have constituents that would precipitate and damage machinery or stain products. Irrigation water is not as restrictive; plants and trees are tolerant of a range of water quality. The presence of boron and of high concentrations of dissolved solids is harmful, and the sodium-calcium balance is important for soil structure.

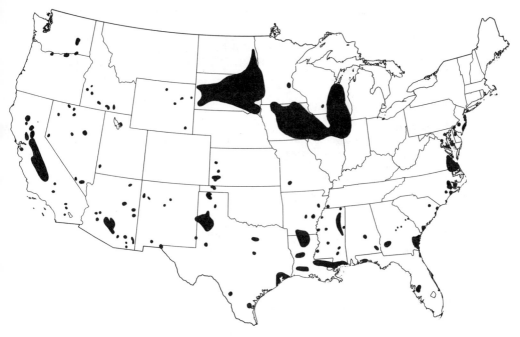

Figure IV-2. Major water table decline. Shaded areas indicate water table decline or artesian water level decline in excess of 40 ft in at least one aquifer since predevelopment. *(Source:* U.S. Geological Survey. 1984. *National water summary 1983.)*

Another water quality problem is acid rain. Chapter 19, "Acid Rain," by the Canadian Embassy (to the United States), discusses a subject that has long been a problem between the upper Midwest states, the Northeast, and Canada and has recently become a problem between the U.S. Southwest and Mexico. Evidence for the relationship between sulfur emissions from coal-fired power plants and smelters and above-normal acidic precipitation downwind is growing. At the same time, however, the effects of this acid rain on crops and trees are not clear.

Ground Water Problems

The major problems and hazards associated with ground water are various types of contamination and overdraft of ground water, leading to land subsidence. Chapter 20, "Ground Water Contamination in the United States" by Veronica I. Pye and Jocelyn Kelly, outlines the sources of ground water pollution and their prevalence. In a recent publication, the American Institute of Professional Geologists points out:

> Contamination of ground water is a severe problem because the contaminant generally travels unobserved until detected in a water-supply well. Once contaminated an aquifer is difficult and expensive to clean up. The contaminant disperses in the ground water, is difficult to remove, and may persist for decades. In almost all cases, *prevention is simpler and cheaper than cure* [original emphasis].

> Contaminants include an almost endless list of inorganic chemicals, biological matter, radioactive compounds, and even physical loads such as heat. The impacts on ground water may range

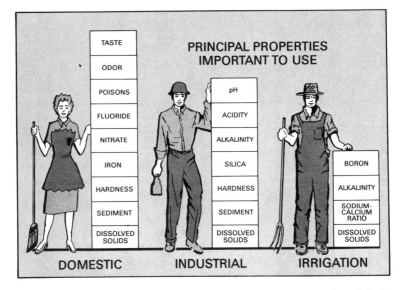

Figure IV-3. Quality determines usability. Water for domestic use should taste and smell good, be free from constituents harmful to health, and should not damage plumbing or appliances. Requirements for industry vary greatly, but generally the water should not be highly corrosive or cause precipitates that would clog equipment. Plants are quite tolerant of a wide range of water quality. They are sensitive to boron, a plant poison, and to dissolved solids, which at high levels make the water unusable. The balance between sodium and calcium is important in maintaining proper soil structure. *(Source:* American Institute of Professional Geologists, 1983. *Ground water issues and answers.* Arvada, Colorado: AIPG. Original source: U.S. Geological Survey, *Primer on ground water.)*

from aesthetic effects (such as unpleasant taste or warm temperature) to imminent hazards to health. Principal sources of pollution, in order of importance nationally, include: industrial wastes, municipal landfills, agricultural chemicals, septic systems and cesspool effluents, leaks from petroleum pipelines and storage tanks, animal wastes, acid mine drainage, oil-field brines, salt-water intrusion, and irrigation return flow (American Institute of Professional Geologists, 1983, *Ground water issues and answers,* 16).

Waste Disposal

Some of the specific problems of waste water generation by industry are discussed in Chapter 21, ''Industrial Waste Water'' by the Environmental Protection Agency. Water used in industrial processes is usually altered and often contains contaminants that degrade water quality and can even be hazardous to health.

The special problem of disposing of hazardous wastes is discussed in Chapter 22, ''Problems of Hazardous Waste Disposal,'' by K. Cartwright, R. H. Gilkeson, and T. M. Johnson. They say:

> In reality, total isolation of wastes in humid areas is not possible; some migration of leachate from wastes buried in the ground will always occur (in abstract of the original paper, p. 357).

Disposal of these wastes is seen by many to be the key environmental problem for the rest of this century. The amounts of waste involved are staggering. Figure IV-4 illustrates total indus-

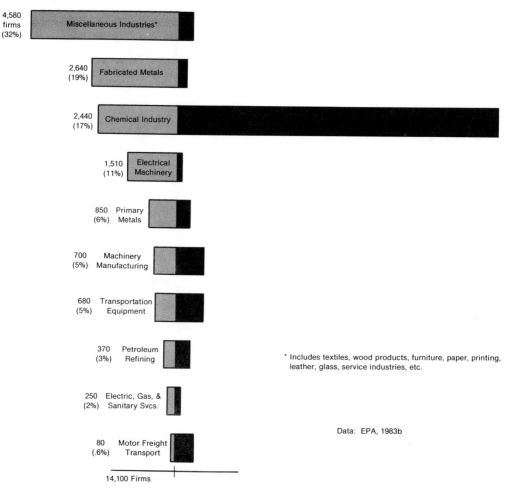

4,580 firms (32%) — Miscellaneous Industries*

2,640 (19%) — Fabricated Metals

2,440 (17%) — Chemical Industry

1,510 (11%) — Electrical Machinery

850 (6%) — Primary Metals

700 (5%) — Machinery Manufacturing

680 (5%) — Transportation Equipment

370 (3%) — Petroleum Refining

250 (2%) — Electric, Gas, & Sanitary Svcs.

80 (.6%) — Motor Freight Transport

14,100 Firms

* Includes textiles, wood products, furniture, paper, printing, leather, glass, service industries, etc.

Data: EPA, 1983b

Figure IV-4. Who generates how much hazardous waste. The EPA has identified 60,000 firms that have the potential to generate hazardous waste. In a 1981 study only 14,000 were found to produce significant waste, and they produced 40 billion gal. One-sixth of the firms (the chemical industry) generate 71 percent of the waste. A third of the firms (diverse, smaller industries) produce only 2.5 percent. *(Source:* American Institute of Professional Geologists, 1984, *Hazardous wastes: Issues and answers.)* Arvada, Colorado: AIPG.

trial waste, with the shaded area the proportion that is poisonous, corrosive, disease-causing, burnable, cancer-causing, or otherwise hazardous. A variety of disposal techniques have been developed, with a tremendous range in costs for applying them (Figure IV-5). The magnitude of these costs indicates why so much waste has been dumped in landfills—both legally and illegally.

Health-related Problems

Health-related problems of water include not only diseases transmitted through drinking water but also diseases transmitted through contact with water. Chapter 23, ''Health Hazards of Water

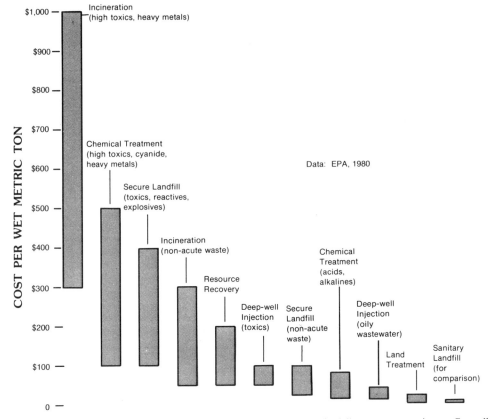

Figure IV-5. Range of costs for each method of waste disposal. Values are in dollars per wet metric ton. Generally, the more toxic the waste, the greater is the cost of disposal. *(Source:* American Institute of Professional Geologists, 1984. *Hazardous Wastes.)* Arvada, Colorado: AIPG.

Pollution'' by the World Health Organization (WHO), is a portion of a famous WHO study on environmental problems. While pollution may be accidental, it more often is caused by the improper disposal of sewage, industrial wastes, and polluted runoff. Pesticides and other chemical compounds may be carried long distances from their area of application, thus spreading health hazards. The wide range of diseases discussed in the WHO article is discouraging. Chapter 24, ''Waterborne-Disease Outbreaks in the United States,'' does not brighten our outlook. Edwin C. Lippy and Steven C. Waltrip illustrate clearly our lack of knowledge about how much disease is actually water-related and what might be causing it. An interesting note is that the water quality standards set for trout streams by the Pennsylvania Fishing Commission are more restrictive than U.S. drinking water standards.

OTHER WATER-RELATED PROBLEMS

Land use and hydrologic-hazard issues are discussed in the next three chapters. In Chapter 25, ''The Desert Blooms—At a Price,'' David Sheridan illustrates the many problems associated

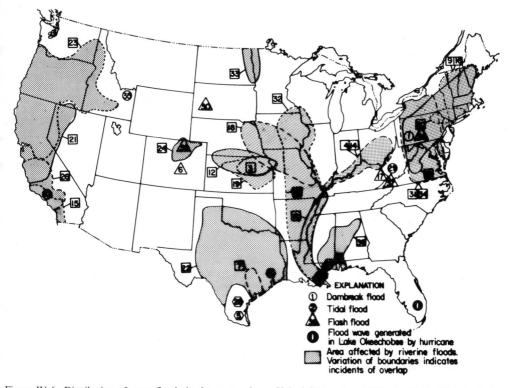

Figure IV-6. Distribution of great floods in the conterminous United States since 1889. See Table IV-1 for further information. (*Source:* G. W. Edelen Jr., 1981, Hazards from floods, in *Facing geologic and hydrologic hazards: Earth-science considerations,* ed. W. W. Hays. U.S. Geological Survey Professional Paper 1240–B.)

with irrigation by examining four case studies. Ground water overdraft, salinization of crop-land, and soil erosion are some of the problems discussed. In a more detailed examination of one of these problems, Chapter 26, ''Subsidence due to Ground Water Withdrawal'' by Joseph F. Poland, clearly shows the cause and effect between the withdrawal of ground water and its subsequent impact on surface topography and features. Subsidence exceeding 1 m has occurred in Arizona, California, Nevada, and Texas. California has over 16,000 km^2 affected, with 84 percent of that in the San Joaquin Valley. The pumping of petroleum often has caused similar subsidence.

Chapter 27, ''Human Responses to Floods,'' by Jacquelyn L. Beyer, emphasizes that floods are a natural occurrence and that they can become disasters when land use by humans does not take their potential damage into account. Figure IV-6 and Table IV-1 indicate the great floods in the United States since 1889 and reveal their impact in terms of lives lost and property damage. These floods have a variety of causes, such as dam breaks, tidal floods, hurricanes with associated storm surges, flash floods, and riverine floods. Flash floods differ from riverine floods in that they are of great volume but short duration, generally caused by torrential rain. Riverine floods, in contrast, take place in river systems that drain large areas, and the duration of riverine floods may range up to days. Tidal floods are overflows of coastal lands. Their

Table IV-1. Great Floods in the United States Since May 1889

Number[a]	Type of Flood	Date	Location	Lives Lost	Estimated Damages (Millions of Dollars)
1	b	May 1889	Johnstown, Pennsylvania, dam failure	3000	—
2	c	September 8, 1900	Hurricane, Galveston, Texas	6000	30
3	d	May–June 1903	Kansas, Lower Missouri, and Upper Mississippi River	100	40
4	d	March 1913	Ohio River and Tributaries	467	147
5	c	September 14, 1919	Hurricane, south of Corpus Christi, Texas	600–900	22
6	b, e	June 1921	Arkansas River, Colorado	120	25
7	d	September 1921	Texas rivers	215	19
8	d	Spring of 1927	Mississippi River Valley	313	284
9	d	November 1927	New England rivers	88	46
10	b	March 12–13, 1928	St. Francis Dam failure, southern California	450	14
11	f	September 13, 1928	Lake Okeechobee, Florida	1836	26
12	d	May–June 1935	Republican and Kansas rivers	110	18
13	d	March–April 1936	Rivers in eastern United States	107	270
14	d	January–February 1937	Ohio and Lower Mississippi River basins	137	418
15	d	March 1938	Streams in southern California	79	25
16	d	September 21, 1938	New England	600	306
17	e	July 1939	Licking and Kentucky rivers	78	2
18	d	May–July 1947	Lower Missouri and Middle Mississippi river basins	29	235
19	d	June–July 1951	Kansas and Missouri	28	923
20	d	August 1955	Hurricane Diane floods, northeastern United States	187	714
21	d	December 1955	West Coast rivers	61	155
22	d	June 27–30, 1957	Hurricane Audrey, Texas and Louisiana	390	150
23	d	December 1964	California and Oregon	40	416
24	d	June 1965	South Platte River Basin, Colorado	16	415
25	c	September 10, 1965	Hurricane Betsy, Florida and Louisiana	75	1420
26	d	January–February 1969	Floods in California	60	399
27	c, d	August 17–18, 1969	Hurricane Camille, Mississippi, Louisiana, and Alabama	256	1421
28	c	July 30–August 5, 1970	Hurricane Celia, Texas	11	453
29	b	February 1972	Buffalo Creek, West Virginia	125	10
30	e	June 1972	Black Hills, South Dakota	237	165
31	c, d	June 1972	Hurricane Agnes floods, eastern United States	105	4020
32	d	Spring 1973	Mississippi River Basin	33	1155
33	d	June–July 1975	Red River of the North Basin	< 10	273
34	c, d	September 1975	Hurricane Eloise floods, Puerto Rico and northeastern United States	50	470
35	b	June 1976	Teton Dam failure, southeast Idaho	11	1000
36	e	July 1976	Big Thompson River, Colorado	139	30
37	e	April 1977	Southern Appalachian Mountains area	22	424
38	b, e	July 1977	Johnstown, western Pennsylvania	78	330
39	d	April 1979	Mississippi and Alabama	< 10	500
40	c	September 12–13, 1979	Hurricane Frederic floods, Mississippi, Alabama, and Florida	13	2000

Source: Adapted from National Oceanic and Atmospheric Administration, 1977, *Climatological Data, National Summary,* vol. 28, no. 13; and by information furnished from the Federal Disaster Assistance Administration.

[a] Numbers correspond to those in Figure IV-6.
[b] Dam break flood.
[c] Tidal flood.
[d] Riverine flood.
[e] Flash flood.
[f] Flood wave generated in Lake Okeechobee by hurricane.

duration is usually short, being dependent in part on the tides, but they are capable of affecting large areas.

Readers might want to look up Perry Rahn's article "Flood-Plain Management Program in Rapid City, South Dakota," (see Appendix C for the complete reference) for a program that was initiated after a disastrous flood (see item 30 in Figure IV-6 and Table IV-1) and that appears to be working. In addition, "Flood Control Failure" by Gary Griggs and Lance Paris gives a case history of a project constructed to prevent floods and the problems that the project itself generated.

To conclude Part IV, Chapter 28, "Deterioration of Water Supply Systems," by Harrison J. Goldin, indicates a growing problem facing municipal water systems—in this case, New York City. In the 1830s and 1840s New York City bought land north of the city and built a water delivery system, parts of which are still functional. The financing and construction of the system was initiated by the Manhattan Company, a group headed by Aaron Burr, which agreed to build a water supply system in exchange for a charter to establish a bank. This bank would thus be able to compete with the Bank of New York, headed by Alexander Hamilton. While the water company is long gone, both banks remain; Hamilton's is still the Bank of New York, and Burr's is now called the Chase-Manhattan Bank. The water system that was established now serves approximately 7 million people; Chapter 28 indicates some of the problems of the deteriorating distribution systems.

Many other water problems exist in every area of the United States. These site-specific problems are dependent not only on the local hydrogeologic characteristics but also on modifications put in place by local human activity.

17

ENVIRONMENTAL PROBLEMS AFFECTING WATER SUPPLY

COUNCIL OF ENVIRONMENTAL QUALITY and
THE DEPARTMENT OF STATE

Five environmental topics related to the projections of water supply and to the consequences of water development and use are explored:

- Environmental developments affecting water supply (deteriorating catchments in river basins, acid rain, climatic change).
- Impacts of hydraulic works.
- Water pollution (of urban and industrial origin and of agricultural origin).
- Water-related disease.
- Extinction of fresh water species.

This range of topics reflects the multiplicity of characteristics, uses, and values of the water resource.

ENVIRONMENTAL DEVELOPMENTS AFFECTING WATER SUPPLY

Worldwide, two environmental developments are likely to have an impact on water supply by 2000: catchment and river basin deterioration and regional or global changes in climate. The trends in catchment and river basin deterioration are clearly discernible and accelerating. Climatic trends are less clear but just as important. Deteriorating catchment and river basin conditions will adversely affect water supplies by increasing variability. Climatic change could further increase variability.

Deterioration of Catchments and River Basins

From the standpoint of both water supply and water quality, the condition of a catchment or river basin is determined largely by the flora on the upper portions of the basin. The high, often steep portions of the basin usually receive a large proportion of the rainfall, and the flora on these slopes are critically important in determining the quality and flows of water throughout the basin.

A continuous mantle of vegetation in the upper portions of a basin has many benefits. The vegetation breaks the fall of the raindrops, absorbing the kinetic energy before it can dislodge soil particles. The vegetation also slows the runoff and enhances the absorptive properties of the soils. Where vegetation is present over the upper portions of a river basin, the basin's water is generally relatively well regulated and clean.

In the absence of vegetative cover rain flows off a basin's steep upper slopes as it would off a tin roof. The full kinetic energy of the raindrops is available to dislodge soil particles. A relatively unobstructed surface accelerates the runoff, producing greater flood peaks downstream. The kinetic energy of the enhanced floods tears away riverbanks, broadens channels, and damages or destroys canals, bridges, and other hydrological developments. Canals and dams are rapidly filled with sediments eroded from upstream, and topsoils are carried far downstream to be deposited ultimately in estuaries and oceans, often adversely affecting biological productivity. Aquifers are not recharged, and in the dry season flows are low. As a result, the removal of vegetation—especially forests—from the upper portions of river basins and catchments increases erosion, reduces water quality, damages hydrologic developments, and reduces the water available during the dryer season. On very steep (therefore unstable) slopes removal of vegetation

This material is excerpted from The water projections and the environment, in *The global 2000 report to the president, the technical report*, vol. 2 (Washington, D.C.: U.S. Government Printing Office, 1980), 334–345.

can trigger landslides and flows of debris. In the Cape Verde Islands, narrow irrigated valleys have been buried meters deep by soil and debris swept down from denuded side slopes by intense rains.

Deforestation, burning, overgrazing, and some cultivation practices all have potential for adversely affecting river basins and catchments. The *Global 2000* forestry and agriculture projections both suggest that by 2000 such practices will have extended much further into the upper portions of river basins and catchments.

Deforestation is one of the most serious causes of deterioration of catchments. In steep, high-rainfall zones such as the midslopes of the Andean and Himalayan mountains, forests are indispensable for protecting catchments and controlling runoff. Removal of trees, however, does not invariably jeopardize water supplies. In regions with moderate relief and low rainfall, removal of forests and the substitution of other soil-binding vegetation that consumes less water (such as grasses) can improve lower-basin water supplies by increasing runoff— at the cost of tree growth. Scientifically controlled cutting of catchment vegetation has been employed in the United States and some other countries to increase runoff, or water yield, in dry regions (1). To be successful, this practice requires strong, enlightened institutional programs for careful land and water management over much of the affected basin. In most of the world there is as yet no such institutional capability, and as a result, the projected deforestation will in virtually all cases lead to adverse water impacts.

Burning, overgrazing, and cultivation practices that expose the soil for long periods can be expected to increase in many areas over the next two decades, contributing further to catchment deterioration. These practices intensify the extremes of flooding and aridity by reducing soil porosity and water storage capacity, by reducing organic matter, and by increasing compaction. In soils that are overgrazed, frequently burned, or continuously cultivated, organic matter (largely mulch from vegetative debris) can become sufficiently depleted to cause soil drought. Without the absorbent properties of these organic materials, soils are less able to retain moisture, and shifts in vegetation occur. The vegetation able to survive in such soils is typical of climates that are more arid than actual rainfall indicates. The intensification of soil drought is already in evidence in the African Sahel and other semiarid regions (2), and much further deterioration of the water-retaining properties of soils can be anticipated on the basis of the population and food projections.

Acid Rain

Acid rain is an environmental problem closely related to energy development. It deserves special note here because of its effects not only on water bodies over much of the world but also on many other parts of the biosphere.

While the *Global 2000* energy projections are not specific enough to permit a detailed analysis of the future prospects for the acid rain problem a few points can be made. Increases in coal combustion in the magnitudes projected (13 percent by 1990) will significantly increase the production of the two primary causes of acid rain—sulfur oxides (SO_x) and oxides of nitrogen (NO_x). The 58 percent increase in oil combustion projected to occur by 1990 will also increase both SO_x and NO_x emissions; the 43 percent increase projected in natural gas combustion will also increase NO_x emissions. Technologies are available to remove sulfur oxides, but their removal is expensive and probably will not be required uniformly throughout the world. There is no practical technology for the removal of oxides of nitrogen from stack gases; the only control now available involves reduced combustion temperatures, which limit efficiencies. The water quality consequences of increased emissions, especially increases of NO_x emissions, need to be considered carefully.

The immediate consequence of both SO_x and NO_x emissions is the acidification of precipitation. These gaseous compounds react in the atmosphere to form sulfuric acid and nitric acid, which, in turn, precipitate out of the atmosphere in both rain and snow. The acidified precipitation falls anywhere from a few hundred to a few thousand miles away from the source, depending on the strength of the prevailing winds (3). As a result, the pH of rainfall is known to have fallen from a normal value of 5.7 to 4.5–4.2 (high acidic values) over large areas of southern Sweden, southern Norway, and the eastern United States (4). In the most extreme case yet recorded, a storm in Scotland in 1974, the rain was the acidic equivalent of vinegar (pH 2.4) (5). Equivalent changes have almost certainly occurred elsewhere, for example, downwind of the German, Eastern European, and Soviet industrial regions. Effects of acid rain are only beginning to be understood but have now been observed in lakes, rivers, and forests, in agricultural crops, in nitrogen-fixing bacteria, and in soils.

The clearest ill effects of acid rainfall observed to date are on lake fisheries. A survey of over fifteen hundred lakes in southwestern Norway, which has acid rainfall problems similar to those of southern Sweden, showed that over 70 percent of the lakes with a pH below 4.3 contained no fish. This was true for less than 10 percent of the lakes in the normal pH range of 5.5 to 6.0 (6). Similar effects have been found in lakes in the Adirondack Mountains of New York (7) and in some areas of Canada (8). Acid rain appears to be the cause of both the low pH and the extinction of the fish. Within the last 20 years salmon disappeared from many Norwegian rivers, and trout soon followed. Measurement in such rivers almost always shows a decline in pH, usually attributable to acid rain. Similar occurrences have been observed in Sweden (9).

Effects of acid rain on forest growth are only beginning to be understood. The effects on tree-seed germination are mixed (10). Reductions in natural forest growth have been observed in both New England and Sweden. One study tentatively attributed a 4 percent decline in annual forest growth in southern Sweden to acid rain (11). Other observers feel that a decline in Scandinavian forest growth has not been conclusively demonstrated but suspect that the even more acidic rainfall expected in the future will cause slower growth (12).

The effects of acid precipitation on leafy vegetation have been studied in the United States in the states of Maryland and West Virginia. While no major damage has yet occurred, one study concludes that current levels of acidity in rainfall present little margin of safety for foliar injury to susceptible plant species, but with the increasing emissions of pollutants that contribute to the formation of acid rain, there is substantial risk of surpassing the threshhold for foliar effects in the future.

Little research has been undertaken on the effects of acid rain on large natural ecosystems, but one interesting study has now been done for the boundary waters canoe area and the Voyageurs National Park (BWAS–VNP) wilderness areas in the north central United States. The study concludes that "as more lakes are eventually impacted, the whole philosophy behind the wilderness experience that forms the basis of the establishment of the BWCA–VNP will be violated and the part of the BWCA which provides recreation will be reduced. Few people who utilize the BWCA–VNP could be expected to enjoy the areas made fishless by pollution from human activity" (13).

The effects of acid rain on nonforest agricultural crops are under study and are beginning to be reported. Shoot and root growth of kidney bean and soybean plants have been found to be markedly reduced as a result of simulated acid rain of pH 3.2. Similarly, nodulation by nitrogen-fixing bacteria on legumes is significantly reduced by simu-lated acid rain. The growth of radish roots has been observed to decline by about 50 percent as the pH of rain falls from 5.7 to 3.0.

The sensitivity of soils to acidification by acid rain varies widely from area to area, depending largely on the amount of calcium in the soil (14). Calcium buffers the soil against acidification but is leached out by acid rain; this leaching of calcium and soil nutrients has been found to increase with decreasing pH, and the pH of soils has been observed to decline more rapidly with more acidic rains (15). The acidic soils that can result from acid rain could be expected to significantly reduce crop production in the affected areas unless large amounts of lime were applied.

In addition to damaging biota and soils, acid rain damages materials extensively over wide areas. Even stone is being severely damaged. A dramatic example of the effects of acid rain and air pollution on stone is provided by the Egyptian obelisk moved from Egypt to New York in the 1890s. While the inscription on the east face of the monument is still legible, the inscription on the west face has been destroyed by chemicals in the city's air, driven by New York's prevailing westerly winds.

The 13 percent increase in coal combustion by 1990 implies that large areas in and near industrial areas will continue to receive highly acidic rainfall. The rainfall in these areas is likely to become increasingly acidic as SO_x and NO_x emissions increase. The areas affected are likely to extend hundreds to thousands of miles downwind from the sources, a total geographic area large enough to include many lakes, watersheds, and farmlands. The combined adverse effects in these areas on water quality (and indirectly on soil quality and plant growth) are likely to become increasingly severe.

Climatic Change

Water supplies and agriculture can be severely affected by climatic changes that are well within the range of historic experience. Changes in global temperatures could lead to either an increase or decrease in both the amount and variability of rainfall. The climate projections therefore have definite significance for water availability in the future.

The *Global 2000* study's climate projections provide little guidance, however, because of disagreement among climatologists on future trends. The experts are more or less evenly divided over the prospects for warming or cooling, and most felt that the highest probability is for no change. Faced with this uncertainty, the *Global 2000* study devised three climatic scenarios of roughly equal probability. There is considerable uncertainty as to the pattern of rainfall to be associated with these climate scenarios, but it is thought by many climatologists that global warming would lead to slight increases in precipitation in many areas and less year-to-year variation. (The central United States, however, might experience more frequent drought.) A cooling trend is thought by many to be associated with less precipitation and increased year-to-year variation.

In short, there is much uncertainty about future global climate because of the present lack of agreement on causes, effects, and trends. Uncertainty over climate—and therefore also over water supplies and agricultural harvests—can be expected to lead to projects for the storage and regulation of water and to the development of food reserves in anticipation of unfavorable years.

Even if it were absolutely certain that the variability and amount of water supply would not deteriorate in the years ahead, there would still be reason to anticipate further projects to increase the storage and regulation of water and to develop food reserves. Population growth, urbanization, and the extension of both agriculture and forestry into more arid and variable regions has made the social and economic impacts of variability of water supplies greater than in the past. In the years ahead the impacts of even present variability can only become greater. As nations attempt to bring more marginal lands into production, fluctuations in water supply will quickly translate into social and economic vulnerability. Therefore even in the absence of any climatic deterioration, incentives will be present to maintain foodgrain reserves, accelerate water conservation efforts, modify macro- and microclimates, and develop hydraulic works to reduce the risk and uncertainty in water availability. What will be the environmental consequences of these efforts?

IMPACTS OF HYDRAULIC WORKS

Both the prospect of destabilizing deforestation in the upper portions of river basins and the certainty of continued (possibly even increased) climatic variability will encourage the development of hydraulic works for flow regulation, electrical generation, irrigation, and flood control. The *Global 2000* study's water projections assume increased withdrawals of water for all uses but make no projections as to how additional water supplies will be developed or where supplies might fall short of future need. So that the projected withdrawals are met, a considerable expansion of engineering works

for water regulation and distribution will be required, especially in regions with highly variable rainfall. By one estimate, 12,000 cubic kilometers (km^3) of runoff will be controlled in the year 2000 by dams and reservoirs—30 percent of the total world runoff and three times the estimated 4000 km^3 now stored in the world's reservoirs (16).

In the less developed countries (LDCs), where most of the world's untapped hydropower potential is located, river basin development schemes that integrate flood control, power production, and irrigation will be implemented for a number of reasons:

· The indispensable role of irrigation in increasing food production.
· The limited amounts of naturally fertile, well-drained, well-watered soils remaining to be brought into production.
· The need to control the floods of large rivers (e.g., Yellow River, Lower Mekong River) where floods have been more or less tolerated in the past.
· The need for electricity in economic development.

The environmental impacts of large river basin development schemes can be great. In the case of large dams the impacts include the following:

· The inundation of farmland, settlements, roads, railroads, forests, historic and archeological sites, and mineral deposits.
· The creation of artificial lakes, which often become habitats for disease vectors such as the mosquitoes that transmit malaria and the snails that transmit schistosomiasis.
· The alteration of river regimes downstream of dams, ending the biologically significant annual flood cycle, increasing water temperature, and sometimes triggering riverbank erosion as a result of an increased sediment-carrying capacity of the water.
· The interruption of upstream spawning migrations of fish.
· Water quality deterioration.

Irrigation systems have their own environmental problems:

· Danger of soil salinization and waterlogging in perennially irrigated areas.
· Waterweeds, mosquitos, and snail infestation of drainage canals, with the danger of malarial and schistosomiasistic infections spreading in areas where these diseases exist, especially in parts of Africa and Latin America.

- Pollution of irrigation return water by a variety of agricultural chemicals, with negative consequences for aquatic life and for the human use of downstream waters.

While the benefits of dams and irrigation development may outweigh the costs, environmental impacts have a definite bearing on the benefit/cost ratios of river basin development schemes. Plans for the development of the Lower Mekong River Basin illustrate this point.

A series of engineering, economic, social, and environmental studies of the Lower Mekong Basin has been carried out under the aegis of the United Nations Committee for Coordination of Investigations. The development plan that has emerged from these studies calls for the construction over a 20-year period of a series of multipurpose dams and associated irrigation works for the basin, which is shared by Thailand, Laos, Cambodia, and Vietnam. In 1974 the portion of the basin downstream from the People's Republic of China supported about 33 million persons. Assuming the adoption of birth control methods at rates based on other South Asian experience, the United Nations studies project this population to grow ultimately to or beyond the Lower Mekong Basin's present food production capacity, which is estimated to be potentially adequate for 123 million persons. For feeding of the expanded population by the end of the century, it will be necessary to expand paddy rice production from 1970s 12.7 million tons to 37 million tons. The studies suggest that this increase of nearly 200 percent cannot be achieved without flood control and new irrigation. It is estimated that multiple dams in the Lower Mekong River system could add up to 5 million hectares (ha) of land for double-cropping of rice and might provide enough food to support an additional 50 million persons in the basin. The dams would generate badly needed power, and the reservoirs could, with proper management, become productive fisheries.

The proposed dams in the Lower Mekong Basin will involve significant social costs. For example, the reservoir behind the Pa Mong—the largest dam proposed for the Mekong River—would force the resettlement of 460,000 persons, mostly in Thailand. Land for resettlement en masse in large communities is not available, and Thailand is faced with the prospect of paying these people an estimated $626 million (approximately $1400 each) to leave without a planned alternative, a situation euphemistically referred to as "self-settlement."

The situation in the Mekong River Basin happens to be relatively well understood because 20 years of internationally coordinated studies have examined the entire river basin as a single planning unit.

Other densely populated river basins in Asia, Africa, and Latin America are the focus of similarly ambitious schemes, but in most cases there are no coordinated studies or even adequate data. Consequently, the full social and economic costs of these proposed projects can scarcely be estimated.

The environmental costs are just as hard to estimate. It is known that large dams produce very considerable ecological impacts on rivers and estuaries in temperate and subtropical areas. The Aswan Dam in Egypt is a case in point.

A considerable list of costly impacts are associated with the High Aswan Dam and the irrigation development that has subsequently taken place in the Nile Delta. They have been documented by Julian Rzoska (17) in a ten-years-later assessment, as well as by earlier researchers such as Kassas, George, and van der Schalie. Here are some of their findings:

- 100,000 people had to be relocated from the reservoir site, which extends into Sudan. The people were mostly floodplain farmers of Nubian origin.
- The ancient Nubian temples were inundated (a considerable portion of them were salvaged intact in a UNESCO-organized emergency operation).
- The dam traps sediments that formerly enriched the floodplain as well as the Mediterranean Sea, with a loss in natural soil productivity and the collapse of the sardine fishery that once provided half of Egypt's fish.
- Waves and tides are now eroding the delta, which formerly was extending into the Mediterranean, and a reduction in the agriculturally important delta is slowly occurring.
- Year-round irrigation in the delta, which represents 60 percent of Egypt's farmland, has elevated the water table and caused salinization, now being remedied through expensive drainage works financed by the World Bank.
- Schistosomiasis is rapidly spreading throughout the rural population as a consequence of the spread of the snail intermediate hosts in the irrigation canals, the lack of sanitary facilities, and the continual exposure of the dense rural population.
- The water hyacinth spread almost uncontrollably throughout the canal systems, where it harbors snails and interferes with water flows.

This controversial project's benefits include an 8000-megawatt (MW) electricity generating potential and a doubling of agricultural potential on perennially irrigated soils.

While the extensive consequences of the Aswan Dam are reasonably well known and established, relatively little is known about the ecological effects of dams in tropical areas, where most of the need and potential is located. The animal species native to tropical rivers, estuaries, and oceans have frequently evolved life cycles that are linked to annual floods and the patterns of salinity and nutrient fluxes that accompany the floods. Regulation of river flows can therefore be expected to significantly affect large numbers of estuarine and oceanic organisms. Similar impacts can be anticipated in fresh water species. Their decline is not likely to be compensated by the development of aquaculture, especially if pollution seriously impairs water quality.

WATER POLLUTION

The *Global 2000* study projections point to worldwide increases in urbanization and industrial growth and in the intensification of agriculture—trends that, in turn, imply large increases in water pollution in many areas.

Water Pollution of Urban and Industrial Origin

By the year 2000 worldwide urban and industrial water withdrawals are projected to increase by a factor of about 5, reaching 1.8–2.3 trillion m³. The higher figure is almost equal to the total annual runoff of 2.34 trillion m³ from the 50 United States. Most of the water withdrawn for urban and industrial use is returned (treated or untreated) to streams and rivers. If 90 percent is returned, the total combined discharge of water flowing through sewers and industrial outfalls by the year 2000 will be on the order of 1.6–2.1 trillion m³.

Urban and industrial effluent will be concentrated in the rivers, bays, and coastal zones near the world's largest urban-industrial agglomerations. In the developing world—where 2 billion additional persons are projected to be living by 2000 and where rapid rates of urbanization continue—urban and industrial water pollution will become ever more serious because many developing economies will be unable or unwilling to afford the additional cost of water treatment.

Few LDCs have invested heavily in urban and industrial waste treatment facilities. As a result, the waters below many LDC cities are often thick with sewage sludge and wastes from pulp and paper factories, tanneries, slaughterhouses, oil refineries, chemical plants, and other industries. One consequence of this pollution is declining fishing yields downstream from LDC cities. For example, the inland catch in the eastern province of Thailand,

696 tons in 1963, fell to 68 tons by 1968, and it is thought that water pollution, particularly from Bangkok, was the main cause of the decline. Similar, though less extreme, declines have occurred around the world in freshwater systems and in bays, lagoons, and estuaries. Frequently, the changes are not measured but become apparent with the appearance of eutrophication, poisonous red tides, and the decline of inland fishing occupations.

Efforts to control the effects of pollution from LDC cities can lead to international disputes. An example is the dispute that occurred in 1976 over India's withdrawal of water from the Ganges to flush out the port of Calcutta during the dry season when the water was needed in Bangladesh for irrigation.

The reuse of urban and industrial waste water is likely to increase as urban populations expand rapidly in the water-short regions of West Asia and in arid portions of Mexico, Africa, and the U.S. Southwest. The use of waste water for irrigation will serve to recycle nutrients that would otherwise overload the absorptive capacity of rivers; however, a careful management and monitoring will be required to avoid pollution of ground water and human exposure to disease pathogens, heavy metals, and other toxic substances (18).

The use of water as the transport media for sewage is being questioned because of its high capital requirements, its potential for pollution, and its energy-intensiveness. Composting toilets that avoid the water medium entirely have been developed and are being used more widely.

Some water pollution problems are linked directly to air pollutants from urban and industrial areas, particularly from emissions of sulfur and nitrogen oxides from electric power plants burning fossil fuels. The increased use of coal (a rich source of both sulfur and nitrogen oxides) promises a growing contribution of acid to rain water and to lakes. There is no known economically practical method for controlling NO_x emissions. Control of SO_x emissions from coal is now technologically possible but is estimated to increase electricity costs by 6 to 15 percent. It seems that the price of "live" lakes will be high.

Urbanization and industrial growth, in addition to increasing various forms of water pollution, will also increase the consumptive uses of water. Evaporative cooling for thermal electric generating facilities is one of the fastest-growing consumptive uses of water.

Large amounts of water are used to remove waste heat from thermal electric (primarily coal and nuclear) power plants, but until recently, relatively little of this water has been consumed (i.e., evaporated). Until the early 1970s in the United States most of the waste heat from electricity generation—which amounts to approximately two-thirds of the total primary energy input to electrical generation—was dissipated by means of once-through cooling. With once-through cooling large quantities of river-, lake-, or ocean water are pumped through condensers and returned to the natural water body approximately 10°C warmer. Between 1950 and 1972 annual water withdrawals in the United States for thermal electric power plant cooling jumped from 50 billion to 275 billion m³—2.5 times the average flow (110 billion m³) of the lower Mississippi River—surpassing irrigation withdrawals in volume.

Concern for biological and ecological damage caused by water intakes and by thermal pollution in the United States resulted in the promulgation in 1974 of standards for levels of thermal discharge to water bodies by power plants, but in 1976 the standards were remanded by court order. At present thermal pollution has a low priority at the Environmental Protection Agency. Thermal pollution impacts are numerous and generally deleterious in mid to low latitudes. In high latitudes waters are naturally so cold that aquatic life processes are slowed, and in these areas heated discharges from power plants can stimulate production of fish and other organisms. In the tropics, on the other hand, where waters are naturally warm and many species live near their upper temperature tolerance, thermal discharges are often lethal. At all latitudes increased temperature reduces the dissolved oxygen in the water, stressing aquatic fauna by speeding metabolic rates while at the same time depleting oxygen supplies. Other impacts include the following:

- Destruction of small organisms such as fish larvae and plankton entrained in the cooling water intake and poisoned by antifouling biocides.
- Reduction of fish abundance, biomass, and species diversity in downstream thermal "plumes."
- Synergistic exacerbation of the stresses caused most organisms by other factors such as increased salinity, biological oxygen demand, and toxic substances.
- Shifting of the balance among algae species to favor blue-green algae, which create taste and odor problems in municipal water supplies.
- Sudden changes of temperature during start-ups and shutdowns, causing death of many sensitive species (19).

The remanding of the 1974 Environmental Protection Agency regulations on thermal pollution left U.S. problems of thermal pollution unresolved. New plants tend to utilize evaporative cooling towers rather than once-through cooling because of insufficient volumes of water available rather than because of ecological considerations. As a result, thermal water pollution in the United States may remain at about 1976 levels, while local atmospheric heat and humidity loadings in areas around new power plants increase. The U.S. Water Resources Council estimates that the consumptive use of water by the country's electrical generating facilities will increase rapidly (650 percent between 1975 and 2000).

The net consumption (i.e., evaporation) of water can also be expected to continue increasing elsewhere in the world during the years ahead as thermal electric generation grows and supply constraints and environmental considerations encourage shifts away from once-through cooling to evaporative cooling towers. A. L. Velikanov has estimated for the United Nations that waste heat discharged from thermal electric plants throughout the world in 1973 was sufficient to evaporate 7 to 8 km³ of water if cooling towers had been in use everywhere. This estimate may be low. European energy specialist Wolf Häfele calculates that if Europe had been using cooling towers exclusively in 1974, Europe alone would have been evaporating water at an annual rate of 16 km³. Häfele expects this consumption to reach 30 km³ per year by 2000. Although this represents only about 1 percent of Europe's yearly runoff of 2800 km³, the additional consumptive demand on Europe's water resources would be significant.

Water Pollution of Agricultural Origin

Extensive pollution from fertilizer runoff can be expected, especially in developed, densely populated regions, if worldwide fertilizer use increases from 55 kg/ha—the average 1971–75 rate—to around 145 kg/ha as projected. The U.S. Department of Agriculture projects fertilizer application rates in Japan, Western Europe, and Eastern Europe to reach 635, 355, and 440 kg/ha, respectively, by 2000. At these application rates it will be difficult to avoid at least some increase in the nitrogen pollution of water supplies and eutrophication of bodies of water.

The LDCs are likely to experience increasing water pollution by pesticides, especially chlorinated hydrocarbon insecticides used in irrigated rice culture and export crop production. The Food and Agriculture Organization (FAO) expects that pesticide use in the LDCs will grow at 10 percent per year for at least the near future. Should this trend continue until 2000, the volume of pesticides used in the LDCs will have increased more than sevenfold (20). Presently, about half the pesticides used in the LDCs are organochlorines, a trend that may continue because organochlorines are substantially less expensive than the more specific, less destructive, and less persistent alternatives.

A sevenfold increase in the use of persistent pesticides in Asia would virtually eliminate the culturing of fish in irrigation canals, rice paddies, and ponds fed by irrigation water. Organochlorine insecticides continue to collect in aquatic systems years after they have been applied and affect waters many miles downstream. At moderately high concentrations they kill fish. Already, many Asian farmers are reluctant to buy fry for their paddies or ponds for fear that pesticide pollution will kill the stock. The amount of protein forfeited could be substantial. Per hectare yields of fish from well-tended ponds can be as high as the per hectare yields of rice, that is, 2500 kg/ha animal protein versus 2500 kg/ha carbohydrate. Cage culture yields are extraordinarily high and show great commercial promise in several developing countries, as long as waters are not poisoned by pesticides. Projected pesticide increase seriously threatens both fresh water and brackish water aquaculture in much of Asia. If pesticide trends continue, aquaculture in Latin America and Africa will eventually face the same threat.

The protein that fish culture could provide is badly needed, especially in the humid tropics where aquaculture can thrive. Moreover, while alternative forms of producing animal protein tend to increase the pressures on already stressed soil systems, fish culture places no strain on terrestrial systems and is complementary to the careful water management schemes required for sustained agricultural production in many parts of the humid tropics. The FAO estimates that culture of fresh water and marine organisms could reach 20–30 million metric tons by 2000—between one-third and one-half of the present marine catch. Further pesticide pollution will sharply diminish this promising prospect.

Increased pesticide use will also create water contamination problems in industrialized nations. To cite but one example: California health officials report that they have found dangerous levels of a pesticide—dibromochloropropane (DBCP)—in half of the irrigation and drinking water wells they have tested in one of the state's major agricultural areas, the San Joaquin Valley. The U.S. Environmental Protection Agency banned the use of DBCP in 1977 on 19 fruit and vegetable crops after tests showed that the pesticide caused sterility in the workers who manufactured it and caused cancers in laboratory animals. Two years after the ban, California health officials found residues averaging 5 parts per billion (ppb) in the wells tested. The state has recommended that all wells showing more than 1 ppb of DBCP be closed to human consump-

tion. At that level one case of cancer is expected for every 2500 persons who use the wells. Arizona health officials have tested 18 wells near Yuma and found 6 wells with concentrations of 4.6–18.6 ppb of DBCP. The Environmental Protection Agency has allowed continued use of an estimated 10 million lb of DBCP annually in the United States on crops such as soybeans, citrus fruits, grapes, and nuts but is now considering restricting this amount.

Other pesticides may not cause as many problems with water. California officials report that DBCP is the only pesticide they tested that shows a tendency to be absorbed into ground water. Nonetheless, the projected increases in pesticide use will create a variety of water contamination problems in the industrialized nations as well as in the less developed countries.

Irrigation will also add large amounts of salt contamination to the waters of many areas. The water use projections reveal that by the year 2000 between 4600 and 7000 billion m^3 water will be withdrawn for irrigation. Approximately 25 to 30 percent will be returned to streams carrying dissolved salts. In very arid areas return water is heavily contaminated with salts, concentrated by high rates of evaporation.

Salt pollution of arid zone rivers draining away from irrigated lands will ultimately make the rivers unfit for further irrigation use in their lower reaches, as has already happened to the Shatt-al-Arab River in Iraq and the Lower Colorado River in the United States. The Shatt-al-Arab was formed by the Tigris and Euphrates rivers, whose delta soils were once covered with extensive date palm and citrus orchards.

One remedy—a very costly and energy-intensive remedy—for the salt pollution of rivers is to desalinate the water. A 104-million-gallon-per-day (mgd) desalting facility will soon be in operation on the Lower Colorado River. Now under construction at Yuma, Arizona, this plant will be the largest desalting plant in the world, costing over $300 million. It is needed to fulfill a U.S. agreement with Mexico to deliver water in the Colorado River with a total dissolved-solids content (including salts) of no more than 115 milligrams per liter (mg/L). Because of the leaching of salts from fields upstream, the dissolved-solids content had increased to 850 mg/L and was expected to reach 1300 mg/L by 2000. Most plant species cannot tolerate water with more than 500 mg/L of dissolved solids (21).

WATER-RELATED DISEASES

Water-related diseases have been an unfortunate accessory to irrigation systems and dams as well as to pollution by human wastes and are virtually certain to become more prevalent during the rest of the century as more of the water environment becomes affected by human activities and wastes.

A wide variety of water developments can increase the incidence of water-related diseases. The creation of ponds, reservoirs, and irrigation and drainage canals in the course of water resource development, and the widespread inadequacy of waste water disposal systems in LDC cities, all favor the persistence or spread of a number of such diseases. In recent years new irrigation systems and reservoirs in Middle and North Africa and West Asia have provided ideal habitats for the intermediate snail host of schistosomiasis, which has spread dramatically among rural populations. This debilitating disease of the intestinal and urinary tract now affects an estimated 250 million people throughout the world, approximately 7 percent of the entire human population. In some irrigation project and reservoir areas, up to 80 percent of the population is affected.

In addition to schistosomiasis there are numbers of other serious water-related diseases. These include malaria, filariasis (elephantiasis), and yellow fever, all of which are transmitted by mosquitos. Onchocerciasis, "river blindness" disease, is transmitted by flies. Paragonimiasis is a disease transmitted by a snail. Poorly managed water resource development projects, as well as the impact of urbanization on aquatic habitats and water quality, contribute to the spread of all of these diseases. Diseases typical of waste water contaminated by human feces—cholera, typhoid fever, amoebic infections, and bacillary dysentery—can become problems anywhere in the world. In LDCs today almost 1.5 billion persons are exposed to these diseases for lack of safe water supplies and human waste disposal facilities. Largely for this reason infant deaths resulting from diarrhea continue at a high rate. Every day 35,000 infants and children under five years of age die throughout the world; most of these deaths occur in LDCs. Schistosomiasis afflicts 200 million people in 70 countries, and elephantiasis is estimated to cripple 250 million more. In parts of Asia where night soil is extensively used as fertilizer, roundworm (Ascaris) infections will continue to be a threat because the roundworm's eggs are not easily killed.

Water-related diseases are not limited to countries that cannot afford sewage treatment. In industrialized countries the treatment of city waste waters with chlorine presents a different kind of water-related health problem—the possibility of cancer. When chlorine reacts with organic compounds in waste water, one of the resulting by-products is chloroform, a carcinogen. Elevated rates of fatal gastrointestinal and urinary cancer are reported by some scientists in communities that utilize water supplies contaminated with chloroform. The U.S. National Academy of Sciences has recommended that strict criteria be applied in setting limits for chloroform in drinking water (22).

EXTINCTION OF FRESH WATER SPECIES

The International Union for Conservation of Nature and Natural Resources notes in the draft of its *World Conservation Strategy* that 274 fresh water vertebrate taxa are threatened by extinction as a result of habitat destruction. This number is larger than the number of similarly threatened vertebrate taxa in any other ecosystem group (23).

It is not surprising that a large number of freshwater species are threatened with extinction through loss of habitat in view of the major changes that are occurring in fresh water systems. Damming, pollution, channelization, and siltation are causing massive alterations in fresh water ecosystems throughout the world. Fresh water species endemic to specific lakes, rivers or upper reaches of river branches are particularly vulnerable because they are often easily extinguished by changes in water chemistry (the effects of acid rain, for example), modification of streambed contours, alteration of water temperature, or the imposition of dams that prevent species from reaching their spawning grounds. Because of the anticipated increase in pollution and in manipulation of fresh water systems, many of the species now threatened may be extinct by 2000, and many now relatively common species may be on the way to extinction.

The trends in fresh water extinctions will be difficult if not impossible to reverse. In many areas political and social realities will stand in the way of installing expensive pollution control systems, of changing dam sites, or of reducing pesticide use or coal combustion in order to save a fish or amphibian whose existence may be known to only a few people and whose value and importance may be perceived by fewer still. As a result, higher rates of extinction among fresh water species are expected to continue.

CONCLUSION

Fresh water, once an abundant resource in most parts of the world, will become increasingly scarce in coming decades for two reasons. First, there will be greater net consumption, by cooling towers and, especially, by irrigation so that the total supply will decline. Second, pollution and the impacts of hydraulic works will effectively limit the uses of fresh water—and therefore, in effect, the supply.

The deterioration of river basin catchments, especially as a result of deforestation, will increase the variability of supply, accelerate erosion, damage water development projects, and degrade water quality. It seems inevitable that the function of streams and rivers as habitat for aquatic life will steadily be sacrificed to the diversion of water for irrigation, for human consumption, and for power production, particularly in the LDCs.

The 1977 U.N. Water Conference served to focus global attention on the critical problems of managing the world's water resources in the coming decades. In the LDCs the development of water resources for irrigation and power is a key to providing for the economic needs of expanding populations. At the same time the ecological impacts of hydraulic works and of pollution from agricultural fields and urban industrial concentrations is greatly diminishing the capacity of water systems to support fish that are sorely needed to supplement meager diets. The lack of safe water supplies and of methods for sanitary disposal of human waste and waste water means that as many as 1.5 billion persons are exposed to fecally related disease pathogens in drinking water. These problems of water supply and quality in LDCs are so severe as to be matters of survival for millions of persons.

In industrial nations water supply and quality will pose more subtle and therefore more complex questions of trade-offs and conflicts among users (or values) of fresh water. Water resources management in such nations is concerned not with human survival but with balancing demands for water resources against considerations of quality-of-life. But scarcities and conflicts are becoming more acute, and by the year 2000 economic, if not human, survival in many industrial regions may hinge upon water quality, water supply, or both.

Perhaps the most underrated aspect of fresh water systems throughout the world is their function as aquatic habitat. At some point high social and economic costs will follow the continued neglect of the water quality needed to maintain ecosystem health. This point may be marked by the failure of fish farms, by a decline of the capacity of streams to accommodate wastes, or by the decline or disappearance of species that may possibly be of great future value. Given the criticality of the other uses of fresh water resources, the future integrity of aquatic habitats is by no means assured. In fact, since aquatic habitats are much more difficult to know and monitor than terrestrial ones, it is in serious doubt.

NOTES

1. H. W. Anderson et al., 1976, *Forests and water: Effects of forest management on floods, sedimentation and water supply*, Pacific Southwest Forest and Range Experiment Station report (Washington: U.S. Forest Service).

2. M. Kassas, 1977, Desertification versus potential for recovery in circum-Saharan territories, in *Arid Lands in Transition*, ed. H. E. Dregne (Washington: American Association for the Advancement of Science), 123–142; Status of desertification in the hot arid regions; climate aridity index map; experimental world scheme of aridity and drought probability, U.N. Conference on Desertification, 1977.

3. A. P. Altschuller, 1978, Transport and fate of sulfur and nitrogen containing pollutants related to acid precipitation, unpublished report to the Environmental Protection Agency; George M. Hidy et al., International aspects of the long range transport of air pollutants, report prepared for the U.S. Department of State (Westlake Village, Calif.: Environmental Research and Technology, Inc.).

4. Commission on Natural Resources, National Research Council, 1977, *Implications of environmental regulations for energy production and consumption: A report to the U.S. Environmental Protection Agency from the Committee on Energy and the Environment*, vol. 6 (Washington: National Academy of Science) 68; Northrop Services, Inc., 1978, Interim Report: Acid precipitation in the United States—History, extent, sources, prognosis," a report to the Environmental Protection Agency (Corvallis, Ore.: Environmental Research Laboratory, Dec. 18); J. N. Galloway et al., 1976, Acid precipitation in the Northeastern United States, *Science*, 194: 722–724; G. E. Likens, 1976, Acid precipitation, *Chemical Engineering News*, 54: 29–44.

5. G. E Likens et al., 1977, Acid rain, *Scientific American* (October): 43–51.

6. D. M. Hendrey et al., 1976, Acid precipitation: Some hydrobiological changes, *Ambio*, 5 (5–6): 224–227. For a good general overview of the impact of acid precipitation in Norway, see Finn H. Braekke, 1976, *Impact of acid precipitation on forest and freshwater ecosystems in Norway*, SNSF-Project, NISK, 1432 AasNHL, Norway.

7. C. L. Schofield, 1975, Acid precipitation: Our understanding of the ecological effects, in *Proceedings of the Conference on Emerging Environmental Problems: Acid precipitation*, U.S. Environmental Protection Agency Region II, New York.

8. M. Whelpdale, 1978, Large scale atmospheric sulphur studies in Canada, *Atmospheric Environment*, 12: 661–670; Roderick W. Shaw, 1979, Acid precipitation in Atlantic Canada, *Environmental Science and Technology*, 13 (4): 406–411; Ross Howard, 1979, 48,000 lakes dying as Ontario stalls, *Toronto Star*, Mar. 10, p. C4.

9. Arid Holt-Jensen, 1973, Acid rains in Scandinavia, *Ecologist* (October): 380–381.

10. Jeffrey J. Lee and David E. Webber, The effect of simulated acid rain on seedling emergence and growth of eleven woody species, unpublished report, Corvallis Environmental Research Laboratory, U.S. Environmental Protection Agency, Corvallis, Ore.; Braekke, op. cit., 53–59.

11. Leon S. Dochinger and Thomas A. Seliga, 1975, Acid precipitation and the forest ecosystem, *Journal of the Air Pollution Control Association* (November): 1105.

12. Holt-Jensen, op. cit., 381–382; Carl Olof Tamm, 1976, Acid precipitation: Biological effects in soil and on forest vegetation, *Ambio*, V (5–6): 235–238; Braekke, op. cit., 53–59.

13. Gary E. Glass, 1979, Impacts of air pollutants on wilderness areas of northern Minnesota, Environmental Research Laboratory-Duluth, U.S. Environmental Protection Agency, March 5, pp. 129–130 (draft).

14. W. H. Allaway, 1957, pH soil acidity, and plant growth, and N. T. Colemean and A. Mehlich, 1957, The chemistry of soil, pH, in *Soil: The 1957 yearbook of agriculture* (Washington: Government Printing Office), 67–71 and 72–79.

15. John O. Reuss, 1975, Chemical/biological relationships relevant to ecological effects of acid rainfall, Corvallis, Ore.: National Environmental Research Center, U.S. Environmental Protection Agency, June; Dale W. Johnson and Dale W. Cole, 1977, Anion mobility and soils: Relevance to nutrient transport from terrestrial to aquatic ecosystems, U.S. Environmental Protection Agency.

16. K. Szesztay, 1973, Summary: Hydrology and man-made lakes, in ed. William C. Ackermann et al., *Man-made lakes: Their problems and environmental effects* (Washington: American Geophysical Union).

17. Julian Rzoska, 1976, A controversy reviewed, *Nature* (June 10): 444–445.

18. Office of Water Program Operations, 1976, Application of sewage sludge to cropland: Appraisal of potential hazards of heavy metals to plants and animals (Washington: U.S. Environmental Protection Agency, no. 15); Bernard P. Sagik and Charles A. Sorber, eds., 1977, *Risk assessment and health effects of land application of municipal wastewater and sludges,* Conference Proceedings, San Antonio, Center for Applied Research and Technology, University of Texas; G. W. Leeper, 1978, *Managing the heavy metals on the land* (New York: Marcel Dekker).

19. P. R. Ehrlich et al., 1977, *Ecoscience: Population, resources, environment* (San Francisco: Freeman), 670.

20. U.S. Agency for International Development, 1979, *Environmental impact statement on the AID pest management program,* vols. 1 and 2 (Washington); also its *Environmental and natural resource management in developing countries: A report to Congress,* vols. 1 and 2 (Washington: 1979).

21. U.S. Bureau of Land Reclamation, 1976, Increasing available water supplies through water modification and desalinization, U.N. Water Conference, New York.

22. Drinking water and health (Washington: National Academy of Sciences, 1977).

23. World Conservation Strategy, Morges, Switzerland, International Union for Conservation of Nature and Natural Resources, January 1978 (draft), p. WCS/Strategy/4.

18

DROUGHT IMPACTS ON PEOPLE

HOWARD F. MATTHAI

"When droughts have occurred in the past, there have been few intelligent plans of action. Actions during a drought often can be characterized as too little and too late, if not actually counterproductive. While droughts cannot be forecast, their effects can be anticipated." The preceding statement was made in June 1978 by the Office of Science and Technology Policy in its report to the President's Committee for the Water Resources Policy Study. Some of the effects that can be anticipated and a few that could not are discussed in general terms in this section.

Often too many people react to a drought in the manner exemplified by the flowchart in Figure 18-1.

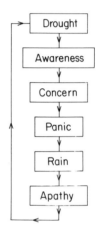

Figure 18-1. Human reaction to drought.

PUBLIC WATER SUPPLIES

Water for domestic, commercial, and industrial uses in the United States under normal conditions was withdrawn at the rate of 168 gal/day per individual served in 1975. About 175 million people are served by public supply systems nationwide. In states affected by the drought in 1976–1977 normal withdrawals ranged from 119 gal/day per individual served in Virginia to 331 gal/day per individual served in Utah (Murray and Reeves, 1977).

The 168 gal/day per individual served is an increase of 16 percent over the use in 1950, about the midpoint of the last drought in the Southwest. And the number of individuals served has increased from 93.5 million to 175 million. Therefore the demand for public water supplies has doubled in 25 years from 14 billion gallons per day (bgd) to 29 bgd. Increased stresses of this magnitude on the available water supplies and distribution systems could cause water shortages locally even during nondrought periods. Then when a severe drought does occur, the problems faced by the public are compounded.

Though the drought started during 1976 in California, the Upper Peninsula of Michigan, and Colorado, public water supplies were not seriously affected until later. Supplemental supplies from wells and the use of carryover storage were the main sources of water used to meet the public's demands.

It was not until the winter of 1976–1977 that many public entities realized that a serious drought was imminent. Water managers appealed to the public to conserve water and to voluntarily reduce water use by 10 percent. Suggestions were offered on how to save water, some utilities furnished flow restrictors for showers and faucets, and the news

Howard F. Matthai is a hydrologist with the U.S. Geological Survey. This material is excerpted from *Hydrologic and human aspects of the 1976–77 drought*, U.S. Geological Survey Professional Paper 1130, 1979, pp. 14–19.

media cooperated by publishing or televising pictures of reservoirs with very little water and by keeping the public informed.

A small percentage of the public seems to be very skeptical of any warnings about hydrologic phenomena that are usually considered natural, whether they are floods or droughts. Some people reacted by claiming that there was no serious drought and that one group or another was putting out propaganda to further its own motives. As the drought worsened, most skeptics became believers.

Sometime during the period February–April 1977, mandatory water rationing was imposed in many areas, and additional water districts established reduced quotas during the summer. An allowance of 75 percent of the amount used during the same billing period in 1976 was fairly common, and other rations ranged from 30 to 90 percent. Users were warned once or twice if they exceeded their allowance. The next time they were penalized or fined for any overuse; and in a few homes flow restrictors were installed so that it took 20 min to draw water for a bath.

Typically, a few people made light of the drought. There were jokes about when or where one should use or not use water and stories of unique situations that no one could foresee. Someone suggested that you should shower with a friend to conserve water. A suburbanite drove to his club in San Francisco to shower before going to work, thus shifting his water use from his allotment to someone else's. And one citizen claimed the drought in California was caused by the rain in California which is only half as wet as rain should be.

In spite of the carefree attitude and the people who would rather pay fines than curtail their water use, water use was reduced significantly and commonly below the ration allowed.

Several public supply systems found it necessary to raise rates because their operating costs remained about the same but revenues were less. The city of Bessemer, in the western end of the Upper Peninsula of Michigan, had to import water by truck starting in December 1976, after their wells went dry.

The Marin Municipal Water District, just north of San Francisco, California, imposed one of the strictest water-rationing programs upon their customers when they set a limit of just 50 gal/day per person. To ease the water shortage, they constructed a 24-in. pipeline across the Richmond–San Rafael Bridge and obtained water from the State Water Project through the facilities of the East Bay Municipal Utility District. In the interim entrepreneurs trucked water from San Francisco and other nearby areas to large estates to save the valuable landscaping and to dairies. This was a thriving business while it lasted.

RURAL WATER SUPPLIES

Rural use of water other than for irrigation is primarily for domestic and livestock use. In 1975, 42 million people depended upon their own supply; and they withdrew 5 bgd (Murray and Reeves, 1977), an increase of 37 percent since 1950. About 95 percent of the rural domestic water and 58 percent of the water for livestock comes from wells.

During the drought shallow wells went dry or yielded meager quantities of water because the additional pumping from deeper wells lowered the water table more than in the past. Solutions to water problems at many individual homes and farms were not easy, and many were expensive. Numerous wells located in alluvium were deepened when one of the very busy well drillers could schedule the work and a loan could be obtained. In the Upper Peninsula of Michigan drilling wells deeper did not always produce sufficient water to alleviate the drought because in most of the area the deeper formations are not good aquifers.

More than a thousand wells in the Upper Peninsula went dry, and the Michigan National Guard and State Police trucked water to some areas. In several northern Michigan counties people had to obtain water for cooking and drinking from schools or community buildings that had wells with more dependable water supplies. Water for sanitary purposes was taken from streams and lakes.

The number of livestock on many farms in several states was reduced to conserve water, and water was hauled to supply those retained. Some animals were moved to places with sufficient water, and ranchers received financial assistance from the Federal Disaster Assistance Administration or the Department of Agriculture for the move.

WATER FOR IRRIGATION

Irrigation water withdrawals amount to roughly twice the water withdrawn for public, rural, and industrial uses combined, excluding water used by electric utilities in power generation. Ground water supplies 40 percent of the irrigation withdrawals nationwide; but in eight western states in the drought areas, ground water withdrawals normally average 32 percent.

The first effect of a drought on agriculture is low soil moisture caused by the below-normal rainfall. The additional irrigation required depletes the reserves in both surface water and ground water reservoirs. Where surface supplies were practically exhausted, additional wells were drilled or existing wells were deepened. The increased use of ground water, though expensive, was enough to produce near-normal crop yields in many areas.

Irrigation generally is considered a lower beneficial use than municipal use; so in some places irrigation diversions were reduced to provide water for municipalities. Conversely, the California Aqueduct was shut down south of the Tehachapi Mountains south of Bakersfield because additional water was available to southern California from the Colorado River. The additional water remaining in the San Joaquin Valley was used mainly for agriculture.

In Idaho and Washington some temporary redistributions of irrigation water were made. To save orchards and vineyards, they were irrigated rather than field crops. Also, irrigated acreage was reduced or crops needing less water were planted in anticipation of decreased water supplies.

Very low soil moisture because of the longer-than-normal periods between rains during the growing season in the Midwest stunted corn and other farm products.

WATER FOR HYDROELECTRIC POWER

Hydroelectric power is generated in 46 of the 50 states; therefore a widespread drought will affect seriously the ability of utilities to generate hydroelectric power. When storage in a reservoir is drawn down, the head on the generator is reduced and less power is produced. The water level was lowered below the intakes to a few powerhouses in California, and power generation ceased. At other sites the number of hours that power was generated had to be reduced.

The reduced hydroelectric generation required increased use of natural gas and oil at steam generating plants, which added millions of dollars to the cost of producing electrical energy. The additional cost was passed on to the consumers when rates were raised.

Rolling brownouts were expected in California, and electric utility customers were advised of the proposed schedules. *Rolling brownouts* are planned periods of a few hours when electric service to different areas or to different classes of customers would be interrupted on a scheduled rotation to reduce the demand for electricity and therefore reduce the water use by hydroelectric plants. However, no brownouts occurred. Reduced hydropower output in the Columbia River Basin affected some of the large consumers such as aluminum plants.

Cutbacks in industry that is dependent on electric power increased the number of unemployed.

WATER FOR FORESTS

Drought conditions were severe in many of the nation's forests; therefore the fire season started earlier than usual—as early as April in Idaho. The larger fires in 1976 in California occurred in June and July, burning 85 percent of the 165,000 acres burned in 1976. Normally, only 25 percent of the acreage burned annually is burned by July 15.

The obvious results of fires can be seen immediately, but the secondary results of fires will not be known for awhile. When rains finally come, erosion of hillsides, head and bank cutting along streams, deposition of sediment and debris, and flooding will occur in various degrees along streams draining the burned-over areas.

Recent research by the U.S. Forest Service has shown that another effect of forest fires, at least in chaparral, is the formation of nonwettable or water-repellent soils. When chaparral plants and the litter on the ground are burned, a complex of waxlike substances is produced, which tends to coat the soil particles and makes them hard to wet. The fire vaporizes these substances, and the gases that are heavier than air sink into the soil layer, where they cool enough to recondense and again coat the soil particles. This second process makes soils that formerly were only hard to wet virtually waterproof. If the water cannot enter the soil, it must run off; therefore the flood potential is increased (Wells, 1978).

Another long-term effect of the drought is the damage to timber and other trees from insects, disease, and smog. Vegetation undergoes additional stress when it does not receive some minimum amount of moisture or does not receive it at the proper time to foster growth. Therefore vegetation is more susceptible to deleterious influences because of a drought.

WATER QUALITY

Whenever stream flow falls below threshold amounts, water quality problems can be expected soon thereafter. The inability of low stream flows to flush and dilute contaminants in stream channels may let concentrations from waste discharges increase to the point that the water is not usable. Higher-than-usual concentrations of dissolved solids, one of the indices of water quality, occurred in North and South Dakota, western Colorado, and

Ohio. Both natural sources and pollutants contributed to the high concentrations.

The lower the flow for a prolonged period, the less water there is to absorb wastes that demand oxygen. A decrease in dissolved-oxygen concentrations, called an *oxygen sag,* can cause unpleasant odors and fish-kills and will reduce the ability of a stream to purify itself naturally. Oxygen sags were reported in Minnesota and California.

Higher water temperatures are associated with low stream flows and shallow depths. Aquatic growths may increase; fish, particularly trout, may die; evaporation from water surfaces will increase, and the efficiency of water-cooled systems is impaired when water temperatures increase.

WATER FOR FISHERIES

Fishery resources are important economic factors in many regions of the United States, and low stream flows resulting from drought conditions can cause serious problems to the fish populations. A few problems are mentioned briefly below.

The combination of low stream flow and very low temperatures in the eastern part of the country increased the ice cover on streams and farm ponds and resulted in fish-kills. A similar result occurred when high temperatures in several streams in Idaho and in the Trinity River in California occurred in 1977. Hydroelectric power generation was suspended at Trinity Dam so that cooler water could be released from lower levels of Clair Engle Lake to preserve the fishery. The level of Lake Tahoe, California, dropped below the outlet channel, and part of the Truckee River dried up. There was a reduction in the fish population locally but no fish-kill, because the fish moved downstream and adjusted to the reduced flow there (California Department of Fish and Game, oral communication, 1978).

So that several races of salmon and steelhead trout would be protected from near extinction, special flow releases and spills were made in 1977 at dams in the Columbia River Basin to augment the low flows and move the juvenile salmonids more rapidly to the ocean. Operations at four reservoirs in northern California were altered to provide the best water temperatures possible and to stabilize the flows during the salmon-spawning seasons.

There are no facilities on many Pacific Coast streams to enhance flow conditions for fish; therefore the number of fish that could successfully spawn was reduced. This condition will affect the fisheries for at least several years (California Department of Water Resources, 1977).

WATER FOR RECREATION

Water-based recreation is a major activity, considering the number of people involved and the economic value to many areas. Skiing, boating, fishing, swimming, and water skiing are pursuits that use water directly, and camps or homes around lakes or along streams are enhanced by the aesthetic values of water.

Skiers and ski resort operators were among the first to feel the effects of the drought. Low snow depths and a short season made most skiing only fair at best. Also, warmer-than-usual weather in the Sierras hampered the production of snow by machines, and resorts with equipment to move snow onto the ski slopes and to pack it did so even though this procedure was costly. The lack of patronage and the short season were enough to cause several resorts to declare bankruptcy.

Boaters and marina operators had drought-related problems during the summer and fall of 1977 when water levels in reservoirs were drawn down to such an extent that marinas were stranded long distances from the water. Temporary expedients were needed to provide launching facilities. Obstacles to boating were exposed in some lakes, and white-water boaters in kayaks or rafts found more rocks showing than usual because of the low stream flows. Scheduled float trips were canceled; and trips by individual parties were about half the number in 1975.

A number of recreational areas in parks and forests were closed to vacationers because of lack of water or because the fire hazard was too great. The ban in northern Minnesota came just prior to the hunting season, and resort owners were faced with a large number of canceled reservations.

WATER FOR NAVIGATION

Major navigation problems did not develop because of drought-induced low flows. Adequate flows were maintained on the Missouri River by releasing water from the main-stem reservoirs. Low flows in 1976 on the Upper Mississippi River were the reason that pleasure-boat operators were requested to reduce their use of locks, because the long time necessary to fill the locks delayed other traffic. Shoaling and dredging on the Lower Mississippi River were very much like numerous other years (Corps of Engineers, oral communication, 1978). However, the combination of lower-than-usual flows in the Mississippi River between St. Louis, Missouri, and Cairo, Illinois, and the severe ice conditions halted navigation for several weeks in January and February 1977.

Two ferries that cross the Missouri River north

of Lewistown, Montana, were taken out of service in June 1977 because of low flow in the river. Usually, any cessation of ferry service does not occur until late fall. Minor navigation difficulties occurred on the Sacramento River in California.

REFERENCES

California Department of Water Resources. 1977. *The continuing California drought.*
Murray, C. R., and E. B. Reeves. 1977. *Estimated use of water in the United States in 1975.* U.S. Geological Survey Circular 765.
Wells, Wade. 1978. Fires producing ''water-proof soils.'' *Sediment Management Newsletter,* no. 3, Spring, California Institute of Technology and Scripps Institution of Oceanography.

19

ACID RAIN

CANADIAN EMBASSY

Acid rain has become a matter of increasing interest and debate. Numerous studies have demonstrated that acid rain is adversely affecting many lakes and rivers in the United States and Canada. There are growing indications that acid rain may be having an impact upon crops and forests as well.

A number of strategies for controlling acid rain have been proposed. The cost and desirability of acid rain control are now being debated in the United States and Canada. In the course of this debate a number of questions about the causes, effects, and controllability of acid rain have been raised. This chapter seeks to provide answers to these questions.

What causes acid rain?

Acid precipitation is rain or snow that contains significant amounts of sulfuric acid or nitric acid. Sulfuric or nitric acid is formed when sulfur dioxide (SO_2) or nitrogen oxide (NO_x) gases emitted by industrial or transportation sources undergo a chemical transformation in the atmosphere.

The U.S. National Commission on Air Quality, in its March 1981 report, found that the process by which man-made pollutants are transformed into acid rain is now "reasonably well known." The SO_2 or NO_x gas released into the atmosphere is first oxidized to sulfate or nitrate particles. If water vapor is present, the particles are further transformed into sulfuric or nitric acid, which contaminates rain, snow, or fog. Otherwise, the sulfate or nitrate particulates can be deposited on the ground in dry form, and later they combine with surface water or ground water to produce sulfuric or nitric acid.

The SO_2 or NO_x emissions can remain aloft for several days. The longer the time aloft, the greater is the quantity of SO_2 or NO_x transformed into particulates or acid. While in the atmosphere SO_2 and NO_x and their transformation products, sulfate and nitrate, can be transported hundreds or even thousands of miles by weather systems.

Man-made pollution is responsible for most of the sulfur dioxide and nitrogen oxides. *Although small quantities of SO_2 are naturally present in the atmosphere, in eastern North America over 90 percent of the sulfur content in the environment is man-made.* According to Environmental Protection Agency (EPA) figures, 65 percent of the man-made SO_2 comes from electric utilities; the remainder is produced by various industrial processes and transportation. Automobiles are the main source of nitrogen oxides (about half), with electric utilities contributing another 30 percent of NO_x emissions.

There is little doubt about the strong link between man-made emissions and acid rain. According to the National Academy of Sciences (NAS) the circumstantial evidence "linking power-plant emissions to the production of acid rain (is) . . . overwhelming."

What effects of acid rain have been observed?

Abnormally high levels of rainfall acidity have been observed in many areas of the United States and in eastern Canada. Normal rainfall is naturally somewhat acidic, due principally to the acid that results from the combination of carbon dioxide and water vapor. While the acidity of natural precipitation varies somewhat, it is normally between pH 5 and pH 5.6 (see Figure 19-1). When rain is abnormally acidic, it also contains unnaturally high quantities of sulfate and nitrate, products of the transformation of sulfur and nitrogen pollutants. This is the kind of high acidity that has been found downwind from industrialized regions.

Rain that is 10 to 40 times as acidic as normal rainfall has been occurring frequently in many parts

This material is taken from a pamphlet published by the Canadian Embassy, Washington, D.C., undated.

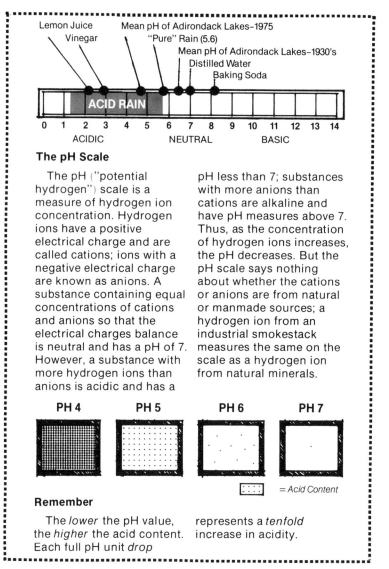

The pH Scale

The pH ("potential hydrogen") scale is a measure of hydrogen ion concentration. Hydrogen ions have a positive electrical charge and are called cations; ions with a negative electrical charge are known as anions. A substance containing equal concentrations of cations and anions so that the electrical charges balance is neutral and has a pH of 7. However, a substance with more hydrogen ions than anions is acidic and has a pH less than 7; substances with more anions than cations are alkaline and have pH measures above 7. Thus, as the concentration of hydrogen ions increases, the pH decreases. But the pH scale says nothing about whether the cations or anions are from natural or manmade sources; a hydrogen ion from an industrial smokestack measures the same on the scale as a hydrogen ion from natural minerals.

Remember

The *lower* the pH value, the *higher* the acid content. Each full pH unit *drop* represents a *tenfold* increase in acidity.

Figure 19-1. The pH scale. (*Source:* Environmental Protection Agency, 1980, *Acid rain*, EPA–600/9–79–036, p. 4. Diagram added by the editors.)

of New England and New York. The average pH of rainfall over substantial portions of the Northeast is 4.3, approximately 10 times normal acidity. Comparable levels of acidity are occurring in many parts of Canada. Highly acidic rainfall has also been observed in a number of southeastern states, particularly Florida, Virginia, Texas and North Carolina, in the Northern Plains states of Minnesota and Wisconsin, and in the Colorado Rockies.

According to estimates by scientists in North America and Europe, many thousands of lakes in the two continents have already been affected by acid rain. There has been extensive damage to aquatic life. In the Adirondack Mountains the New York State Department of Environmental Conservation has found that fish populations are endangered in more than half of all the lakes and ponds in the region, and more than 200 of the lakes have become totally fishless. (When the pH of a lake falls much below 5, most fish cannot survive.) In

Ontario 48 percent of 2600 lakes sampled to date have been identified as very sensitive to acid rain. In Sweden and Norway fish life has been destroyed in more than 6500 lakes. The danger to fish populations is particularly acute during the spring, when acid stored in melting snow causes rapid decreases in pH values, a condition known as *spring shock*.

The National Academy of Sciences concluded in its report issued September 1981 that acid rain is causing "widespread damage" to our aquatic systems, including "higher concentrations of toxic metals." Already, according to the Academy, "several important species of fish and invertebrates have been eliminated" in acidified lakes and streams.

Other than the impact on aquatic life, have any other effects of acid rain been demonstrated?

Recent observations in Europe and the United States indicate that acid rain may be reducing soil fertility and forest growth. A recent German study has shown that acidified water running through the soil tends to pick up increased amounts of aluminum; the aluminum interferes with the absorption rate of water and nutrients by tree roots. Acid waters also tend to remove important nutrients such as calcium and magnesium from the soil. Studies at the University of North Carolina suggest that acid deposition increases the susceptibility of some plants to drought. The National Academy of Sciences report notes that while the effects on soils, forests, and plants have not been proven, "long-term permanent damage to the ecosystem may result" from the leaching out of necessary nutrients.

There is also some indication that acid rain and its precursors could have an indirect impact on human health. This can occur in several ways. In addition to the nutrient minerals, acidification causes leaching of heavy metals such as mercury and cadmium. These metals accumulate in lakes and streams and in the tissues of fish that are present. In sufficient amounts these metals may become toxic. The National Commission on Air Quality took note of studies showing that fish caught in acid waters in the United States, Canada, and Sweden have higher concentrations of mercury than fish taken from waters that are not acidic.

Acidification also can cause leaching of lead and copper from the plumbing that supplies drinking water systems—such an effect was observed recently in Clarion and Indiana counties in western Pennsylvania. In New York the Department of Environmental Conservation has found that ground water in seven upstate counties is sufficiently corrosive to leach copper and lead from pipes and soldered joints, causing the water flowing through the pipes to exceed federal drinking water standards.

Finally, because the sulfate and nitrate particles that make up the dry form of acid deposition are extremely fine, they are capable of causing serious respiratory problems. These are set forth in the "Northeast Damage Report," a paper prepared for a consortium of northeastern states.

The terrestrial and health-related impacts of acid rain have not been studied as extensively as the impact on aquatic systems. The evidence that exists suggests grounds for concern in these areas.

Where is most of the acid rain coming from? Is acid rain principally of local or long-range origin?

Although some limited acidification is due to local sources, data compiled by the National Commission on Air Quality confirm that acid rain is principally an interstate and international problem. The data show that 70 percent of the sulfur deposited in New York and New Jersey each year is attributable to sources outside the region, mostly from sources in Illinois, Indiana, Ohio, and Mid-Atlantic states. On the average, between 75 and 90 percent of sulfate concentrations in any state in the eastern half of the United States originate outside the state.

The data also demonstrate that sources in the Ohio River Valley alone account for 70 percent of the sulfate in central Pennsylvania and 35 percent of the sulfate in the Adirondack Mountains and northeastern New York state. Four states in the Ohio River Valley—Ohio, Indiana, Illinois, and Kentucky—are responsible for nearly one-fourth of all the SO_2 produced in the United States each year. Sources in Ohio alone emit as much SO_2 as New York and the six New England states combined.

Tall stacks constructed by electric utilities and smelters to disperse emissions have compounded the transboundary pollution problem. Prior to 1970 only two stacks exceeding 500 ft were in existence; today in the United States there are more than 175 such stacks, and in Canada there are 25. These stacks were built during the 1970s in order to reduce ambient air pollutant concentrations in the vicinity of the stacks (and thus to comply with the Clean Air Act's ambient air requirements). The net effect has been the export of enormous quantities of air pollution to neighboring states and provinces.

What regions are most affected by acid rain?

There is little question that the regions hardest hit to date by acid rain have been New York, the New

Figure 19-2. Distribution of acid rain in North America; precipitation-weighted, average annual pH for January–December 1981. Light lines are lines of equal pH value, contour interval in 0.2 and 0.5 pH units. Heavy lines are arbitrary boundaries for region: NE = northeastern and midwestern United States and southeastern Canada; W = western United States; SE = southeastern United States. (*Source:* John T. Turk, 1983, *An evaluation of trends in the acidity of precipitation and the related acidification of surface water in North America,* U.S Geological Survey Water Supply Paper 2249. Diagram added by the editors.)

England states, and the provinces of Ontario, Quebec, and Nova Scotia in Canada (see Figure 19-2). These areas lie generally downwind from major SO$_2$ emitters in the Ohio Valley and the industrial regions of the Mid-Atlantic United States and Canada. Aquatic systems in these receptor areas also have poor buffering capacities—the rocks and soils lack natural alkalinity that might otherwise help to offset the damaging consequences of acid rain.

The effects are showing up in other areas as well. In the Northern Plains, particularly Minnesota and Wisconsin, several thousand lakes have been identified as acid-sensitive. In the Southeast, from Florida through Kentucky, rainfall with pH values between 4.6 and 4.2 (five to ten times more acidic

than pure rainfall) has been observed with increasing frequency. In Raleigh, North Carolina, there have been occasions when the rainfall has been more acidic than vinegar.

In the western states a major culprit is NO$_x$, primarily from automobile exhausts. The NO$_x$ from metropolitan centers along the Pacific Coast is being deposited hundreds of miles to the east in the Sierras and Rockies in the form of nitric acid–contaminated rain or snow. Studies published in *Science* magazine show that precipitation with 4.6 pH (at least five times normal acidity) is occurring frequently in parts of Colorado. Mountain lakes in Colorado and California are becoming acidic, with local residents concerned about potentially adverse consequences for the tourism and recreation industries. Metal smelters also play a role in the West; nonferrous smelters accounted for more than 2 million tons of SO_2 emissions in 1980.

To what extent is Canada responsible for causing acid rain, and what is Canada doing about it? How strict are Canadian clean air laws compared with those in the United States?

Approximately 90 percent of the SO_2-based acid rain in the northeastern United States is of domestic origin, and 10 percent is of Canadian origin. Of the acid rain affecting Canada, about 50 percent originates in the United States and 50 percent in Canada. In regions of particular concern in Canada, such as the Muskoka-Haliburton tourist and recreation area in Ontario, as much as 75 percent of acid rain originates in the United States. Overall, according to the National Commission on Air Quality, the United States "exports" about four times as much SO_2 as it "imports" from Canada each year.

Canada thus receives far more acid rain than it exports. But both countries are responsible for generating SO_2 and NO_x emissions that cause acid rain. The problem is a mutual one, requiring cooperative efforts to control it.

The structure of Canadian clean air laws differs from those in this country. Unlike the U.S. Clean Air Act, neither mandatory criteria nor air quality standards are set at the federal level in Canada. The Canadian federal government issues national ambient air quality objectives for the major air contaminants. However, each province establishes emissions levels for sources within its jurisdiction to satisfy provincial ambient air quality criteria. This system has resulted in standards that are generally as strict—and in many cases stricter—than standards under the Clean Air Act. For example, the Canadian ambient air standards for SO_2 (in permissible annual average micrograms per cubic meter) are Ontario, 55; Saskatchewan, 30; Quebec, 55; Alberta, 30; Manitoba, 60. The comparable levels under the U.S. Clean Air Act are 80 micrograms (μg) for the primary standard (health-based), and 60 μg for the secondary standard (welfare-based). Thus for provinces with significant emission levels, Canadian standards for SO_2 are as tough or tougher than either the primary or secondary U.S. standards.

With respect to certain specific sources, Canadian controls are also being tightened. The largest single source of SO_2 on the North American continent is the INCO Ltd. smelter at Sudbury, Ontario. In the mid 1960s the smelter emitted SO_2 at

a rate of about 7000 tons per day. Under Ontario regulations, this had been cut to a current emissions level of 2500 tons per day. A further cut—to 1950 tons per day—took effect in 1983. New sources have even stricter controls; the zinc and copper smelters built by Texas Gulf, Inc., at Timmins, for example, are achieving 97 percent and 95 percent sulfur containment, respectively.

Under a recently promulgated regulation Ontario Hydro, the province's major utility, is required to reduce SO_2 and NO_x emissions 43 percent below current levels by 1990. This will put Ontario Hydro below pre-1970 emission levels, despite a projected increase in power generation of about 30 percent by 1990.

Tight emission standards for major stationary sources in Ontario represent the first significant effort to control air pollution beyond the traditional locally based ambient standards. This approach is specifically designed to address the problems associated with long-range pollution transport—including acid rain.

In the area of mobile sources, however, Canada's requirements are weaker than the U.S. standard. The U.S. Clean Air Act specifies limits of 1.0 grams per mile (g/mi) of NO_x for 1981 gasoline-powered automobiles. In Canada, the comparable standard for NO_x is 3.1 g/mi. Canadian officials readily acknowledge the need to tighten the mobile source NO_x standard, and the standard is currently under review.

Canada is actively participating in the dialogue with the United States as set out in the August 1980 Memorandum of Intent on acid rain. In the Canadian government's view this means above all else pursuing simultaneously twin goals as set out in that document: "to improve scientific understanding of the long-range transport of air pollutants and its effects *and* to develop and implement policies, practices and technologies to combat its impact."

Will acid rain controls make it easier for Canadian utilities to sell electricity to customers in the United States?

Business arrangements between U.S. and Canadian utilities have traditionally encouraged sales and exchanges of electricity between the two countries. Between 1958 and 1978, for example, more than thirty licenses to Canadian utilities for electricity exports to the United States were approved by the National Energy Board.

In recent years a variety of factors has created increased demand in the United States for Canadian electricity. Canadian utilities have not been able to fulfill all requests; the demand has outstripped the supply. Because of the availability of cheap hydroelectric power and the operation of efficient nuclear reactors, Canadian power is very competitively priced. If an acid rain control program results in an average increase of less than 2 percent in U.S. utility rates, as the November 1981 ICF, Incorporated, study indicates, it will have no significant impact on Canadian energy exports.

Canadian utilities have on occasion been requested to supply additional power to several American utilities. For example, Ontario Hydro has responded to a request by General Public Utilities (GPU) to assist in overcoming the power deficits caused by the accident at Three Mile Island.

The supply of this power is subject to a nonappealable regulation issued by the Ontario cabinet. The regulation requires a 43 percent reduction in Ontario Hydro's SO_2 emissions by 1990. Ontario Hydro has assured Canada's National Energy Board that the GPU contract will in no way alter its schedule for complying with the 43 percent overall reduction required by the Ontario government.

Thus the sale of this power by Ontario Hydro in no way conflicts with an effective acid rain control strategy, and the Canadian utility is already on a compliance schedule to meet the reduction targets for the end of the decade.

Hasn't acid rain been around for centuries?

Normal rain and snow are naturally mildly acidic, due mainly to the presence of carbonic acid (combination of CO_2 and water vapor). The pH of natural rain or snow is in the vicinity of 5.0 to 5.6 (pure distilled water, with no acid whatever, would have a pH of 7). The term *acid rain* does not refer to such mild, natural acidity; it is generally used to denote acid levels 10 to 100 times more acidic. It refers to rain that has been contaminated with sulfuric or nitric acid, which is almost entirely of man-made origin.

Ice cores taken from glaciers in Antarctica, which are thousands of years old, show higher levels of acidity. The reason for this is that the natural process of glacier formation tends to concentrate the carbonic acids naturally present in rainfall and snowfall. There is no evidence whatsoever in the glaciers of any sulfur or nitrogen compounds, which are chemical precursors of the acids that are responsible for damaging aquatic systems.

Therefore, although most natural precipitation is slightly acid, it is not true that acid rain has been around for centuries. There is no evidence, either historical or geological, of such wide-ranging and damaging acidity; this is a uniquely modern phenomenon.

Is there any evidence that acid rain is increasing? Why are coal-fired utilities blamed for acid rain, if coal use has remained relatively constant in the United States since the 1940s?

A number of studies prepared for the National Commission on Air Quality conclude that acid rain has been increasing over the eastern half of the United States for the past quarter century.

A Florida study, for example, demonstrated that in that state the concentration of sulfates in precipitation has doubled, and the concentration of nitrates quadrupled, over the past 25 years. Scandinavian studies indicate a strong correlation between concentrations of sulfates and nitrates in precipitation and acidity.

Several of these studies have been challenged by the utility industry. What is important, however, as the National Commission on Air Quality points out, is that *precipitation much more acid than normal continues to be observed over large parts of North America, and currently observed levels are causing increasing numbers of lakes to become acidified.* In effect, the areas exposed to acid rain have been increasing.

With respect to coal, although total use has not increased significantly, seasonal use has changed dramatically. Coal consumption is now much greater in the summertime, due to increased use of air conditioning. Summertime heat and humidity hasten the transformation of SO_2 and NO_x emissions into acid rain. Also, as noted earlier, the construction of tall stacks over the past decade has meant much more diffusion of SO_2 and NO_x emissions into the atmosphere. According to the NAS, the height at which gases are injected into the atmosphere "has increased the probability of oxidation to sulfuric acid." In addition, auto exhausts have increased markedly over the past several decades, and these play a catalytic role in speeding up the oxidizing of SO_2 and NO_x into sulfates and nitrates. Thus even though overall coal consumption has not changed greatly, sulfates and nitrates that result from coal combustion and a variety of other sources have increased significantly.

How can acid rain have differing impacts upon lakes situated within a few miles of each other?

All lakes have different capacities to deal with incoming acid. The buffering capacity of a lake, more than any other factor, determines its ability to neutralize or absorb acidity. Buffering capacity varies with the nature of the underlying rocks, surrounding soils, and vegetation. Lakes with low buffering capacity within regions heavily impacted by acid rain can become acidified within a decade. A lake within the same region can withstand acidic precipitation for a much longer period if the buffering capacity is higher.

In addition, a lake's size and depth, and the rate of flushing capacity, can affect its response to acid loading. Whether a lake is surface water fed or ground water fed also is a major determinant in how it responds to acid rain.

Thus acid rain can induce very different levels of acidity in lakes situated within several miles of each other.

Can the effects of acid rain be neutralized by liming the lakes?

Liming may be a temporary solution if just a few bodies of water were involved. But tens of thousands of lakes are involved, and they are spread over vast areas of North America.

In addition, liming does not necessarily undo existing damage. In a number of lakes where liming has been tried, and fish restocked, the fish have died. *There is evidence that chemical imbalances other than pH occur which render aquatic systems inhospitable for fish for many years even after the excess acid has been neutralized.* As one scientist observed, "If you take an acid lake and lime it, you do not now have a normal lake; you now have a limed, formerly very acid lake, with a very peculiar water chemistry and a very peculiar biota as a result" (Dr. Harold Harvey, University of Toronto).

Also, acid rain may be having deleterious effects on our terrestrial systems. These cannot possibly be handled with liming.

What steps can be taken to reduce acid rain, and what would they cost? Would the benefits be worth the costs?

According to the National Academy of Sciences, "only the control of emissions of sulfur and nitrogen oxides" can significantly reduce the deterioration caused by acid rain.

To implement the NAS recommendation, legislation has been introduced in the Senate and House to impose a 10 million ton annual reduction in SO_2 by 1990. The reductions would be required in the 31 states in the eastern United States and would be allocated on a proportional basis among states in the region. The NO_x emission reductions could be substituted for SO_2 emission reductions on a 2-for-1 basis. The controls would bring about a 43 percent reduction in emissions in the region by 1990.

Technologies to achieve these reductions are available. These include scrubbers to remove in excess of 90 percent of sulfur from stack gases, coal washing, and swapping low-sulfur fuels for high-sulfur fuels. (Coal washing by itself could only achieve 25 percent of the needed reductions.) In addition, new technologies such as fluidized bed combustion methods and limestone injection through multistage burner (LIMB) are under active development.

The Congressional Office of Technology Assessment estimates the cost of a 10-million-ton SO_2 reduction program at $3.4 billion per year. The National Commission on Air Quality arrived at a figure of $2.2 billion a year, based on a 10-year reduction program of 7.6 million tons. The Commission stated that the *emissions reductions can be "achieved at reasonable cost."* The cost estimates vary, depending upon the mix between technologies and coal washing/substitution and whether interstate trading of emissions is permitted.

ICF, Incorporated, in an analysis based on work done for the EPA, was asked to estimate the utility rate increases that would result from a 10-million-ton reduction program. ICF estimated the control program would result in an average utility rate increase of 1.9 percent in the eastern United States by 1990. In states that are particularly dependent upon coal-fired energy sources, the costs would be somewhat higher. According to ICF estimates, rate increases would be 7.4 percent in Indiana, 6.6 percent in Ohio, 5.6 percent in Wisconsin, and 2.8 percent in Illinois.

If new LIMB technology came to be employed, costs could decrease sharply—a recent EPA study suggests that LIMB methods may have as much as an 80 percent cost advantage over other technologies.

Benefits of acid rain reduction are somewhat more difficult to estimate. The 1980 Crocker study at the University of Wyoming, prepared at the direction of the EPA, puts the value of reducing acid rain in the eastern one-third of the United States at $5 billion a year. This figure includes $2 billion in effects on materials, $1.75 billion in damage to forest ecosystems, $1 billion in direct effects on agriculture, $250 million in effects on aquatic ecosystems, and $100 million in other effects, including damage to water supply systems.

How many jobs might be affected if acid rain controls were initiated?

Coal industry witnesses have testified before the Senate Environment and Public Works Committee that 98,600 direct mining jobs would be lost in the Appalachian and Midwest coal regions if the 10-million-ton SO_2 reduction became law.

This estimate is faulty in at least two respects. First, it assumes that the reduction would be achieved entirely by swapping low-sulfur coal with high- and medium-sulfur coal, and it calculates the worker loss impact by considering only the impact upon high- and medium-sulfur coal-producing regions. Even if the assumption were valid, the National Commission on Air Quality points out that the decline of 67 million tons of coal production in northern Appalachia and the Midwest (where most of the high- and medium-sulfur coal is now produced) would be partially offset by an increase of 34 million tons of coal production in the low-sulfur coal regions of central Appalachia. Thus the increased demand for low-sulfur coal would probably generate at least half the number of jobs that might be lost in the higher-sulfur coal-producing regions.

Second, to the extent that the law requires new technology to be used (e.g., "best available control technology"), or to the extent that utilities choose to use new technology to meet statutory emission levels, coal-mining jobs need not be affected, and new manufacturing jobs will actually be created in pollution control industries. Jobs may also be created in research and development, to the extent that the law succeeds in forcing new technologies to be developed.

There should also be taken into account the job loss or dislocation that could result if acid rain is not controlled. Thousands of jobs in Canada, New England, New York, and Colorado are dependent upon the tourist industry. To the extent that tourism suffers if lakes and rivers become fishless, or forest damage occurs, job losses and dislocation could be considerable.

Undoubtedly some job dislocation could result if effective acid rain control legislation is enacted. But *the employment impact on the coal industry is likely to be far more modest than industry figures suggest, and this impact may be partially or totally offset by gains in other sectors of the economy.*

Doesn't the Clean Air Act already provide for reductions in sulfur dioxide emissions?

The principal mechanism for controlling SO_2 emissions in the Clean Air Act is the New Source Performance Standards (NSPS) program. Under NSPS the emission standard for "new" sources (plants constructed after the act's provisions went into effect) is approximately seven times as strict as the standard for existing sources.

The vast majority of the SO_2 emissions are produced by older plants (utilities and industrial sources)—plants that are categorized as "existing"

sources by the Clean Air Act. The difference in standards for old versus new plants gives utilities and industrial sources a strong incentive to keep cheaper-to-operate existing sources on-line as long as possible—in many instances well beyond the time when the old plants would otherwise be retired. Thus the rate of turnover to newer, cleaner plants is very low.

EPA estimates that at the current turnover rate under existing law, SO_2 emissions will remain at the same level—or even increase slightly—between now and the mid 1990s. The EPA does not foresee any significant reduction in total SO_2 emissions before the year 2000 under current law. Moreover, the EPA estimates were made using 30- to 35-year retirement estimates. The emissions increases will be larger if plants are not retired according to this schedule. This makes control of present plants all the more important.

According to the National Academy of Sciences, if current emission rates of SO_2 and NO_x continue, the number of lakes affected by acid rain is expected to "more than double" by 1990. *Controlling only new electrical generating plants, says the NAS, is "insufficient" to accomplish any significant reductions of SO_2 or NO_x emissions in the foreseeable future.*

To what extent are nitrogen oxides responsible for acid rain? Can they be controlled?

Nitrogen oxides are produced both by stationary sources (utilities and industrial processes) and by automobiles. Nitrogen oxides currently are responsible for approximately one-fourth to one-third of the acid rain—but this proportion is expected to increase over the next two decades. In parts of the West, NO_x is already the major contributor to acid rain. *If current trends continue, by 1990 NO_x caused acid rain could equal or exceed the acid rain caused today by SO_2.*

Nitrogen oxide pollution also is associated with the production of ozone. High levels of ozone cause crop damage, forest damage, and a number of respiratory problems. Ozone, like acid rain, is a product of atmospheric chemistry acting on pollutants. It, too, is principally a transboundary pollutant; most of its damage is done outside the state or province where the NO_x originates.

Available technology to control NO_x from stationary sources is less effective than SO_2 removal technology. Technology is available to control automobile NO_x emissions. The 1982 automobile fleet is meeting the current Clean Air Act requirement for NO_x of 1 g/mi. This requirement—if it stays in the law—would provide for a steady reduction in NO_x emissions over the next ten years.

However, the automobile industry is asking that the NO_x standard be relaxed to 2 g/mi. If this occurs, EPA estimates that there would be no decrease in NO_x emissions, and possibly a slight increase, over the next ten years.

How do we know that controlling SO_2 or NO_x emissions will lead to reduced levels of acid rain?

Most scientists who have reviewed the issue believe that a strong correlation exists between SO_2 and NO_x emissions and acid rain. This is certainly the view of the National Academy of Sciences, which believes that man-made emissions of sulfur and nitrogen compounds provide "the only plausible explanation" for acid rain.

It is a law of nature that material is conserved. In this case the amount of sulfur emitted into the atmosphere must be directly related to the amount that comes down, since the atmosphere, on a global scale, does not store SO_2 and its oxidation products. Thus the material that is injected into the atmosphere must come down eventually.

On the average, SO_2 has a life expectancy in the atmosphere of less than five days. The atmospheric sulfur budget for eastern North America, prepared for Environment Canada in 1979, demonstrated that about 70 percent of the total SO_2 emissions in eastern North America are deposited on the continental land mass before the air moves off to the east over the Atlantic.

On a regional scale it is clear that those areas with the most acidic precipitation and the highest concentration of air pollutants are co-located. The hot spot regions where the highest levels of acid precipitation have been observed in North America and Europe are consistently located downwind of major SO_2 and NO_x emissions—in the majority of cases many hundreds of miles downwind. (See the Report of Work Group I of the U.S.–Canadian Work Groups established by the August 1980 Memorandum of Intent, February 1981.)

Thus while a precise relationship between a single SO_2 source and rain or air quality downwind cannot be conclusively established, nor probably will ever be established, *significant reductions in emissions in general must lead to significant reductions in deposition of both acidic precipitation and dry materials.*

Shouldn't we postpone acting on acid rain until the scientific uncertainties are resolved? Isn't more study needed on the causes and effects of acid rain?

Title VII of the Energy Security Act passed by Congress in 1980 created an Interagency Task Force on Acid Precipitation to conduct a 10-year study on the causes and effects of acid rain. Congress authorized $20 million in fiscal year 1982 for the task force.

In authorizing this effort, Congress expected the study primarily to help provide more precise data on the factors that contribute to the production of acid rain. These include the proportionate contribution of SO_2 and NO_x, the relative importance of wet deposition (acid rain or snow) versus dry deposition (sulfate or nitrate particles), and the relative contribution of stationary sources (utilities, smelters) and mobile sources (automobiles, trucks, etc.) to acid rain. The study seeks additional data on the specific roles of weather factors (e.g., sunlight and humidity), atmospheric chemistry and other chemical factors (e.g., vanadium, iron, and manganese).

The existence of these uncertainties, however, should not obscure the fact that a great deal of evidence (both concrete and circumstantial) is currently available about the causes and effects of acid rain. This evidence was analyzed by the National Commission on Air Quality, which issued its report in March 1981 after a four-year study, and by the National Academy of Sciences, which issued its report in September 1981.

Unquestionably, more study will be helpful in resolving the uncertainties about relative contributions to acid rain. But *what we know now is clearly sufficient to begin to remedy the situation.* While there is risk in taking action based on imperfect knowledge, there is much greater risk—environmental, social, and economic—of delay. As the NAS noted, allowing current emissions to continue would be "extremely risky" to parts of the earth's environment.

The responsible course is to do both. We must expand our research efforts, and we must begin now to fashion ways to significantly reduce SO_2 and NO_x emissions into the atmosphere.

SUMMARY: WHAT WE KNOW ABOUT ACID RAIN

- The "broad dispersion of acid rain over large parts of the European and North American continents" represents a "major anthropogenic perturbation" of the environment; that is, it is man-made and it is serious.
- The waters and soils over "extensive areas of North America" are "susceptible to acidification".
- Acid rain has caused the destruction of "many species of fish and their prey." It has also caused toxic trace metals to reach concentrations in surface waters and ground waters that are "undesirable for human consumption".

- Acid rain has led to "severe degradation of many aquatic ecosystems" in the United States, Canada, the United Kingdom, and Scandinavia. "Many thousands of lakes" have been affected.
- There has been an increase in both acidity and toxic substances in many lakes and rivers over the past several decades. This has been observed particularly in New England and southeastern Canada.
- Fish taken from acid waters show high concentrations of mercury and other heavy metals.
- Stone buildings, monuments, and other building materials are eroded by a number of pollutants, including acid rain.
- In the United States and Canada the sources of acid rain are almost entirely man-made.
- The conditions that lead to the formation and long-range transport of acid rain are "reasonably well known."
- The SO_2 and NO_x emissions are transformed in the atmosphere to sulfuric acid and nitric acids, transported great distances, and deposited on vegetation, soils, and surface waters.

- The circumstantial evidence relating power plant emissions to acid rain is "overwhelming."
- Between 75 and 90 percent of sulfate depositions in any state in the eastern half of the United States originate outside the state.
- Provisions of the Clean Air Act that seek to control interstate pollution "have proved ineffective."
- A reduction in SO_2 and NO_x emissions should lessen acid deposition, improve visibility, and help eliminate fine particles that are hazardous when inhaled.
- Only the control of SO_2 and NO_x emissions can significantly reduce the rate of deterioration of sensitive fresh water ecosystems.
- Continued emissions of SO_2 and NO_x at "current or accelerated rates" pose clear evidence of "serious hazard to human health."
- The needed reductions in SO_2 emissions can be achieved "at reasonable cost" (National Commission on Air Quality, 1981) and should result in less than a 2 percent average increase in utility rates.

REFERENCES

Abrams. 1981. *Evidence summary: Sulfates transported into New York State—Impacts and origins.* Submission to the Environmental Protection Agency (Docket No. A–81–09, October 7).

Altschuller and McBean. 1979. *A preliminary overview on the long range transport of air pollutants.* Prepared for the U.S.–Canada Bilateral Research Consultation Group.

Bolin. 1972. *Air pollution across national boundaries.* Sweden's Case Study for the United Nations Conference on the Human Environment.

Bridge and Fairchild. 1981. *Northeast damage report of the long range transport and deposition of air pollutants.* Northeast Regional Task Force on Atmospheric Deposition, Boston.

Cowling. 1981. *A status report on acid precipitation and its ecological consequences.*

Drablos and Tollan. 1980. *Ecological impact of acid precipitation.*

Galloway and Whelpdale. 1979. An atmospheric sulfur budget for eastern North America.

Hutchinson and Havas. 1980. *Effects of acid precipitation on terrestrial ecosystems.*

ICF, Incorporated. 1981. *Cost and coal production effects of reducing electric utility sulfur dioxide emissions.* Prepared for National Wildlife Federation and National Clean Air Coalition.

Likens. 1976. Acid precipitation, *Chemical and Engineering News,* 48.

National Academy of Sciences. 1981. *Atmosphere-biosphere interactions: Toward a better understanding of the ecological consequences of fossil fuel combustion.*

National Commission on Air Quality. 1981. *To breathe clean air.*

Office of Technology Assessment. 1981. *Impact of atmospheric alterations.* Draft report submitted November.

Ulrich, Mayer, and Khanna. 1980. Chemical changes due to acid precipitation in a loess-derived soil in central Europe, *Soil Science,* 193–199.

U.S.–Canada Work Groups. 1981. Established pursuant to August 1980 U.S.–Canada Memorandum of Intent, Report of Work Groups 1, 2, 3A, 3B.

20

GROUND WATER CONTAMINATION IN THE UNITED STATES

VERONICA I. PYE and JOCELYN KELLY

The mobility of our society and the distribution of industry and agriculture depend on an available supply of clean water; nevertheless, instances of ground water contamination have been found in most sections of the country [Kerns, 1977; U.S. Environmental Protection Agency (U.S. EPA), 1980a]. For the purpose of this chapter ground water contamination will be defined as the addition of elements, compounds, and/or pathogens to water that alter its composition.

One of the major difficulties with ground water contamination is that it occurs underground, out of sight. The pollution sources are not easily observed nor are their effects often seen until damage has occurred. There are no obvious warning signals such as fish-kills, discoloration, or stench that typically are early indicators of surface water pollution. Where contamination affects pumping wells, some indications may occur, although many commonly found contaminants are both colorless and odorless and occur in low concentrations. The tangible effects of ground water contamination usually come to light long after the incident causing the contamination has occurred. This long lag time is a major problem.

Ground water can be contaminated by a variety of compounds, both natural and man-made. Contamination due to man has occurred for centuries, but industrialization, urbanization, and increased population have greatly aggravated the problem in some areas (Figure 20-1).

A contaminant usually enters the ground water system from the land surface, percolating down through the aerated soil and unsaturated (vadose) zone. The root zone may extend 2 or 3 ft into the soil, and many reductive and oxidative biological processes take place in this zone that may degrade or biologically change the contaminants. Plant uptake can remove certain heavy metals; microbial fixation and other biological processes can also remove a fraction of the contaminants—the size of the fraction being dependent on the nature of the contaminant.

Deeper below the root zone, which consists mainly of humus and weathered rocks, there is a reduction in such biological processes. Attenuation may occur by surface adsorption, cations in the contaminant being attracted to the negative charge on clay particles. Soils have a cation exchange capacity. Other contaminants may be removed by complexing with insoluble organic matter, which gives rise to complexed humic acids. Microbial action may influence redox potentials and cause the release of inorganic ions during decomposition (Braids, 1981). The susceptibility of different contaminants to differential attenuation varies.

Once in the aquifer, a contaminant will move with the ground water at a rate varying between a fraction of an inch to a few feet per day, forming, under certain idealized conditions, an elliptical plume of contamination with well-defined boundaries. This dispersion process causes a spreading of the solute in a longitudinal flow direction and also transverse to the flow path (Freeze and Cherry, 1979); thus the plume will widen and thicken as it travels. Attenuation of the contaminants in the aquifer may

Veronica I. Pye and Jocelyn Kelly are with the Academy of Natural Sciences, Philadelphia. The project on ground water contamination was part of the program of the Environmental Assessment Council of the Academy.
This material was originally published as "The extent of ground water contamination in the United States," in *Studies in geophysics: Ground water contamination* (Washington, D.C.: National Research Council, 1984), Chapter 1, 23–33. Used with permission of the authors and National Academy Press.

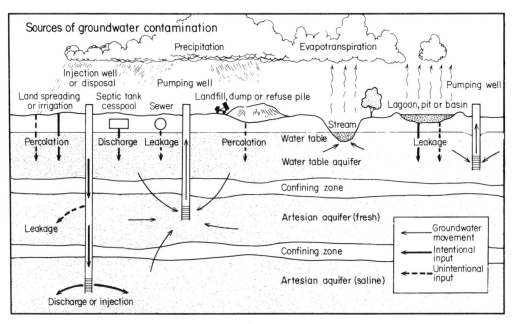

Figure 20-1. Sources of ground water contamination. (*Source:* David W. Miller, 1984, Sources of ground water-pollution, *EPA Journal,* 10 (6):17–19. Diagram added by the editors.)

take place through dilution, volatilization, mechanical filtration, precipitation, buffering, neutralization, and ion exchange. Diffusion and dispersion will bring contaminants into contact with material that may retard their progress; thus attenuation may vary with the time and distance traveled. Unless the plume is blocked, it will usually reach points of ground water discharge such as streams, wetlands, lakes, and tidal waters (Miller, 1981). The shape of the plume will vary according to the continuity and duration of the source of contamination. Dispersion tends to dilute the contaminants; however, concentrations of contaminants are typically much higher in ground water than in surface water (Miller, 1981).

Ground water is a major natural resource in the United States and is often more easily available than surface water. Between 40 and 50 percent of the population depends on ground water as its primary source of drinking water (U.S. Water Resources Council, 1978a; U.S. EPA, 1977). Ground water delivered by community systems supplies 29 percent of the population, and a further 19 percent has its own domestic wells. About 95 percent of the rural population is dependent on ground water for drinking purposes. Approximately 75 percent of American cities derive all or some of their supplies from ground water (Leopold, 1974). The states vary in their dependence on ground water.

Approximately 92 percent of New Mexico's population uses ground water for drinking water, as compared with 30 percent of Maryland's population (U.S. EPA, 1977). West of the Mississippi, in the area where irrigated agriculture is prevalent, the states depend heavily on ground water, whereas the humid eastern portions of the country are less dependent. In 1980 the fresh ground water withdrawals for the United States totaled 88.5 billion gallons per day (bgd), of which 68 percent was used for agricultural irrigation (U.S. Geological Survey Water Information Service, unpublished data).

SOURCES OF GROUND WATER CONTAMINATION

There are three main ways in which the chemical composition of ground water may be changed. The first is due to natural processes. Mineralization can result from leaching, especially in arid areas. Evapotranspiration can further concentrate salts in the remaining water. In the arid Southwest and south central areas of the United States, natural leaching has been identified as the most prevalent source of contamination (Fuhriman and Barton, 1971; Scalf et al., 1973). Chlorides, sulfates, nitrates, fluoride, and iron commonly occur in localized deposits, and their concentration in ground water may exceed U.S. EPA standards. Arsenic and radioactivity from uranium ore also may cause local problems.

The second category of sources of contamination is that due to man's waste disposal practices. The 1977 *Report to Congress* on waste disposal practices and their effects on ground water (U.S. EPA, 1977) provided estimates of the sources and the extent of ground water contamination. At the time the report was compiled, definitive data on waste disposal practices often were not available, and indeed this is frequently the case today. Implementation of the Resource Conservation and Recovery Act of 1976 would be expected to affect waste disposal practices, but such recent data or estimates are not available. Sources of contamination involve all aspects of our lives, including manufacturing and service industries, agriculture, and government. It is estimated that over thirty thousand chemicals are now being used and distributed through the environment and that an additional one thousand are being added each year (Weimar, 1980). Besides the 1977 *Report to Congress,* a report by the Environmental Assessment Council of the Academy of Natural Sciences of Philadelphia (Pye et al., 1983) assessed the various sources of ground water contamination. Contaminant sources from waste disposal practices include individual sewage disposal systems; land disposal of solid wastes; collection, treatment, and disposal of municipal waste water; industrial and other waste water impoundments; land spreading of sludge; brine disposal associated with the petroleum industry; disposal of mine wastes; deep-well disposal of liquid wastes; disposal of animal feedlot waste; and disposal of high- and low-level radioactive wastes resulting from a variety of activities. All of these potential sources of contamination are direct effects resulting in natural and synthetic substances entering the ground water because of human activities (Matthess, 1982).

The third category of sources is also the direct result of human activities but is unrelated to waste disposal practices. It includes accidental spills and leaks, agricultural activities, mining, highway deicing salts, atmospheric contaminants and acid rain, surface water, improperly planned ground water development leading to saltwater intrusion, and improper well construction and maintenance.

Ground water pollution problems and their sources have been the object of numerous studies. An incomplete, but illustrative, listing would include Fuhriman and Barton (1971), Scalf et al. (1973), Miller et al. (1974), van der Leeden et al. (1975), Keeley (1976), Miller et al. (1977), U.S. EPA (1978a, 1978b, 1980a), U.S. General Accounting Office (1978, 1980); U.S. Water Resources Council (1978a, 1978b, 1978c), Jackson (1980), and Pye et al. (1983).

SEVERITY OF GROUND WATER CONTAMINATION

In a determination of the overall magnitude of the national problem, defining severity poses some difficulty. The definition may be approached in several ways.

- If the contaminants in the ground water exceed the interim standards set for drinking water, then the problem could be said to be severe, depending on the extent to which the contamination exceeds the standard, if the intended use is for drinking water. If the ground water was not intended for drinking, then the problem need not be severe. It should be noted that the interim standards do not cover the many synthetic chemicals that can often be found in water. (See proposed rule making for volatile synthetic organic chemicals published in the *Federal Register,* March 4, 1982).
- The number of persons affected by contamination might be taken into account. Thus the contamination of an aquifer in the vicinity of a municipal well field would be of more concern than contamination occurring in an isolated, sparsely populated area.
- In terms of a single aquifer, the severity of contamination may be related to the percentage of the aquifer contaminated by point or non-point sources.
- Nationwide, the severity of the problem may be indicated by the percentage of the available ground water that is affected.
- A different measure of severity might be obtained if the volume of known and suspected contaminated plumes of ground water is expressed as a percentage of the nationwide ground water reserves.
- The degree of hazard posed by the contaminants varies according to the volume discharged, toxicity, persistence, and concentration and would be affected by how the contaminants move in the aquifer.

Thus severity could depend on one or a combination of the following parameters: concentration, persistence, and toxicity of the contaminants; the number of people affected if the contaminated aquifer is a source of drinking water; and the percentage of the available ground water (both locally and regionally) affected by such contamination. Interwoven with each of these factors would be the economic cost of finding an alternative source of water if the contamination renders the ground water unfit for its previous or future uses or if treatment of the water before use is not possible.

Many of the data required for these methods of assessing severity of contamination in quantitative terms simply are not available. The U.S. EPA (1980b) in the appendices to the planning workshops to develop recommendations for a ground water production strategy accurately summarized the sort of information that is available from existing reports and studies, namely documentation of a large number of contamination incidents, identification of important sources of contamination, determination of the mechanisms of contamination, in-depth studies of some contamination incidents, and surveys of the number of certain contamination sources nationwide. The U.S. EPA (1980a) recognized the usefulness of conducting a nationwide survey at randomly selected sites to obtain an estimate of contamination but ruled out this possibility because of the expense of drilling and water sampling and the long time required. Instead, the U.S. EPA adopted a second approach of estimating the number of sources of contamination and the amount of contamination per source. The agency obtained an order-of-magnitude estimate of the extent of the problem and utilized existing information, both qualitative and quantitative. Such assessments serve a useful purpose but only when their inherent flaws are kept in mind. Using information and estimates for only two sources of contamination—landfills and surface impoundments—and whether they are sited over usable aquifers, the length of time they have been operating, and the volume of available ground water in storage, the U.S. EPA (1980b) estimated that between 0.1 and 0.4 percent of the usable "surface" aquifers are contaminated by industrial impoundments and landfill sites. The U.S. EPA cautioned that this is a nationwide estimate and that the two types of disposal sites used are usually found in areas of significant industrial and domestic water use and that the problem could be exacerbated by the area's dependence on ground water. The U.S. EPA, although considering landfills and impoundments to be the most important sources, also evaluated secondary sources such as subsurface disposal systems (septic tanks) and petroleum exploration and mining and concluded that such sources had contaminated about 1 percent and 0.1 percent, respectively, of the nation's usable aquifers. The U.S. EPA (1980b) therefore concluded that at present nearly 1 percent by area of the usable "surface" aquifers in the United States may be contaminated by these four activities and that the areas of contamination will increase with time. It did not include an estimate of the percentage of available ground water contaminated by man's activities unrelated to waste disposal.

Lehr (1982) recently completed another independent estimate, assuming a total of 200,000 point sources. Using faster rates of contaminant spread over a longer period of time, Lehr considered this to be a worst-case estimate. He arrived at a range of between 0.2 and 2 percent. He concluded by stating that no matter how liberal or conservative the estimate, the fraction of polluted ground water is small. This raises the question of how to determine the salience of the problem. Should contamination of 1 to 2 percent of our usable "surface" aquifers be considered a minor or a serious problem? Obviously, for those immediately affected, it is a serious problem. Also, there are many regions where the contamination could exceed 2 percent as the contamination is not uniform geographically. Lehr (1982) optimistically predicted that the initiation of new sources of pollution from waste disposal could be eliminated in ten years by careful siting and facility design and operation and that if these steps are taken, 98 percent of our available ground water could remain unpolluted. However, this does not mean that existing sources will not continue to produce contamination.

Little information is readily available concerning the size of the population affected by well closings due to ground water contamination. There are well-documented cases of well closings in South Brunswick, New Jersey (Geraghty and Miller, Inc., 1979); New Castle County, Delaware (Frick and Shaffer, undated); and Long Island, New York, and California (Council on Environmental Quality, 1981). About 3 million people have been affected by well closings, both municipal and domestic, in these four states alone. Thus at present it is difficult to make useful judgments on the severity of ground water contamination based on the number of people affected.

The types of chemicals that emanate from anthropogenic sources of contamination are varied. They range from simple inorganic ions such as nitrate (from septic tanks, feedlots, and fertilizer use), chlorides (from highway deicing salt, salt water intrusion, certain industrial processes), radioactive materials and heavy metal ions (e.g., chromium from plating works) to complex organic compounds resulting from manufacturing and industrial activities and some of which are found in household cleaning fluids. The chemical composition of wastes deposited in landfills or surface impoundments is often known. Nevertheless, when the constituents of such wastes interact, new compounds may be formed that would not appear on the original waste content inventory. Many industrial waste disposal practices involve stabilization of waste, thereby rendering it less chemically active, but leachate production may still alter some of these chemicals. The degradation of contaminants by microbial organisms in the soils also changes the chemical composition, thus making it almost impossible to predict precisely what contaminants reach the aquifer.

REGIONAL ASSESSMENTS OF GROUND WATER CONTAMINANTS IN THE UNITED STATES

In the 1970s the U.S. EPA commissioned five regional assessment reports. They summarized the geology, major aquifers, natural ground water quality, and major pollution problems for each region. The reports were completed for Arizona, California, Nevada, and Utah (Fuhriman and Barton, 1971), the south central states (Scalf et al., 1973), the Northeast (Miller et al., 1974), the Northwest (van der Leeden et al., 1975), and the Southeast (Miller et al., 1977). Three additional reports were to have covered Alaska and Hawaii, the Great Lakes area, and the north central region, but these were not completed. The findings for the principal sources of contamination in the five regions are shown in Table 20-1. The assessment of the relative importance of the sources of contamination may have changed recently owing to a greater knowledge of the occurrence of toxic organic chemicals in ground water. Nevertheless, at the time the U.S. EPA reports were completed, contamination resulting from natural processes unrelated to man's activities was considered to be of most concern in the arid south central and southwestern states. The extensive irrigation that transformed parts of the Southwest into a major crop-producing area was considered to have caused problems from irrigation return flow. Such problems would involve an increase in nitrate concentrations from fertilizer use, an increase in pesticide content, and leaching of salts. The slow rate of recharge of many of the aquifers in this area, coupled with extensive ground water withdrawals, has led to ground water overdrafting and associated changes in ground water quality and increased pumping costs. The south central region, the most important area for petroleum production, had a major problem with oil field brines contaminating ground water. Because of their sheer numbers and density, individual septic tank systems were considered the main cause for concern in the Northwest, Northeast, and Southeast. Contamination from septic tanks would result in elevated levels of nitrate. More recently, widespread use of septic

tank cleaners containing degreasing agents such as trichloroethylene has resulted in ground water contamination by synthetic organic chemicals (Council on Environmental Quality, 1981). The U.S. EPA assessments further identified the importance in the Northeast and the Southeast of problems associated with industrial development, namely, leakage from buried pipelines, storage tanks, landfills and impoundments. Most of the U.S. EPA assessments were completed before the attendant publicity of Love Canal and the Valley of the Drums had made public the potential threat of such dump sites and thus spurred further investigations. All the reports were completed before the U.S. EPA's Surface Impoundment Assessment (U.S. EPA, 1978a) made available the number of such potential sources of contamination.

Keeley (1976) concluded from the U.S. EPA regional assessments that problems indigenous in one area may not occur in another but that several sources of ground water contamination occur at a high or moderate degree of severity in each area studied. He noted that the four pollutants most commonly reported were chlorides, nitrates, heavy metals, and hydrocarbons but that this may merely be a reflection of the monitoring practices. Sampling for organic chemicals is not routine and is usually expensive, although these chemicals are almost always associated with municipal and industrial wastes. Keeley (1976) also made the point that the rank-ordered problems in the five regions were not selected on the basis of statistical information, as such information was not available. The priorities were established empirically on the basis of the experience of the authorities and individuals who had worked in the five regions.

SUMMARY AND CONCLUSIONS

The summaries of ground water pollution sources completed by the U.S. EPA (Table 20-1) and the Environmental Assessment Council (EAC) (Table 20-2) show that contamination problems from several sources have been reported from all parts of the United States and that the problems vary from one region to another, depending on climate, population density, intensity of industrial and agricultural activities, and the hydrogeology of the region. A comprehensive national survey might well uncover other important sources of contamination or different frequencies of the same sources. Neither the U.S. EPA nor the EAC summaries can be considered complete, as they do not result from comprehensive national surveys. It is difficult to estimate severity from these summaries as there is no established method for doing so. Essentially,

Table 20-1. Sources of Ground Water Pollution Throughout the United States
and Their Prevalence in Each Region

Source	Northeast	Northwest	Southeast	South Central	Southwest
NATURAL POLLUTION					
Mineralization from soluble aquifers				1[a]	1
Water from fault zones, volcanic origin					2
Evapotranspiration of native vegetation				2	2
Aquifer interchange					3
GROUND WATER DEVELOPMENT					
Connate water withdrawal					3
Overpumping/land subsidence				1	4
Underground storage/artificial recharge	4		1	4	
Water wells	4				
Salt water encroachment	3	4	3		1
AGRICULTURAL ACTIVITIES					
Dryland farming		1			
Animal wastes, feedlots		4		3	
Crop residues, dead animals		4			4
Pesticide residues	4	3	2		4
Irrigation return flow		1		2	1
Fertilization	4	3	2		2
MINING ACTIVITIES					
Mining activities	2	2	3		2
WASTE DISPOSAL					
Septic tanks/cesspools	1	1	2	2	1
Land disposal, municipal and industrial wastes			3	2	2
Landfills	1	3	1		
Surface impoundments		2	1	3	
Radioactive waste disposal		3			
Injection wells		2		4	2
Disposal of oil field brines				1	1
MISCELLANEOUS					
Accidental spills	2	3	1	3	2
Urban runoff					3
Highway deicing salts	1	4	4		
Seepage from polluted surface waters	3		4		3
Buried pipelines and storage tanks		1			
Abandoned oil and test wells	1				
Petroleum exploration and development	3		4		

Note: Northeast includes NY, NJ, PA, MD, DE, and New England; Southeast: AL, FL, GA, MS, NC, SC; Northwest: CO, ID, MT, OR, WA, WY; Southwest: AZ, CA, NV, UT; South central: AR, LA, NM, OK, TX. Reports not completed for Great Lakes and north central regions, AK, and HI.

[a] Numbers indicate degree of contamination: 1, high; 2, medium high; 3, medium low; 4, low.

the information contained in the EAC state summaries presents a best-case scenario; the situation can only change as new cases of contamination are discovered.

The important sources of contamination identified in the EAC state summaries (Table 20-2) differ somewhat in order of importance from those identified by the U.S. EPA summaries (Table 20-1), partly because the EAC summarized information from individual states whereas the U.S. EPA did

Table 20-2. Most Frequently Reported Sources of Ground Water Contamination
in the Ten States Reviewed by the Environmental Assessment Council

	Ground Water Use (mgd)[a]	Natural Quality of Ground Water[b]	Most Frequently Reported Sources of Contamination	Total Number of Known Contamination Incidents	% Affecting or Threatening Water Supply	% for Which Remedial Actions Have Been Undertaken
ona	4800	Generally good; mineralization problems	1. Industrial wastes 2. Landfill leachate 3. Human and animal wastes	23	100	26
ornia	13,400–19,000	Good	1. Saltwater intrusion 2. Nitrates from agricultural practices 3. Brines and other industrial and military wastes	Not known	Not known	Not known
ecticut	116	Good to excellent	1. Industrial wastes 2. Petroleum products 3. Human and animal wastes	64	59	30
da	3000	Generally good	1. Chlorides from salt water intrusion and agricultural return flow 2. Industrial wastes 3. Human and animal wastes	92	63	39
▪	5600	Good	1. Human and animal wastes 2. Industrial wastes 3. Radioactive wastes	29	97	45
is	1000	Generally good	1. Human and animal wastes 2. Landfill leachate 3. Industrial wastes	58	76	70
aska	5900	Generally good	1. Irrigation and agriculture 2. Human and animal wastes 3. Industrial wastes	35	34	26
Jersey	790	Generally good	1. Industrial wastes 2. Petroleum products 3. Human and animal wastes	374	50	41
Mexico	1500	Fair to good; mineralization problems	1. Oil field brines 2. Human and animal wastes 3. Mine wastes	105	83	29
Carolina	200	Suitable for most uses	1. Petroleum products 2. Industrial wastes 3. Human and animal wastes	89	74	45

: Pye et al. (1983).
(1981).
er Leeden et al. (1975); Miller et al. (1974, 1977); Fuhriman and Barton (1971); and Scalf et al. (1973).

regional summaries and because the methods of assessment were different. The U.S. EPA summaries are empirical assessments relying on the expertise of professionals who had worked in the regions studied (J. W. Keeley, Kerr Environmental Research Laboratory, personal communication, 1982). Data for the EAC state summaries were based on anecdotal reports of case histories supplied by state agencies, and the categories for types of contaminants are fewer and broader than those in the U.S. EPA surveys.

Some generalizations can be made from the combined results of the two surveys. It is clear that human and sometimes animal wastes are a high-priority source of contamination throughout the country. In the EAC survey human and animal wastes are among the top three contaminants in every state surveyed except California (for which individual cases were not available). Human wastes are ranked as highest-priority contaminants in three of the regions surveyed by the U.S. EPA—Northeast, Northwest, and Southwest. They are of medium-high priority in the southeastern and south central regions. Industrial wastes are also common and high-priority contaminants in most regions of the country. In the Northeast both surveys show industrial waste disposal as the biggest source of ground water contamination. In the southeastern (Florida, South Carolina) and northwestern (Idaho) states surveyed by EAC, industrial wastes are the second most prevalent contaminants of ground water, which concurs with the U.S. EPA regional survey. And in the north central states (Illinois, Nebraska), where agricultural activities are important, industrial wastes are ranked third in reported frequency of contamination. In the southwestern (Arizona, California) and south central (New Mexico) states, disposal of oil field brines accounts for a high percentage of industry-related contamination. Land disposal of general industrial wastes is of secondary importance.

Beyond industrial wastes and human and animal wastes, the sources of contamination vary considerably from one region of the country to another depending on the intensive activities of a particular state (e.g., mining, agriculture, industry) and the geographic and geologic location of the state (some coastal states have severe problems with salt water intrusion into ground water; states located in snowbelts have problems with chloride contamination from road salts; states having soluble aquifers have problems with mineralization). The major sources of contamination are nearly all induced by human activity, apart from naturally poor-quality water caused by dissolution of natural compounds in the strata through which ground water flows.

Although contamination of ground water has occurred throughout the United States, and is likely to continue to some extent in the future, we are still in a position to make choices on how best to use, manage, and protect this valuable resource.

REFERENCES

Braids, O. C. 1981. Behavior of contaminants in the subsurface, in *Seminar on the fundamentals of ground water quality protection.* Presented by Geraghty and Miller, Inc., and American Ecology Services, Inc., Cherry Hill, N.J., October 5–6.

Council on Environmental Quality. 1981. *Contamination of ground water by toxic organic chemicals.* Washington, D.C.: CEQ.

Frick, D., and L. Shaffer. Undated. Assessment of the availability, utilization, and contamination of water resources in New Castle County, Delaware. Department of Public Works, Office of Water and Sewer Management, New Castle County, Delaware (for U.S. EPA Office of Solid Waste Management Programs, Contract No. WA–6–99–2061–J).

Freeze, R. A., and J. A. Cherry. 1979. *Ground water.* Englewood Cliffs, N.J.: Prentice-Hall.

Fuhriman, D. K., and J. R. Barton. 1971. *Ground water pollution in Arizona, California, Nevada, and Utah.* Report #1600ERU for the Office of Research and Monitoring, U.S. EPA.

Geraghty and Miller, Inc. 1979. *Investigations of ground water contamination in South Brunswick Township, N.J.* Syosset, N.Y.: Geraghty and Miller.

Jackson, R. E., ed. 1980. Aquifer contamination and protection, in *Project 8.3 of the international hydrological programme.* Paris: UNESCO.

Keeley, J. W. 1976. Ground water pollution problems in the United States, in *Proceedings of a water research conference, Ground Water Quality, Measurement, Prediction and Protection.*

Kerns, W. R., ed. 1977. Public policy on ground water protection, in *Proceedings of a national conference, April 13–16.* Virginia Polytechnic Institute and State University, Blacksburg, Va.

Lehr, J. H. 1981. Ground water in the eighties. *Water and Engineering Management,* 123(3): 30–33.

———. 1982. How much ground water have we really polluted? (editorial). *Ground Water Monitoring Review* (Winter): 4–5.

Leopold, L. B. 1974. *Water: A primer.* San Francisco: Freeman.

Matthess, G. 1982. *The Properties of ground water.* New York: Wiley.

Miller, D. W. 1981. Basic elements of ground water contamination, in *Seminar on the fundamentals of ground water quality protection.* Cherry Hill, N.J.: Geraghty and Miller, Inc. and American Ecology Services.

Miller, D. W., F. A. DeLuca, and T. L. Tessier. 1974. *Ground water contamination in the northeast states.* EPA-660/2–74–056. Office of Research and Development. Washington, D.C.: U.S. EPA.

Miller, J. C., P. S. Hackenberry, and F. A. DeLuca. 1977. *Ground water pollution problems in the southeastern United States.* EPA-600/3–77–012. Washington, D.C.: U.S. EPA Office of Research and Development.

Pye, V. I., R. Patrick, and J. Quarles. 1983. *Ground water contamination in the United States.* Philadelphia: University of Pennsylvania Press.

Scalf, M. R., J. W. Keeley, and C. J. LaFevers. 1973. *Ground water pollution in the south central states.* EPA-R2-73-268. Washington, D.C.: U.S. EPA.

U.S. Environmental Protection Agency. 1977. Waste disposal practices and their effects on ground water, *Report to Congress.* Washington, D.C.: U.S. EPA.

———. 1978a. *Executive summary: Surface impoundments and their effects on ground water in the United States—A preliminary survey.* EPA–570/9–78–005. Washington, D.C.: U.S. EPA.

———. 1978b. *Surface impoundments and their effects on ground water quality in the U.S.—A preliminary survey.* EPA-570/9–78–004. Washington, D.C.: U.S. EPA.

———. 1980a. Ground water protection, *U.S. Environmental Protection Agency water quality management report.* Washington, D.C.: U.S. EPA.

———. 1980b. *Planning workshops to develop recommendations for a ground water protection strategy.* Appendixes. Washington, D.C.: U.S. EPA.

U.S. General Accounting Office. 1978. *Waste disposal practices—A threat to health and the nation's water supply.* CED-78-120. Washington, D.C.: GAO.

———. 1980. *Ground water overdrafting must be controlled, a report to Congress of the United States by the comptroller general.* CED–80–96. Washington, D.C.: GAO.

U.S. Water Resources Council. 1978a. *The nation's water resources, 1975–2000, second national water assessment.* Vol. 1: *Summary.* Washington, D.C.: WRC.

———. 1978b. *The nation's water resources, 1975–2000, second national water assessment.* Vol. 2: *Water quantity, quality and related land considerations.* Washington, D.C.: WRC.

———. 1978c. *The nation's water resources, 1975–2000.* Vol. 3: *Analytical data survey.* Washington, D.C.: WRC.

van der Leeden, F., L. A. Cerillo, and D. A. Miller. 1975. *Ground water problems in the northwestern United States.* EPA-660/3-75-018. Washington, D.C.: U.S. EPA.

Weimar, R. A. 1980. Prevent ground water contamination before it's too late. *Water and Wastes Engineering,* 30–33: 63.

21

INDUSTRIAL WASTE WATER

UNITED STATES ENVIRONMENTAL PROTECTION AGENCY

More than 300 billion gal of water are withdrawn from our nation's lakes, rivers, and streams each day. Of this quantity 91 percent is devoted to industrial use, an amount of water roughly equal to 75 percent of the daily flow of the Mississippi River at its mouth.

While some water is evaporated, or is incorporated into the product itself, most is discharged back to its source, Figure 21-1. The U.S. Department of Commerce estimates that major industrial water users discharged approximately 285 billion gal of waste water daily in 1975.

This water, usually altered considerably in the industrial process, may contain contaminants that degrade water quality and pose a threat to human health. Degradation of water quality comes about with the addition of large amounts of nutrients, suspended sediments, bacteria, and oxygen-demanding matter. The possible addition of toxic pollutants is even more serious. These pollutants are particularly important because of their persistence, harmful effects at low concentrations, and ability to enter the food chain.

Past water use and discharge practices have levied their toll on the nation's water supply. A particularly severe example, illustrating the potential hazards associated with uncontrolled discharge, occurred in South Charleston, West Virginia, in 1977. A 5000-lb discharge of carbon tetrachloride into an Ohio River tributary contaminated the river for more than 600 mi—from Gallipolis, Ohio, to Paducah, Kentucky. Since carbon tetrachloride is a carcinogen and can cause damage to the liver, kidneys, lungs, and central nervous system, this contamination was a matter of great concern. The Environmental Protection Agency (EPA), through its sampling and analysis program, was able to detect the contamination and notify cities on the Ohio River to close their water treatment plant intake gates, thereby protecting drinking water supplies along the river's course.

The effects of past use and discharge practices are increasingly in evidence—from the contamination of New York's Hudson River by polychlorinated biphenyls (PCBs), to the presence of Kepone in the James River in Virginia. Such incidents are particularly alarming in light of documented evidence that some organisms can ingest, accumulate, and bioconcentrate toxicants such as these to lethal levels.

REGULATION OF INDUSTRIAL WASTE WATER

In recognition of the need for industrial waste water control measures, Congress enacted the 1972 Federal Water Pollution Control Act Amendments identifying several major environmental goals for the nation. The elimination of pollutant discharges or "zero discharge" into navigable waters by 1985 highlighted the act, with the preservation of water quality providing for fishable and swimmable waters by 1983 as an interim goal.

To meet these goals with respect to industrial waste water, the act mandates the establishment and imposition of discharge limitations based on protection of receiving water quality, toxicity, and technological practicability. The latter regulatory effort requires the definition and achievement of best practicable control technology currently available (BPCTCA) and best available technology economically achievable (BATEA). Best practicable control technology levels requires industries to use the best broadly demonstrated technology available, while the best available technology levels require the use of the best technology available.

This material is excerpted from *Research summary—Industrial waste water*, EPA–600/8–80–026 (Washington, D.C.: U.S. Environmental Protection Agency, 1980), 1–25.

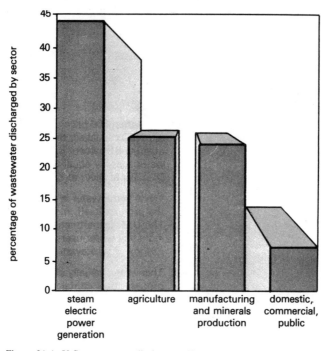

Figure 21-1. U.S. wastewater discharges. (*Source:* U.S. Department of Commerce.)

NRDC Consent Decree

From 1972 to 1976 EPA concentrated on developing best practicable control technology and best available technology limitations for conventional pollutants such as suspended solids, biochemical oxygen demand (BOD), and chemical oxygen demand (COD). The BOD is a measure of the oxygen required to biologically decompose organic matter in water, while the COD is a measure of the oxygen required to chemically oxidize both organic and oxidizable inorganic compounds in water. As such, both BOD and COD are used to determine the degree of pollution in an effluent.

In June 1976 in settlement of a suit with the National Resources Defense Council (NRDC), EPA agreed to devote more attention to potentially toxic substances in industrial waste water. The resulting NRDC Consent Decree required EPA to promulgate regulations for 65 classes of toxic pollutants associated with 21 industrial categories—updated in 1979 to 34 industrial categories (Table 21-1). The 65 classes represent 129 specific substances referred to as "consent decree priority pollutants" or simply "priority pollutants." The Clean Water Act of 1977 (PL 95–217), which further amends the Federal Water Pollution Control Act, incorpo-

rates substantial portions of the NRDC settlement and broadens regulations to improve water quality and the control of potentially toxic pollutants.

Other federal legislation related to industrial waste water treatment and control includes the National Environmental Policy Act, the Toxic Substances Control Act, the Ocean Dumping Act, the Safe Drinking Water Act, the Resource Conservation and Recovery Act, and the Clean Air Act.

Regulatory Responsibility

Federal responsibility for promulgating and enforcing industrial waste water control regulations is held by EPA. The National Pollution Discharge Elimination System (NPDES), a national permit program administered through EPA, was created under the Federal Water Pollution Control Act Amendments of 1972 to control the discharge of pollutants into waterways from all point sources including industrial, municipal, and commercial facilities. Under the law it is illegal to discharge pollutants into the nation's waterways without a permit. Through this system EPA regulates what may be discharged and in what quantity by imposing the discharge limitations described previously.

Table 21-1. NRDC Consent Decree for 34 Industrial Categories and 65 Toxic Pollutant Classes

INDUSTRIES WHOSE WASTE WATER DISCHARGE IS REGULATED

Adhesives	Plastics processing	Pesticides
Leather tanning	Porcelain enamel	Pharmaceuticals
and finishing	Gum and wood chemicals	Plastic and synthetic
Soaps and detergents	Paint and ink	materials
Aluminum forming	Printing and publishing	Rubber
Battery manufacturing	Pulp and paper	Auto and other laundries
Coil coating	Textile mills	Mechanical products
Copper forming	Timber	Electric and electronic
Electroplating	Coal mining	components
Foundries	Ore mining	Explosives manufacturing
Iron and steel	Petroleum refining	Inorganic chemicals
Nonferrous metals	Steam electric	
Photographic supplies	Organic chemicals	

PRIORITY INDUSTRIAL WASTE WATER POLLUTANTS

Acenapthene	DDT and metabolites	Nitrobenzene
Acrolein	Dichlorobenzenes	Nitrophenols
Acrylonitrile	Dichlorobenzidine	Nitrosamines
Aldrin/dieldrin	Dichloroethylenes	Pentachlorophenol
Antimony and compounds	2, 4-Dimethylphenol	Phenol
Arsenic and compounds	Dinitrotoluene	Phthalate esters
Asbestos	Diphenylhydrazine	Polychlorinated biphenyls
Benzene	Endosulfan and metabolites	(PCBs)
Benzidine	Endrin and metabolites	Polynuclear aromatic
Beryllium and compounds	Ethylbenzene	hydrocarbons
Cadmium and compounds	Fluoranthene	Selenium and compounds
Carbon tetrachloride	Haloethers	Silver and compounds
Chlordane	Halomethanes	2,3,7,8,-Tetrachlorodibenzo-
Chlorinated benzenes	Heptachlor and metabolites	p-dioxin (TCDD)
Chlorinated ethanes	Hexachlorobutadiene	Tetrachloroethylene
Chloralkyl ethers	Hexachlorocyclopentadiene	Thallium and compounds
Chlorinated phenols	Hexachlorocyclohexane	Toluene
Chloroform	Isophorone	Toxaphene
2-Chlorophenol	Lead and compounds	Trichloroethylene
Chromium and compounds	Mercury and compounds	Vinyl chloride
Copper and compounds	Naphthalene	Zinc and compounds
Cyanides	Nickel and compounds	

The Research Program

EPA's Office of Research and Development (ORD) supports the agency's regulatory activities by producing the scientific data and technology necessary for development of effective pollution control strategies and environmental standards. This research falls under two major categories:

- Treatment.
- Reuse and recycling and other process modifications.

Researchers at ORD's Industrial Environmental Research Laboratories in Cincinnati, Ohio, and Research Triangle Park, North Carolina, and the Robert S. Kerr Environmental Research Laboratory in Ada, Oklahoma, perform EPA's in-house research. The majority of the industrial waste water control research, however, is performed extramurally through contracts, grants, and cooperative agreements with universities, research foundations, trade associations, professional societies, and industrial companies.

WASTE WATER TREATMENT TECHNOLOGIES

Treatment technologies reduce or eliminate industrial waste water pollutants, thereby producing an effluent that either can be discharged into a nearby waterway without threatening environmental quality or can be accepted by a local municipal treatment plant without disturbing treatment processes.

In March 1979 the EPA opened a waste water Test and Evaluation Facility in Cincinnati, Ohio. The facility both increases EPA's capacity for in-house research on pollution control technologies and enhances its capability to evaluate the health and environmental impacts of these controls.

Through a cooperative agreement with EPA, the city of Cincinnati is providing the land for the facility for 20 years at no cost. A sewage treatment plant, located adjacent to the facility, provides the industrial and municipal waste waters and sludges needed for research efforts. The Test and Evaluation Facility is especially suited for research on techniques to remove or treat toxic and hazardous materials in industrial waste waters.

Three technologies for removal of toxic materials from waste water currently being tested are carbon adsorption, activated sludge, and steam stripping. Carbon adsorption removes toxic organic compounds through adsorption, or attraction and accumulation, onto the surface of activated carbon. Activated sludge processes essentially duplicate natural stream purification mechanisms through biological degradation of water pollutants by bacteria and other microorganisms. The principal difference is that treatment is carried out in a controlled environment with high concentrations of bacteria and microorganisms, thereby speeding up and increasing the efficiency of the process. Steam stripping involves the removal of certain volatile organic pollutants from waste water through distillation. Limited information exists concerning the efficiency of carbon adsorption, biological treatment, and steam stripping in removing priority pollutants from industrial plant waste water systems. Research, development, and demonstration of these treatment technologies will answer important questions, such as: Can these technologies be made more energy efficient and cost effective? And can they be broadly applied?

Due to the wide variability in industrial waste water discharges, waste water treatment research must be performed on an industry-by-industry basis.

Organic Chemicals

The waste water produced during the manufacture of synthetic detergent bases, fuel additives, solvents, plastics, resins, and synthetic fiber bases is frequently contaminated with toxic substances that require specialized treatment. Both activated carbon treatment and wet air oxidation treatment are receiving particular attention. As stated earlier, activated carbon is a highly adsorbent form of carbon to which specific pollutants adhere. After

pollutants are removed from water, the carbon can be changed or cleaned and used again.

A second process, wet air oxidation, involves exposing toxic organic waste waters to high temperatures and pressures while they are confined in a reactor. These conditions cause the oxidation and conversion of toxic organic substances into nontoxic forms. The resulting effluent is suitable for discharge directly to a waterway or to a nearby municipal waste water treatment facility.

Petrochemicals

The petrochemicals industry produces waste waters rich in chemicals that can be extremely hazardous. One chemical that has an especially deleterious ecological impact is caprolactam, used in the manufacture of nylon. Biological treatment of waste streams containing caprolactam has not been successful. Steam stripping, while somewhat more successful, requires large amounts of energy and is costly to apply. In order to develop a more inexpensive, energy-conserving technology to treat chemical waste streams, ORD has been examining the use of a new solvent to extract caprolactam from industrial waste waters. After removal, the caprolactam can be extracted from the solvent and recovered for reuse. Research is also being conducted on the use of powdered activated carbon in combination with biological treatment to treat petrochemical industry waste waters.

Petroleum Refining

The petroleum refining industry, one of the five largest industrial waste water dischargers in the nation, requires large volumes of water to produce its products. Sour water, which results from some of its refining processes, contains high concentrations of sulfur and cyanide compounds and ammonia. Treatment techniques employ a form of steam stripping known as *sour water stripping*. This process offers the benefit of recovering ammonia and sulfur, which can later be sold. Unfortunately, the sour water–stripping process has not been as efficient in removing high concentrations of ammonia as predicted. Apparently, other constituents in the sour water are preventing removal of the ammonia.

ORD, along with the American Petroleum Institute, is conducting studies to determine the feasibility of using waste water from several industrial stripping operations to determine what constituents in sour water prevent efficient stripping. The results of this project will assist engineers in predicting the concentrations of sour water constituents that can be stripped and in designing more effective stripping systems.

Research efforts in the petroleum refining industry are also being directed at the catalytic cracking process, a major source of sour water. Catalytic cracking involves splitting large crude oil molecules into smaller molecules.

EPA, in cooperation with the Oil Refiners Waste Control Council and Oklahoma State University, is characterizing and treating effluents from catalytic cracking operations. Through intensive treatment of the effluents from these operations, the cost of end-of-pipe treatment for waste waters from the entire petroleum refining process can be reduced.

Pilot plant testing of treatment methods is underway at an operating petroleum refinery that is currently using biological treatment. If these treatment methods prove to be effective in treating waste waters from catalytic cracking operations, then recycle/reuse techniques may be applied throughout the petroleum refining industry.

Pesticides

The manufacture of pesticides involves substantial production of toxic chemicals. In conjunction with treatability research, the EPA is developing a guideline document to review best available technology to support 1984 effluent limitations and standards. The document will concentrate on identifying pesticides industries and evaluating waste water control and treatment technologies. In support of this effort, the Environmental Monitoring Systems Laboratory in Research Triangle Park (EMSL–RTP) is developing analytical procedures for measuring pesticide and priority pollutant levels in waste waters.

Inorganic Chemicals

The inorganic chemical industry comprises about 1600 plants, produces 110 million metric tons of products annually, and generates 40 million metric tons of wastes per year. Large quantities of water are used for cooling, processing, product washes, waste transport, and other production purposes. Resulting waste waters contain heavy metals and cyanide, suspended solids, fluoride, iron, ammonia, and have a high chemical oxygen demand. Although many plants operate sophisticated treatment systems capable of producing high-quality effluents, the inorganic chemical industry continues to have significant water pollution problems, which are difficult and expensive to solve.

Projects in support of research and development in the inorganic chemical industry are conducted at IERL-Cincinnati. Researchers are currently assessing the industry to identify research and development needs for solving major air, water, and land pollution problems.

Battery Manufacturing

Low concentrations of toxic metals such as lead, arsenic, cadmium, and antimony are often contained in battery-manufacturing waste water discharges. Waste water treatment technology is expensive, both in terms of high costs and large land requirements. Since most battery-manufacturing operations are inside cities, neither land nor money to purchase high-priced land is available for building large treatment facilities.

Metal Finishing

Metal-finishing operations daily produce more than 1 billion gal of waste waters containing toxic heavy metals and cyanide. Metal-finishing processes add a protective metal coating or plating to metal surfaces such as precision instruments and tools or to nonmetallic surfaces such as plastics.

An object is plated by being submerged first in a plating tank and then in a series of rinse tanks. The concentration of heavy metals declines in each consecutive bath until the product is essentially clean. As the concentration of metals increases in the rinse baths after extended use, the bath water must be discharged and replaced.

Over the past several years ORD has assisted in developing several methods for the cost-effective treatment of waste rinse waters from metal-finishing plants. The most promising methods involve membrane and electrochemical techniques and centralized waste treatment.

Membrane techniques concentrate rinse water pollutants on the membrane through which the rinse water is passing while simultaneously generating an effluent stream that is relative pollutant-free. The effluent can then be discharged or reused in the process, while the concentrated pollutants can be returned to the plating bath or treated by chemical means. Major benefits of membrane technologies are that they recover reagents used in the process, do not require the addition of treatment chemicals, do not generate sludge, and are low in energy consumption.

An electrochemical reactor at an operating metal-plating plant removes heavy metals and cyanide from the rinse tank water through electrical attraction, thereby preventing accumulation. In the process of removing toxic pollutants from waste water, heavy metals may also be recovered for possible reuse.

Although inexpensive treatment technologies for metal-finishing waste waters are being developed, the total cost of pollution control will be high because of the enormous number of metal-finishing plants in the United States. It is estimated that there are currently about twenty thousand electroplating plants in operation. Although some plants will not suffer financially from implementing waste water treatment technologies, EPA predicts that some plants in the metal-finishing industry may be forced to close because of the high cost of installing and operating treatment systems. Clearly, the need exists for an alternative to individual-plant on-site treatment that will remedy waste water problems without debilitating a significant portion of the industry.

One method is centralized waste treatment. Centralized treatment of wastes is possible where industries producing similar types of waste are located in close enough proximity to allow the economical transportation of waste materials. Either wastes can be treated at one centralized location, or individual plants can exchange wastes and treat those for which they are best technologically suited. The Federal Republic of Germany has successfully used this concept of industrial wastes treatment for many years, and the Office of Research and Development is examining its feasibility in the United States.

Iron and Steel

Water pollution is among the major environmental problems associated with the iron and steel industry. The largest single industry in the United States, it has an annual production capacity of approximately 150 million metric tons. Waste waters from steel plants contain such pollutants as suspended and dissolved solids, oils and greases, phenols, cyanides, ammonia, sulfide, and have a high biochemical oxygen demand. Initial studies using mobile waste water treatment systems have focused on coke plant and blast furnace waste waters, two of the most contaminated waste waters found in steel plants.

Steam Electric Power

Large quantities of water are used by the steam electric power industry, but most is used for steam generation, or cooling, and is returned to its source containing few pollutants. However, boiler-cleaning operations at these power plants result in waste water containing toxic pollutants. During operation corrosion products accumulate in the boiler tubing and, if uncontrolled, cause the power plant to become less efficient. These corrosion products contain heavy metals such as iron, cooper, zinc, and nickel. Strong chemicals are used to dissolve the corrosion products in the cleaning process. As a result, the heavy metals previously contained in the boiler tubing are put into a solution that ultimately becomes a toxic industrial discharge.

The Industrial Environment Research Laboratory in Research Triangle Park (IERL–RTP), in cooperation with the Utilities Water Act Group, which represents the steam electric power industry, is conducting numerous laboratory studies to assess the effectiveness of lime treatment in precipitating and removing heavy metals in boiler-cleaning solutions. Boiler waste waters from six steam electric power plants using different boiler tube–cleaning methods are presently being treated and analyzed. The results of this project will demonstrate which boiler-cleaning methods used in conjunction with lime treatment can achieve the highest-quality waste water and will provide EPA with the data on which to base effective effluent limitations and standards.

Textiles

Waste water from textile mills presents an environmental problem in several regions of the United States due to the large volume and the widely varying and specialized character of the wastes. Textile-processing plants utilize a variety of dyes and chemicals such as acids, bases, salts, detergents, and finishes. Many of these are not retained in the final textile product but are discarded in waste waters after they have served their purpose.

Leather Processing

Water is used extensively to process leather in cleaning, tanning, and dyeing operations. Cleaning is required to remove flesh and fat from the inside of animal skins. Tanning involves soaking the skins in chemicals to produce a flexible, long-lasting product. And dying involves numerous chemicals to color the hides. Consequently, the process effluents are high in BOD and suspended solids, rich in process chemicals, and noticeably colored.

Waste water from leather processing is treated to settle suspended solids. The resulting water does not, however, meet effluent guidelines. In 1976 EPA initiated a full-scale demonstration of the use of an oxidation ditch in which many of the pollutants in waste water can be biologically degraded by bacteria and other microorganisms. After extended full-scale operation, it has been shown that this treatment method is successful in producing water that meets effluent limitations. This treatment technique represents not only the best available technology for the leather-tanning industry but the most cost-effective method available as well.

Paper

More than 7 billion gal of waste water are discharged daily by pulp and paper mills across the country. Soaplike resins and the fatty acids of pulp and paper manufacturing effluents are suspected of contributing to foam problems on rivers, streams, and lakes receiving mill effluents. In Maine the threat to the aesthetic quality of the Androscoggin River due to unsightly mounds of foam was first brought to the public's attention by local duck hunters. Four pulp and paper mills occupying sites along the river were the target of complaints by area residents. However, since municipalities and other industries line the river's banks, and since the river has its own natural foam, the blame could not be fairly placed entirely on the pulp and paper industry.

EPA, in cooperation with Maine's Department of Environmental Protection, recently initiated a study to look into ways of eliminating foam from pulp and paper mill effluents. Three river surveys are being conducted on the Androscoggin River to determine which chemicals are causing the foams and to develop foam removal techniques and mill effluent treatment systems.

Treatment technologies involving the elimination of foam through recovery of resin and fatty acids, technologies currently employed in southern mills, are a prime consideration for use in northern manufacturing plants. Foam separation and chemically assisted treatment techniques are also being evaluated in this study.

Environmental Fate Studies

The fate of priority pollutants in the biological treatment of waste waters from the wood-preserving, pharmaceuticals, petrochemicals, petroleum refining, pesticides, and rubber industries is the focus of much research at ORD's Robert S. Kerr Environmental Research Laboratory in Ada, Oklahoma. A major objective of this research is to determine whether the priority pollutants that are removed from waste water in biological treatment are degraded to an innocuous state or if they are transferred to another medium, such as the air, or to sludge, where they create new pollution problems.

Observations from sampling and analysis at industrial sites have shown that biological treatment is generally very successful in removing priority pollutants from waste waters. However, some priority pollutants are removed from the waste water but appear in significant quantities in air emissions and sludges. Compounds not originally present in the untreated waste water are also appearing in air emissions and sludges. Apparently,

additional compounds are generated in the biological treatment of industrial waste waters.

REUSE, RECYCLE, AND PROCESS MODIFICATIONS

Water use practices in industry have undergone radical changes in recent years. Whereas early water pollution control measures focused on end-of-pipe treatment to reduce the hazardous potential of industrial discharges, recent legislation, which has a goal of zero pollution discharge by 1985, has encouraged recycle/reuse alternatives. As shrinking fresh water supplies become an increasing problem, recycle/reuse systems are gaining attention as water-conserving techniques. In California, for example, repeated cycles of drought and flooding have prompted the state to enact a law allowing allocations of fresh water only if manufacturing facilities can prove they cannot operate on, or cannot locate, used water. Industrial waste water recycle/reuse systems offer the additional advantage of allowing reagent and by-product recovery and reuse.

Major benefits of recycle/reuse systems include reduction of waste water volume; reduction of intake water and related harmful effects to aquatic life; improved treatment efficiency; conservation of water, raw materials, and other natural resources; and containment of conventional and toxic pollutants. Water reuse can bring about intake reductions by reusing the effluent of one process or plant in the same, or another, process or plant. Treatment of each effluent must be tailored to its contaminants and its next use. Effluent from another process may be used directly in processes that do not require pure water. In contrast to water reuse, water is recycled in a closed loop and is continuously applied to one production process. Such a cycle usually includes a treatment system to remove contaminants from the process water to a degree suitable for reuse.

Continued research on waste water treatment technology is essential because treatment will not only remain a major method for pollution control for many years to come but is also an inherent part of any recycle/reuse system. Industry involvement in EPA's program is encouraged for the benefit of the program and for the benefit of industry as well. The fact that industrial water intake has been falling dramatically while gross water use has been steadily rising is indicative of this increasing industrial awareness of recycle/reuse technologies.

Recycling in Steam Electric Power Plants

Recycle/reuse efforts are gaining attention in the steam electric power industry. Water-cooled power plants are well known for their abundant water use.

The water coming into the plant must first be treated to ensure that suspended and dissolved solids will not settle, clog the cooling system, and reduce its efficiency. In a coal-fired power plant, fly ash produced in the combustion process is sluiced or flushed from the system. As a result, the treated water regains dissolved solids from the fly ash as it passes through the cooling system. In the past, water involved in the cooling process was used on a once-through basis to prevent the buildup of fly ash solids and subsequent cooling system clogging.

Closed-Loop Fiberglass Production

Large volumes of water are used in the production of fiberglass, particularly in the cooling process. To melt the glass and form the fibers, high temperatures are required. Once the fibers are formed, they are cooled with water. One fiberglass textile industry plant, through an EPA grant, has implemented a closed-loop system for the reuse of process water. Prior to the initiation of this project in 1973, approximately 350 gal of water per minute were discharged from the plant following primary and biological treatment. So that water of suitable quality for reuse could be obtained, a three-stage advanced treatment system was added to the facility. The first stage, sand filtration, removes biological solids; the second stage, carbon adsorption, removes organic chemicals; and finally, chlorination provides adequate disinfection. Under present operating conditions, an 80 percent discharge reduction to approximately 70 gal of water per minute has been achieved. It is expected that zero waste water discharge will be achievable in the very near future.

When the system is complete, the only intake water required by the industry will be that needed to compensate for evaporation and other small losses. This intake water should amount to approximately 130 gal/min—a considerable reduction from the 1000 gal/min requirements of many fiberglass plants prior to implementation of recycle/reuse programs.

Closed-Cycle Dyeing

In textile operations, dyes and chemicals used to set dyes are typically diluted with effluent from the rest of the plant. The effluent is then treated and discharged. Since pure water for dyeing is necessary for obtaining high-quality colored fabrics, dye water is typically used on a once-through basis. As a result, industries must make large expenditures to treat and remove the color from water that they never use again.

In an effort to eliminate dye water discharges from textile plants, reduce intake water needs, recover materials, conserve energy, and save money, EPA has funded a major project to evaluate and demonstrate closed-cycle dyeing. The technology that is applied is called *reverse osmosis* or *hyperfiltration*.

Reverse osmosis is a method of reversing nature's osmosis process in which a dilute or less concentrated solution passes spontaneously through a semiporous membrane into a more concentrated solution. In reverse osmosis sufficient pressure is applied to the concentrated solution to reverse the flow through the membrane. Membrane filters are designed to prevent molecules or dissolved solids larger than a designated size from passing through. In this way the quality of the recycled water is strictly controlled.

Nontreatment Alternatives

For those industries at which recycle/reuse is not possible, EPA is examining nontreatment alternatives. The theory behind nontreatment is to modify the industrial process so that there is no pollution produced in the first place. Modifications include process changes, such as water reduction techniques, and material substitutions, which involve using less polluting materials in manufacturing. These methods of industrial waste water control may prove to be more efficient and economical than end-of-pipe waste water treatment.

Activated Carbon Regeneration

One million gallons or more of waste waters saturated with pesticides are produced daily at many pesticide manufacturing plants. Biological treatment alone often does not successfully remove toxic organics from waste water, and in many instances the pesticides are poisonous to microorganisms used in the treatment process. For this reason activated carbon treatment has been applied extensively in the pesticides industry. When the carbon becomes ''loaded'' through continuous adsorption of soluble components in waste water, it is either discarded or regenerated. Due to the high cost of activated carbon, many industries have implemented regeneration technologies. Thermal regeneration is frequently employed. However, this process is often not considered to be a desirable alternative because it is expensive, highly energy consuming, and is not applicable for many types of waste water.

EPA, in cooperation with private industry, is developing and laboratory-testing the use of high-pressure carbon dioxide, or supercritical fluid CO_2 for regenerating activated carbon in an inexpensive and energy-conserving manner. As supercritical fluid CO_2 passes through the activated carbon, it removes the adsorbed pesticide components by putting them into solution. The pesticide components are later separated out of solution and, with further purification, can be sold as dry pesticides or recycled back into the process. In this way resource recovery is also made possible. The success of this technology has inspired future ORD research. Plans to initiate programs involving pilot and full-scale demonstrations of supercritical fluid CO_2 at pesticide-manufacturing plants are under consideration.

Water Reduction in Food Processing

The development of techniques to reduce the amount of water used in food processing has been a major focus of ORD research for several years. Dry peeling, recently initiated in potato processing, represents an important breakthrough for the food-processing industry. Prior to dry peeling, large volumes of water directed at high pressure forced the peels off the potatoes. In the dry peeling process potatoes and other fruits and vegetables are pre-treated to soften the skin and then moved over rubber rollers, which wear off the peels. In 1969 EPA funded the first commercial plant in the world, in North Dakota, to implement full-scale dry peeling. The success of that project has encouraged the widespread use of this technique. Currently, about two hundred fifty food-processing plants around the world use dry peeling in the preparation of beets, tomatoes, pears, peaches, carrots, and so on. Because food-processing wastes contain such high concentrations of organic matter, water pollution problems can also be reduced through water reduction processes.

FUTURE RESEARCH

Over the next five years ORD's research will continue in three major areas: determining and analyz-

ing industrial waste water sources, evaluating and developing control methods, and developing recycle/reuse alternatives.

Any meaningful pollution control strategy requires the ability to determine which pollutants are, and will be, created, as well as the ability to accurately measure those pollutants. Analyses will be performed to determine probable future pollutant problems and their environmental impacts. So that those impacts can be assessed, dependable methods of determining the presence and concentration of pollutants will be developed. One such method calls for the determination of indicator organisms or substances that have characteristics similar to problem pollutants and are usually found in the same contaminated water but are easier to detect.

A second major focus of future ORD research will be the evaluation and development of pollutant control techniques. Continued research will be performed on the treatability of specific pollutants by a wide range of conventional biological, physical, and chemical methods. Since individual pollutants increasingly disrupt, or are incompatible with, traditional treatment processes, more emphasis will be placed on developing new technologies or non-treatment alternatives to meet changing pollution control problems. This may involve either process changes or raw material substitutions to eliminate or minimize the production of toxic wastes.

In light of national energy concerns, a final major area of focus will be the continued development of reuse/recycle alternatives. Research in this area has previously focused on industries producing highly toxic chemicals or on industries for which alternative waste water solutions are not being brought into full-scale operation. Future research efforts will emphasize the development of reuse/recycle techniques that are applicable to a variety of installations and reuse/recycle alternatives that will result in more efficient pollution control.

22

PROBLEMS OF HAZARDOUS WASTE DISPOSAL

K. CARTWRIGHT, R. H. GILKESON, and T. M. JOHNSON

The geology and hydrogeology of a site must be carefully considered in planning the disposal of wastes and in assessing the potential problems that may result from some current regulatory criteria. An evaluation process is necessary that considers both the specific character of the wastes for disposal and the specific geologic conditions at the proposed disposal site. Wastes are frequently categorized as nonhazardous and hazardous. Nonhazardous wastes vary from construction debris to municipal waste and sludges, and each affords a very different hazard to the environment. Hazardous waste has not yet been satisfactorily defined in total by the U.S. Environmental Protection Agency (U.S. EPA), and none is offered here—only suggestions on an approach. This chapter deals with wastes that are liquid or will produce a leachate that must be controlled to prevent environmental degradation.

The term *hazardous wastes* includes the general categories of toxic chemical, biological, radioactive, flammable, and explosive wastes. Such wastes, if improperly managed, obviously threaten public health and welfare. There are many sources of hazardous wastes. Approximately 10 percent of all nonradioactive wastes generated by industry are considered hazardous wastes. About 90 percent of the hazardous wastes are in liquid form, of which about 40 percent are inorganic and 60 percent are organic (Garland and Mosher, 1975). The quantity of hazardous wastes has been growing at the rate of 5 to 10 percent annually; the volume of solid wastes, sludges, and liquid concentrations of pollutants from industry that will be disposed of in landfills or lagoons is expected to double during the next ten years.

The problems of disposal of wastes into geologic materials are a relatively recent subject of research. Research has, until recently, concentrated on the disposal of municipal refuse (e.g., Andersen and Dornbush, 1967; Apgar and Langmuir, 1971; Hughes et al., 1971; Palmquist and Sendlein, 1975; Gilkeson et al., 1978; Baedecker and Back, 1979). Although criteria for disposal of hazardous wastes may differ in some detail from those for disposal of general refuse, the principles developed from this research are applicable to both. Current practices of landfill disposal of wastes, particularly in humid climates, usually generate a noxious liquid leachate. An extensive bibliography on the subject can be found in Cartwright et al. (1981).

Studies of the hydrologic systems in fine-grained geologic materials, into which current practices direct most wastes, were almost nonexistent 20 years ago; however, during the past two decades procedures for study of fine-grained materials have been developed to provide the data required to study waste disposal sites (e.g., Farquhar and Rovers, 1975; Griffin et al., 1976, 1977; Cartwright et al., 1981). Case histories also have been developed to help predict chemical and physical changes that may occur as a result of the burial and surface disposal of waste.

Investigation of the waste attenuation characteristics of geologic materials is now one of the major areas of research. Attenuation capacity is the material's ability to remove contaminants from per-

Dr. Keros Cartwright is a geologist and head of the General and Environmental Group; Robert H. Gilkeson is a geologist in the Hydrogeology Section; Thomas M. Johnson is a geologist and head of the Hydrogeology Section; all are with the Illinois State Geological Survey, Champaign, Illinois 61820. This material is taken from Geological considerations in hazardous-waste disposal, *Journal of Hydrology*, 54 (1981): 357–369. Reprinted by permission of the authors and Elsevier Publishing Company. Copyright © 1981, Elsevier Scientific Publishing Company.

colating fluids. Approximations of the attenuation capacities of some geologic materials are known for some contaminants in leachates. These approximations provide the general relationships and principles on which judgments can be made; however, it probably will be a number of years before the mechanisms of attenuation are fully understood and the attenuation characteristics of most geologic materials for various contaminants and combination of contaminants are known.

PRESENT CRITERIA USED BY REGULATORY AGENCIES

Each U.S. state has different rules, regulations, and guidelines concerning the siting of waste lagoons and landfills. Many categorize sites according to the type of wastes to be received. As an example, landfill sites in Illinois (IPCB, 1973) are divided into five classes on the basis of geologic and ground water conditions. Hazardous wastes may only be accepted at sites that have the strictest requirements. Illinois regulations require disposal of hazardous wastes in class I and possibly class II sites, the distinction being made on the basis of the hydraulic conductivity at the bottom and sides. Class I sites require 3 m of material with a hydraulic conductivity of 10^{-8} centimeter per second (cm/s) or less, and class II sites require the same thickness but a hydraulic conductivity of 5×10^{-7} cm/s or less. There are no standard specifications for making this measurement, although ASTM (1970) does have a suggested laboratory method. Laboratory measurement of such low values is difficult; the error in measurement may be quite large, possible greater than the difference that distinguishes the sites. Field measurement is also difficult, time-consuming, and costly and may be no more accurate.

The current guidelines require, in addition to a permeability barrier, a minimum distance of commonly 150 m from the nearest water well. For protection of surface water, siting on a floodplain is prohibited, surface runoff must be controlled, and the site must be at least 150 m from any body of surface water. If a site does not meet these criteria, engineering modifications in accordance with certain guidelines may be implemented to enhance site conditions. Such modifications frequently involve a liner and a leachate collection system.

U.S. EPA (1977) recently published criteria for disposal of hazardous wastes in *Secure Chemical Waste Landfills*. The rules provide two alternatives for landfill design: natural geologic containment or artificial containment using engineered features.

Conditions in at least one-half to two-thirds of the United States will not allow natural geologic containment of hazardous wastes in accordance with U.S. EPA rules for two reasons: (1) the mean annual precipitation is too great, and (2) the water table is too high, especially in geologic materials with low hydraulic conductivity. Therefore, the federal guidelines require an artificial liner and a leachate collection system for such a facility. However, on the basis of our understanding of the behavior of leachate movement in geologic materials, natural containment, with attenuation of leachates by earth materials, is preferable to artificial containment; the reasons for this statement are explained later in this chapter.

Additional U.S. EPA requirements for all disposal sites include low hydraulic conductivities (10^{-7} cm/s), a minimum of 1.5 m between the base of the artificial liner and the water table, and a 150 m separation from any functioning public or private water supply. The rules also prohibit direct contact between the wastes and surface water, location on a wetland, on a floodplain, in a fault zone, or in the recharge zone of a single-source aquifer.

PERFORMANCE STANDARDS FOR DISPOSAL

The objective of regulations governing hazardous waste disposal is protection of surface water and ground water resources. Regulations with rigid specifications of geological and hydrological criteria for sites cannot be applied over the entire United States or even a region the size of most states. Strict application of some criteria, such as depth to water table, can actually lead to selection of less suitable sites. Rather, regulations should provide performance standards that the site design must meet to be acceptable and should be applied on a site-by-site basis. In the evaluation of a site it is the possible effect upon the environment that must be considered. The specific character of the wastes, the geologic materials at the proposed site, and the interaction between the two must be carefully examined.

A performance standard should stipulate the effect that a disposal site can have on the surrounding land. The standard must be written to limit the amount (volume and concentration) of contaminant allowed to be discharged by specifying: (1) the degree of water quality that is required, such that it must be fit for human consumption or that it must meet specific ion concentrations; (2) the degree to which the quality of ambient water can be altered; and (3) the exact area that must meet these requirements, such as within certain property lines and including the nearest aquifer or surface water body. If a mixing zone (as for point source discharge to surface water) is acceptable, the size of that zone should be specified. These specifications must be realistic; that is, a specification of "zero discharge measured at the waste boundary" cannot be accomplished.

These performance standards are an alternative approach to the use of liners and total containment; they permit the selection of disposal sites that will provide for an acceptable amount of attenuation of the toxic constituents from the leachate by interaction with geologic materials at the site. The sites must be carefully designed and engineered to minimize differential compaction that may occur, and trench covers must be constructed to control infiltration so that it is equal to or less than the possible migration rate from the site. An understanding of both unsaturated and saturated ground water flow is required by those people who both design and review sites. To use this approach will minimize the "bathtub" effect (see the section on site hydrogeology) and may allow the refuse in the landfill to leach and compact sooner and shorten the required monitoring time. Lagoons and landfills designed to meet performance standards should take into account five factors: (1) the type of waste to be disposed; (2) the site hydrogeology that governs the direction and rate of contaminant travel; (3) the attenuation of contaminants by geochemical interactions with the geologic materials; (4) the release rate of unattenuated pollutants to surface water or ground water; and (5) the character of the receiving waters.

CONSIDERATION OF THE WASTES

Regulatory agencies commonly classify wastes as hazardous and nonhazardous, a categorization that is not easily made and often must be arbitrary. In addition, mixed wastes, such as building debris and general municipal refuse, may contain some hazardous materials. The hazardous wastes in building debris may present a problem because current regulations regarding the disposal of building debris are lenient. The hazardous materials that are frequently present in general municipal refuse also may pose problems; general municipal refuse is a mixture of different types of wastes that may promote reactions that enhance the mobility of certain toxic constituents. Presently, the disposal of industrial wastes and sewage sludges into sanitary landfills constructed to receive general municipal refuse is common.

Segregation of Wastes

The authors presently believe that hazardous wastes should be segregated by type where possible; this may sometimes be accomplished by designating sites for particular types of waste disposal. Segregation of wastes allows for better prediction of attenuation characteristics of the geologic material for geochemically similar wastes. This procedure is less complicated than for mixed wastes, and it prevents interaction among incompatible wastes. Chemical reactions between some mixed wastes may increase the mobility of certain toxic constituents. For example, the mobility of most heavy metals is directly related to the pH of the solution; and polychlorinated biphenyls (PCBs), which are nearly immobile in aqueous solution, become highly mobile in organic solvents (such as carbon tetrachloride). In some instances an immobile ion may complex with a more mobile ion and migrate with it. Presently, the kinds of complex species that can form and their types of mobility are not very well understood for many hazardous wastes.

Degradation of Wastes

The nature of degradation of wastes must be considered, that is, whether by some natural process the waste may change from its present form to some less complex chemical compound and, hopefully, a less noxious form. Categorization into degradable and nondegradable wastes is desirable for all types of waste; it allows the addition of a time factor to geological and geochemical considerations. The decay/decomposition rate governs the duration of time that is required for isolating the hazardous waste from the environment; the time can range from a few days to thousands of years. The decay/decomposition process may be the result of radioactive decay, organic decomposition or some other process. Wastes that require a long decay/decomposition period (thousands of years) generally should, from a practical hydrogeological point of view, be considered nondegradable. Obviously, this distinction is arbitrary and perhaps should differ from site to site, depending upon the site characteristics.

Some geologic materials provide both containment time and attenuation of the wastes; however, wastes buried in landfills ultimately will return byproducts to the environment in some form and concentration.

Toxicity of Wastes

Because of the ultimate return of waste by-products from disposal sites to the environment, consideration of the toxicity of the wastes is essential. The toxicity of many wastes is not well known, and the assigned values are often arbitrary. One approach to this problem of waste classification is to consider the level at which toxicity occurs, that is, as parts per thousand, million, billion, and so on. In the evaluation of waste for disposal by landfill, the toxicity of the waste should also be related to its decomposition/decay rate. Geologic conditions in many areas may be unsuitable for landfill disposal of some wastes that have slow decomposition/decay rates and contain constituents that are toxic in low concentrations. These wastes may require destructive treatment, deep-well disposal, or shipment to a site with unique geologic conditions that may make it suitable for landfill disposal of the wastes.

CONSIDERATION OF SITE

Site Hydrogeology

The objective of existing regulations that require disposal of hazardous wastes in trenches or lagoons in natural clay materials or with artificial clay liners of very low hydraulic conductivity is to contain the wastes and thereby protect ground water resources. This approach is valid; however, it can create problems in humid climates where natural precipitation and infiltration of water from the surface exceeds the hydraulic conductivity of the surrounding natural material or liner. When this excess infiltration occurs, the disposal site fills with leachate and overflows, a phenomenon called the *bathtub effect* (Hughes et al., 1976; Cartwright et al., 1977). The water table rises into the disposal excavations, even if the original water table was located well below the bottom, eventually filling the trenches or lagoons and sometimes spilling out the sides as springs. Thus a site that on the basis of standards designed for ground water protection was suitable for disposal of hazardous wastes may become a hazard. The leachate will then have to be collected, treated, and redisposed of.

The bathtub effect occurs, in part, because most wastes have much higher hydraulic conductivities than the natural material into which they are placed; they may also have very different unsaturated soil–moisture characteristics. The hydraulic conductivity of some wastes can be reduced by compaction. The bathtub effect also occurs because more infiltration enters the disposal excavation than would under normal undisturbed conditions. Trench covers may be constructed to achieve the desired hydraulic conductivity and to limit infiltration for the required period of containment or until compaction of the wastes occurs; however, it is difficult

to maintain the trench covers. The covers must withstand attack by plants, weather (freeze–thaw, wet–dry), erosion, and strains caused by consolidation within the trench. Most trench covers are not capable of meeting these demanding requirements without costly long-term maintenance programs. The cover should be designed to allow for expected consolidation and to utilize hydrogeological concepts of saturated and unsaturated flow systems present at the site.

The authors presently believe that the importance of the water table is exaggerated in most regulations (Hughes and Cartwright, 1972). With the goal of protecting ground water, regulations commonly require that the base of the disposal trenches and lagoons must be situated a specified distance above the water table; therefore, a relatively deep water table is required. However, as pointed out earlier, the water table may be altered by the disposal operations.

In much of the humid areas of the United States deep water tables usually occur in the coarse-grained deposits with relatively high hydraulic conductivities; this is especially true in areas of low topographic relief. These materials may be a potential ground water resource rather than a suitable medium for burial of wastes. In these materials the location of the wastes above the water table does not ensure protection for ground water from leachate contamination. Research has shown that infiltration through refuse buried in these materials rapidly moves contaminants down to the water table. Shallow water tables generally occur in fine-grained geologic materials having low hydraulic conductivity; however, the fine-grained material is not a ground water resource. In the proper hydrogeologic setting fine-grained materials are well suited for the disposal of hazardous wastes because they are more effective than coarse-grained materials in containing and attenuating wastes and isolating them from aquifers.

Site Geology

The geologic setting at the disposal site determines whether the leachate will discharge near the trench or flow for great distances through natural geologic materials (Bergstrom, 1968). The pathway to a point of concern and the materials through which the contaminants must pass are critical.

Fine-grained materials of low hydraulic conductivity have been found to be the most suitable medium for burial and attenuation of wastes. These materials will be more effective in slowing the movement of the leachate and removing contaminants than coarse-grained or fractured materials. Water wells in such materials are usually unsuccessful because the materials, though saturated, do not transmit water fast enough to supply the pump. Where substantial deposits of these materials are present and water-yielding deposits (aquifers) are absent or isolated, conditions are most suitable for attenuation by physicochemical processes for disposal of wastes. Fine-grained geologic materials are frequently found in glacial drift, alluvium, colluvium, and other unconsolidated deposits; also, shale, claystones, mudstones, and other materials of the bedrock characterized by their fine grain. In addition, many crystalline rocks, evaporites, and such have extremely low hydraulic conductivities. These materials may protect aquifers, a primary objective when establishing criteria for disposal facilities. However, coarse-textured materials have been shown to be acceptable under certain circumstances. It may be possible to dispose of wastes with very low solubility in sand or other coarse-textured materials. Disposal may also be acceptable in areas where migration of contaminants is over moderate distances and attenuation by dilution and limited cation exchange can occur prior to public contact with the contaminants via ground water development.

The geology of the site should be studied in sufficient detail to provide the information required for the site design and to predict the fate of the waste by-products. For some sites areal geological mapping with limited or no exploratory borings may be sufficient, whereas other sites may require considerable drilling and laboratory testing. Exploration methods making use of piezometers, lysimeters, tensiometers, and drilling cores have been developed to provide the required data. Although some drilling will generally be necessary at most hazardous waste disposal sites, we believe it should be held to a reasonable minimum, because each boring represents a possible man-made conduit for the waste by-products to follow. (Strict plugging specifications should be required.) As was mentioned earlier, the hydrology of the site may be altered by disposal trenches. Such consequences should be considered in the proposed site design and operation plan.

Waste Attenuation Capacity of Site

Individual constituents in waste leachate may have markedly different mobilities in different geologic materials (Griffin et al., 1976, 1977; Fuller, 1978). Geochemical mechanisms that strongly inhibit the migration of one constituent may have little or no effect on other constituents. The chemical composition of waste leachates vary widely (U.S. EPA, 1974), and the interactions with geological materials are complicated; however, the leachates can

be considered in terms of their basic constituents. This permits the evaluation of factors affecting attenuation and their disposal in landfills with regard to their impact on the environment and the public health. Although only generalizations can be made, it should be recognized that the particular combination of leachate and site will be unique.

The attenuating characteristics of fine-grained geologic materials are considered favorable for waste disposal. The properties of geologic materials that are considered most important for attenuation are texture (grain size and structure), pore size distribution, clay composition, and chemical composition. The high percentage of clay content and the distribution of small pores in the fine-grained materials provided a low flux of solution and gases and long contact times and extensive contact areas between the earth materials and the contaminants.

The chemical composition of the geologic material includes its soluble-species adsorption capacity and matrix composition. The surficial materials are generally low in soluble species that could contribute to the pollutant load. This condition differs from that of arid regions where in some instances the natural salt content of surficial materials is sufficient to degrade the ground water below waste disposal sites. Surficial geologic materials in the eastern half of the United States frequently contain moderate to high amounts of hydrous oxides and carbonate minerals, which, along with clay minerals, provide good adsorption properties for a wide range of chemical species.

Some important geochemical mechanisms that attenuate waste leachates are exchange processes in which contaminants are selectively removed from the leachate and replaced by nontoxic constituents from the enclosing geologic materials. Since this is an exchange process, the total concentration of dissolved solids in the leachate does not change greatly. This is a reversible process; adsorbed ions may later be released. Thus this process may be considered as dilution.

The capacity of the geochemical mechanisms in the geologic materials to renovate contaminants from leachates is finite and, if exceeded, will allow the leachate to pass with little change. Therefore the attenuation capacity of the site's geologic materials must be the limiting factor for volume of wastes for disposal.

Release Rate for Unattenuated Contaminants

Determining the release rate of unattenuated or poorly attenuated contaminants from the disposal site to surface water or ground water (aquifer) is a necessary step in evaluating a waste disposal site. A decision must be made as to which ions must be attenuated "totally" and which could be released to the environment in the primary movement of contaminants (Cartwright et al., 1977). A properly designed and operated site promotes the dilution of contaminants by restricting the rate of their release into the environment to some acceptable level at which the concentrations in the receiving natural waters will remain below an acceptable maximum.

The calculation of the release rate of leachate from the bottom and sides of a landfill or lagoon and its flow path presents a complex problem. The use of a high-speed digital computer to predict the rate and path of fluids may be required; however, preliminary estimates of leakage can be made by use of the Darcy equation. If we consider the accuracy of most of the input data, this may be almost as reliable as more complex models. Alterations in the hydrogeologic system caused by the presence of the landfill must be considered, and the leakage calculated for the landfill must then be compared in volume to the receiving waters.

The time required for contaminants leached from the wastes to arrive at some point away from the source may be important under certain circumstances. The rate of water movement provides an estimate of the rate of travel of nonattenuated contaminants. The rate calculation must take into account the pore volume and structure of the geologic materials in which flow occurs. The attenuation characteristics of the geologic materials will retard this rate for a specific contaminant by an amount related to its attenuation factor. This factor is of great importance for wastes that undergo decay or decomposition during flow through a porous medium.

Geochemical mechanisms that attenuate waste by-products as they migrate through fine-grained geologic materials have been discussed briefly. If the composition of the leachate is known, an approximation of the attenuation factor can be made in the laboratory. The distribution coefficient, or the less complicated retardation factor, can be measured for the samples of geologic materials from the disposal site and for the leachate from the waste. The values measured in the laboratory, in practice, present some difficulty, as they are not constant but vary as the concentration of waste by-products changes in the leachate; nevertheless, these factors do provide data upon which a judgment can be made. However, the composition of the leachate is rarely known prior to disposal. An approximate attenuation factor for the by-products of some wastes may also be calculated from known general relationships and principles. More research will be necessary before the mechanisms of attenuation

can be understood and the attenuation characteristics for various contaminants by most geologic materials are known.

DISCUSSION

For both hazardous and municipal wastes there has been a trend in recent years from numerous, widely dispersed, small disposal sites to few and larger sites. This strategy should be used with extreme caution, especially if both large and small sites are judged by the same design standards. The authors believe that the use of performance standards rather than design standards are essential under these circumstances. The attenuation capacity of any geologic materials has a limit that, if exceeded by the volume of leachate that enters the material, will allow contaminants to pass almost unretarded. Unfortunately, there are insufficient data on the attenuation capacities of geologic materials for most leachate constituents to clearly define this limit. Larger land disposal sites are more likely than the smaller sites to exceed this limit.

For slowly degradable or nondegradable wastes, we view the trend to engineered sites with leachate collection systems as an interim disposal technique. Eventually, the leachate collected will have to be redisposed of at a final disposal site and perhaps at great expense. Such engineered sites may be suitable for the disposal of degradable wastes where isolation of wastes from the environment is not necessary for long periods of time. Sites may be engineered so as to reduce the volume of wastes that need to be transferred for final disposal; such a site may be appropriate in densely populated regions. The leachate collected from these sites eventually presents a disposal problem. Destructive treatment may be difficult, and processing the leachate in a standard municipal waste treatment plant may only dilute hazardous substances, possibly causing the sludge from the treatment plant to become hazardous.

This discussion has considered waste in general. Radioactive materials, mentioned several times in the chapter, represent a special type of hazardous waste that is often given special consideration. In the author's opinion, such special consideration is not always necessary; the principles discussed in this chapter apply equally to low-level radioactive waste materials and some short-lived high-level wastes.

An adequate monitoring system is essential to the operation of hazardous waste disposal sites. The monitoring system should test the extent to which the operation of the site meets performance standards of the site design. Also, the monitoring system should provide sufficient warning of potential pollution problems so that remedial measures, called for in a contingency plan, can be instituted. A contingency plan should be part of all disposal site designs. These measures should be specified in the site design to ensure against environmental degradation in the event that operation of the disposal site fails to meet performance standards.

Most regulations prohibit burying wastes in floodplains of rivers; however, these regulations do not recognize that often the floodplain does not occupy the entire river valley. The term *floodplain* and its application should be clearly defined. Where river valleys are underlain by shallow high-capacity aquifers, geologic conditions are generally not suited for the disposal of hazardous wastes. Sites that are located in the valley, well out of the reach of erosion by flood waters and are not underlain by coarse sands and gravel with high conductivities, would, under some circumstances, be suited to the disposal of certain types of hazardous wastes. These sites should be underlain by fine-grained geologic materials that have adequate attenuation capacity and should permit only the slow release of contaminants to the environment at an acceptable rate. The migration of contaminants from disposal sites in this hydrogeologic setting will follow along short, well-defined flow paths, and consequently there will be a limited area of contamination that may be relatively easy to monitor. At these sites the slow release of poorly attenuated or unattenuated contaminants to the environment where there is a comparatively high volume of receiving water will provide very high dilution rates. The proper operation of a waste disposal site in the floodplain setting is very similar in concept to the current operation of waste treatment plants that discharge directly to surface water.

Deep bedrock disposal wells were mentioned briefly in this chapter as a possible method for disposal of some wastes. These disposal wells are drilled into deep bedrock formations containing saline ground water (greater than 10^4 ppm of total dissolved solids), and the wastes for disposal are pumped into these formations. Engineering aspects of constructing deep disposal wells requires careful consideration of the nature and volume of wastes for disposal and the geologic and hydrologic conditions of the disposal zone. With thorough site investigation, careful operation, and adequate monitoring, deep-well disposal has the potential to provide excellent isolation from the environment for limited quantities of highly toxic wastes.

REFERENCES

Andersen, J. R., and J. N. Dornbush. 1967. Influence of sanitary landfill on ground water quality. *Journal of Water Works Association,* 59: 457–470.

Apgar, M. A., and D. Langmuir. 1971. Ground water pollution potential of a landfill above the water table. *Ground Water,* 9: 6.

ASTM (American Society for Testing Materials). 1970. *Special procedures for testing soil and rock for engineering purposes.* ASTM Special Technical Publication, 479, 141–145.

Baedecker, M. J., and W. Back. 1979. Hydrogeological processes and chemical reactions at a landfill. *Ground Water,* 17(5): 429–437.

Bergstrom, R. E. 1968. *Disposal of wastes: Scientific and administrative considerations.* Illinois State Geological Survey, Environmental Geology Note 20.

Cartwright, K., R. A. Griffin, and R. H. Gilkeson. 1977. Migration of landfill leachate through glacial tills. *Ground Water,* 15(4): 294–305.

Cartwright, K., R. H. Gilkeson, R. A. Griffin, T. M. Johnson, D. E. Lindorff, and P. B. DuMontelle. 1981. *Hydrogeologic considerations in hazardous-wastes disposal in Illinois.* Illinois State Geological Survey, Environmental Note 94.

Farquhar, G. J., and F. A. Rovers. 1975. *Leachate attenuation in undisturbed and remoulded soils.* Proceedings of the Research Symposium on Gas and Leachate from Landfills: Formation, Collection, and Treatment, New Brunswick, N.J., March 15–16, 1975. Cincinnati: U.S. Environmental Protection Agency, National Environmental Research Center.

Fuller, W. H. 1978. *Investigation of landfill leachate pollutant attenuation by soils.* EPA–600/2–78–158. Cincinnati: U.S. Environmental Protection Agency.

Garland, G. A., and D. C. Mosher. 1975. Leachate effects from improper land disposal. *Waste Age,* 6: 42–48.

Gilkeson, R. H., K. Cartwright, L. R. Follmer, and T. M. Johnson. 1978. *Hydrogeologic investigation of ground water contamination from land disposal of toxic wastes in Ogle County, Illinois.* Illinois State Geological Survey, Reprint 1978–D (reprint from Proceedings of the 15th Annual Engineering Geology Soils Engineering Symposium, Pocatello, Idaho, 1977, 17–28.

Griffin, R. A., K. Cartwright, N. F. Shimp, J. D. Steele, R. R. Ruch, W. A. White, G. M. Hughes, and R. H. Gilkeson. 1976. *Attenuation of pollutants in municipal landfill leachate by clay minerals. Part 1. Column leaching and field verification.* Illinois State Geological Survey, Environmental Geology Note 78.

Griffin, R. A., R. R. Frost, A. K. Au, G. Robinson, and N. F. Shimp. 1977. *Attenuation of pollutants in municipal landfill leachate by clay minerals. Part 2. Heavy metal absorption.* Illinois State Geological Survey, Environmental Geology Note 79.

Hughes, G. M., and K. Cartwright. 1972. Scientific and administrative criteria for shallow waste disposal. *Civil Engineering,* 42(3): 70–73.

Hughes, G. M., R. A. Landon, and R. N. Farvolden. 1971. *Hydrogeology of solid waste disposal sites in northeastern Illinois.* U.S. Environmental Protection Agency Report, Solid Waste Management Series, SW–12d.

Hughes, G. M., J. A. Schleicher, and K. Cartwright. 1976. *Supplement to the final report on the hydrogeology of solid waste disposal sites in northeastern Illinois.* Illinois State Geological Survey, Environmental Geology Note 80.

IPCB (Illinois Pollution Control Board). 1973. Rules and regulations. *Solid Waste.* Springfield, Ill.: Pollution Control Board, Chapter 7.

Palmquist, R., and L. V. A. Sendlein. 1975. The configuration of contamination enclaves from refuse disposal sites on floodplains. *Ground Water,* 13(2): 167–181.

U.S. EPA (U.S. Environmental Protection Agency). 1974. *Summary report: Gas and leachate from land disposal of municipal solid waste.* Cincinnati: Solid Hazardous Waste Research Division, Municipal Environmental Research Laboratory.

———. 1977. *Waste disposal practices and their effects on ground water.* U.S. Environmental Protection Agency, Office of Water Supply—Office of Solid Waste Management Programs, Report to U.S. Congress, edited by D. W. Miller. Report also published in book form as *Waste disposal effects on ground water.* Berkeley, Calif.: Premier Press, 1980.

23

HEALTH HAZARDS OF WATER POLLUTION

WORLD HEALTH ORGANIZATION

Water, as a part of the human environment, occurs in four main forms—as ground water, in fresh water surface masses, in the sea, and as vapor in the atmosphere. Human health may be affected by ingesting water directly or in food, by using it in personal hygiene or for agriculture, industry, or recreation, and by living near it. Two main categories of water-associated health hazards are considered here: (1) hazards from biological agents that may affect man following ingestion of water or other forms of water contact, or through insect vectors; and (2) hazards from chemical and radioactive pollutants, usually resulting from discharges of industrial wastes.

BIOLOGICAL HAZARDS

Water-Associated Hazards from Ingestion of Biological Agents

The principal biological agents transmitted in this way can be grouped into the following categories: pathogenic bacteria, viruses, parasites, and other organisms.

The contamination of water by pathogenic bacteria, viruses, and parasites can be attributed either to the pollution of the water source itself or to the pollution of the water during its conveyance from source to consumer. The pollutants may include the excretions, fecal and urinary, of man and animals, sewage and sewage effluents, and washings from the soil. Infections are spread both by patients and by carriers who shed the pathogen in feces or urine. Carriers may be patients who have recovered but still harbor the infective agent without suffering any further ill health themselves, or patients with mild or asymptomatic disease that has neither been

discovered nor diagnosed. The prevention of pollution and the purification of water are largely concerned with, and were developed to bring about, the eradication of waterborne infections.

Pathogenic Bacteria

Pathogenic bacteria transmitted directly by water or indirectly through water to food constitute one of the principal sources of morbidity and mortality in many developing countries. They include the causative agents of the great epidemic diseases—cholera and typhoid—and of the less spectacular but far more numerous cases of infantile diarrhea, dysenteries, and other enteric infections that occur continuously, and often with fatal results, among rural and urban populations, particularly in the developing countries.

Bacterial infections, especially those caused by the *Salmonella* group, may also be transmitted by shellfish grown in contaminated waters, unless a sufficient period is allowed for self-cleansing in tanks containing pathogen-free water treated with chlorine or by ultraviolet light.

During the past decade classical cholera caused by *Vibrio cholerae* has receded remarkably, even in such areas as Calcutta. However, cholera "El Tor," which emerged in 1961 from its endemic foci in Indonesia, has spread to many countries in the western Pacific and in the Southeast and central Asia. During 1970 a series of outbreaks of cholera "El Tor" occurred in areas not normally affected, for example, the eastern Mediterranean region and the Soviet Union, as well as a number of African countries [*Official Record of the World Health Organization (Official Record WHO)* 1971]. In 1971 cholera spread to nine more African countries, and small outbreaks or individual cases of cholera occurred in six European countries. Person-

This material is excerpted from Health hazards of water pollution and water-related diseases, in Chapter 2, Water, of *Health hazards of the human environment* (Geneva, Switzerland: World Health Organization, 1972), 50–59. Reprinted by permission of the World Health Organization. Copyright © 1972 by World Health Organization.

Table 23-1. Bacterial Diseases Capable of Being
Transmitted Through Contaminated Water
or Food Prepared with Such Water

Disease	Causative Organism
Cholera	*Vibrio cholerae,* including biotype "El Tor"
Bacillary dysentery	*Shigella* species
Typhoid fever	*Salmonella typhi*
Paratyphoid fever	*Salmonella paratyphi* A, B, and C
Gastoenteritis	Other *Salmonella* types: *Shigella, Proteus* species, and so on
Infantile diarrhea	Enteropathogenic types of *Escherichia coli*
Leptospirosis	*Leptospira* species
Tularaemia (rarely)	*Pasteurella (Brucella* or *Francisella) tularensis*

to-person transmission of cholera does occur, but by far the most important mode of dissemination is through the environment, especially water (Barua et al., 1970). Typhoid and paratyphoid fevers are still widely disseminated throughout the world; in Europe the explosive outbreak of typhoid fever in Zermatt in 1963 was a salutary warning (Bernard, 1965). Outbreaks of salmonellosis, although usually foodborne, may occasionally be spread by water (Greenberg and Ongerth, 1966).

Viruses

Certain viruses that multiply in the alimentary tract (including the oropharynx) of man, and may be excreted in considerable amounts in feces, can be found in sewage and polluted waters, but their mere presence is not necessarily evidence of significant risk to man. The viruses most commonly present in polluted waters and sewage are the enteroviruses (poliovirus, coxsackieviruses, and echoviruses), adenoviruses, reoviruses, and the virus (not yet identified) of infectious hepatitis (Chang, 1968). Of the enteroviruses, the spread of poliovirus by water has rarely if ever been demonstrated, because of the extremely high dilution of the virus and the consequent difficulty of isolating it, while the more direct fecal-oral route is the most likely mode of spread of the echoviruses and coxsackieviruses; adenoviruses and reoviruses are usually transmitted to other persons from the oropharynx (respiratory route).

Although the virus of infectious hepatitis has not yet been isolated and identified, there is ample epidemiological evidence that outbreaks of this infection, which has a global distribution, are caused by polluted waters (Mosley, 1967; Taylor et al., 1966; Koff, 1970). A striking example was the

epidemic of infectious hepatitis in Delhi (1955–1956), in which more than 28,000 cases were identified, with a case fatality rate of 0.9 per 1000 and an estimated total case incidence of 97,600 (Viswanathan, 1957).

Infectious hepatitis can also be spread by shellfish contaminated with sewage effluent (WHO Expert Committee on Hepatitis, 1964).

Parasites

Of the parasites that may be ingested, *Entamoeba histolytica* is the causal agent of both intestinal amoebiasis (e.g., amoebic dysentery and its complications) and extraintestinal forms of the disease, such as amoebic liver abscess. It is widespread throughout the warm countries of the world and wherever sanitary conditions are poor. Fine filtration, as practiced for the removal of bacteria, is both effective and essential against vegetative and ecysted amoebae, since amoebic cysts are resistant to chlorine in the doses normally applied in water treatment (WHO Expert Committee on Amoebiasis, 1969). The guinea worm, which causes dracontiasis, is common among the rural populations of many developing countries. This parasite is transmitted principally through open village wells and ponds infested with the copepod intermediate host.

Some intestinal helminths, such as *Ascaris lumbricoides* and *Trichuris trichiura,* may also be waterborne, although ingestion of contaminated soil is the normal means of transmission. Distomatosis is another parasitic disease that may be contracted by swallowing contaminated water containing, in this case, cysts of species of *Fasciola* or *Dicrocoelium.*

Hydatid disease (hydatidosis), a zoonosis that usually involves a dog-sheep-dog cycle in maintaining the reservoir of infection (cattle, pigs and other animals, including wildlife, may act as intermediate hosts), is occasionally transmitted to man through drinking water or foods contaminated with the excreta of the primary hosts.

Hazards from Biological Agents Transmitted Through Water Contact Other Than Ingestion

In economically advanced countries direct contact with water, other than in personal hygiene, occurs mostly in recreational activities—swimming, waterskiing, and so on—and ordinarily involves only minimal hazards. In many developing countries, on the other hand, water in rivers, ponds, canals, and so forth, is used for a variety of purposes—ablutions, washing of clothes, disposal of human excreta, domestic uses—so that these waters become highly polluted and serve as an important vehicle for the transmission of enteric infections, such as cholera, typhoid fever, and the dysenteries, and of certain parasitic infections.

Of the communicable diseases spread by the penetration, by parasites, of the skin and certain mucous membranes, the most widespread is schistosomiasis. Certain bacteria also cause disease in this way.

Schistosomiasis and Other Communicable Diseases

Schistosomiasis is a chronic, insidious, debilitating disease that may cause serious pathological lesions, saps energy, lowers resistance, and reduces output of work. In certain endemic areas it may be classed not only as an important health problem but also as a major social and economic one. Accurate figures for morbidity and mortality are lacking, but a conservative estimate is that the number of people infected with the parasite at any given time amounts to some 200 million. In some endemic areas the prevalence may exceed 50 percent (WHO Expert Committee on Bilharziasis, 1965; WHO Expert Committee on the Epidemiology and Control of Schistosomiasis, 1967). The accelerated construction of artificial lakes, reservoirs, and water impoundments to meet the needs of agriculture and industry in developing countries must inevitably lead to an increase in fresh water snail populations and a corresponding increase in schistosomiasis if preventive measures are not taken.

The disease in man is chiefly due to three species of trematodes, namely, *Schistosoma mansoni, S. japonicum,* and *S. haematobium.*

Of these, the first two give rise to intestinal manifestations, and the third is the causative agent of genitourinary or vesical schistosomiasis. The eggs, released in feces or urine, hatch as miracidia on reaching a free body of water, where they penetrate snails (such as *Biomphalaria, Bulinus,* and *Oncomelania*). They emerge from the snails in the form of cercariae ready to infect man. Penetration is through the skin while wading, bathing, and so on. Water flowing with only a very gentle current, as in slow-moving canals, is preferred by all species of aquatic snail hosts.

In many parts of the world bathers in lakes may be affected by "swimmer's itch." This dermatitis is caused by the penetration of the skin by cercariae of schistosomes from animals, such as birds and rodents.

Other parasitic diseases where entry is through the skin are ancylostomiasis and strongyloidiasis. *Necator americanus* and *Ancylostoma duodenale* are the hookworms (roundworms) responsible for most cases of ancylostomiasis. The eggs, passed in the feces of the infected person, hatch and larvae emerge that develop into a filariform stage infective to man. Strongyloidiasis is caused by the nematode *Strongyloides stercoralis,* which inhabits the submucous tissue of the small intestine. The mode of transmission is identical to that of ancylostomiasis. While water may serve as the medium whereby these infective agents are swallowed, infection is usually acquired by skin penetration from soil or by autoinfection, in the case of *Strongyloides.*

Leptospirosis is the principal bacterial infection transmissible from vertebrate animals to man through direct contact with water. The natural hosts are wild and domestic animals that excrete leptospires in urine, and man is infected through the skin and mucous membranes from contact with the water of contaminated ponds, canals, rivers, and so forth.

Health Hazards from Bathing Beaches and Coastal Waters

With the increased use by large groups of people of beaches for recreational purposes and because of the possible consumption of marine fish and shellfish from polluted waters, it is understandable that coastal pollution has received increased attention in many countries.

Epidemiological studies on coastal pollution have given equivocal findings, and no definite conclusions have yet been reached (Brisou, 1968, 1971; Moore, 1954, 1970).

In a large-scale survey of over forty bathing beaches around the English coast that were known to be contaminated with sewage, only four cases of paratyphoid fever probably due to bathing were recorded; these were from two beaches with median coliform counts of more than 1000 per 10 milliliters (mL); both beaches also showed gross macroscopic fecal pollution. There was no evidence that bathing in polluted coastal waters played any part in the occurrence of 150 cases of poliomyelitis among children living near the sea (Committee on Bathing Beach Contamination of the Public Health Laboratory Service, 1959). More information is needed about the possibility of young children acquiring superficial mycotic diseases, such as thrush, or becoming infected with enteroviruses other than poliovirus as a result of bathing in such waters.

There is no agreed basis for establishing a limiting bacteriological quality for coastal water, above which there is a danger of infection from bathing. Maximum acceptable counts in different countries for coliforms may be as high as 10,000 per liter of water; for *Streptococcus faecalis* the corresponding figure is 200 per liter (Brisou, 1968). There are no internationally accepted criteria for the quality of coastal water with respect either to microbial contamination or to chemical pollution.

Hazards of Diseases Transmitted by Water-Associated Insect Vectors

Of diseases caused by water-associated vectors, the most widespread is malaria, transmitted by the anopheline mosquito. The habitats of different vector species of anophelines, their ecology, and their bionomics are well known.

Onchocerciasis, or river blindness, is associated, under natural conditions, with clear springs running over rocky beds. In artificial works of water resources development, the same conditions may arise on dam spillways or concrete-lined channels if suitable precautions are not taken. The disease, transmitted by blackflies (*Simulium* species), is of considerable health, social, and economic importance in the vicinity of breeding grounds for the vector, where a large proportion of the resident population may become partially or completely blind. The creation of new breeding areas, as a result of water resources development, may outweigh in significance any potential benefits of the works concerned (WHO Expert Committee on Onchocerciasis, 1966). An indirect effect of onchocerciasis on the environment may follow the application of control measures against the vector if persistent insecticides, such as DDT, are used for this purpose. However, it is now recommended that these insecticides should be replaced in onchocerciasis vector control by newly developed biodegradable larvicides.

Other diseases of a similar nature include yellow fever (transmitted by a water-breeding mosquito); trypanosomiasis, or sleeping sickness, which is widespread in Africa and is transmitted by the tsetse fly (the form of the disease caused by *Trypanosoma gambiense* is particularly associated with waterside vegetation); and filariasis.

Filariasis, which affects more than 250 million people throughout the world, continues to be one of the major parasitic infections. The public health importance of the causative organisms, *Wuchereria bancrofti* and *Brugia malayi,* is widely recognized, but control programs, although successful to some extent, have not succeeded in containing the infection significantly on a world basis. Indeed, in many of the developing countries the amount of infection and disease may in fact be increasing. This trend is due primarily to the rapid urbanization now taking place in many of the new countries of Africa and Asia, a phenomenon characterized by vast population movements and the growth of urban ghettos in which breeding sites for the major vector, *Culex pipiens fatigans,* have been greatly increased [WHO Expert Committee on Filariasis (*Wuchereria* and *Brugia* infections), 1967].

Culex pipiens fatigans may occupy the first place

as the mosquito that has profited most by increasing urbanization and industrialization. In Kaduna, Nigeria, for instance, a survey carried out in 1942 failed to detect the presence of this species at all. However, in 1958 another survey found as many as 760 per room (Mattingly, 1963). The significance of this rather spectacular development is that it occurred unrecognized. There seems little doubt that similar rapid increases in numbers are taking place in other areas. In addition to its well-developed ability to shelter in houses, *C. p. fatigans* possesses a remarkable genetic adaptability in developing resistance to insecticides; it was noted in Malaya that the urban strain of this species was about twenty times as efficient a vector of *W. bancrofti* as a rural strain.

Nuisance Organisms

Other organisms constitute, so far as is known at present, only indirect health hazards in that they may make a safe water unpalatable or unattractive or interfere with treatment and distribution processes. Such organisms include the biological slimes that accumulate on the interior surfaces of mains and upon which methane-utilizing bacteria may grow; algae and bryozoal growths, such as *Plumatella,* which may interfere with the operation of filters; molluscs, such as *Dreissena,* which may choke mains; crustacea, such as *Asellus* (water louse or sow bug), and nematodes, which while not in themselves pathogenic, may harbor bacteria or viruses in their intestines, thus protecting possible pathogens from destruction by chlorine; and certain algae, which may impart bad tastes and odors to water. Table 23-2 lists the nuisance organisms normally found in drinking water.

Nuisance organisms, which, once they have entered a distribution system, may multiply there,

Table 23-2. Nuisance Organisms Normally Found
in Drinking Water

Organism	Effects
Biological slimes	Choking of treatment plants and distribution systems. Support for methane-utilizing bacteria. Risk of rendering water unacceptable
Molluscs (*Dreissena*)	Choking of water mains
Plumatella, algae	Interference with filtration
Asellus	Risk of rendering water unacceptable
Nematodes	Possible concentration of pathogens

can be kept to insignificant levels by regular cleansing of mains and storage tanks and by the use of harmless pesticides, such as pyrethrin, to which some of these organisms (e.g., *Asellus*) are particularly sensitive. Research is needed on the influence of organic matter in treated water on bacterial growths, leading to the production of bad tastes and odors.

HAZARDS FROM CHEMICAL AND RADIOACTIVE POLLUTION

If present above a certain level, some chemical pollutants (e.g., nitrates, arsenic, and lead) may constitute a direct toxic hazard when ingested in water. Other water constituents, such as fluorides, are beneficial and may be essential to health, if present in small concentrations, though toxic if taken in larger amounts. Certain other substances or chemical characteristics may affect the acceptability of water for drinking purposes. They include substances causing odors or tastes; acidity or alkalinity; anionic detergents; mineral oil; phenolic compounds; and naturally occurring salts of magnesium and iron, as well as sulfate and chloride ions, if present in excessive concentrations. Both international and national criteria and standards have been established to provide a basis for the control of human exposure to many of these substances through ingestion of polluted water (World Health Organization, 1970, 1971).

Ingestion is, however, only one possible pathway to exposure. Man can be exposed to water pollutants through other types of direct contact, for example, in recreation or the use of water for personal hygiene. The possible health implications of these nondrinking uses of water (including agricultural and industrial uses) are less well understood, and no international criteria or guidelines exist for the control of such exposure.

In addition to the possible effects of ingestion and other direct water contacts, chemical water pollutants may influence man's health indirectly by disturbing the aquatic ecosystems or by accumulating in aquatic organisms used in human food. For some pollutants, at the levels now existing in water bodies, these effects may be the most important public health aspects of water pollution and should be considered particularly in respect of such substances as compounds of toxic metals and organochlorine pesticides.

The various chemical and biochemical transformations that pollutants may undergo in the aquatic environment also deserve attention. Chemical change may affect their biological availability or toxicity, which may be either enhanced or reduced. Degradation or transformation products may appear that are more toxic than the original pollutant. Little is known of these physical, chemical, and biological processes and their mechanisms, yet they are essential to the understanding of the health implications of chemical water pollution.

Many water pollutants also appear in air and food, which are often more important sources of intake than water. Such pollutants include metals, organic substances resistant to biodegradation, and radionuclides. The assessment of pollutant levels in water should always be made in relation to the actual intake of drinking water and to the body burden resulting from other sources in a given locality (World Health Organization, 1971).

REFERENCES

Barua, D., W. Burrows, and J. C. Gallut. 1970. *Principles and practice of cholera control.* Geneva: World Health Organization (Public Health Paper, number 40).

Brisou, J. 1968. *Bulletin of World Health Organization,* 38: 79–118.

———. 1971. *La pollution des mers par les microorganismes.* Geneva: World Health Organization (Unpublished document WHO/EP/71.1).

Chang, S. L. 1968. *Bulletin of World Health Organization,* 28:410–414.

Committee on Bathing Beach Contamination of the Public Health Laboratory Service. 1959. *Journal Hygiene* (London), 57:435–472.

Greenberg, A. E., and Ongerth, J. H. 1966. *Journal of American Water Works Association,* 58:1145–1150.

Koff, R. S. 1970. *CRC Critical Reviews of Environmental Control,* 1:383–442.

Mattingly, P. F. 1963. *Bulletin of World Health Organization,* 29 Supplement: 135–139.

Moore, B. 1954. *Bulletin of Hygiene,* 29:689–704.

———. 1970. *Review Oceanography of the Mediterrean,* 18–19:1–26.

Mosley, J. 1967. Transmission of viral diseases by drinking water. *Transmission of viruses by the water route.* New York: Interscience.

Official Record of the World Health Organization. 1971. No. 193, p. 144.

Taylor, F. B., et al. 1966. *American Journal of Public Health,* 56:2093–2105.

Viswanathan, R. 1957. *Indian Journal of Medical Research,* 45 (Supplement no. 1). WHO Chronicle. 1972. 26:51.

WHO Expert Committee on Amoebiasis. 1969. *Report*. Geneva (World Health Organization Technical Report Series, no. 421).

WHO Expert Committee on Bilharziasis. 1965. *Third report*. Geneva (World Health Organization Technical Report Series, no. 299).

WHO Expert Committee on the Epidemiology and Control of Schistosomiasis. 1967. *Report*. Geneva (World Health Organization Technical Report Series, no. 372).

WHO Expert Committee on Filariasis (*Wuchereria* and *Brugia* infections). 1967. *Report*. Geneva (World Health Organization Technical Report Series, no. 359).

WHO Expert Committee on Hepatitis. 1964. *Second report*. Geneva (World Health Organization Technical Report Series, no. 285).

WHO Expert Committee on Onchocerciasis. 1966. *Second report*. Geneva (World Health Organization Technical Report Series, no. 335).

World Health Organization. 1970. *European standards for drinking-water*. 2nd edition. Geneva.

———. 1971. *International standards for drinking-water*. 3rd edition. Geneva.

24

WATERBORNE-DISEASE OUTBREAKS IN THE UNITED STATES

EDWIN C. LIPPY and STEVEN C. WALTRIP

Information on waterborne disease outbreaks has been collected since 1920 and published in summary fashion for the periods 1920–1936,[1] 1938–1945,[2] 1946–1960,[3] 1961–1970,[4] and 1971–1978.[5] Outbreak information on file by year dating back to 1946 was recently entered into an automatic data-processing system that provided rapid recall of data for analyses of occurrence, distribution, and trends and for other statistical interpretation. The data for 1946–1980 are presented in this chapter to give a 35-year perspective on waterborne illness associated with drinking water.

Information on waterborne illness has been submitted voluntarily to the U.S. Public Health Service (U.S. PHS) since 1920 by state agencies concerned with intestinal illness. What originally began as a concern for morbidity and mortality resulting from typhoid fever and infant diarrhea and their relationship with food, milk, and water led to the development of important public health measures that had a profound influence on decreasing the incidence of enteric diseases. The voluntary reporting procedure was changed somewhat after the responsibility for it was transferred in 1966 from the National Office of Vital Statistics to the Centers for Disease Control (CDC), which began contacting state health agencies on a yearly basis for reports of disease outbreaks. In 1971 the U.S. Environmental Protection Agency (U.S. EPA) joined with CDC in a collaborative effort to improve reporting of waterborne illness by contacting state regulatory agencies responsible for water supply activities. A form was developed to provide detailed information on waterborne outbreaks, and its use was initiated in 1976 after two years of field testing. Reporting was further refined in 1978 when annual summaries of waterborne outbreaks were

separated from those caused by food and published under the title *Water-Related Disease Outbreaks.*[6]

Authors involved in writing summaries and articles on waterborne disease outbreaks have been critical in their appraisal of the effectiveness of voluntary reporting. They indicate that it is a passive surveillance system that detects only a fraction of the outbreaks and cases of illness that occur each year in the United States. In support of this criticism are the statistics developed from an improved surveillance system in one state that accounted for 48 of the 189 outbreaks, or 25 percent of the total reported in the period from 1976 to 1980. Craun and McCabe,[7] in a review spanning 1946–1960, estimated that one-half of the outbreaks in community systems and only one-third in noncommunity systems are reported. Information from a study done in a Rocky Mountain state under auspices of a U.S. EPA contract showed that under intensified surveillance efforts the annual number of outbreaks increased nearly fourfold in a three-year period compared with the occurrence in the previous three years. Surveillance and reporting efforts improved from 1971 to 1980 (Figure 24-1), but much more needs to be done.

The analyses that follow must be interpreted with caution because the intensity of surveillance varies among states. Agencies that conscientiously report outbreaks may cast doubt on the safety of drinking water in their states although this may not be warranted. If all states placed equal emphasis on surveillance, a more accurate representation of occurrence, distribution, and trends could be presented.

The data presented in this chapter are from outbreaks associated with water obtained from public or individual water systems and intended for

Mr. Lippy is a water resources and sanitary engineer, and Mr. Waltrip is a biological technician, both at the A. W. Breidenbach Environmental Research Center, U.S. Environmental Protection Agency, Cincinnati, Ohio 45268. This material is taken from Water disease outbreaks—1946–1980: A thirty-five year perspective. 1984. *Journal AWWA*, 76(2):60–67. Reprinted by permission of the authors and AWWA. Copyright © 1984 by The American Water Works Association.

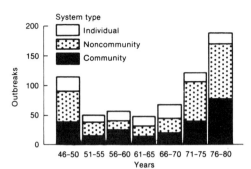

Figure 24-1. Occurrence of waterborne disease by five-year increments.

drinking or domestic purposes. At least two cases of acute infectious disease or one case of an acute intoxicating illness must occur and originate from a common source to qualify as an outbreak. An infectious disease that produces symptoms affecting the upper and lower gastrointestinal tract is caused by a microbiological agent and is accompanied by an incubation period ranging from about 24 hr to several weeks. An intoxicating illness that produces symptoms associated with the upper gastrointestinal tract is caused by a chemical agent, and the incubation period from exposure to illness is brief—usually a matter of a few hours. For one case of illness to qualify an an outbreak associated with a chemical agent, the causative agent must be identified in a sample of water collected from the system.

A public water system normally consists of (1) a water supply from a surface source (e.g., river, lake, reservoir), a ground source (e.g., well, spring, or a mixture of the two); (2) treatment facilities (e.g., filtration, disinfection); and (3) a distribution network of mains and pipes that deliver water to the consuming public. Unfortunately, many systems have no treatment facilities. By definition, public water systems also have at least 15 service connections and provide water to 25 persons. There are two types of public water systems. A community water system serves year-round residents; a noncommunity water system, which does not meet the residence requirement, serves the traveling public at institutions, camps, parks, or motels. Where public water systems are not available, people develop their own household water supplies—individual systems.

Public water systems currently number about 215,000, with 65,000 community services serving a resident population of 195 million persons. Noncommunity systems number 150,000, and about 10 million individual systems serve 35 million persons.

OCCURRENCE, DISTRIBUTION, AND TRENDS

Outbreaks

Figure 24-2 illustrates the annual frequency of the 672 waterborne outbreaks reported for the 35-year period commencing in 1946. The trend in number of outbreaks has been increasing since 1966. The trend is even better exemplified in Figure 24-1 and is especially notable since 1971, when the current system of reporting was established. Figure 24-2 shows a peak in 1980; the 50 reported outbreaks were the third highest number of outbreaks since record collection began in 1920. The record of 60 outbreaks was reported in 1941. Cursory examination of Figures 24-1 and 24-2 indicates that outbreaks were under control for the 1951–1970 period, whereas the opposite may be said about the outbreaks for 1971–1980. A four-year cycle in annual occurrence began in 1972, with peaks following in 1976 and 1980. There is no obvious scientific explanation for the cycle; however, if it and the current trend continue, 95 outbreaks can be expected in 1984. Again, Figure 24-2 as well as subsequent analyses must be tendered with cautious interpretation. Although Figures 24-1 and 24-2 show large increases in the number of reported outbreaks for the 1971–1980 period compared with those in previous years, relatively few outbreaks were reported by the 50 states. The 50 outbreaks reported in 1980 averaged one per state. If a few states become aggressive in investigating and re-

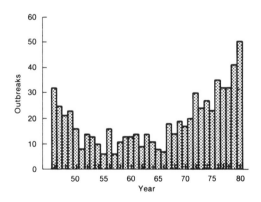

Figure 24-2. Annual occurrence of waterborne-disease outbreaks.

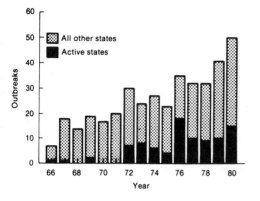

Figure 24-3. Waterborne-disease outbreaks in two active states versus those in all other states.

porting waterborne outbreaks, the trend can easily be affected. The influence of two such states is shown in Figure 24-3.

The geographic distribution of outbreaks is shown in Figure 24-4. Delaware was the only state that did not report an outbreak. States in which the reported outbreaks were low are clustered in the upper Midwest or Plains states, and those in which the number was high are clustered in the eastern United States (New York, Pennsylvania, New Jersey, and Ohio) and in the western United States (Washington, Oregon, and California). The combination of a low population density and a long distance between sources of water supply and waste water discharge locations may explain the low-outbreak cluster, whereas opposite conditions may explain the high-outbreak clusters. The clusters may also be due to the effect devoted to surveillance of outbreaks. New York has been active in investigating and reporting since the 1940s, and Pennsylvania began a more active effort in the 1970s (Figure 24-3).

The gross distribution shown in Figure 24-4 was corrected for population density by expressing the information as a rate—outbreaks per number of community water systems—for each state (Figure 24-5). Outbreaks in noncommunity and individual systems were not included in the distribution. In Figure 24-5 the rate of occurrence is based on the

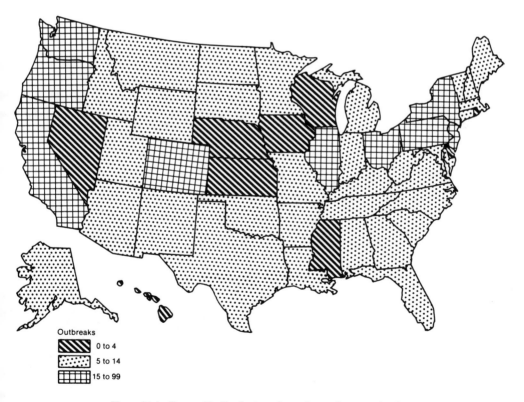

Figure 24-4. Geographic distribution of waterborne-disease outbreaks.

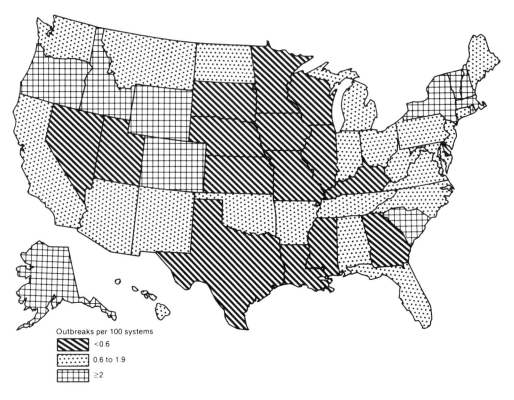

Outbreaks per 100 systems

▨ <0.6

▦ 0.6 to 1.9

⊞ ≥2

Figure 24-5. Geographic distribution of waterborne-disease outbreaks per 100 community water systems.

number of outbreaks that occurred over the 35-year period for each state and the number of water systems that were counted during a nationwide inventory in 1963.[8] The 1963 date represents a near midpoint in the set of outbreak data.

Of those states in the high-outbreak category of the gross distribution shown in Figure 24-4, only New York, Colorado, and Oregon remain in the high-outbreak category of the distribution based on the number of water systems (Figure 24-5). These three remaining states share the high category in the rate distribution with the states of Vermont, New Hampshire, Connecticut, Rhode Island, South Carolina, Wyoming, Idaho, and Alaska. The preponderance of the states in the low-rate category are located in the mid-continental United States. The topic of geographic outbreak clustering is also considered in the section on water system deficiencies.

A temporal distribution of outbreaks, which is given in Figure 24-6, shows that a peak is evident in the summer months. Summer is also the period of the year for vacations, including visits by many people to recreational areas that are primarily served by noncommunity water systems. Figure 24-7 was

developed to compare the monthly distribution of outbreaks by type of water system. The pronounced peaks shown in this figure indicate that noncommunity systems experience problems in the summer. A similar, clearly defined, temporal distribution is not apparent for community systems.

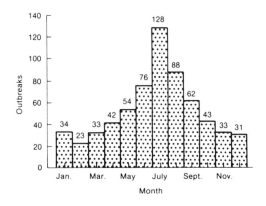

Figure 24-6. Waterborne-disease outbreaks by month (month not known for 25 outbreaks).

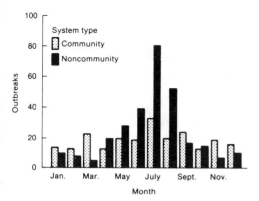

Figure 24-7. Waterborne-disease outbreaks for community and noncommunity systems.

Table 24-1. Occurrence of Waterborne-disease Outbreaks by Type of Water System

System Type	Outbreaks		Occurrence per 1000 Systems[a]
	1946–1980	1971–1980	
Community	237	121	1.9 (121/65)
Noncommunity	296	157	1.0 (157/150)
Individual	137	36	
Unknown	2	0	

[a] 1971–1980 data.

Outbreaks in noncommunity systems outnumbered those in community systems, as shown in Table 24-1 and Figure 24-8. However, when the rate of occurrence was based on the number of outbreaks per 1000 community or noncommunity systems, the rate for community systems was nearly twice that for noncommunity systems (Table 24-1). Rates were computed only for the 1971–1980 period, to take advantage of information recently collected on the number of water systems in the United States through the Federal Reporting Data System (FRDS). Because national inventories prior to the FRDS did not discriminate between community and noncommunity systems, rates for 1946–1980 could not be calculated.

Compared with the outbreak rate for noncommunity systems, the greater rate for community systems is probably due to better reporting of

outbreaks; clusters of illness would be more readily recognized and reported in a resident population than in a nonresident population. Travelers exposed to contaminated water from a noncommunity system may become ill some time after exposure (the incubation period) and at a distance from the point of exposure. Associating their illness with contaminated water served at a motel, campground, or restaurant that they visited two to three days previously would be difficult. If an association would be perceived, the separation by time and distance might tend to make people want to forget the event and not become involved with reporting an illness potentially related to contaminated drinking water. Therefore although the outbreak rates are greater for community systems than for noncommunity systems, the recognition and reporting of outbreaks in noncommunity systems must be acknowledged as affecting the rates.

Illness

The annual occurrence of cases of illness associated with waterborne-disease outbreaks are shown in

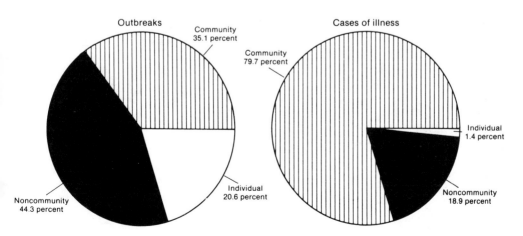

Figure 24-8. Waterborne-disease outbreaks and cases of illnes by type of system.

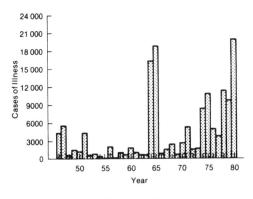

Figure 24-9. Cases of illness.

Figure 24-9. The total number of cases during the period was 150,475. More than 20,000 cases occurred in the peak year of 1980, with four major outbreaks accounting for 14,000 of these cases. The prominent peaks in two other years, 1964 and 1965, each resulted from one major outbreak that caused 16,000 cases of illness. Excluding these

two years, illness could be labeled as insignificant for a 25-year period. Since 1974, however, the number of cases has increased dramatically and been accompanied by a similar increase in outbreaks (Figures 24-1 and 24-2).

The geographic distribution of illness is shown in Figures 24-10 and 24-11. Figure 24-10 represents a gross distribution of all cases. Figure 24-11 shows a distribution that was corrected for population density by expressing illness as a rate based on the number of people served by community water systems. Those cases attributable to outbreaks in noncommunity and individual systems were not considered in the rate. In Figure 24-11 the rate is based on the number of cases per 10,000 service population counted during the U.S. PHS national inventory in 1963.

The gross distribution emphasizes the large outbreaks that occurred in California (Riverside with 16,000 cases in 1965), Texas (Georgetown with 8000 cases in 1980), Florida (Gainesville with 16,000 cases in 1964), Pennsylvania (Sewickley with 5000 cases in 1975 and Bradford with 3500 cases in 1979), and New York (99 outbreaks, the

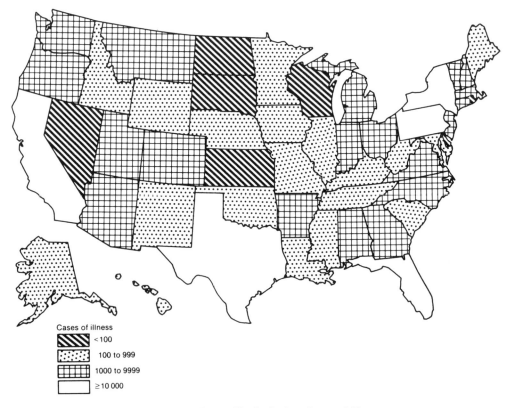

Figure 24-10. Geographic distribution of cases of illness.

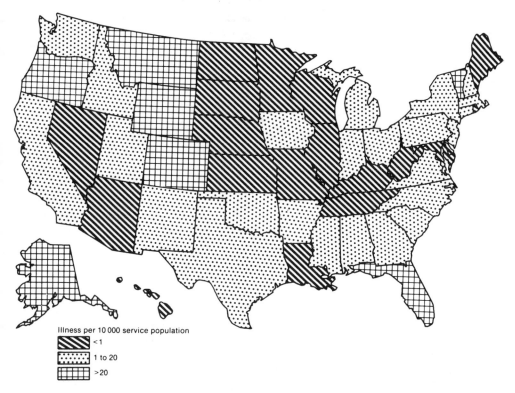

Illness per 10 000 service population

- <1
- 1 to 20
- >20

Figure 24-11. Geographic distribution of cases of illness per 10,000 service population in community water systems.

largest of which was in Rome, with 4800 cases in 1974). In Figure 24-11 the use of a rate to portray illness changed the distribution pictured in Figure 24-10, with only Florida remaining in the high-rate category and being joined by Oregon, Montana, Wyoming, Colorado, Vermont, and Alaska.

Figure 24-12 indicates that bacterial agents were a major factor among outbreaks. In more than half of the outbreaks, however, an agent was not determined. Microbiological agents are rarely identified in water (<1 percent of outbreaks) but are determined through collection and assay of stool or blood specimens. The lack of success in identifying agents in water stems mainly from the lack of local capability to isolate, culture, and identify microorganisms from an environment in which they are normally reduced in number by dilution or die away as opposed to medical specimens collected from a human subject in which agents are propagated. Another reason for poor success with environmental samples is that outbreaks are normally recognized one to two weeks after the exposure, which allows ample time for the water system to be purged of contamination through normal usage of water. Conversely, opportunities for identifying

the agent in medical specimens are better because of secondary transmission of illness through person-to-person pathways. A person originally exposed to the agent in his drinking water may

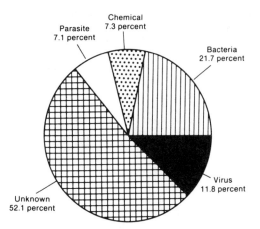

Figure 24-12. Waterborne-disease outbreaks by causative agent.

Table 24-2. Causative Agents of Waterborne Disease

Agent	Outbreaks	Cases of Illness
Bacterial		
Campylobacter	2	3,800
Pasteurella	2	6
Leptospira	1	9
Escherichia coli	5	1,188
Shigella	61	13,089
Salmonella	75	18,590
Total	146	36,682
Viral		
Parvoviruslike	10	3,147
Hepatitis	68	2,262
Polio	1	16
Total	79	5,425
Parasitic		
Entamoeba	6	79
Giardia	42	19,734
Total	48	19,813
Chemical		
Inorganic	29	891
Organic	21	2,725
Total	49	3,616
Unknown	350	84,939
Grand total	672	150,475

transmit the disease some days later to a contact, who then becomes a positive candidate for investigators. Because some agents survive freezing, contaminated water drawn from the tap and frozen as ice cubes for later use can also be analyzed and used to supplement the analysis of medical specimens.

Chemical agents can affect aesthetic qualities of water. Unusual taste, odor, or discoloration leads to complaints that prompt collection of samples and investigation by local authorities. Outbreaks and illness caused by chemicals are probably better recognized and reported than are those caused by microbial agents. Table 24-2 lists microbiological and chemical agents responsible for outbreaks from 1946 to 1980.

Cases of illness are classified according to type of water system in Figure 24-8. Although noncommunity systems have the largest percentage of the total number of outbreaks, cases of illness in community systems greatly outnumber those occurring in other water systems. Simply stated, outbreaks in community systems are more serious in terms of illness because of the availability of a resident population to become infected. The severity of illness in community or noncommunity systems is further exemplified by comparing cases per out-

break. Community systems averaged 506 cases of illness per outbreak, and noncommunity systems averaged 96 cases, which gives a severity factor of five.

The number of deaths resulting from waterborne outbreaks has greatly diminished over the years. From 1920 through 1945, 960 deaths were reported (mostly from typhoid fever), for an annual rate of 37. During the 35-year period since 1946, the rate has been reduced to 1 death per year. An underreporting problem probably exists in terms of deaths associated with gastrointestinal illness, which can stress a weakened cardiovascular system or other critical body function. That is, even though mortality occurs in the weakened and distressed population affected by gastrointestinal illness, the death is not attributed to gastrointestinal illness.

WATER SYSTEM DEFICIENCIES

Deficiencies in water system design, operation, and maintenance that contribute to waterborne outbreaks are of importance to regulatory agencies and to the water utility industry, including purveyors, equipment manufacturers, consulting firms, and laboratories. Overwhelming percentages of a certain deficiency are indicative of a breakdown in an approach by a regulatory agency or the industry. Unfortunately, the analyses that follow show that the causes of outbreaks for 1946 to 1980 do not markedly differ from those presented in summaries for 1920 to 1945, which may indicate that drastic changes are needed in attitudes of regulatory agencies and the industry.

Water system deficiencies that caused or contributed to outbreaks were categorized under five major headings as follows:

1. Use of contaminated, untreated surface water.
2. Use of contaminated, untreated ground water.
3. Inadequate or interrupted treatment.
4. Distribution network problems.
5. Miscellaneous.

Use of contaminated, untreated surface water and ground water is self-explanatory. Inadequate or interrupted treatment includes breakdown or failure of equipment, insufficient chlorine contact time, and an overloaded process. Distribution network problems include cross-connections, improper or inadequate main disinfection, and contamination of open distribution reservoirs. The miscellaneous category encompasses deliberate sabotage, events that systems are not expected to cope with, and undetermined causes of outbreaks.

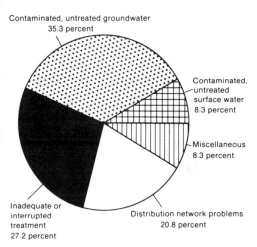

Contaminated, untreated groundwater
35.3 percent

Contaminated, untreated surface water
8.3 percent

Miscellaneous
8.3 percent

Distribution network problems
20.8 percent

Inadequate or interrupted treatment
27.2 percent

Figure 24-13. Waterborne-disease outbreaks by deficiency in public water systems.

Water system deficiencies that caused or contributed to outbreaks are graphically depicted by the five categories in Figure 24-13. Outbreaks in individual systems were omitted so that deficiencies in public water systems, which are regulated by a federal law, could be emphasized. Use of untreated, contaminated ground water and poor practices in treatment and distribution of water are deficiencies that deserve attention. More than 80 percent of the outbreaks were associated with these deficiencies. The deficiencies in community and noncommunity systems are compared in Figure 24-14. Two-thirds of the outbreaks in community systems were caused by deficiencies in treatment

and distribution of water, whereas three-fourths of the outbreaks in noncommunity systems were related to use of untreated ground water or to poor treatment practices.

The overwhelming statistic gleaned from Figure 24-14 and Table 24-1 is that the 229 outbreaks in noncommunity systems were caused by use of untreated, contaminated ground water or by inadequate or interrupted treatment. Disinfection is generally the only treatment provided in these systems. The 229 outbreaks represent nearly one-half of all outbreaks reported for public water systems that are regulated by law; that is, 237 outbreaks in community systems plus 296 in noncommunity systems equals 533 outbreaks, and 229/533 is approximately one-half. If disinfection was in place where needed and applied properly, many outbreaks could be prevented. Some of the common problems that deserve attention from regulatory agencies and the water utility industry include the following:

1. Noncommunity systems suffer from lack of proper design, construction, and operation and receive minimal attention from regulatory agencies to correct these problems.
2. One microbiological sample is required during each quarter a noncommunity system is operational. A sample collected from the system during start-up after the off-season may not show the same results as one collected during heavy visitation and system overload.
3. Coliform samples that are collected periodically from a challenged system with problems and show negative results provide a false sense of security. Though it is easy to inactivate coli-

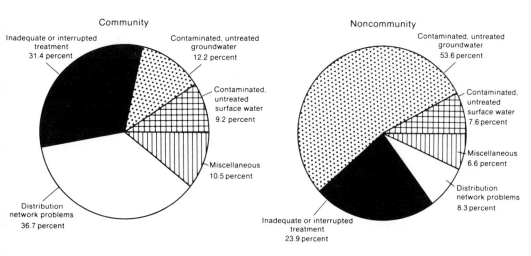

Community

Inadequate or interrupted treatment
31.4 percent

Contaminated, untreated groundwater
12.2 percent

Contaminated, untreated surface water
9.2 percent

Miscellaneous
10.5 percent

Distribution network problems
36.7 percent

Noncommunity

Contaminated, untreated groundwater
53.6 percent

Contaminated, untreated surface water
7.6 percent

Miscellaneous
6.6 percent

Distribution network problems
8.3 percent

Inadequate or interrupted treatment
23.9 percent

Figure 24-14. Waterborne-disease outbreaks by deficiency in community and noncommunity water systems.

forms with chlorination, more resistant pathogens not detected by the testing procedure may survive.

4. The key to disinfection is reliability. Reliability is enhanced by yoked-up cylinders; switchover devices; auxiliary power; dosage applied in response to output; automatic residual recording; loop-controlled feed; adequate contact time; pH, temperature, and turbidity considerations; standby equipment or spare parts; and reports to substantiate operation.

A simple correction can often enhance reliability. For example, during investigation of a recent outbreak, examination of the chlorination facility showed that hypochlorite was pumped from a day tank in response to float operation in a reservoir. The operator duly noted in the record every day that the chlorine solution level in the day tank was "O.K." After an unsuccessful wait for the pump to cycle during the inspection, it was discovered that the electrical contacts in the relay box were so corroded that the pump was not receiving the signal. The solution to this problem was a simple correction in record keeping: the recorded level of the chlorine solution would be used to indicate whether the pump was operating. It took only a few minutes for the operator to realize that if he regularly measured a 250–350 millimeter (mm) (10–12-in.) decrease in the solution level on a daily basis and suddenly found only a 75-mm (3-in.) decrease, something was wrong.

Cases of illness attributed to deficiencies in public water systems are shown in Figure 24-15 and are further classed according to community and noncommunity systems in Figure 24-16. Inade-

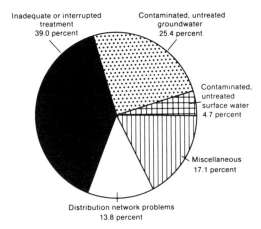

Figure 24-15. Cases of illness by deficiency in public water systems.

quate or interrupted treatment was responsible for the greatest number of cases of illness in community systems, whereas use of contaminated, untreated ground water caused the most illness in noncommunity systems.

The states with low and high outbreak rates (Figure 24-5) were tabulated, and comparisons were made based on the source of water supply and treatment provided. The comparisons were limited to community water systems identified in the 1963 U.S. PHS inventory. In Table 24-3 ratios of systems using ground water to those using surface water were computed for states with low and high outbreak rates, and the medians of the ratios were compared. The results indicate that water systems

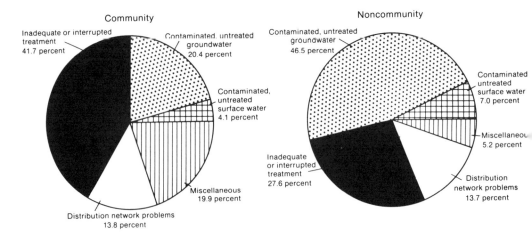

Figure 24-16. Cases of illness by deficiency in community and noncommunity water systems.

Table 24-3. Comparison of Sources of Water Supply[7] for Community Water Systems with Low
and High Outbreak Rates

Outbreak Rate	State	Systems Using Surface Sources	Systems Using Ground Sources	Ratio of Ground to Surface Sources
Low	Delaware	5	36	7.2
	Georgia	80	327	4.1
	Illinois	118	786	6.7
	Iowa	33	666	20.2
	Kansas	73	400	5.5
	Kentucky	130	123	0.9
	Louisiana	43	235	5.5
	Minnesota	23	578	25.1
	Mississippi	3	230	76.7
	Missouri	87	341	3.9
	Nebraska	3	427	142.3
	Nevada	9	43	4.8
	South Dakota	11	237	21.5
	Texas	172	960	5.6
	Utah	32	249	7.8
	Wisconsin	21	404	19.2
	Total	843	2661	a
High	Alaska	25	11	0.4
	Colorado	92	196	2.1
	Idaho	32	147	4.6
	Massachusetts	54	168	3.1
	New Hampshire	50	52	1.0
	New York	296	653	2.2
	Oregon	123	197	1.6
	Rhode Island	8	35	4.4
	South Carolina	63	148	2.3
	Vermont	41	116	2.8
	Wyoming	27	69	2.6
	Total	811	1792	b

[a] Median ratio = (6.7 + 7.2)/2 = 6.95.
[b] Median ratio = 2.3.

in states with a low outbreak rate depended more on ground water than on surface water as a source of supply. This is a reasonable expectation in that ground water generally has better microbiological quality than does surface water. This does present a conflict, however, in that more outbreaks are attributed to use of ground water (for noncommunity systems). The conflict may be explained by the premise that community water systems are better operated, with disinfection in place where it is required.

Table 24-4 shows a comparison between states with low and high outbreak rates and the number of systems providing treated and untreated surface water. Treatment was defined in the 1963 U.S. PHS inventory as "any action taken upon the water"; therefore, for example, communities that treated surface water with aeration were included.

Comparison of the ratios indicates that states with a low rate of outbreaks had a much higher ratio of systems treating water to those not treating than did states that had a high rate of outbreaks. The fourfold difference in the ratios for low and high rates reinforces the need for adequate treatment of surface sources.

A similar comparison for the treatment of ground water is shown in Table 24-5. Comparison of the ratios indicates more systems provided treatment of ground water in the states with low outbreak rates.

SUMMARY

In the 35-year period from 1946 to 1980, 672 waterborne-disease outbreaks affecting more than

Table 24-4. Comparison of Community Water Systems Providing Treated and Untreated Surface Water[7]
in States with Low and High Outbreak Rates

Outbreak Rate	State	Systems Providing Treated Surface Water	Systems Providing Untreated Surface Water
Low	Delaware	5	0
	Georgia	79	1
	Illinois	118	0
	Iowa	33	0
	Kansas	73	0
	Kentucky	127	3
	Louisiana	43	3
	Minnesota	23	0
	Mississippi	3	0
	Missouri	87	0
	Nebraska	3	0
	Nevada	9	0
	South Dakota	11	0
	Texas	166	6
	Utah	21	11
	Wisconsin	21	0
	Total[a]	822	24
High	Alaska	13	12
	Colorado	77	15
	Idaho	22	10
	Massachusetts	48	6
	New Hampshire	47	3
	New York	275	21
	Oregon	118	5
	Rhode Island	8	0
	South Carolina	63	0
	Vermont	21	11
	Wyoming	25	2
	Total[b]	717	85

[a]Low rate = 822/24 = 34.2.
[b]High rate = 717/85 = 8.4.

150,000 persons were reported. The increase in the annual occurrence of outbreaks that began in 1967 was probably due to better reporting from a few states. If this is true, the statistics for the United States during this 35-year period are greatly understated. The reporting of outbreaks is improving, if an increase in the number of outbreaks is used as an indicator. If the data for 1976–1980 are used as an attainable level of sensitivity under the current reporting procedure, the average annual frequency of outbreaks is 38. Applying this value for annual occurrence to the 35-year period increases the period total to 1330, or nearly double the number reported. A similar analogy applied to illness shows a frequency of 10,000 cases per year, which applied to the 35-year period results in 350,000 cases of illness—considerably more cases than the number reported.

Clustering was evident when outbreaks in community systems were geographically distributed and rate-based according to the number of water systems in a respective state. Water systems in states with low outbreak rates were compared with those in states with high outbreak rates by using source of supply and treatment as indicators. Those states with low outbreak rates had a greater ratio of systems that depended on ground water as a source of supply and that provided treatment of surface water and ground water.

Outbreaks exhibit temporal distribution, with peak occurrences in June, July, and August. This distribution was more pronounced when outbreaks in community systems were compared with those in noncommunity systems. The number of outbreaks in noncommunity systems greatly exceeded those in community systems for the summer months.

Table 24-5. Comparison of Community Water Systems Providing Treated and Untreated Ground Water[7]
in States with Low and High Outbreak Rates

Outbreak Rate	State	Systems Providing Treated Ground Water	Systems Providing Untreated Ground Water	Ratio of Treated to Untreated Ground Water
Low	Delaware	22	13	1.6
	Georgia	155	171	0.9
	Illinois	575	211	2.7
	Iowa	387	272	1.4
	Kansas	399	1	399.0
	Kentucky	92	31	3.0
	Louisiana	112	123	0.9
	Minnesota	287	291	1.0
	Mississippi	123	107	1.1
	Missouri	150	191	0.8
	Nebraska	39	387	0.1
	Nevada	13	30	0.4
	South Dakota	75	148	0.53
	Texas	524	436	1.2
	Utah	68	181	0.4
	Wisconsin	217	173	1.3
	Total	3238	2767	[a]
High	Alaska	9	2	4.5
	Colorado	98	98	1.0
	Idaho	40	106	0.4
	Massachusetts	63	101	0.6
	New Hampshire	14	38	0.4
	New York	279	366	0.8
	Oregon	63	128	0.5
	Rhode Island	19	16	1.2
	South Carolina	49	99	0.5
	Vermont	10	105	0.1
	Wyoming	30	39	0.8
	Total	674	1098	[b]

[a] Median ratio $= (1.0 + 1.1)/2 = 1.05$.
[b] Median ratio $= 0.6$.

Noncommunity systems serve recreational areas that experience heavy visitation during the summer, causing maximum demand on water systems and overloads to sewerage-sewage facilities. Noncommunity systems suffer from design, construction, operation, and maintenance problems and normally receive less attention from regulatory agencies in terms of monitoring and inspection to correct the problems.

Although outbreaks in noncommunity systems outnumber those in community systems, the frequency of outbreaks is almost twice as great in community systems. However, outbreaks in noncommunity systems are more difficult to detect because they serve the traveler who may become ill 1600 kilometers (1000 mi) from the point of

exposure to contaminated drinking water and would not associate the illness with a place he visited two days earlier. Although underreporting of outbreaks occurs for community systems, it is thought to be a much more serious problem for noncommunity systems.

Cases of illness occurring in community water systems far outnumber those in noncommunity systems. Regulatory agencies should concentrate their efforts on protecting the source of water supply, ensuring adequate and reliable treatment, and improving distribution network practices in community systems to achieve the greatest overall reduction of waterborne illness.

Microbiological agents that cause waterborne outbreaks are rarely isolated from the water system.

Investigators are more successful with identification of agents in specimens from cases where the agent is propagated and excreted in large numbers. For more than half of the outbreaks, no agent is identified, probably because of the lapse in time between occurrence of illness and the beginning of an investigation that includes collection of specimens.

The deficiencies in water systems that caused and contributed to waterborne outbreaks during this 35-year period differed little from those reported for the previous 26 years (1920–1945). The glaring deficiencies were that disinfection was not in place where it was needed and not properly operated where it was in place. Changes in regulatory and industry approaches are indicated and may include scrutiny in approval, frequent inspection, operator training, and expanded monitoring from regulatory agencies with improvements in design, operation, and equipment from industry.

NOTES

The authors thank V. Tilford, secretary, Health Effects Research Laboratory, Cincinnati, Ohio, and B. Lippy, journalism student, Ohio University, Athens, Ohio, for typing, proofreading, and editing the manuscript.

1. A. E. Gorman and A. Wolman, 1939, Waterborne outbreaks in the United States and Canada and their significance, *Journal AWWA*, 31(2):225.

2. R. Eliassen and R. H. Cummings, 1948, Analysis of waterborne outbreaks, 1938–45, *Journal AWWA*, 40(5):509.

3. S. R. Weibel et al., 1964, Waterborne-disease outbreaks, 1946–60, *Journal AWWA*, 56(8):947.

4. A. Taylor et al., 1972, Outbreaks of waterborne disease in the United States, 1961–1970, *Journal of Infectious Diseases*, 125(3):329.

5. G. F. Craun, 1981, Outbreaks of waterborne disease in the United States: 1971–1978, *Journal AWWA*, 73(7):360.

6. *Water-related disease outbreaks–Annual summary*, (Atlanta: Centers for Disease Control, Department of Health and Human Services, 1978).

7. G. F. Craun and L. J. McCabe, 1973, Review of the causes of waterborne-disease outbreaks, *Journal AWWA*, 65(1):74.

8. *Statistical summary of municipal water facilities in the United States*, Publication 1039 (Washington, D.C.: U.S. PHS, 1965).

25

THE DESERT BLOOMS—AT A PRICE

DAVID SHERIDAN

SANTA CRUZ BASIN, ARIZONA

Standing atop the bank of the Santa Cruz River a few miles northwest of Tucson, one finds it almost impossible to imagine what this floodplain looked like a hundred years ago. Then water flowed through an unchanneled river that wound sluggishly across a flat marshy area. Trout were abundant. Beavers built dams. There were giant cottonwood, mesquite, willow, sycamore, and paloverde, and grass—grass tall enough to "brush a horse's belly," to shelter wild turkeys. Meandering, ungullied tributary creeks fed the river.[1]

Today the river channel is dry, a broad trench filled with nothing but gravel and sand. The river's bank is a bare dirt wall. Mesquite trees, 4 to 6 ft tall, grow along the trench. Some of the mesquite clumps are so thick they are impassable. Where the mesquite have not taken hold, the ground is bare except for a rare patch of grass. Farther back from the trench the mesquite give way to desert shrubs, especially white-thorn (*Acacia species*) and creosote, as well as cacti such as ocotillo. At irregular intervals dry gullies—the river's tributaries—intersect the trench wall.

There is no question that ground water overdraft in the Santa Cruz Basin is as severe as anywhere in the United States. In the Lower Santa Cruz, where some 552,000 acre · ft of ground water are overdrafted every year, agriculture is the prime water consumer. The federally subsidized Central Arizona Project, now under construction, will deliver a yet to be determined amount of Colorado River water to this desert sometime in the 1980s (see Figure 25-1). But even this imported water may only temporarily decrease the level of ground water overdraft. If present water use patterns continue, the annual overdraft will again exceed half a million acre-feet per year by the year 2020.[2]

Thirsty Tucson

The ground water situation in the upper Santa Cruz Basin is, if anything, even more tenuous because the booming city of Tucson competes with agriculture and the area's copper industry, and the amount of water available for pumping is less. Tucson draws its water from the Upper Santa Cruz and Avra Valley basins to the west of the city. At present rates of consumption the Upper Santa Cruz's aquifers will be exhausted, for practical purposes, within a hundred years. The annual ground water overdraft now totals about 236,000 acre · ft of water.[3] The Avra Valley aquifers will also be exhausted within a hundred years at current rates of consumption.[4]

Tucson, which averages about 11 in. of rain per year, is the largest city in the United States to rely entirely on ground water.[5] There are wells in the Tucson area in which the water level has dropped 110 ft in the last ten years.[6] Tucson currently pumps water out of the ground at five times the rate nature puts it back in.[7] The city also consumes some water that was deposited more than five thousand years ago, so-called fossil water.[8]

And yet Tucson continues to grow and attract new industry. In 1977 the voters of Tucson recalled the then-existing city council, which favored controlling growth through increased city water rates.[9] At its current growth rate the population of the Tucson metropolitan area, now almost 450,000, will rise to about 652,000 by 1990.

The city of Tucson has purchased numerous irrigated farms in the Upper Santa Cruz Basin and

David Sheridan is a Washington, D.C.–based free-lance writer and consultant who specializes in natural resource and environmental issues. His articles have appeared in *Life, Saturday Review, Smithsonian,* and *Sierra*. This material is taken from *Environment*, 23(3) (April 1981): 6–20 and 38–41. Reprinted by permission of the Helen Dwight Reid Educational Foundation. Copyright © 1981 by Helen Dwight Reid Educational Foundation.

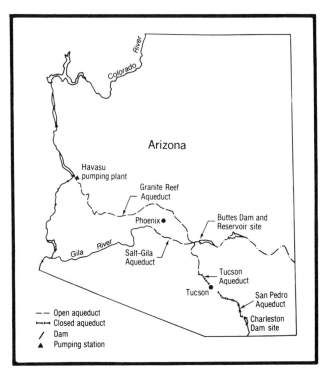

Figure 25-1. The Central Arizona Project. Currently under construction, this project will carry 1.2 million acre·ft of Colorado River water annually through a 250-mi-long pipe to the desert land and cities of central Arizona. This project is being heavily subsidized by federal taxpayers. (*Source:* U.S. Water and Power Resources Service.)

Avra Valley in order to gain control of the farms' wells. To date, the city has retired about 12,000 acres of farmland, and mining companies have bought up and retired another 8000 acres.[10] Tucson officials anticipate the need to purchase about 36,000 acres by the mid-1980s and have budgeted about $20 million for that purpose.[11] This purchase will effectively end irrigated agriculture, mostly pecan trees and cotton, in the Avra and Upper Santa Cruz valleys.[12] Agriculture, notes an Arizona geographer, is "expendable in this area."[13] As a consequence, the acres of once-plowed fields that are "retired"—that is, abandoned to wind and weeds—will grow right along with Tucson.

Water conservation—the more efficient use of water— has not been pursued as a major policy alternative by Tucson in dealing with its ground water overdraft problem. The city's voluntary water conservation effort is geared toward keeping water consumption in the peak summer months within the water system's capacity—151 million gal a day—rather than bringing about long-term and fundamental changes in water-conserving habits.[14]

Tucson has begun to recycle water. Effluent from one of the city's three water treatment plants is now used to water a municipal golf course. But the area's 16 other golf courses still use ground water to keep the desert green.[15] Apparently, Tucson sees the acquisition of additional ground water sources through displacement of agriculture and the importing of Colorado River water through the Central Arizona Project as the means for stretching its resources.

Uncertainties loom ahead, however. Legal problems have arisen. Tucson relies, in part, on water pumped from wells south of the city on the edge of the Papago Indian San Xavier Reservation. The Papagos are now suing the city of Tucson and other non-Indian interests (including a mining company), claiming the pumping has caused some wells on the reservation to dry up and the water levels in others to drop precipitously. They are seeking to restrain all ground water withdrawals off the reservation that affect ground water levels on the reservation. This case could have far-ranging implications for Tucson's future water supply.[16]

In addition, whether Tucson will actually get Colorado River water remains open to question. A recent National Science Foundation–sponsored report observes that "it is neither certain that the Central Arizona Project will ever reach Tucson nor that the requested allocations will or can be granted or delivered."[17]

Even assuming that Tucson gets the Colorado River water it has requested, the city's and the Upper Santa Cruz Basin's long-term water problems are far from over. The University of Arizona's Water Resources Center made the following assessment of the area's water prospects:

Central Arizona Project water will counter urban and mining depletions of ground water and for a brief period of time there will be a dependable water supply in the basin. However, by the year 2005, an estimated dependable supply of 191,000 acre-feet will be exceeded by non-agricultural demands for 237,000 acre-feet.

The Basin supply-use picture could be further clouded if the Department of Justice and Indian residents on the San Xavier Reservation are successful in their lawsuit.[18]

Reasons for Concern

One may ask, so what? Why should the water problems of the Upper Santa Cruz Basin trouble anyone but the residents of that area? The agricultural output that will be lost is insignificant in national terms.

Two reasons come to mind. In the first place, the United States depends on the Upper Santa Cruz Basin for roughly one-fourth its supply of copper, an essential metal. There are five major open-pit copper mines in the basin and another under development. These mines produce an estimated 200,000 tons of sulfide-copper ore daily. It takes roughly 210 gal of water per ton to mine and concentrate this ore. The concentration of the ore through a flotation process is especially water-intensive and getting more so because it takes more water to recover copper from the increasingly lower-grade ores being mined. In 1960 the ore mined here contained, on average, 1 percent copper. Today, it contains 0.5 percent.

At present production the copper industry in the Upper Santa Cruz Basin consumes about 50,735 acre·ft of water per year.[19] To date, water scarcity has caused only minor disruptions in copper production. The worry is that as ground water supplies become further depleted, a time will come when water scarcity will prevent the copper industry here from responding to the nation's copper demand, particularly if political events interfere with the export of copper from Chile, Zambia, or Zaire, or if there is a surge in domestic demand like that which occurred during the Korean and Vietnam wars.

At present, mining accounts for 27 percent of the ground water depletion in the Upper Santa Cruz Basin, while urban users account for 29 percent and agriculture 41 percent.[20] Mineral industry officials see agriculture as the chief culprit in the area's water plight. For example, Tom Chandler, attorney for Anamax (a mining partnership between Anaconda and Amax), is quoted as saying:

The farmer leases land from the state at a cheap rate, plus he gets power breaks, a cheap tax rate, government subsidies. He's pumping the copper mining industry dry; he's pumping the state dry; he's pumping the Indian people dry; and when he's got that done, he's going to move to La Jolla and raise martinis.[21]

A second reason for concern about this arid area's massive ground water overdraft problem is that the U.S. taxpayers are being called on to finance its solution. Currently under construction, the Central Arizona Project will divert 1.2 million acre·ft of water per year from the Colorado River near Havasu, Arizona. The water will be lifted 2000 ft and transported to metropolitan areas and irrigators in the middle of the state by means of a 250 mi-long pipe.

About $270 million has been spent to date. And it will take $1.5 billion to complete the project, according to current estimates. But the cost to the U.S. taxpayer may eventually run much higher. A recent study, sponsored by the Andrew Mellon Foundation and under the auspices of the National Audubon Society, reportedly estimates that the Central Arizona Project will, over the next 50 years, cost U.S. taxpayers $5.4 billion.[22]

The pattern is a familiar one in the arid West. A local economy is built and thrives on the depletion of ground water; then when it becomes apparent the resource will not last, an expensive water import project is launched. In his book *Killing the Hidden Waters*, Charles Bowden put the water problem of areas such as the Santa Cruz Basin into an illuminating perspective:

Water is energy, and in arid lands it rearranges humans and human ways and human appetites around its flow. Ground water is a nonrenewable source of such energy. . . . Humans build their societies around consumption of fossil water long buried in the earth, and these societies, being based on a temporary resource, face the problem of being temporary themselves.[23]

Environmental Consequences

The environmental consequences of ground water overdraft, while not so dramatic as the economic consequences, are also worth noting. One of the most obvious consequences is the drying up of once-perennial streams and rivers where they are in "hydraulic contact" with the ground water that is being overdrafted.[24] For example, the Santa Cruz River ran dry in the 1930s because of "the advent of deep-well turbines which are capable of pumping water in excess of the recharge rate."[25]

Land subsidence caused by ground water overdraft is occurring in various parts of central Arizona. Land near Eloy, Arizona, for example, has subsided as much as 10 ft in the last 30 years. Cities such as Casa Grande and Tucson have also experienced subsidence.[26]

Accompanying subsidence in central Arizona are earth fissures and faults. They vary in size, but some fissures measure as much as 25 ft wide and 60 ft deep.[27] More than seventy-five cracks in the earth have been found in central Arizona. A U.S. Department of Agriculture (USDA) report describes the fissures in this area (the lower Santa Cruz and Willcox basins) as ruptures of valley soil (alluvium) that have "disrupted local drainage and irrigation water application, damaged wells and canals, caused misalignment of highways and endangered homes."[28]

They have also caused gullying. Researchers from the University of Arizona observed in this area, for example, "the creation overnight of a gully 5 feet deep, 6 feet wide, and 25 feet long where an existing drainage was breached by the surface opening or fissure after (or during) a [rain] storm."[29]

Keith concludes: "The amount of land lost to the use of man by fissuring, faulting, and subsequent gullying can only be speculated on."[30] One thing does seem certain, however—the fissuring and faulting of the earth is expected to continue as long as the ground water overdraft continues.[31]

Perhaps the most serious environmental consequence of ground water overdraft in the Santa Cruz Basin is the abandonment of irrigated cropland. A recent analysis of Arizona agriculture reported:

Long-term intensive pumping in this area has lowered water tables to a point at which production of some crops is now marginal. Minor fluctuations in fuel or power costs of pumping and commodity prices are sufficient to cause financial losses and have forced some abandonment of fields or shifts to high value crops. Farmers throughout southern Arizona who use ground water face similar situations.[32]

Of the 549,100 acres of irrigated land in the Santa Cruz–San Pedro Basin, 369,800 acres are in production, and 157,800 are "idle" but may be returned to production. Some 53,000 acres of formerly irrigated land have been abandoned.[33]

Lacking any natural cover, these abandoned fields and the "idle" fields with sandy and loamy soils fall quick prey to wind erosion. Blowing dust from abandoned or "idle" fields in the Lower Santa Cruz Basin has been so severe at times that nearby interstate highways have had to be closed.[34]

The amount of irrigated acreage in this area is expected to decline in the years ahead, but by how much remains uncertain. A 15 to 20 percent decrease by the year 2000, as projected by some economists, could mean that an additional 82,000 to 110,000 acres of cropland end up abandoned, that is, producing dust and tumbleweed.[35]

To see one of the abandoned fields in the lower Santa Cruz Basin of Arizona, its desiccated surface scoured by the wind, its irrigation ditches choked with sand, is to be reminded that arid land can be a merciless place for those who try to domesticate it.

Controlling Ground Water

Given the long-term economic and environmental consequences of massive depletion of ground water, it is remarkable that the state of Arizona has done so little to manage and conserve this resource. Until very recently, that state's water code treated ground water as essentially a property right rather than a public resource. Aside from designating basins where the withdrawal rate exceeded the replenishment as "critical ground water areas," the state had done nothing substantive to control its use.

Secretary of the Interior Cecil D. Andrus warned Arizona that unless the state took effective action, the Central Arizona Project would be delayed. This threat apparently provided the impetus for the competing water interests in the state—agriculture, the mining industry, and the cities—to negotiate an agreement on how ground water should be allocated and conserved. A consensus was hammered out, and the state enacted a comprehensive water management and conservation law in June 1980.

Aimed at achieving a "safe yield" in ground water use by the year 2225, the new law requires a statewide registration of all wells; mandatory water conservation in the state's three major urban areas (Phoenix, Tucson, and Prescott) and its major agricultural area (Pinal County), a large portion of which comprises the Lower Santa Cruz Basin; and empowers the director of a new Department of Water Resources to set per capita consumption limits for cities and to purchase and retire the water rights of irrigated farms in this area after the year 2006. The law prohibits new growth in areas where the developer cannot assure that a water supply will exist for at least 100 years and new irrigated agriculture in ground water problem areas.

Arizona Governor Bruce Babbit, who signed the law, states:

> In the old West, we're going overnight from a laissez-faire system, a system where everybody used whatever they wanted wherever they wanted, to the most comprehensive ground water management system of any state in the American West.[36]

Obviously, many questions remain unanswered about Arizona's new law. It is too soon to tell whether or not the state will be able to accomplish the exceptionally difficult task it has undertaken—the allocation and conservation of its most important scarce resource, ground water. But by passing such a law, Arizona has acknowledged that ground water is a finite resource and should be managed accordingly. This realization alone is historic.

GAINES COUNTY, TEXAS

On a spring morning, as you drive east of Hobbs, New Mexico, out across the southern High Plains, with the wind at your back, you can see a brownish haze hugging the land ahead. Soon it envelops you. The sky is cloudless, but you cannot see the sun. Visibility is diminished to about a quarter of a mile. You have entered Gaines County, Texas, and the substance you see in the air is topsoil, blown off fields that are bare and dry because they have been plowed in preparation for planting cotton.

Forty years ago this was grassland where ranchers grazed cattle. Today it is the ninth most productive county in terms of cash crop output in Texas—a state whose crop output ranks third in

the nation.[37] In 1977, a drought year here, Gaines County farmers produced $73 million worth of crops, mostly cotton. In 1976, a wetter year, they produced $76.2 million.[38] The land has paid dearly for this production, however.

Soil Erosion

Gaines has a soil erosion problem. It also has a serious ground water depletion problem.

A General Accounting Office (GAO) study team evaluated 39 farms in Gaines County and found that 31 of them were suffering an annual soil loss of 40 tons or more per acre. Of the 10 counties (283 farms) across the country that GAO studied, Table 25-1 indicates that Gaines County had the worst soil erosion.

In the spring of 1979 indications were that the soil erosion in Gaines County was every bit as bad as it was at the time of the study. Soil could be seen blowing along the ground and over roads even during moderate wind conditions–5 to 10 mi/h. Soil drifts all but covered 4-ft-high fences and strips of unharvested wheat and alfalfa. "Blowout" areas in fields were a common sight, and piles of dirt leaned against telephone poles and walls.

There are approximately 750,000 acres of cropland in Gaines County—400,000 of which is dryland-farmed and 350,000 irrigated with ground water. Cotton is grown on about 90 percent of this cropland, and wheat and alfalfa account for much of the remainder. There are about 150,000 to 175,000 acres of natural grassland left, but it is being plowed up at a rate of about 10,000 to 15,000 acres per year in order to plant more cotton.[39] About 25 percent of this rangeland appears overgrazed. Particularly noticeable is the relative pau-

Table 25-1. Soil Erosion in Six Counties
in the Arid West

County	Farms in Sample	Estimated Annual Soil Loss (tons/acre)				
		0–5	5.1–10	10.1–20	20.1–40	Over 40
Gaines, Texas	39	1	0	2	5	31
Roosevelt, New Mexico	28	2	7	9	10	0
Finney, Kansas	35	1	23	2	9	0
Benton, Washington	20	0	11	8	0	0
Whitman, Washington	30	5	14	11	0	0
Burleigh, North Dakota	11	7	4	0	0	0

Source: U.S. General Accounting Office, *To protect tomorrow's food supply, soil conservation needs priority attention* (Washington, D.C.: U.S. Government Printing Office, 1977), 5.

city of native grasses and the abundance of mesquite. One soil conservation service official estimates that most of the county's grassland produces approximately one-third to one-fourth as much grass as it did in its native state—short-grass and mid-grass prairie.[40]

Gaines County's average annual rainfall is about 16 in., but the rainfall varies erratically from year to year. Since 1923 annual rainfall has ranged from a low of 6.6 in. in 1956 to 37.6 in. in 1941.[41]

The land here is relatively flat—this is high plains.[42] Prevailing winds are strong and southwesterly from November through April, the period when so much of the ground is bare of vegetation. The fine sand soils that predominate are fine-grained and do not hold moisture well. Hence they are highly erodible.[43]

According to experts in the Soil Conservation Service (SCS) and the Agricultural Stabilization and Conservation Service (ASCS) who are familiar with Gaines County, ignorance is not the problem. District conservationist Walter Bertsch reports:

> The great majority of farmers know what has to be done to stop the soil from blowing, but they can't afford to do it. In the short run, they've got a bank loan to meet.[44]

Controlling Soil Erosion

What is being done to control soil erosion in Gaines County? In the main, the time-tested conservation practices are little used.

Crop Rotation
The Gaines County Soil and Water Conservation District's *Program and Plan of Work* states: "Crop rotations of milo, wheat, or other crops high in organic matter are needed to maintain the fertility of the soil and help protect if from erosion."[45] Nonetheless, most farmers here plant cotton year in and year out. Prices on the commodity market are such that a farmer can earn more per acre planting cotton here than wheat, alfalfa, milo (sorghum), or any other crop. Also, cotton requires less water per acre to grow in this climate. Last and by no means least, USDA policies encourage cotton over wheat or feed grains in arid areas such as Gaines.

The ASCS' formula for computing disaster benefits favors cotton:

$$\text{Established crop yield per acre} \times 75\%$$
$$\times \text{ planted crop acreage.}$$

Moreover, to be eligible for disaster payments or federal cost share programs such as the ASCS Agricultural Conservation Program, a farmer who plants wheat or feed grains must set aside 10 percent of his normal crop acreage in that crop. For example, let us look at a Gaines County farmer who did practice crop rotation in 1977 on his 640 acres of land, with 320 acres in grain sorghum and 320 in cotton. In 1979 he can harvest only 288 acres of grain sorghum, setting aside 10 percent or 32 acres. For cotton, however, there is *no* set-aside requirement. He can plant all 640 acres of his land in cotton. Obviously, it pays to do so in the short term, and that is exactly what most Gaines County farmers do.

In the long term, one-cropping this land year after year in cotton will lead to reduced soil fertility just as it did in the southeastern United States. According to Jim McGehee, the ASCS executive director in Gaines County, "A Gaines County farmer really has no choice under current market conditions and government policies; he has to plant cotton if he is going to have any chance at all of making ends meet."[46]

Conversion to Grass
The trend in Gaines County is in the opposite direction, even though a good share of the land that is being plowed up is class VI (highly erosive) land. What about the federal government's Great Plains Conservation Program? It was specifically designed to treat the kind of land found in Gaines County, that is, high-erosion cropland that needs permanent vegetative cover. McGehee of ASCS reports:

> The Great Plains Conservation Program has been a big bust here. It simply is not economically feasible for a farmer to convert his cropland to range under this program. The average farm size here is about 525 acres. You cannot support yourself and your family today raising cattle on that amount of land in this arid area.[47]

James Abbot of SCS vigorously disagrees that the Great Plains Conservation Program has been a bust. It has been successful, he reports, in revegetating private rangeland and in developing management plans with the ranchers so the land is no longer overgrazed. "It would have been more successful at converting high erosion cropland from cotton monoculture to a crop system that provides better cover and residues if the government's commodity adjustment programs and disaster relief had not made this abuse of the land financially advantageous."[48]

Irrigation Practices
Much of the federal soil conservation money spent in Gaines County through such programs as the ASCS Agricultural Conservation Program has, in

fact, gone for the installation of irrigation systems or for deep plowing.[49] The Great Plains Conservation Program has spent 37.5 percent of its cost share funds for grass plantings, 53 percent on irrigation practices, and 9.5 percent on livestock-watering facilities and fencing.[50]

A moist soil is less likely to blow than a dry one; hence irrigation does qualify as a soil-conserving practice even though its primary purpose is to increase production. However, even the huge, quarter-mile-long, central-pivot irrigation systems used in Gaines County cannot cover all of the land all of the time; furthermore, the soils dry quickly. Therefore the unwatered soil on irrigated land blows. Indeed, during the windswept winter and spring months, the irrigated cropland is virtually indistinguishable from the unirrigated cropland. Furthermore, irrigation has become increasingly expensive.

In the previously mentioned GAO study, 10 of the 30 Gaines County farms sampled were participating in some federal cost share, soil conservation program. The annual average soil erosion from these farms was 40 tons per acre, compared with 50 tons per acre from those farms not participating in the federal programs. So the federal effort is having some effect, especially when it comes to minimum tillage, but it falls far short of solving the county's terrible soil erosion problem.[51]

The Ogallala Aquifer

James Abbott of SCS warns that at the current rate of soil erosion in Gaines and adjoining counties, "we will eventually exhaust the resource."[52] Gaines County farmers, however, may run out of water before they run out of soil (that is, the water will become too expensive to use for agriculture). They are using up their only water source—the ground water—at more than twice the rate of natural recharge; they are "mining" ground water.[53]

Gaines County sits on the southern end of an underground reservoir known as the Ogallala Aquifer that stretches all the way to Nebraska. Vast quantities of water are stored in this layer of sand and gravel laid down during the late Miocene and Pliocene epochs. Beneath Gaines County alone there are 9.2 million acre·ft of water.[54] The entire billion-dollar-plus agricultural economy of the Texas High Plains is built upon the overdraft of water from the Ogallala.[55] (For further discussion of the use of the Ogallala Aquifer for irrigation see "The Irrigation Revolution," *Environment,* October 1979.)

How long the water will last is a tantalizing and difficult question. Its answer depends, in large part, on energy prices, because it takes energy to pump that water out of the ground, and the more you pump, the more energy it takes. Ten years ago this underground water cost Gaines County farmers about $1.50 per acre-foot to pump. Today it costs about $60 per acre-foot.[56] The increase is due to the increased price of the fuel used to power the pumps and to the fact that during this time the water level in Gaines County wells dropped on average about 12.8 ft.[57]

The Texas Department of Water Resources reports that the overdraft of the Ogallala Aquifer in Gaines County is "expected to continue, ultimately resulting in reduced well yields, reduced acreage irrigated, and reduced agricultural production."[58] The department projects that the amount of water stored in the aquifer beneath Gaines County will decline to 7.9 million acre·ft by 1990 and 5.6 million acre·ft by 2020. The water level in wells is projected to drop at an average rate of 1.26 ft per year in the 1980s.[59]

These projections are, if anything, too conservative. They do not take into account the recovery of oil from inactive oil fields in Gaines County. The process currently being used involves pumping large quantities of water out of the Ogallala and injecting it into the inactive oil wells. As the price of oil continues to climb, recovery of this sort becomes increasingly lucrative. To date, oil recovery has accounted for the consumption of several hundred acre-feet of ground water in Gaines County where total ground water pumpage is about 241,000 acre·ft per year.[60] It is feared, however, that oil recovery will soon become a significant factor in the depletion of the area's ground water supply.[61]

Soil conservationists are concerned that rising energy costs and lowering water levels in wells will lead to the abandonment of once-irrigated cropland that then will become a prime source of dust storms and weeds.

The overdraft of ground water resources has a profound, long-lasting implication for Gaines County and the 38 other counties that comprise the arid High Plains of Texas. Several other counties are undergoing even more rapid rates of depletion than Gaines.[62] Charles Bowden predicts:

> By the 1980s water declines should make serious inroads in irrigated agriculture; thirty or forty years hence this commerce of pumped water should be over. The humans of the High Plains will be staring down tens of thousands of dry holes.[63]

Though somewhat less apocalyptic, the Texas Department of Water Resources also sounds a note of warning:

> If this overdraft continues, the aquifer ultimately will be depleted to the point that it may not be economically feasible to produce water for irrigation. . . . The

actions of the water users will determine whether the projections of this study come to pass.[64]

Given such concerns, Texas' continued nonregulation of ground water is difficult to fathom. The only method used to regulate the amount of ground water pumped on the Texas High Plains is well spacing. John Graves observes:

resource is ground water rather than pasture. Noting the prosperity of the resource depletors, the others increase their use of the common resource. The long-term consequence is, of course, ruinous.

Finally, it should be noted that the federal government has subsidized the rapid depletion of the Ogallala Aquifer—first by price supports for commodities such as cotton, then by crop disaster

---

FEDERAL DISASTER RELIEF

In recent years the federal government has spent several times more money on disaster relief in Gaines County than on soil conservation. In 1978 federal disaster relief payments to Gaines County farmers soared to $10 million, or about $13.33 per acre of cropland. While $10 million is a substantial sum, it is hardly unique in arid land agriculture on the High Plains of west Texas. For example, Dawson County, immediately to the east of Gaines, received $11 million in federal crop disaster payments in 1978, though it had less cropland but more dryland farming than Gaines.

There are 39 counties on the High Plains of west Texas. In 1977 they produced $1.4 billion worth of crops, and the farmers in them received $32 million in crop disaster relief—feed grains, wheat, and cotton—from the federal government. In Gaines County it is not at all unusual for a farmer to take out a $75,000 to $100,000 loan to cover his operating expenses for the coming year. And in a dry year like 1978, that farmer might not break even. For example, a farmer harvests 500 acres of cotton, but because of lack of moisture in the soil and the destructive effects of wind erosion, his yield is only 295 lb per acre. Assuming he sold his cotton for about 53 cents a pound, his balance sheet might look something like this:

Costs per acre	$221.50
Output per acre	156.35
Net per acre	− $ 65.15

In other words, he incurred an operating loss of $32,575 for the year. Under such circumstances, how is he going

to repay his $100,000 loan with the bank? His crop disaster payment from the federal government—about $15,000—will cover the $10,000 interest on the loan and leave $5000 for beginning to repay the principal. In this situation he will probably then renegotiate the loan, that is, go deeper into debt, and pray for rain. If it comes, he could earn $59,250 at current cotton prices (68 cents per pound) and begin to reduce his total indebtedness. If not, the farmer may seek additional low-cost financing from federal institutions such as the Farmers Home Administration (FmHA) or the Small Business Administration (SBA).

The main point is that the erratic climate and fine sand soils of arid Gaines County make the planting of any crop, even cotton, a high-risk endeavor. There are few if any private lending institutions that would consistently bet on the farmer in this farmer-versus-the-elements contest, and without their ante there would not be 750,000 acres of cropland in Gaines County. The private lending institutions are willing, however, to bet on the farmers' federal disaster relief versus the elements. In other words, federal disaster payments should be viewed, therefore, as subsidized insurance for banks as well as for arid-land farmers. The west Texas bankers profit and the farmers stay in business, for the most part. What is *not* at all clear is whether the general public gains or loses from this arrangement. Are the benefits—presumably lower cotton prices due to the increase in supply—greater than the costs—the taxpayer-supported crop disaster payments?

Texas law continues to regard most ground water as a mysterious blessing legitimately subject to capture and use in unlimited quantities by any property owner who digs or drives a well.[65]

As a consequence, there exists what economists call a "negative incentive" to conserve the resource. A recent report noted:

If one farmer does not practice water conservation while those around him do, the one who is profligate will benefit from the water belonging to those who are conservative.[66]

It is, in fact, the "tragedy of the commons" all over again.[67] Only this time the commonly held

payments, the various cost share "soil conservation" programs, and low-interest loans provided by the SBA and the FmHA. In addition, federal tax policy encourages the depletion of this resource. High Plains farmers are granted a depletion allowance on pumped ground water, thereby enjoying a tax break similar to that which the oil industry has had for many years and which the mineral extraction industry currently enjoys. The more water they consume, the less tax they pay!

The cost of all these various subsidies paid for by the American taxpayer has never been tallied; however, they might someday seem insignificant compared to what it will cost to rescue the agricultural economy of the Texas High Plains when the ground water becomes too expensive to pump. In our society billion-dollar-plus private economic interests do not lose their investments meekly. They seek aid from the federal government, that is, the general public. When the Ogallala Aquifer water runs out, the farmers, bankers, irrigation system manufacturers, fertilizer producers, and others who have built their livelihoods on the overdraft of this resource will form a powerful lobby.

THE SAN JOAQUIN VALLEY

Only a few hundred acres of land in the San Joaquin as yet wear a glistening mantle of salt. But salinization of the topsoil could spread to large stretches of this rich valley during the next 30 years. And although salinization is the most serious threat to the San Joaquin's productivity, it is not the only

one. In fact, all the major forces of "desertification" are at work: poor drainage of irrigated land; overgrazing; cultivation of highly erodible soils; overdraft of ground water; off-road vehicle damage to soil and vegetation.

The San Joaquin is the southern half of the great Central Valley of California, lying between the Coast Ranges and the Sierra Nevada (see Figure 25-2). The San Joaquin Basin encompasses 18.2 million acres—mountains, rolling foothills, and a flat valley floor. The valley floor and foothills total about 10 million acres.[68]

The San Joaquin is an arid land. Annual precipitation in the north averages about 14 in. per year and declines with movement southward, averaging about 5 in. in the southernmost Tulare subbasin. Nonetheless, this basin is one of the most productive agricultural areas of the world.[69]

The eight San Joaquin counties produced $4.76

Figure 25-2. The Central Valley Project. First authorized by Congress in 1933, the Central Valley Project is a "complicated concrete jigsaw" that is constantly being expanded. Financed by the federal government, it provides water for a half million people and irrigates crops that had a gross value of $13 billion in 1976. (*Source:* U.S. Water and Power Resources Service.)

billion worth of farm products in 1977, which is more than most states produce.[70] In fact, the San Joaquin outproduced all but three states—Iowa, Texas, and Illinois.[71] Major crops grown in the San Joaquin are cotton, grapes, tomatoes, barley, alfalfa, and sugar beets, as well as a variety of tree crops—including walnuts, almonds, oranges, and apricots. It also has a sizable livestock industry.

The San Joaquin is so productive for several reasons. Material eroded over millennia from the mountains on either side have accumulated in the valley to form a thick, rich soil. The growing season is long—most of the valley is frost-free for at least eight months. It possesses a Mediterranean climate with hot, dry summers and mild, moist winters. Of the basin's 4.8 million acres of cultivated cropland, 97 percent are irrigated.

Where does the water come from? Twenty percent is imported from outside the basin, mostly from the northern part of California.[72] This water is stored in reservoirs behind government-built dams and is delivered by government-built aqueducts (state and federal). The cost to the irrigator varies, but it is relatively inexpensive—ranging from about $12 to $35 per acre-foot. The state charges more for water from its projects than does the federal government.

Forty percent of the irrigation water comes from aquifers within the basin and 40 percent from streams. The San Joaquin's agricultural prosperity rests in part on the very shaky foundation of ground water overdraft. About 1.5 million acre·ft more water is pumped from the basin's aquifers each year than is naturally replenished. This overdraft fills 12.5 percent of the San Joaquin's average annual water supply.[73]

Agriculture dominates the San Joaquin, particularly the 8.5 million acres of valley floor. Over the past 150 years, first cattle and sheep grazing and then crop production (especially wheat, at the beginning) transformed the valley's natural ecosystem beyond recognition. Desert shrubs occupied portions of the valley between the coastal ranges and the valley trough, although most of the San Joaquin was grassland dotted with oak trees; surprisingly, large marshes and shallow lakes once existed there.

Today the native perennial grasses are gone, and the wetlands have almost disappeared. Gone also is most of the native wildlife. Major portions of the valley have become a crop-producing factory. Except for livestock grazing, development of the San Joaquin foothills came mostly after World War II. The combination of cattle and sheep overgrazing and the introduction of European plant species such as filaree desolated the native perennial grasses of both the foothills and the valley floor. Later, or-

chards, vineyards, and subdivisions were planted in the foothills.[74]

Salinization of Cropland

Today about 400,000 acres of irrigated farmland in the San Joaquin are affected by high, brackish water tables. Ultimately—by the year 2080—*1.1 million acres* of San Joaquin farmland *will* become unproductive unless subsurface drainage systems are installed.[75]

The poorly drained area runs along the west side of the valley for almost its entire length. A few of the highly saline "perched" water tables occur naturally, but many have been created by irrigation water percolating down from the surface.

All water, including fresh irrigation water, contains some salt. When the water is applied to a field for irrigation, some of it evaporates, some is consumed by the plants, and the remainder trickles down into the ground. The difficulty is that the sun and plants extract almost pure water from any water supply, and the water that is left and that trickles downward has a higher content of dissolved salts than when it was first applied.

Some irrigated areas, such as west Texas, are fortunated because underlying the fields is a thick stratum of permeable material that allows the salty unused irrigation water to drain deep, far below crop roots. Other irrigated areas, such as the Tigris-Euphrates Valley, California's Imperial Valley, and the San Joaquin, are not so fortunate. Not too far beneath the surface of the fields is a tight layer of material that blocks the water's downward passage. Hence the salty water builds up, perhaps adding to a water deposit that has already collected naturally or creating a new underground water deposit. In either case, as the deposit's volume increases, its level rises toward the surface—toward the roots of the crops.

The salty perched water table need not actually rise to root level to hurt the plants. If it comes within 5 ft of the roots, it will cause damage because some of the water will continue to rise through capillary action. When the salty water does reach a crop's roots, it inhibits the plants' ability to absorb moisture and oxygen. As a result, the plants either become stunted or die, depending on how concentrated the salt is in the water.

If the salty ground water reaches the surface of a field, it will evaporate and leave salt crystals behind. If enough salty water reaches the surface and evaporates, a salt crust will form over the soil, a crust that is relatively impermeable, thereby diminishing the natural leaching power of water falling on the field.

There is no technical mystery to solving the age-old irrigation problem of salinization. The solution is clear—drain off the excess ground water. First, the irrigator needs to install an on-farm drainage system to collect the saline ground water. To date, only about 40 percent of the San Joaquin's farms have on-farm drainage systems.

Next, the irrigator needs to dispose of the drained water. In some cases on-farm drainage can be combined with on-farm disposal by providing a few vertical wells or deep ditches on the perimeter of the fields so that the unused irrigation water drains into a water-holding stratum far beneath root level and beneath the layer of material that has been blocking its downward flow. The big danger of this procedure is that the salty water will drain into the aquifer system from which water is being pumped for irrigation. The great advantage to this approach is that there is no off-farm drainage water disposal problem.

An alternative, the one that is apparently most practical for farms in the San Joaquin's poorly drained areas, is to lay perforated pipes in parallel lines 6 to 10 ft beneath the fields. (Historically, such drains were made of clay tile, and today they are still referred to as "tile drains," although they are now usually made of concrete or plastic.) After collection in an on-farm sump, the saline water must then be pumped along a lined ditch to a master drain. From the master drain the saline water can be dumped into a natural salt sink, a salt lake, or an ocean, or into an evaporation pond created for the purpose. In a large shallow evaporation pond, the saline water evaporates, leaving a salt bed behind. Another option is to release irrigation drainage water into a naturally low-lying area, thus creating a saltwater marsh and wildlife refuge.

Drainage Disposal

The San Joaquin's natural sink is the Pacific Ocean, but there is no way that a master drain from the San Joaquin can reach the Pacific without causing political, economic, or environmental problems. The San Joaquin's drainage problems are in essence a drainage *disposal* problem. The San Joaquin drains northward into the Pacific through the Delta, a 1150-mi^2 area northeast of the San Francisco–Oakland area, where the Sacramento and San Joaquin rivers converge.[76] The fresh water from the Central Valley meets the salty ocean water in the area between the western Delta and San Pablo Bay.

One solution to the San Joaquin's drainage disposal problem would be to build a concrete-lined ditch some 290 mi along the valley's natural drainage course. This master drain would carry the salty ground water drained off farms all the way north to the Delta area, discharging it into Suisun Bay. An 82-mi segment of this drain, known as the San Luis drain, has already been built.

Such a project has indeed been proposed by a task force comprised of the federal Water and Power Resources Service (formerly the Bureau of Reclamation), the California Department of Water Resources, and the California State Water Resources Control Board.[77] But for their proposal to achieve a political consensus at both the state and federal levels, some very difficult questions will have to be resolved:

- Who will pay for the project? Should the farmers who will directly benefit pay? Should all farmers who use irrigation water in the San Joaquin pay? Should the state of California and the federal government subsidize some of the costs and, if so, to what extent?
- What impact will the annual discharge of some 250,000 acre·ft of salty drainage water into the tidal waters of Suisun Bay have on the Delta environment? Will it endanger the drinking water supply of the people who live in the Delta area? Will the arsenic, boron, and mercury[78] present in the drainage water reach toxic levels? Will the salts in the drainage water alter the subtle salt–freshwater balance of the Delta's complex ecosystem? In sum, will the project endanger one of California's last remaining great wetlands?

One alternative to discharging the drainage water into the Delta area would be to pipe it west over the Coast Range and discharge it directly into the Pacific Ocean. To do so, however, would be extremely costly both in dollars (capital and operating costs) and in energy. A tunnel through the mountains would reduce the operating costs but greatly increase the initial capital and energy expenditures. The proposed valley-long drain has the important advantage of letting gravity do much of the work of moving the drainage water.[79]

Another alternative would be to pump drainage water from the southern San Joaquin up into the Carrizo Plain, a valley within the Coast Range. There, a huge (80,000 acres) evaporation pond could be formed, or the entire valley could be turned into a salt lake. The drainage water from the northern San Joaquin would go into local evaporation ponds (9400 acres). A big advantage of this plan is that the Carrizo Plain is already a natural salt sink and no agricultural land would be taken out of production. The big disadvantage is that the energy costs of pumping the drainage water up 2000 ft to the plain would be high, ten times higher than for the master drain–Suisun Bay discharge alternative.[80]

There are other possibilities as well. For example, the drainage water could be run through strategically located desalinization plants and then reused for irrigation. The difficulty with this approach is that desalinization, even using the most up-to-date technology available, is costly and energy-intensive. It currently costs about $300 per acre·ft to desalinate salty water, not including disposal of the brine.[81] Of course, radical improvements in the technology will lead to radically reduced costs, but these improvements may not occur in time to solve the San Joaquin's drainage problem.

A strictly local solution to the drainage problem would also be possible. Resource conservation districts cover about 75 percent of the drainage problem areas, and each one could develop its own disposal system. They could then dispose of the drainage water in local evaporation ponds or low-lying areas suitable for the development of saltwater marshes.

There are two major difficulties with the localized approach, however. First, it is uncertain whether some of the resource conservation districts or whatever local entities took on the job could afford the initial capital investment needed to build drainage disposal systems. Secondly, because there will be such an enormous quantity of salty water to dispose of over the coming years, the local evaporation ponds could eventually (by 2060) take over 150,000 acres of San Joaquin Valley farmland.[82] The advantages of the local alternative, on the other hand, are that it leaves the Delta wetlands alone and it avoids the political hurdles that confront a big public works project in this era of tight budgets.

Conservation as an Alternative

More *efficient use* of irrigation water would mitigate the San Joaquin's drainage water disposal problem. Logically, the less water farmers apply to their fields in the first place, the less drainage water there will be to dispose of. It appears that water use per acre could be reduced without reducing agricultural productivity in the San Joaquin.

Incentives exist for San Joaquin farmers to use water more efficiently. If farmers are pumping ground water for irrigation, more efficient water use will cut their energy costs. And if they are buying imported water, more efficient use will cut their water costs as well. Of course, if the federal government raised its water rates to cover the full costs of its projects, the irrigator's incentive to conserve water would be even greater.

There is, however, one drawback to more efficient water use—less tailwater would be available for reapplication. So it is conceivable that some farmers might have to buy *more* imported water as a result of more efficient use. In other words, every acre-foot of consumption reduced through more efficient use is not necessarily an acre-foot saved.

The major advantages of more efficient water use are decreased energy costs, now a significant factor in the overall cost of irrigated agriculture, and less buildup of salty ground water.

Why, then, has not more efficient water use, sometimes termed water conservation, been achieved? Lack of technical information is probably one major reason.

Irrigators need to know what the most efficient irrigation method—drip, sprinkler, or surface flooding—is for particular crops and soils. They need to know the specific water requirements of each crop they cultivate and to have a water application system that can be controlled so as not to exceed those requirements. They need to know how to take an accurate reading of their root zone's moisture content so they do not overirrigate.

Irrigators also need to know the best time to irrigate so as to minimize evaporation. They need advice on field leveling for more even and efficient distribution of water during surface flooding. In short, the farmers are not conserving water because their irrigation practices are scientifically imprecise.

The prime government agency for transmitting technical information and advice to the farmer about water conservation is the U.S. Soil Conservation Service (SCS). Gylan Dickey, water management engineer with the SCS's state office in Davis, California, reports, however, that "the water conservation job is not getting done because we don't have the staff or the money to do it."[83]

The Extension Service, which receives about 30 percent of its funding from the U.S. Department of Agriculture (USDA) and the rest from states and counties, is supposed to provide farmers with technical information on soil and water conservation. A recent study in California found, however, that in that state the Extension Service's administrative mandate for soil and water conservation "is incidental to an *over-arching concern for crop productivity*"[84] [emphasis added]. It appears that conclusion accurately describes the Extension Service's role in other arid areas as well. In other words, the Extension Service does not seem to be a major force for soil or water conservation in the arid United States.

Water conservation often entails additional financial outlays by the farmer—whether it be for installing a drip irrigation system, for field leveling, or whatever. The main source of federal cost-sharing assistance for such improvements is the Agricultural Conservation Program (ACP), which is administered by the Agricultural Stabilization and Conservation Service (ASCS). Under the ACP, owners of farms may apply to their county ASCS committee for funding up to 80 percent of the cost of soil or water conservation measures and may receive up to $3500 in a year.

The problem is that the ACP funds available to any given county in a year are so paltry that many eligible farmers do not bother to apply. In fact, the ACP funds available to a county increase very little or not at all from year to year and have not kept pace with either inflation or ground water salinization rates. In addition, since the county ASCS committees are elected by farmers, they mirror the farmers' preoccupation with immediate cash flow, so that measures promising a quick return, such as the installation of irrigation pipe or the drilling of a new well, receive the preponderance of the funds. One recent assessment concluded: "ASCS provides a funding function which is limited by available funds and priority biases." [85]

Any future analysis of the alternative solutions for the San Joaquin's salty-water drainage disposal problem should include a calculation of the net benefits of a reinvigorated water conservation program by the SCS, ASCS, and Extension Service. It would be particularly interesting to see whether more efficient water use cuts the problem down to a size that makes the local alternative more practical than a valleywide public works project.

Who Pays?

The master drain project recommended by the task force would cost the federal government an estimated $258 million and the state another $89.1 million.[86] These figures, however, do not include the cost of installing on-farm drainage systems, which are essential whether the ultimate disposal is local or valleywide. Various federal programs are available to help farmers finance these systems. The figures also fail to include the cost of facilities for collecting water from farms in a given area for injection into the master drain. These costs would be the responsibility of local entities such as the water districts, drainage districts, or resource conservation districts. Here again, however, a number of federal programs are available to assist on a cost-sharing basis.[87] Therefore the total federal expenditure would be greater than $258 million, but how much greater has not been determined.[88]

Under the proposed plan, the cost of the master drain would be repaid, primarily by farmers in the drainage problem areas, over a 40- to 50-year period through a surcharge on each acre-foot of water applied and a charge as well on each acre-foot of water discharged. The annual cost to a farmer served by a federal water project would

total about $44 per acre and for the farmer served by a state water project $75 per acre.[89]

In calculating the costs for such a water project, the federal government uses a very low interest rate–6.62 percent. Hence the proposed master drain represents a partially subsidized solution to the salinity problem of the San Joaquin.

At present the task force proposal is stalled. Political opposition from environmentalists and residents of the Bay-Delta area exists, of course, but at this stage it is the lack of support from farmers in the San Joaquin Valley that is the major political obstacle. The farmers have balked at paying $44 or $75 per acre per year—even though the task force estimates the benefits to farmers in the drainage problem areas at about $130 an acre per year.[90] Farmers with no drainage problem now— that is, in areas where the saline ground water has not yet reached crop roots—are more concerned with their immediate cash flow problems. Rising costs, especially energy costs, are more of a priority than costs that will be incurred sometime in the future due to poor drainage. In addition, farmers who contribute to salinity problems downslope and do not suffer directly the consequences of poor drainage are reluctant to pay for the remedy.

The farmers experiencing salinity problems right now are employing a variety of very short-term remedies. Many are, for example, converting affected fields from deep-rooted, salt-sensitive crops to shallow-rooted, salt-tolerant crops or from crops to pasture. In either case irrigation is still necessary, so the saline ground water continues to rise. A few farmers are using already highly salinized fields as evaporation ponds for salty ground water drained from elsewhere on their farms. A very few have simply taken salt-encrusted fields out of agricultural production entirely and are trying to recoup their losses through intensified cultivation and irrigation of other fields.[91]

Future Prospects

If the master drain–Suisun Bay discharge alternative or one of the other alternatives is *not* adopted, that is, if no significant remedial action is taken either locally or valleywide, what will happen to the San Joaquin? For one thing, the agricultural yield from poorly drained land will drop precipitously. Today, on the 400,000 acres of San Joaquin farmland that already have a drainage problem, crop yields have declined 10 percent, or $31.2 million, annually since 1970. With no action the amount of poorly drained land will increase to about 700,000 acres by the year 2000, and the annual crop yield loss will climb to $321.3 million.[92]

A certain percentage of these 700,000 acres will be taken out of agricultural production entirely—it is impossible to tell exactly how much from current data. If the land taken out of production was once desert, it will slowly revert to desert shrubs, unless it has become a salt flat. The process will take hundreds of years, and in the meantime the lack of vegetation and desert pavement (a thin layer of rocks of various sizes) will make this land highly vulnerable to wind erosion, especially during periods of drought.

If the land taken out of production was once grassland, then vegetation will return to it more swiftly but in a much debased form—with invader weed species such as Russian thistle (tumbleweed) and filaree dominating. This land too will be vulnerable to wind erosion during times of drought and, if it slopes, to water erosion during rains. Overall, more than 1 million acres of land in the San Joaquin could undergo desertification during the next 100 years if the ground water salinization problem goes untreated.

Overgrazing

In terms of acres affected today, overgrazing is the second most serious land degradation force at work in the San Joaquin. The major interagency study done of the entire San Joaquin concluded: "A *significant portion* of the Basin's rangeland has problems" [emphasis added].[93] An all-too-familiar chain of events is outlined. Overgrazing reduces forage plant cover; this reduction, in turn, leads to both increased soil erosion, which means lower soil fertility, and to the invasion of weeds and bushes. The result is a land that produces still less forage *and* that is especially vulnerable to the big erosion events—windstorm and flood.

Of the 4 million acres of private rangeland, 3.2 million acres, or 80 percent, have problems. Of the public rangeland managed by the Forest Service, 20 percent, or 102,000 acres, were found to have problems. The rangeland in the basin managed by the Bureau of Land Management—about 400,000 acres, was not assessed by this study, but it is thought to be in roughly the same condition as the private land.[94]

The study observed that "many of the range problems in the Basin can be traced to ineffective management techniques."[95] James Clawson, Extension Service range specialist at the University of California at Davis, suggests that absentee ownership of private grazing land—a condition encouraged by federal, state, and local tax laws—contributes to rangeland abuse. This condition is a problem, he suggests, equal in proportion to the mismanagement of public rangelands.[96] W. O.

Beatty, area conservationist for the SCS in Fresno, voices a similar view and adds that rangeland along the western rim of the San Joaquin "is still deteriorating."[97]

The interagency study found that on some 338,000 acres of basin rangeland, the forage vegetation was so badly overgrazed that it could not revegetate; and on some of this land woody and noxious plants were replacing forage vegetation.[98] The land is, in other words, undergoing desertification.

Soil Erosion

Some 2.2 million acres in the foothills and mountains of the San Joaquin Basin are undergoing moderate to severe water (sheet and gully) erosion of the soil. Rangeland is a prime victim of erosion. According to the estimates of USDA, erosion causes an annual loss of $1.2 million of forage in the San Joaquin.[99] Erosion is also costly because of the sediment it deposits in watersheds. The San Joaquin Basin study explains:

Sedimentation has a number of repercussions. The capacity of streams, channels and reservoirs is reduced which causes flooding. Floods destroy cropland and deposit sediments and other debris which are expensive to remove. Sediment also destroys fish spawning by covering gravel beds.[100]

Overgrazing also plays a major role in the loss of soil by wind erosion. Such soil loss was dramatically demonstrated on December 20, 1977, in the southern San Joaquin in the Bakersfield vicinity. Early that day a windstorm struck the crescent of foothills and canyons that form the southern border for the valley. Within a 24-hr period that windstorm moved more than 25 million tons of soil in a 373-mi^2 rangeland area. As much as 23 in. of soil was stripped from some foothills. And as the wind moved down onto the valley floor, it scoured the recently plowed fields, and millions more tons of soil were displaced. A gigantic plume of dust—that is, soil—formed over the San Joaquin and extended northward to at least the far end of the Sacramento Valley, some 360 mi away.[101]

The wind removed 167 tons of soil *per acre* from the affected rangeland.[102] That such huge soil losses occurred in so short a time was because of both the terrific velocity of the wind—up to 186 mi/hr—and the poor condition of the land. To quote the U.S. Geological Survey study of the windstorm's impact: "This land was particularly vulnerable to wind erosion because the vegetative cover had deteriorated seriously under the combined stresses of drought and grazing and because of low soil moisture due to drought.[103]

Windstorms of this intensity, although rare, have occurred before in this part of the San Joaquin Basin and will no doubt occur again.

Recreation and Urbanization

The previously mentioned San Joaquin Basin study also cites "off-road vehicles and other recreational pursuits" as another major cause of rangeland deterioration because they "destroy vegetative cover and accelerate erosion." Some 521,000 acres of San Joaquin rangeland has suffered reduced productivity due to recreational "overuse" of the land, the study reported.[104]

Soil experts also worry about the effect of urban sprawl on the San Joaquin. "Urbanization is a terrible thing to see as it takes over the best farmland and the best soils," Says Morris A. Martin, district conservationist, SCS Fresno field office.[105] It also causes severe erosion problems when the bulldozers move into the foothills, stripping away the natural vegetation and exposing unstable soils to sheet and gully erosion.

The San Joaquin has three major, fast-growing urban areas: Modesto, Fresno, and Bakersfield. Urbanization is projected to take over some 407,100 acres of irrigable farmland in the San Joaquin between 1972 and the year 2000.[106] A by-product of the loss of farmland to urbanization is that poorer-quality land is pressed into cultivation. Over the past five years California has lost approximately 55,000 acres per year of prime agricultural land to urban development; during the same period 75 percent of the newly irrigated acres brought into production has been on medium- and low-potential land (class III and IV land under the SCS land classification system).[107]

Depletion of Ground water

The depletion of ground water resources could also lead to the abandonment of irrigable farmland in the San Joaquin. Whether this, in fact, will occur depends ultimately on a number of factors, including the ability of the San Joaquin to import still more water from other basins through new federal or state water projects. When the Water and Power Resources Service sought authorization for a project to serve the Westlands Water District, one of the San Joaquin's more recent water import projects, the agency stated that the ground water level in the area was declining an average 10 ft per year and, in some spots, 20 ft per year. Advocates of the project claimed that without the imported water this 72,000-acre area would soon be fit only for growing sagebrush.[108]

Before an aquifer is totally depleted of water,

the energy costs of pumping water from it becomes prohibitive. These steeply rising costs can lead to the abandonment of irrigated cropland. As yet, however, there is no record of farmers in the San Joaquin abandoning cropland because of increased ground water pumping costs, but this certainly is a distinct possibility in the future.

The deeper the well, of course, the higher the energy costs per acre-foot of water become. In the western San Joaquin today some farmers are pumping water for irrigation from 3500 ft beneath the surface.[109] Dropping ground water levels and rising energy prices will certainly make the pumping of ground water an increasingly significant cost item in San Joaquin agriculture.

The major physical effect on the San Joaquin of ground water overdraft has been land subsidence. When water is mined from an aquifer system of fine-grained, unconsolidated sediments, the aquifer system compacts. As a result, the land surface above it sinks. In the San Joaquin, about 5200 mi of land had subsided as of 1972, with about 4200 mi of this land subsiding more than 1 ft. In the western San Joaquin some areas have sunk as much as 29 ft.[110]

The major cost of subsidence is the damage it does to irrigation and drainage facilities, particularly canals and underground pipes. For example, between June 1975 and September 1976 the Bureau of Reclamation spent about $3.7 million to rehabilitate federal irrigation projects damaged by subsidence.[111] On-farm costs can also be high. In some cases in the San Joaquin, tilting of the land surface has changed the flow pattern on farms, and irrigators have had to realign their entire irrigation systems. In addition, land subsidence damages homes and other buildings. South of Fresno there is a small community that had to be entirely abandoned because of land subsidence.[112]

One of the long-term consequences of ground water overdraft and subsidence that has not received much attention is the loss of water storage capacity. As an aquifer system compresses with the mining of its water, the amount of pore space within it shrinks. Since it is this very pore space that enables the system to store water, its storage capacity is therefore greatly diminished.

Water storage is vital to areas such as the arid West, which are subject to periodic droughts, and aquifer systems are by far the most efficient means of storing water. (Reservoirs lose tremendous quantities of water to evaporation; for example, the San Luis Reservoir in the western San Joaquin loses about 120,000 acre·ft of water per year to evaporation.[113]) Aquifer systems that have subsided because of overdraft will never again be able to hold as much water as they did before overdraft began. The result is the partial loss of a valuable nonrenewable resource. This is already occurring in the San Joaquin Basin as well as in other parts of the arid West.

THE WELLTON-MOHAWK

In 1947 Congress authorized the Wellton-Mohawk Project. Built by the Bureau of Reclamation and completed in 1952, the project diverts water from the Colorado River, northeast of Yuma, Arizona, and pumps it about 30 mi east to irrigate 60,000 acres of desert.

Situated on the floodplain and mesas along the lower Gila River, the Wellton-Mohawk only gets about 4 in. of rainfall a year, but the sun shines more than 90 percent of the time during daylight hours, and killing frosts are shortlived. The area's soil is naturally fertile except for a lack of organic matter that is being replenished by crop residues from years of farming.

With a steady supply of cheap water from the federal government, Wellton-Mohawk has prospered. The area now produces $1082 worth of crops per acre, up from $145 per acre in 1955. Iowa crop production, by comparison, is about $125 per acre.[114]

Farm net income figures are not available for Wellton-Mohawk but they are thought to be comparable, if not somewhat higher, than for Arizona as a whole. From 1970 to 1976 net income per farm in Arizona averaged $39,679, compared with $10,102 for Iowa and $7589 for the entire nation.[115]

The key to Arizona agriculture's high profits, aside from the long growing season, has been cheap water. For example, Wellton-Mohawk farmers pay the federal government between $6.25 and $21.50 per acre-foot of water depending on the amount used.[116] These prices reflect only the operating costs of delivering the water to Wellton-Mohawk and not the capital costs of the project. If there were a free market in water, the water might sell for *$100 to $500 per acre-foot.*[117]

Wellton-Mohawk's chief money crops are lettuce, cotton, alfafa, wheat, cantaloupes, grass, and oranges. Its yields are impressive. Wellton-Mohawk produces 1142 lb of cotton per acre, compared with the national yield of 120 lb of cotton per acre. It produces 87 bushels of wheat per acre, compared with 32 for the nation, and 8.8 tons of alfalfa per acre, compared with 3.1 tons for the nation.[118]

And yet, political uncertainty and controversy cloud Wellton-Mohawk's future. The Wellton-Mohawk's problem is saline ground water and what to do with it. In the Wellton-Mohawk the substra-

tum that effectively blocks downward drainage of water is relatively close to the surface and underlies a large percentage of the irrigated land. Hence irrigation water that does not evaporate or transpire soaks into the ground and rises rapidly into the root zone. The solution to Wellton-Mohawk's drainage problem is made especially complicated by the fact that the Colorado River, its natural depository or "sink," is "perhaps the most overdeveloped river in the world."[119]

The Colorado River

In 1961 the Wellton-Mohawk Irrigation District began to operate a system of drainage wells that discharged into the Colorado River. At that time the Wellton-Mohawk's drainage water had a salinity of about 6000 parts per million (ppm), and the Colorado River water taken in by Wellton-Mohawk was about 800 ppm.

The salinity of the Colorado River water flowing into Mexico increased sharply. In 1960 it averaged 800 ppm. By 1962 it had increased to over 1500 ppm. Wellton-Mohawk drainage water was the primary cause. It was not, however, the only cause of the dramatic rise in the lower Colorado's salinity; beginning in 1961 the flow of Colorado River water into Mexico was sharply reduced by the United States in anticipation of storage in Lake Powell behind the newly constructed Glen Canyon Dam.

Mexico raised strenuous objections to the Colorado's increased salinity and charged that the saline water was damaging crops and soils in the Mexicali Valley, a major agricultural area. The salt tolerance of crops cannot be defined in any absolute sense, but the U.S. Salinity Laboratory established a general classification of the *salinity hazard* to crops of irrigation water in parts per million: low, 100–250; medium, 250–750; high, 750–2250; very high, 2250.

In 1962 the United States and Mexico entered into negotiations over the problem. A five-year agreement was signed in 1965. Under the terms of this agreement the United States undertook remedial measures that cost $12 million, but only a marginal improvement in the quality of water delivered to Mexico (1500 ppm in 1962 to 1240 ppm in 1971) resulted. The two nations negotiated a new agreement in July 1972. The United States agreed to undertake additional mitigative measures, but again only marginal improvement in water quality resulted—the average annual salinity of water delivered to Mexico dropped from 1240 ppm in 1971 to 1140 ppm in 1973.[120]

In 1973 the two nations signed an agreement known as Minute No. 242, "The Permanent and

Definitive Solution to the International Problem of the Salinity of the Colorado River.'' Under this agreement the United States for the first time committed itself to a specific level of water quality for the Colorado River water that it released into Mexico. The United States agreed to release to Mexico water with an average salinity of not more than 115 ppm over the salinity at the Imperial Dam in the United States. The salinity of the water at Imperial Dam was about 809 ppm in 1979.[121]

To implement this agreement, Congress passed the Colorado River Salinity Control Act of 1974. It is currently estimated that a $333 million federal effort will be required under Title I of the law, with much of the money allocated to solving the Wellton-Mohawk drainage problem.[122] In other words, Wellton-Mohawk's ground water salinity problem has become a very expensive one.

A Structural Solution

The most expensive item in the ''permanent and definitive solution'' to the Colorado River's salinity problem is a desalinization plant to be built near Yuma, Arizona. Under current Water and Power Resources Service plans, the plant will cost $178 million to construct and $12 million per year to operate; it will desalinate about 120,000 acre·ft of water per year drained from the Wellton-Mohawk.[123] The U.S. Department of the Interior reportedly expects to award the main construction contract in the mid 1980s if Congress appropriates the money. Before completion, however, inflation might result in the plant costing as much as $300 million, driving up the total cost to the taxpayer to over $500 million. In addition, as energy costs continue to climb, so will the plant's annual operating costs because of the energy-intensive nature of desalinization.

Are there alternatives to this capital-intensive, energy-intensive solution to the Wellton-Mohawk's drainage problem? At present a U.S.-built, concrete-lined drainage ditch is siphoning all of the Wellton-Mohawk's drainage water directly into the Santa Clara Slough in Mexico on the Gulf of California, about 70 mi south of Wellton-Mohawk.

Why not simply continue with this arrangement? The drainage water moves along the concrete-lined drainage ditch primarily by means of gravity, and its emptying into the already salty Santa Clara Slough has created a splendid wetlands-wildlife habitat. The difficulty is that none of this water is credited to the United States as part of the U.S. treaty commitment to release 1.5 million acre·ft per year of Colorado River water to Mexico. If the drainage water is desalinated, however, before being

sent to Mexico, then it can be used for irrigation, and the United States would get credit for it.

To implement a no-desalinization-plant alternative, therefore, the United States would have to either: (a) stop supplying water to the Wellton-Mohawk District in order to increase the flow of water to Mexico or (b) reduce the water allotments of one or more states along the Colorado by the amount that is drained into the Santa Clara Slough and then send that amount of water to Mexico. Any attempt to do the latter would stir strong opposition. Indeed, it would be somewhat like trying to take food away from a nest of angry rattlesnakes. Water is too precious in the arid Colorado River Basin for any state to accept willingly a reduction, even a relatively modest reduction, in its allotment.

The only reason the federal government has been able to continue to meet its commitment to Mexico while still draining water into the Santa Clara Slough is that the Colorado's flow in recent years has been higher than normal. For the years ahead, however, the Colorado is already *overbooked*, and continued supply at the current levels will present a problem.

The U.S. government cannot simply stop supplying the Wellton-Mohawk with water; it has a contractual obligation to the farmers. The government could, however, buy out the farmers and then close down the Wellton-Mohawk District. Referring to the Colorado River Salinity Control Act of 1974, Rafael Mosses, counsel to the Colorado Water and Conservation Board of the Colorado River Commission, commented:

> We could have bought up the Wellton-Mohawk Project and retired the whole thing for a lot less than it's going to cost, but politically, of course, it is not feasible.[124]

In fact, the government has purchased and retired from production 5000 acres of land in the Wellton-Mohawk where water use was particularly high—citrus trees grown on sandy soils—in an effort to reduce the district's drainage water outflow, although no districtwide purchase plan is underway.

The people who farm the Wellton-Mohawk do not want to sell. They insist they will fight in the court any federal effort to buy them out and retire the project. Many years of litigation are threatened.

Another alternative is suggested by Jan van Schilfgaarde, director of the U.S. Salinity Laboratory: ''The time has come to realize we can't continue to use huge capital and energy intensive solutions where management and social solutions will work.'' Van Schilfgaarde suggests that a desalting plant might not be necessary if Wellton-Mohawk farmers used irrigation water more efficiently and reused some of the drainage waste to grow salt-tolerant crops.[125]

Efficiency in irrigation—that is, using less water without reducing crop yields—is important for two reasons: It reduces demand for a scarce resource—water; and it means less buildup of salty water in the ground and therefore less to be drained. The federal government is working with the district and individual farmers to reduce their water losses. It is subsidizing irrigation efficiency—providing capital on a cost share basis for lining of irrigation ditches, leveling fields, and installing water control and measurement devices, as well as low-pressure drip irrigation systems.

Sol Reznick of the University of Arizona's Water Resource Center thinks the potential for more efficient water management is great. Resnick points to Israeli irrigation projects as a model of efficiency for Wellton-Mohawk and other subsidized irrigation projects in the United States. "At Wellton-Mohawk, they are using 13 feet of water per acre in some places to grow citrus crops. In the Negev Desert, the Israelis are using 2.5 feet, and the citrus yields per acre are higher," he reports. "They have much more sophisticated water control systems."[126]

In 1973 when Minute No. 242 was signed, on-farm irrigation efficiency in Wellton-Mohawk was about 56 percent, that is, 56 percent of the water applied to the land was consumed by crops, and the rest was lost to the sun or the ground. The current federal program has set 72 percent efficiency as the goal.[127] But there is strong disagreement among the federal agencies involved over whether this goal can be achieved. The U.S. Environmental Protection Agency and the USDA believe that on-farm efficiency in excess of 72 percent can be achieved within ten years. The Water and Power Resources Service questions whether an overall efficiency greater than 64 percent can be achieved.[128]

These differing projections are important because the government is trying to determine what size desalting plant it needs to build. At 64 percent efficiency the district's drainage outflow would total about 167,000 acre·ft per year; at 72 percent it would be 136,000 acre·ft; and at 82 percent, 94,000 acre·ft.

But even if 72 percent efficiency is achieved, it still leaves the government with a big problem—what to do with 136,000 acre·ft of saline drainage water. The government agencies involved in the problem see no alternative to a desalting plant of some size. What they have not considered seriously is whether there is some alternative that would *not* require building a desalting plant at all; something less drastic than buying out the entire district and

more politically realistic than draining all of the water outflow from the district into the Santa Clara Slough. Would, for example, a continuation of the current program to increase irrigation efficiency, combined with a limited reduction in the irrigated acreage and with the development of solar salt ponds, eliminate the need for a desalting plant?

Future Prospects

Despite its relatively insignificant size, Wellton-Mohawk bears scrutiny because it exemplifies the issues involved in the future of irrigated agriculture in the Colorado River Basin. First, high-yield, federally subsidized irrigated agriculture is already straining the basin's water supply. Under the circumstances it will be difficult to justify subsidies for new projects. In other words, federally supported reclamation of additional desert land in the basin may be over. A more realistic prospect might be increased desertification of land now under cultivation. The 5000 acres in Wellton-Mohawk that the federal government purchased and retired are reverting to desert; this fate awaits other acreage in this land of scant rainfall and poor irrigation.

Secondly, "a permanent and definitive" solution to the river's salinity problem does not yet exist. Salinity is one of the major external costs of irrigated agriculture. Bringing it under control will require federal outlays as well as the resolution of a number of technical issues.

Thirdly, economic logic plays little or no role in the resolution of the basin's central problem—a scarcity of water. For example, the U.S. government has sunk a series of wells in the Yuma Mesa 5 mi from the Mexican border and will soon begin pumping water from this aquifer for delivery to Mexico. This action will give the United States credit for water as part of its 1.5 million acre·ft delivery requirement per year.

But meanwhile, the Mexicans in the San Luis area immediately across the border continue to pump water from the same aquifer. They are already pumping out water faster than it is being replenished by nature and irrigation runoff.[129] The United States' pumping will, of course, hasten the aquifer's depletion and eventual exhaustion.

Finally, as a recent study notes:

> It appears that salinity control, like water resource development in general, prefers structural solutions. . . . Nonstructural remedies . . . would have invoked conflict and delayed the implementation of Minute No. 242 and the Basin-wide salinity control program.[130]

But this conclusion raises a further question: How long will the other regions of the country, espe-

cially those such as the upper Midwest and North-
east, that have experienced a net outflow of dollars
to the federal government agree to underwrite the
arid West's phenomenal growth by subsidizing
costly remedies to the region's perennial problems—water scarcity and salinity?

NOTES

1. James Rodney Hastings and Raymond M. Turner, 1972, *The changing mile* (Tucson: University of Arizona Press), 3–5, 32–38; Andrew W. Wilson, 1963, Tucson: A problem in uses of water, in Hodge and Duisberg, eds., *Aridity and man* (Washington, D.C.: American Association for the Advancement of Science), 4–5.

2. University of Arizona, Water Resources Research Center, 1978, Groundwater projections for 11 basins, *Arizona Water Resources News Bulletin*, 78 (3): 2.

3. Ibid., 3.

4. Martin Fogel and Gerald D. Harwood, 1978, *Identification and analysis of major water problems of arid and semi-arid lands,* prepared for the National Science Foundation, University of Arizona, Tucson, p. 15.

5. U.S. General Accounting Office, 1977, *Ground water: An overview,* CED–77–69 (Washington, D.C.), p. 7.

6. U.S. General Accounting Office, *Reserved water rights for federal and indian reservations: A growing controversy in need of resolution,* CED–78–176 (Washington, D.C.), 34.

7. GAO, note 5, p. 7.

8. Charles Bowden, 1977, *Killing the hidden waters* (Austin: University of Texas Press), 30, 120.

9. Fogel and Harwood, note 4, p. 17.

10. GAO, note 6, pp. 33–34; Kenneth E. Foster, undated, *Alternatives for managing a finite ground water supply in an arid region* (Tucson: University of Arizona), 19.

11. GAO, note 6, pp. 33–34.

12. Foster, note 10, p. 19.

13. A. W. Wilson, 1977, Technology, regional interdependence and population growth: Tucson, Arizona, *Economic Geography* 53: 389.

14. Fogel and Harwood, note 4, pp. 16–18, 53–54, 57.

15. Sol Resnick, Water Resources Center, University of Arizona, Tucson, interview, April 17, 1979.

16. GAO, note 6, pp. 34–35.

17. Fogel and Harwood, note 4, p. 15.

18. University of Arizona, note 2, p. 3.

19. Foster, note 10, pp. 9–10, 12.

20. University of Arizona, note 2, p. 3.

21. Michele Strutin, 1979, Pulling the plug on Arizona, *Mother Jones* (August):20.

22. U.S. Bureau of Reclamation, 1978, *Central Arizona project,* Information Paper No. 3, Phoenix, Arizona.

23. Bowden, note 8, p. 7.

24. Susan Jo Keith, 1977, *The impact of ground water development in arid lands* (Tucson: University of Arizona), 11–12.

25. Foster, note 10, p. 20.

26. Thomas Holzer, U.S. Geological Survey, Menlo Park, California, interview with author, October 1, 1980.

27. Ibid.

28. U.S. Department of Agriculture, in cooperation with the Arizona Water Commission, 1977, *Santa Cruz–San Pedro River basins, Arizona—Main report* (Portland, Ore), 3.26.

29. Keith, note 24, p. 26.

30. Ibid.

31. USDA, note 28, p. 3.26.

32. Melvin E. Hecht et al., 1978, Agriculture: Its historic and contemporary roles in Arizona's economy, *Arizona Review,* 27 (11):10.

33. USDA, note 28, p. 3.33.

34. Jack Johnson, Office of Arid Land Studies, University of Arizona, Tucson, interview, January 19, 1979; USDA, note 28, p. 3.36.

35. Maurice M. Kelso et al., 1973, *Water supplies and economic growth in an arid environment: An Arizona case study* (Tucson: University of Arizona Press), 105.

36. *New York Times,* June 15, 1980, p. 35.

37. Texas Crop and Livestock Reporting Service, *1977 Texas agricultural cash receipts statistics* (Austin: Texas Department of Agriculture), 10: USDA, *Economics, statistics, and cooperative service* (Washington, D.C.).

38. Texas Crop and Livestock Reporting Service, note 37, pp. 11, 26.

39. Walter Bertsch, District Conservationist, Soil Conservation Service, Seminole, Texas, interview, April 19, 1979.

40. D. B. Polk, State Resource Conservationist, Soil Conservation Service, Temple, Texas, interview, April 20, 1979.

41. Gaines County Soil and Water Conservation District, 1979, *Program and plan of work* (Gaines, Texas), p. 21.

42. Gaines belongs to the physiographic region known as the Llano Estacado or Staked Plains.

43. Gaines County, note 41, pp. 22–24; Bertsch, note 39.

44. Bertsch, note 39.

45. Gaines County, note 41, p. 9.

46. James McGehee, Gaines County Executive Director, U.S. Agricultural Stabilization and Conservation Service, Seminole, Texas, interview, April 20, 1979.

47. Ibid.

48. James Abbott, Soil Conservation Service, Temple, Texas, interview, August 29, 1980.

49. Bertsch, note 39; McGehee, note 46.

50. Abbott, note 48.

51. U.S. General Accounting Office, 1977, *To protect tomorrow's food supply, soil conservation needs priority attention*, CED–77–30 (Washington, D.C.), 16.

52. James Abbott, Soil Conservation Service, Temple, Texas, interview, April 20, 1979.

53. Anne E. Bell and Shelly Morrison, 1979, *Analytical study of the Ogallala Aquifer in Gaines County, Texas* (Texas Dept. of Water Resources), 1.

54. Ibid., 7.

55. Bowden, note 8, pp. 119–120.

56. Bertsch, note 39.

57. Bell and Morrison, note 53, p. 6.

58. Ibid., 1.

59. Ibid., 6–7.

60. Ibid., 8; Bertsch, note 39.

61. Bertsch, note 39.

62. Fogel and Harwood, note 4, p. 23.

63. Bowden, note 8, p. 112.

64. Bell and Morrison, note 53, pp. 1, 9.

65. John Graves, 1971, You ain't seen nothing yet, *The Water Hustlers* (San Francisco: Sierra Club), 39.

66. Robert M. Sweazy et al., 1979, Ground water resources of the high plains: State-of-the-art and future research needs, *Summaries of reports: Arid and semiarid lands* (Washington, D.C.: National Science Foundation), 15.

67. Garrett Hardin, 1968, The tragedy of the commons, *Science,* 162: 1243–1248.

68. U.S. Department of Agriculture (River Basin Planning Staff, Soil Conservation Service, Forest Service, Economic Research Service), in cooperation with the California Department of Water Resources, 1977, *San Joaquin Valley Basin study* (Washington, D.C.), 37–44.

69. U.S. Bureau of Reclamation, California Department of Water Resources, California State Water Resources Control Board, 1979, *Agricultural drainage and salt management in the San Joaquin Valley,* San Joaquin Valley Interagency Drainage Program (Fresno, Calif.), 2.1–2.3.

70. Ibid.

71. U.S. Department of Agriculture, 1979, *Economics, statistics and cooperative service* (Washington, D.C.).

72. Bureau of Reclamation, note 69, pp. 2.3, 4.1–4.2.

73. Ibid.

74. Ibid., 2.4–2.5: Raymond F. Dasmann, 1973, *The Destruction of California* (New York: Macmillan), 68–72.

75. Bureau of Reclamation, note 69, pp. 1.2, 6.3.

76. The Lower San Joaquin normally drains into the enclosed Tulare sub-basin, although in especially wet years it also drains northward.

77. Bureau of Reclamation, note 69, pp. 11.1–11.11.

78. Unusually high concentrations of boron and arsenic are found in the soils in certain parts of the western and southern San Joaquin. The mercury in the drainage water would come from pesticides used in the San Joaquin.

79. Bureau of Reclamation, note 69, pp. 9.1–9.9.

80. Ibid., 7.4, 9.4, 17.5, 17.6.

81. Ibid., 8.4.

82. Ibid., 17.5.

83. Gylan Dickey, U.S. Soil Conservation Service, Davis, California, interview, April 10, 1979.

84. Gary Weatherford et al., 1979, *Erosion and sediment in California watersheds: A study of institutional controls* (Napa, Calif.: John Muir Institute), 99.

85. Ibid., 103.

86. Bureau of Reclamation, note 69, p. 13.4.

87. Ibid., 13.2–13.3.

88. Craig Stroh, U.S. Bureau of Reclamation, Fresno, California, interview, April 11, 1979.

89. Ibid.

90. Ibid.

91. Louis A. Beck, Director, San Joaquin Valley Drainage Program, Fresno, Calif., April 11, 1979.

92. Bureau of Reclamation, note 69, pp. 6.3, 9.3.

93. USDA, note 68, p. 55.

94. Robert Webb, Stanford University, interview, September 9, 1979.

95. USDA, note 68, p. 55.

96. James Clawson, 1979, quoted in California Department of Conservation, California soils: An assessment, draft, April, p. IV–38.

97. W. O. Beatty, U.S. Soil Conservation Service, Fresno, California, interview, April 11, 1979.

98. USDA, note 68, p. 55.

99. Ibid.

100. Ibid.

101. U.S. Geological Survey, undated, *Field observations of the December 1977 windstorm, San Joaquin Valley, California,* pp. 1, 8, 10, 18–19.

102. Howard Wilshire, U.S. Geological Survey, Menlo Park, California, interview, November 20, 1979.

103. USGS, note 101, p. 19.

104. USDA, note 68, p. 55.

105. Morris Martin, U.S. Soil Conservation Service, Fresno, California, interview, April 11, 1979.

106. USDA, note 68, p. 107.

107. California Department of Conservation, note 96, pp. IV–39 to IV–40.

108. GAO, note 5, p. 7.

109. Ibid., 15.

110. Ibid.

111. Ibid., 16.

112. Beck, note 91.

113. Dickey, note 83.

114. Wellton-Mohawk Irrigation and Drainage District, *Crop production report 1978 and Crop production report 1955* (U.S. Bureau of Reclamation); USDA, 1979, *Economics, statistics and cooperative service* (Washington, D.C.: U.S. Department of Agriculture).

115. Ibid.

116. Jim Naguin, Wellton-Mohawk Irrigation and Drainage District, Wellton, Arizona, interview, April 14, 1979.

117. Ibid.; Herb Guenther, Senior Biologist, U.S. Bureau of Reclamation, Boulder City, Nevada, interview, April 13, 1979; Edward M. Hollenbeck, Project Manager, Bureau of Reclamation, Yuma, Arizona, interview, April 14, 1979.

118. Wellton-Mohawk, note 114; Crop Reporting Board, 1979, *Annual crop summary,* (Washington, D.C.: USDA), B–22, B–28, B–30.

119. Milton Jamail et al., *Federal-state water relations in the American West: An evolutionary guide to future equilibrium* (Tucson: University of Arizona), 13.

120. Myron B. Holbert, 1978, International problems, in Dean F. Peterson and A. Berry Crawford, eds., *Values and choices in the development of the Colorado River Basin* (Tucson: University of Arizona Press), 225–227.

121. Letter from Bruce Blanchard, U.S. Department of the Interior, Washington, D.C., to the Council on Environmental Quality, August 14, 1980.

122. Holbert, note 120; Jamail, note 119, pp. 37–39; Hollenbeck, note 117.

123. Hollenbeck, note 117.

124. Quoted in Jamail, note 119, p. 38.

125. Quoted in William Curry, 1979, The desert blooms, but at an ever higher price, *The Washington Post,* June 25, p. A3.

126. Resnick, note 15.

127. Harold Pritchett, U.S. Soil Conservation Services, Wellton, Arizona, interview, May 15, 1979.

128. W. K. Sidebottom, Chairman, Technical Field Committee on Irrigation Efficiency, 1979, Wellton-Mohawk Irrigation and Drainage District, memo to the Advisory Committee on Irrigation Efficiency, Wellton-Mohawk Irrigation and Drainage District, May 11.

129. Hollenbeck, note 117.

130. Resnick, note 15.

26

SUBSIDENCE DUE TO
GROUND WATER WITHDRAWAL

JOSEPH F. POLAND

This chapter is a summary of land subsidence due to ground water withdrawal in the United States. The chief source of data is a casebook being prepared for the United Nations Educational, Scientific, and Cultural Organization (UNESCO) by an international working group, of which the writer of this chapter is the chairman. Other sources of data are: (1) The Questionnaire on Land-Subsidence Occurrence, Research, and Remedial Work that was distributed worldwide in 1975–1978 by Ivan Johnson, in behalf of International Association of Hydrological Sciences (IAHS); and (2) the *Proceedings of the 2nd International Symposium on Land Subsidence* held at Anaheim, California, in December 1976.

The goal of the casebook is to produce a guide to assist engineers and geologists in the study, evaluation, and control of land subsidence, especially in developing countries.

The casebook contains a manual covering the occurrence, measurement, mechanics, prediction, and control of subsidence, and 16 invited case histories of land subsidence due to ground water withdrawal. The case histories cover a wide range of conditions and magnitude of subsidence; four are for areas in the United States. The casebook also includes a tabulation of 41 subsidence areas, of which 17 are in the United States (exclusive of sinkhole areas). Figure 26-1 shows the location of the 17 areas on a map of the conterminous United States. Subsidence of the land surface in the 17 areas ranges from 0.3 m at Savannah, Georgia, to

9.0 m on the west side of the San Joaquin Valley (Los Banos–Kettleman City area) in California (Figure 26-2). Subsidence exceeding 1 m occurs in four states: Texas, Arizona, Nevada, and California. The areal extent ranges from 10 km² in San Jacinto Valley, California, to 13,500 km² in the San Joaquin Valley. California is the state ranking number one for the dubious honor of having the largest area of subsidence—about 16,000 km². Close behind is Texas with 12,000 km² and Arizona is third with 2700 km².

Principal problems caused by the subsidence are: (1) differential changes in elevation and gradient of stream channels, drains, and water transport structures; (2) failure of water well casings due to compressive stresses generated by compaction of aquifer systems, and (3) tidal encroachment in lowland coastal areas.

Subsidence due to ground water withdrawal develops principally under two contrasting environments and mechanics. One environment is that of carbonate rocks overlain by unconsolidated deposits, or old sinkholes filled with unconsolidated deposits, that receive buoyant support from the ground water body. When the water table is lowered, buoyant support is lost and the hydraulic gradient increased; the unconsolidated material may move downward into openings in the underlying carbonate rocks, sometimes causing catastrophic collapses of the roof. In Alabama an estimated 4000 man-induced sinkholes have formed since 1900, in contrast to less than 50 natural collapses

Mr. Poland is a research hydrogeologist with the U.S. Geological Survey in Sacramento, California 95825. He has worked for over forty-five years with the USGS, including ten years as district geologist for California. Since 1956 he has specialized in studies of land subsidence due to ground water withdrawal, also serving as consultant to UNESCO on subsidence problems worldwide. This material is taken from *Journal of the Irrigation and Drainage Division 107 (IR2)*, ASCE, 115 (June 1981). Reprinted by permission of the author and ASCE. Copyright © 1981 by American Society of Civil Engineers.

Figure 26-1. Areas of land subsidence from ground water withdrawal in the United States.

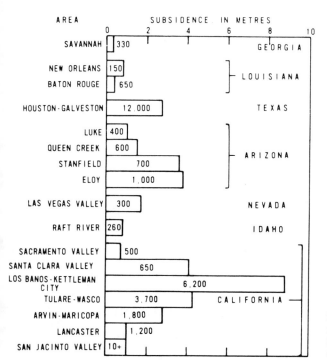

Figure 26-2. Magnitude of land subsidence from ground water withdrawal in the United States. Numbers in the columns represent area in square kilometers.

(16). In the United States the occurrence of man-made sinkholes is common in carbonate terrane from Florida to Pennsylvania, numbering many thousands. The individual sinkhole area is small, however, usually within the range of 10,000 m².

The other and by far the most extensive occurrence is that of young unconsolidated or semiconsolidated sediments of high porosity laid down in alluvial, lacustrine, or shallow marine environments. All areas are underlain by semiconfined or confined aquifer systems containing aquifers of sand or gravel, or both, of high permeability and low compressibility interbedded with clayey aquitards of low vertical permeability and high compressibility under virgin stresses. Most of the compacting deposits were normally loaded, or approximately so, before man applied stresses exceeding preconsolidation stress (maximum past stress). These aquifer systems compact in response to increased effective stress caused by artesian-head decline in the coarse-grained aquifers and time-dependent pore pressure reduction in the fine-grained compressible aquitards, causing land surface subsidence. Note that from Savannah, Georgia, to Houston, Texas (Figure 26-1), the ground water withdrawal and subsidence are occurring in coastal-plain and shallow marine deposits, whereas from Arizona to northern California the withdrawal and subsidence are occurring in valley fill, chiefly alluvial-fan and floodplain deposits.

STRESSES THAT CAUSE SUBSIDENCE

The idealized pressure diagram of Figure 26-3 can be utilized to show the stresses that cause subsidence.

In 1925 Terzaghi introduced the theory of effective stress:

$$p' = p - u_w \qquad (1)$$

in which $p' =$ effective stress (effective overburden pressure or grain-to-grain load); $p =$ total stress (geostatic pressure); and $u_w =$ pore pressure (fluid pressure or neutral stress). The withdrawal of water from wells reduces the head in the tapped aquifers and increases the effective stress borne by the aquifer matrix. For purposes of computing the geostatic pressure in Figure 26-3, a porosity n of 40 percent, an average specific retention R_s of 0.20 for the moisture above the water table, and an average specific gravity G of 2.7 for the mineral grains were assumed. The lowering of artesian head in a confined aquifer system, for example, from depth z_1 to z_3 (Figure 26-3) does not change the geostatic pressure appreciably. Therefore the increase in effective stress in the confined aquifers is equal to the decrease in fluid pressure. The compaction of these aquifers is immediate and is chiefly recoverable if fluid pressure is restored, but the amount is small.

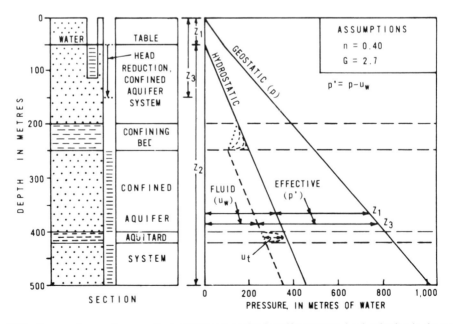

Figure 26-3. Pressure diagram for unconfined aquifer and confined aquifer systems; head reduction in the confined system only.

The decrease in head in the aquifers creates a hydraulic gradient from the confining clays and the aquitards to the aquifers. However, these fine-grained beds have low hydraulic conductivity and high compressibility. Therefore the vertical escape of water and the consequent decrease in pore pressure and increase in effective stress are slow and time-dependent, but the ultimate unit compaction is large and chiefly permanent. It is the time-dependent nature of the stress increase in these fine-grained beds that complicates the problem of predicting subsidence. In these aquitards the stress increase applied by the head decline in the confined aquifers becomes effective only as rapidly as pore pressures decay toward equilibrium with the pressures in adjacent aquifers. (See dashed pore pressure lines of Figure 26-3, where u, represents the excess pore pressure at time t.) Attainment of pore pressure equilibrium (dotted lines) may take months or years; the time required to reach equilibrium varies directly as the specific storage and the square of the draining thickness and inversely as the vertical hydraulic conductivity of the aquitard.

The stress relations of Figure 26-3 show the principle of effective stress but not the hydrodynamic cause of compaction. Actually, the downward hydraulic gradient developed across the confining bed by the head decline induces downward movement of water through the pores and exerts a viscous drag on the clay particles. This downward force due to viscous drag is exerted from particle to particle throughout the aquifer system. The force transferred to the particles at any depth is measured by the head loss to that depth. The stress so exerted on the particles in the direction of flow is a seepage stress. Lofgren (9) has pointed out that in treatment of complex aquifer systems, it is quantitatively convenient to compute effective stresses and stress changes in terms of gravitational and seepage stresses, which are algebraically additive. Gravitational stress is caused by the effective weight of overlying deposits, transmitted downward through the grain-to-grain contacts. Any change in the position of the water table changes the gravitational stress on underlying deposits.

Seven years ago the writer prepared a paper reviewing subsidence in the United States caused by ground water withdrawal (18). The remainder of this chapter is chiefly an update of subsidence developments.

SUBSIDENCE AREAS

Savannah, Georgia

Information available on the subsidence of the Savannah area, Georgia, has been reported by Davis, Counts, and Holdahl (5). Subsidence to 1975 was 0.3 m; it has not been sufficient to be recognized as a serious engineering problem to date.

Louisiana

Information available on subsidence due to ground water withdrawal in Baton Rouge and New Orleans, Louisiana, as of 1972 was summarized briefly by the writer (18). Subsidence in Baton Rouge was 0.4 m as of 1969 and in New Orleans was 0.55 as of 1964. Total subsidence in the Baton Rouge industrial area in 1935–1976 was 0.51 m, of which 0.38 m was attributed to ground water withdrawal and about 0.13 m due to natural regional subsidence, assumed to be at a rate of 3 millimeters per year (mm/year) (23). R. G. Kazmann reported (written communication, December 1975) that maximum subsidence in New Orleans had increased to 0.8 m by 1975. So far as is known, no remedial steps have been taken to date.

Houston-Galveston Area, Texas

The Houston-Galveston area had experienced maximum subsidence of 2.75 m (9 ft) by 1975 (R. K. Gabrysch, written communication, December 1975). The benchmark net in the subsiding area was resurveyed in 1978–1979. Figure 26-4 shows the subsidence from 1943 to 1978. As much as 2.6 m (8.5 ft) of subsidence occurred in the Pasadena area in the 350-year period. From 1890 to 1954 all the water supply for the Houston-Galveston area was ground water. In 1954 surface water first became available from Lake Houston (Figure 26-4). By 1972 pumping of ground water for municipal supply (46 percent), industrial use (33%), and irrigation (21%) totaled 2 cubic hectometers per day (525,000,000 gal/day). In 1975 pumping of ground water for all uses was 1.9 hm³/day.

In the Pasadena and Baytown areas the long-term increasing withdrawal of ground water caused maximum water level declines from 1943 to 1973 of 61 m in wells completed in the Chicot (shallower) aquifer and 99 m in wells completed in the Evangeline aquifer (6). Both are confined aquifer systems. This long-term drawdown of water levels (increase in effective stress in the aquifers) has been the cause of the land surface subsidence.

The Harris-Galveston Coastal Subsidence District was created in 1975 by the Texas legislature to provide for the regulation of the withdrawal of ground water within the district boundaries for the purpose of ending subsidence. Water from Lake Livingston on the Trinity River (Figure 26-4) became available in 1976. Substantial reduction of ground water withdrawal in 1977 and 1978 has

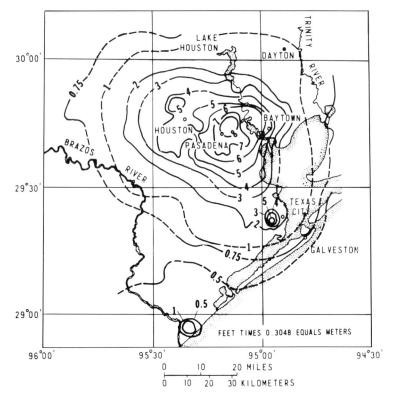

Figure 26-4. Land subsidence, Houston-Galveston region, Texas, 1943–1978 (ft). (*Source:* Modified from R. K. Gabrysch, 1980, *U.S. Geological Survey open-file report,* 80–338, Figure 2.)

resulted in artesian head recoveries in excess of 30 m in both the Chicot and Evangeline aquifer systems (R. K. Gabrysch, oral communication, August 1979). Additional rises are anticipated as the imports from Lake Livingston increase.

South-Central Arizona

Ground water has been mined in Arizona for several decades. In 1975 about 6200 hm³ of ground water was pumped; about 95 percent for irrigation use (15). Probably two-thirds of this was mined. The overdraft, which has persisted for at least two decades, has lowered water levels 100 m or more in several heavily pumped areas. This water level decline has caused subsidence exceeding 1 m in four areas of south central Arizona. Figure 26-5 shows water level declines from 1923 to 1972, earth fissures, and subsidence areas. Table 26-1 lists number, name, and size of subsidence areas, as well as the maximum known subsidence, and the water level decline from 1923 to 1977 in each area (8).

Table 26-1. Subsidence Areas

Name (1)	Area (km²) (2)	Water Level Decline, 1923–1977 (m) (3)	Maximum Subsidence (m) (Year) (4)
Luke	400	90–100	>1 (1967)
Queen Creek	600	90–120	1.5 (1976)
Stanfield	700	90–140	3.6 (1977)
Eloy	1,000	60–90	3.8 (1977)

Water levels are continuing to decline, land subsidence is increasing in these areas of increasing stress, and new earth fissures are developing, all as a result of the severe continuing overdraft on ground water resources. The Central Arizona Project of the U.S. Bureau of Reclamation, when

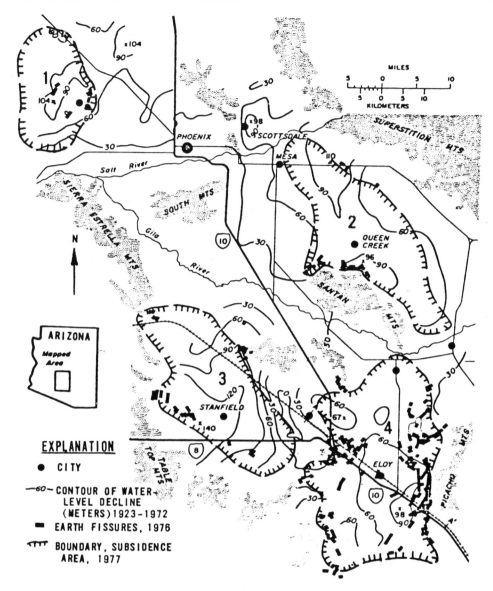

Figure 26-5. Water level declines, earth fissures, and subsidence areas in south central Arizona. Subsidence areas are numbered as follows: 1, Luke; 2, Queen Creek; 3, Stanfield; 4, Eloy. (*Source:* Modified from Winikka and Wold, 1977, Figure 1). Areas are as shown on IAHS questionnaires prepared by C. C. Winikka, January, 1978.)

completed, will alleviate subsidence in some basins, but the additional water supply of 1480 hm³ is less than one-half of the estimated overdraft.

Las Vegas Valley, Nevada

For a summary of data relating to land subsidence in Las Vegas Valley as of 1970, the reader is referred to a report by Anthony Mindling (14). He reported maximum subsidence of 1.2 m from 1935 to 1963, based on leveling of the National Geodetic Survey.

A ground water pumpage inventory made by the State of Nevada Division of Water Resources indicated withdrawal of 85 hm³ in Las Vegas Valley during 1978. The first unit of the Southern Nevada

Water Project, to bring Colorado River water to the Las Vegas area, was completed in 1971 with a design capacity of 163 hm³/year. In 1978 about 114 hm³ of Colorado River water was used in Las Vegas Valley, of which roughly 17 hm³ of Colorado River water was used in the vicinity of Henderson in the southeast part of the valley. The second unit of the Southern Nevada Water Project was completed in late 1981. The design capacity is 206 hm³/year. After completion of the second unit the total design capacity of the project to bring Colorado River water to Las Vegas Valley was about 370 hm³/year. The total of the long-term permits to pump ground water is about 62 hm³/year. After the shorter-term permits expire, it is anticipated that water available from the Colorado River will be adequate to permit holding ground water withdrawal to the 62 hm³/year and slow down or stop the artesian head decline. This should alleviate

the problem of subsidence and fissuring in the Las Vegas area in the 1980s, except in local areas of intensive ground water extraction.

California

The two principal areas of land subsidence due to ground water withdrawal in California are the San Joaquin Valley and the Santa Clara Valley (Figure 26-6). Case histories of these two areas have been included in the UNESCO casebook. Two series of Geological Survey Professional Papers have been published. Professional Papers in the 437 number series represent products of land subsidence studies in cooperation with the California Department of Water Resources; those in the 497 number series are products of the federal program on mechanics of aquifer systems. Three other areas of small subsidence, not considered in this chapter, are the

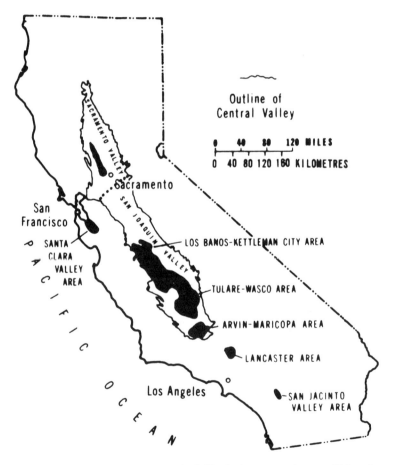

Figure 26-6. Area of Land subsidence in California due to ground water withdrawal.

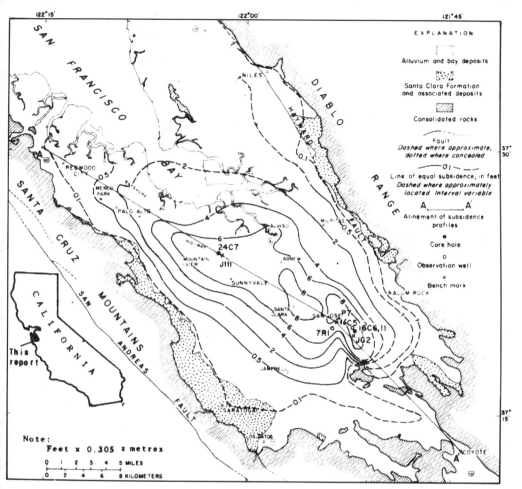

Figure 26-7. Land subsidence from 1934 to 1967, Santa Clara Valley, California; compiled from leveling of National Geodetic Survey in 1934 and 1967.

Sacramento Valley (0.7 m), the Lancaster area (1 m, 1955–1976), and the San Jacinto Valley (1 m, 1950–1974).

Santa Clara Valley

Land subsidence in the Santa Clara Valley was first noted in 1933. By 1969 the central part of the city of San Jose had subsided about 4 m. Figure 26-7 shows the extent and magnitude of subsidence in the 33-year period 1934–1967. It ranged from 3 ft (1 m) or more beneath San Francisco Bay to 8 ft (2.5 m) in San Jose.

In the central two-thirds of the valley below a depth of about 60 m, ground water is confined. The extent of the subsidence defines in a general way the extent of the confined system. Specifically,

in Figure 26-7 the boundary of the confined system is between the 0.5-ft and the 2-ft lines, about along the 1-ft subsidence line (interpolated).

The historical development of land subsidence was in response to a major decline in artesian head (19). From 1916 to 1966 the artesian head in index well 7R1 in San Jose declined 58 m. The two factors that caused this major decline in artesian head are shown in Figure 26-8. The upper line is the cumulative departure of the seasonal rainfall at San Jose from a 50-year mean. In the 50 years between 1916 and 1966 the cumulative departure was generally negative, representing about 300 percent deficiency. As shown by the bottom graph, the pumpage increased fourfold from 1915 to 1965—from 60 hm^3 to 228 hm^3. The 50-year decline in artesian head plainly was caused by generally de-

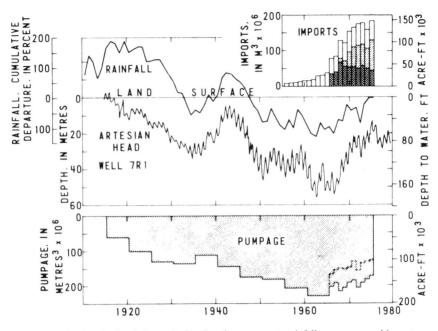

Figure 26-8. Artesian head change in San Jose in response to rainfall, pumpage, and imports.

ficient rainfall and constantly increasing pumping draft.

The subsidence record for benchmark P7 in San Jose is shown in Figure 26-9, together with the artesian head in nearby index well 7R1, taken from Figure 26-8. The fluctuation of artesian head reflects the change in stress on the confined aquifer system: The subsidence is the resulting strain. The cause-and-effect relationship is clear. Between 1912 and 1967 benchmark P7 subsided 3.86 m.

The recovery of artesian head from 1967 to 1975 was dramatic; the head in two index wells recovered about 30 m in the 8 years. As documented in Figure 26-8, recovery of artesian head was due to a fivefold increase in surface water imports from 1965 to 1975, favorable local water supply from rainfall, and decreased pumpage of ground water. Also, the elastic (recoverable) storage coefficient was functional in both aquifers and aquitards during this recovery stage, whereas during the long-term

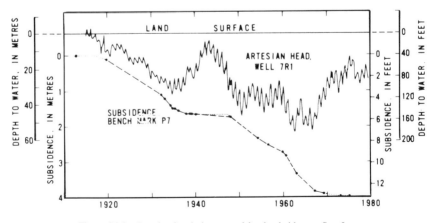

Figure 26-9. Artesian head change and land subsidence, San Jose.

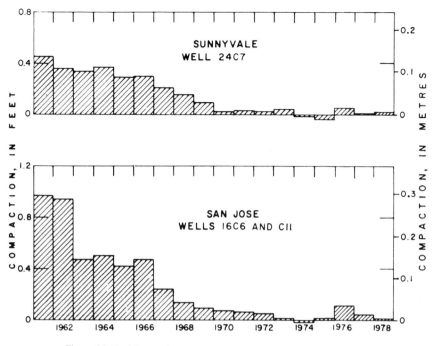

Figure 26-10. Measured annual compaction to a 305-m (1000-ft) depth.

drawdown about 620 hm³ of water was derived from aquitard compaction in response to the much larger virgin storage coefficient.

Although the benchmark net in the subsiding area has not been resurveyed since 1967, two vertical extensometers installed by the Geological Survey to a depth of 305 m have furnished continuous measurements of compaction of the confined aquifer system since 1960 (approximately equal to subsidence).

Figure 26-10 shows the annual compaction at the two sites from 1961 through 1978. In San Jose (wells 16C6 and C11) the rate decreased from 30 cm in 1961 to 0.3 cm in 1973. Net expansion (land surface rebound) of 0.6 cm occurred in 1974.

The successful management of a highly variable water supply to achieve a balance with an ever-increasing demand for water in Santa Clara County (not shown on map) has been remarkable for several reasons. First, maximum development of local water supplies and importation of water from two sources have momentarily brought supply and demand into balance. Secondly, by a building up of the ground water storage in the recharge area and the artesian head in the confined system, land subsidence was stopped, at least temporarily, by 1975. Thirdly, all this has been accomplished by bond issues, revenue from taxes, and water charges,

thus avoiding a drawn-out, expensive legal adjudication of the ground water supply, such as occurred in southern California in the Raymond Basin (17).

Recently, the Santa Clara Valley Water District was awarded historical landmark status by ASCE for its major contributions to the development of the region. It was acknowledged that the district's system is "the first and only instance of a major water supply being developed in a single ground water basin involving the control of numerous independent tributaries to effectuate almost optimal conservation of practically all of the sources of water flowing into the basin."

San Joaquin Valley

The U.S. Geological Survey, in cooperation with the California Department of Water Resources, has prepared several reports describing land subsidence in the three principal subsiding areas in the San Joaquin Valley (Figure 26-6). These include reports on the Los Banos–Kettlemen City area (1, 2, 3, 4), on the Tulare-Wasco area (12), and on the Arvin-Maricopa area (10). A report by the writer, Lofgren, Ireland, and Pugh (20) is a summary of land subsidence in the valley as of 1972.

In most of the central and southern San Joaquin

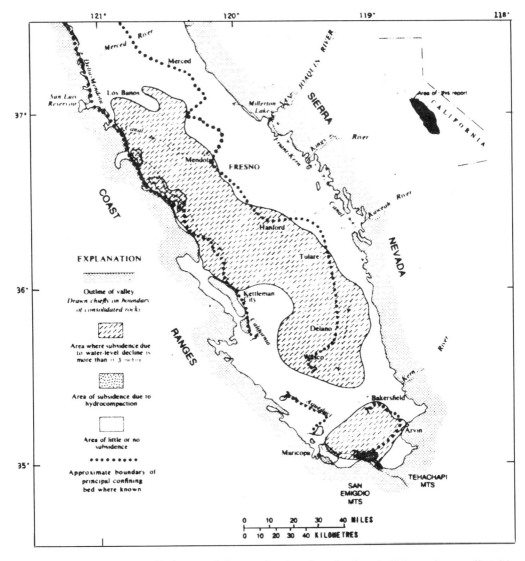

Figure 26-11. Pertinent geographic features of the central and southern San Joaquin Valley and areas affected by subsidence.

Valley, the continental freshwater-bearing deposits can be subdivided into two principal hydrologic units. The upper unit, a semiconfined aquifer system with a water table, extends from the land surface to the top of the Corcoran Clay Member of the Tulare Formation, at a depth ranging from 0 to 275 m. The lower unit, a confined aquifer system, extends from the base of the Corcoran Clay Member down to the saline water body. In the central west-side area the thickness of this confined system ranges from 60 m to more than 600 m. The Corcoran Clay Member, which ranges in thickness from a featheredge to 48 m, is the principal confining bed beneath at least 13,000 km² of the San Joaquin Valley. The dotted line in Figure 26-11 defines the general extent of this principal confining bed in the valley.

Yearly extraction of ground water for irrigation in the San Joaquin Valley increased slowly until 1940. Then during World War II and the following two decades, the rate of extraction increased more than threefold to furnish irrigation water to rapidly expanding agricultural demands. By 1966 pumpage of ground water was 12,000 hm³/year.

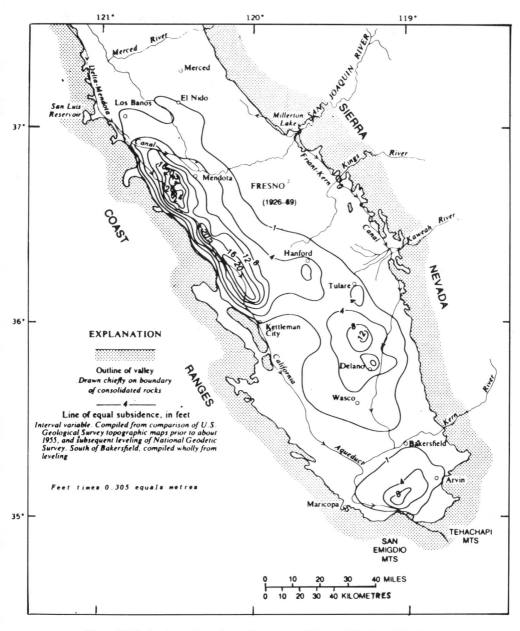

Figure 26-12. Land subsidence in the San Joaquin Valley, California, 1926–1970.

This very large withdrawal caused substantial overdraft on the central west side and in much of the southern part of the valley, essentially within the shaded area of Figure 26-11. The withdrawal in these overdraft areas in the 1950s and early 1960s was at least 5000 hm³/year. As a result, the head (potentiometric surface) in the confined aquifer system between Los Banos and Wasco was drawn down 60 to 180 m. South of Bakersfield the head decline was more than 100 m.

Importation of surface water to these areas of serious overdraft began in 1950 when water from the San Joaquin River was brought south through the Friant-Kern Canal, which extends to the Kern

River (Figure 26-11). About 80 percent of the average annual deliveries of 1250 hm³ of water from this canal is sold to irrigation districts south of the Kaweah River, mostly in the Tulare-Wasco subsidence area.

Large surface water imports from the northern part of the state to overdrawn areas on the west side and south end of the valley beginning in 1967 are being supplied through the California Aqueduct (Figure 26-11). The joint-use segment of the aqueduct between Los Banos and Kettleman City serves the San Luis project area of the U.S. Bureau of Reclamation and transports state-owned water south of Kettleman City. By 1973 the California Aqueduct delivered 1920 hm³ to the central and southern parts of the San Joaquin Valley.

As a result of these large surface water imports, the rate of ground water withdrawal decreased sharply and the decline of artesian head was reversed in most of the areas of overdraft. By the early 1970s many hundreds of irrigation wells were unused, artesian heads were recovering at a rapid rate, and rates of subsidence were sharply reduced.

Subsidence in the San Joaquin Valley is of three types. In descending order of importance these are: (1) subsidence due to the compaction of aquifer systems caused by the excessive withdrawal of ground water; (2) subsidence due to the compaction of moisture-deficient deposits when water is first applied—a process known as hydrocompaction;

and (3) local subsidence caused by the extraction of fluids from several oil fields.

Subsidence due to the compaction of aquifer systems in response to excessive decline of water levels had affected about 13,500 km² of the San Joaquin Valley by 1970. Figure 26-12 depicts the distribution and magnitude of subsidence exceeding 1 ft (0.3 m) that had occurred by 1970—affecting an area of 11,100 km². Three centers of subsidence stand out on this map. The most prominent is the long narrow trough west of Fresno that extends 140 km from Los Banos to Kettleman City referred to subsequently as the west-side area). Maximum subsidence in this area to 1972 was 29 ft (8.8 m), 16 km west of Mendota. The second center, between Tulare and Wasco, is defined by two closed 12-ft (3.7-m) lines of equal subsidence, 32 and 48 km south of Tulare, respectively. The third center, south of Bakersfield, has subsided a maximum of 8 ft (2.4 m), mostly since World War II.

The cumulative volume of subsidence in the San Joaquin Valley soared to 12,350 hm³ by 1960 and reached 19,250 hm³ by 1970. This very large volume is equal to one-half the initial storage capacity of Lake Mead. The volume of subsidence represents water of compaction derived almost wholly from compaction of the fine-grained highly compressible clayey interbeds (aquitards), in response to the increase in effective stress as artesian head in the confined system declined.

NOTES

1. W. B. Bull, 1964, Alluvial fans and near-surface subsidence in western Fresno County, California, *U.S. Geological Survey Professional Paper 437–A.*

2. W. B. Bull, 1975, Land subsidence due to ground-water withdrawal in the Los Banos–Kettleman City area, California. Part 2. Subsidence and compaction of deposits, *U.S. Geological Survey Professional Paper 437–F.*

3. W. B. Bull and R. E. Miller, 1975, Land subsidence due to ground-water withdrawal in the Los Banos–Kettleman City area, California. Part 1. Changes in the hydrologic environment conducive to subsidence, *U.S. Geological Survey Professional Paper 437–E.*

4. W. B. Bull and J. F. Poland, 1975, Land subsidence due to ground-water withdrawal in the Los Banos–Kettleman City area, California. Part 3. Interrelations of water-level change, change in aquifer-system thickness, and subsidence, *U.S. Geological Survey Professional Paper 437–G.*

5. G. H. Davis, H. B. Counts, and S. R. Holdahl, 1977, Further examination of subsidence at Savannah, Georgia, 1955–75, Publication 121, International Association of Hydrological Sciences, 347–354.

6. R. K. Gabrysch and C. W. Bonnet, 1975, Land-surface subsidence in the Houston-Galveston region, Texas, *Report 188,* U.S. Geological Survey, Texas Water Development Board, and the cities of Houston and Galveston, Texas.

7. T. L. Holzer, S. N. Davis, and B. E. Lofgren, 1979, Faulting caused by ground-water extraction in south-central Arizona, *Journal of Geophysical Research,* 84 (B2) (February): 603–612.

8. R. L. Laney, R. H. Raymond, and C. C. Winikka, 1978, Maps showing water-level declines, land subsidence, and earth fissures in south-central Arizona, *U.S. Geological Survey Water-Resources Investigations 78–83,* in cooperation with Arizona Water Commission and U.S. Bureau of Reclamation, Open-File Report, Tucson, Arizona, June.

9. B. E. Lofgren, 1968, Analysis of stresses causing land subsidence, *U.S. Geological Survey Professional Paper 600–B,* B219–B225.

10. B. E. Lofgren, 1975, Land subsidence due to ground-water withdrawal, Arvin-Maricopa area, California, *U.S. Geological Survey Professional Paper 437–D.*

11. B. E. Lofgren, 1979, Changes in aquifer-system properties with ground-water depletion, *Evaluation and prediction of subsidence,* S. K. Saxena, ed. (New York: ASCE), 26–46.

12. B. E. Lofgren and R. L. Klausing, 1969, Land subsidence due to ground-water withdrawal, Tulare-Wasco area, California, *U.S. Geological Survey Professional Paper 437–B*.

13. O. E. Meinzer, 1923, The occurrence of ground water in the United States, with a discussion of principles, *U.S. Geological Survey Water-Supply Paper 489*.

14. A. Mindling, A summary of data relating to land subsidence in Las Vegas Valley, Nevada University System, Desert Research Institute, Center for Water Resources Research.

15. C. R. Murray and E. B. Reeves, 1977, Estimated use of water in the United States in 1975, *U.S. Geological Survey Circular 765*.

16. J. G. Newton, 1977, Induced sinkholes—A continuing problem along Alabama highways, *Publication 121* (International Association of Hydrological Sciences), 453–463.

17. Pasadena *v.* Alhambra (33 CAL 2d 908 207 PAC. 2d 17) 1949; certiorari denied (339 U.S. 937) 1950.

18. J. F. Poland, 1974, Subsidence in United States due to ground-water overdraft—A review, *Agricultural and urban considerations in irrigation and drainage* (New York: ASCE), 11–38.

19. J. F. Poland, 1977, Land subsidence stopped by artesian-head recovery, Santa Clara Valley, California, *Publication 121* (International Association of Hydrological Sciences), 124–132.

20. J. F. Poland, B. E. Lofgren, R. L. Ireland, and R. G. Pugh, 1975, Land subsidence in the San Joaquin Valley as of 1972, *U.S. Geological Survey Professional Paper 437–H*.

21. F. S. Riley, 1969, Analysis of borehole extensometer data from central California, *Publication 89* (International Association of Hydrological Sciences), 423–431.

22. H. H. Schumann, 1974, Land subsidence and earth fissures in alluvial deposits in the Phoenix area, Arizona, *U.S. Geological Survey Miscellaneous Investigation Series, Map I–845–H*.

23. C. G. Smith and R. G. Kazmann, 1978, Subsidence in the capital area ground water conservation district—An update, *Bulletin No. 2* (Baton Rouge, La.: Capital Area Ground Water Conservation Commission), 31.

24. C. C. Winikka and P. D. Wold, 1977, Land subsidence in central Arizona, *Publication 121* (International Association of Hydrological Sciences), 95–103.

27

HUMAN RESPONSES TO FLOODS

JACQUELYN L. BEYER

DEFINITION

All streams are subject to flooding in the hydrological sense of inundation of riparian areas by
stream flow that exceeds bank full capacity. In arid
regions the channel itself, not usually filled with
water, is "flooded" at times of high runoff. The
point at which the channel discharges an overbank
surplus is the flood stage. This may not, however,
coincide with the amount of water outside the
normal channel that will cause damage to human
works. It is also possible to calculate the stage of
high water that is the threshold for damage to
property or dislocation of human activities. Frequently, the use of the term *flood stage* is based
on such a perception of the event and is therefore
a definition subject to change as conditions of
floodplain occupance change. This chapter is not
concerned with coastal flooding.

SPATIAL EXTENT

Floods are the most universally experienced natural
hazard, tend to be larger in spatial impact, and
involve greater loss of life than do other hazards.
Floods can occur on both perennial and ephemeral
streambeds or in an area where no defined channel
exists, such as in an arid region subject to cloudburst types of storms. The problem is compounded
for human adjustment by the fact that few other
hazards present the ambivalent Janus-like aspect of
good and evil. Humans are attracted to settlement
in floor hazard areas by the very characteristics—

water supply and floodplain terrain—that contribute
to the damage potential.

For this reason it is not surprising to find that
historic attempts have been made to resolve the
conflict between the need for riparian occupance
and the inevitable damage, as Wittfogel describes
it: "Thus in virtually all major hydraulic civilizations, preparatory (feeding) works for the purpose
of irrigation are supplemented by and interlocked
with protective works for the purpose of flood
control" (Wittfogel, 1957, p. 24). Less elaborately
organized preindustrial societies have also worked
out ecological adjustments to flooding. Familiar
examples of peasant adaptation to periodic flooding
include the traditional agricultural organization along
the Lower Nile, now altered by the construction of
the high Aswan Dam, and the village rice culture
of the Lower Mekong, which will eventually be
affected by flood control components of the Mekong Basin development. Another such example is
the people of Barotseland in northwest Zambia
where migration to higher ground is the organized
response to the annual seasonal inundation of the
reaches of the Upper Zambezi, which mark the
coreland of Barotse occupance. Changes in socioeconomic patterns as such societies industrialize
will undoubtedly accelerate the damage from floods.
Familiar adjustments such as migration will fall
outside the range of choice. Alternative workable
adjustments may be inhibited by lack of knowledge, technology, and/or capital.

For industrial societies the twentieth-century
concept of multiple-purpose river basin planning,
now widely diffused (United Nations, 1969a), in-

Dr. Beyer is professor of Geography and Environmental Studies, University of Colorado, Colorado Springs, Colorado. She has done
fieldwork in South Africa and North America on resource management problems while teaching at Montana, Capetown, and Rutgers.
Currently, her interests are resource development impacts in Colorado, geographic education, and the geography of women, particularly
lesbian, communities. This material is taken from Global summary of human responses to natural hazards: Floods, in *Environmental
geology*, ed. Ronald W. Tank (New York: Oxford University Press, 1983), 234–249. First published in *Natural hazards, local, national, global*, ed. G. F. White (New York: Oxford University Press, 1974). Reprinted by permission of the author and Oxford
University Press. Copyright © 1974 by Oxford University Press.

volves the consideration of flood damage reduction along with planning for beneficial use of water.

In summary, the potential for flooding is global in nature and can occur, with the proper combination of factors, whenever there is precipitation. This precipitation may range from uniform and general to sporadic and highly localized (see Table 27-1). Adjustments to hazards must be made in the context of both universality and randomness. In addition, there needs to be sensitivity to beneficial uses for floodplains and watercourses.

DAMAGE POTENTIAL

Types of floods are so varied in origin, duration, strength, timing, volume, depth, and seasonality that it is difficult to identify damage potential except in the most general terms. The amount of damage and the damage potential in any flood hazard area is very closely related to the nature of occupance and to the stage of economic development as well as to the physical parameters. There also seems to be an inverse relationship between property damage as measured in monetary terms and loss of life. Societies that have much to lose in terms of structures, utilities, transportation facilities, and so on, also have the technological sophistication to ensure better monitoring, warnings, evacuations, and rehabilitation—all of which contribute to lowering the human costs. Conversely, preindustrial societies, especially with dense rural populations, do not suffer large property losses but are less well equipped to provide preventative or rescue measures for people.

Clearly, the main damage agent is the water itself, overflowing normal channels and inundating land, utilities, buildings, communications, transportation facilities, equipment, crops, and goods that were never meant to operate in or withstand the effects of water. In addition, high velocity of running water operates as a damage agent either directly or indirectly. In the latter case debris carried by the water or dislodged materials batter structures, people, and goods. Debris and silt carried by the water and left behind as the water recedes operate as further damage agents.

Damage is considered to be either direct or indirect (primary or secondary). Such a classification is useful in any assessment that attempts to clarify the benefits of a damage reduction program. Loss of human life is the most dramatic and certainly the easiest to identify as a direct result of flood events. Loss of livestock may be especially costly in rural zones.

In agricultural areas damage involves inundation of land accompanied by erosion and/or loss of crops. It is in such cases that the season of flooding is especially significant. Water damages farm equipment, stored materials (seed, fertilizer, feed), disrupts irrigation systems and other water supply, and disrupts communication.

Urban facilities are all subject to water and force damage—buildings of all kinds, public facilities, utilities, transportation, waterway facilities, and open space. Machinery, manufacturers, goods in retail establishments, household furniture can all be damaged by water, debris, and silt.

Indirect damages are generally associated with health and general welfare although such amenities as scenic values, recreational services, and wilderness preservation may also be taken into account. Normal public health services are subject to greater pressures in the face of disruption of transportation and utilities, especially water supply. Contamination and pollution are more probable, epizootics emerge, stagnant water is left as a flood legacy, and general morbidity increases. Flooding affects the normal sources of food and shelter and hence adversely affects health conditions. Opposed to these considerations is the possibility that emergency relief operations might provide better health care and food than is normally available to some communities, counter-acting, to some extent, the effects of both direct and indirect damages (White, 1945).

BENEFITS

Rivers in flood clearly are hazardous for many kinds of human use, but a complication in planning for amelioration of the hazard is presented by the benefits of naturally flowing rivers, including overbank flooding. "Control" of flooding by protective works, especially, may negate benefits from soil nutrient renewal and fisheries. Restriction on use of floodways will provide for damage reduction and also enhance community values through preservation of open space. In some parts of the world, exemplified by India and Bangladesh, the rhythms of agricultural production are dependent upon water brought by major storms and renewal of fertility through siltation. In such cases serious weather modification efforts should proceed only after careful assessment of the total benefit-cost pattern.

DAMAGE ASSESSMENT

Assessment includes the costs of repair, including temporary repairs; replacement and cleanup; and loss of improvements and inventories. Emergency costs are also involved. Loss of business and em-

Table 27-1. Significant Historical Flood Events

Date	Place	Deaths	Property Damage
June 1972[a]	Eastern United States	100+	$2 billion
June 1972[a]	Rapid City, South Dakota	215 (est.)	$100 million
May 11–23, 1970[b]	Oradea, Rumania	200	225 towns destroyed
January 25–29, 1969	Southern California	95	
July 4, 1969	Southern Michigan and Northern Ohio	33	
August 23, 1969	Virginia	100	
May 29–31, 1968	Northern New Jersey	8	$140 million
August 8–14, 1968	Gujarat, India	1,000	
January–March 1967	Rio de Janeiro and São Paulo states	600+	
November 26, 1967	Lisbon	457	
January 11–13, 1966	Rio de Janeiro	300	
November 3–4, 1966	Arno Valley, Italy	113	Art treasures in Florence and elsewhere destroyed
June 18–19, 1965	Southwest United States	27	
June 8–9, 1964	Northern Montana	36	
December 1964	Western United States	45	
October 9, 1963	Belluno, Italy	2,000+	Vaiont Dam overtopped
November 14–15, 1963	Haiti	500	
September 27, 1962	Barcelona	470+	$80 million
December 31, 1962	Northern Europe	309+	
May 1961	Midwest United States	25	
December 2, 1959	Frejus, France	412	Malpasset Dam collapsed
October 4, 1955	Pakistan and India	1,700	5.6 million crop acres at loss of $63 million
August 1, 1954	Kazvin District, Iran	2,000+	
January 31– February 1, 1953	Northern Europe	2,000+	Coastal areas devastated
July 2–19, 1951	Kansas and Missouri	41	200,000 homeless, $1 billion
August 28, 1951	Manchuria	5,000+	
August 14, 1950	Anhwei Province, China	500	10 million homeless; 5 million acres inundated
July–August 1939	Tientsin, China	1,000	Millions homeless
March 13, 1928	Santa Paula, California	450	St. Francis Dam collapsed
March 25–27, 1913	Ohio and Indiana	700	
1911	Yangtze River, China	100,000	
1903	Heppner, Oregon	250+	Town destroyed
May 31, 1889	Johnstown, Pennsylvania	2,000+	
1887	Honan, China	900,000+	Yellow River overflowed; communities destroyed
1642	China	300,000	

[a] Press reports.
[b] Adapted from Table of Disasters/Catastrophes, *New York Times Encyclopedic Almanac (1970)*, p. 1228; (1972), pp. 322–33.

ployment during flood disruption is also a direct cost associated with the physical damage. Similarly the transfer of public economic development funds to flood emergency and control programs represents a deferred opportunity that may be especially significant as a social cost in developing nations.

There are no comprehensive calculations of global damage from floods. In many cases, local flooding in areas remote from communications may not even be recorded. There are immense difficulties for any inventory of flood consequences including those of cost, technical expertise, comparability of data collected, allocation of losses to proper causes, and time. A pilot survey of global natural disasters by

Sheehan and Hewitt (1969) provides some information for the 20 years 1947–1967 for most of the world—the Soviet Union is the major country excluded. During this period Asia (excluding the Soviet Union) led in loss of life, with 154,000 deaths. Europe (excluding the Soviet Union) followed with 10,540 deaths. Africa, South America, and the Caribbean area each recorded 2000–3000 deaths. During the same period 680 lives were lost to floods in North America, and 60 in Australia, totaling for all these regions 173,170 deaths.

This total can be compared with the 269,635 deaths attributed to all . . . other hazards . . . and if it is considered that many of the other categories—for example, tornadoes, typhoons, hurricanes, and tidal waves—also involve flooding, the death loss is the most impressive comparison. Table 1 will give some idea of the magnitude of the hazard in terms of area affected, people involved, and property damage.

DAMAGE FACTORS

The factors which should be taken into account in assessing the damage potential for flooding in any basin include the following.

Frequency

Flood flows can occur on the average as often as once every 2 years in temperate climates and as infrequently as once in 1000 years elsewhere. The recurrence of a particular flood flow can be predicted with reasonable accuracy over a long time span if sufficient stream-flow data are available over a long period. This emphasizes the certainty of the event, not its timing, which is not predictable since flood flows are assumed to be random events.

Frequency is a physical parameter clearly related both to a perception and adjustments. The greater the frequency the more accurate is the perception of the hazard by floodplain occupants and the greater is their willingness to consider a wider range of adjustments, including alternative sites for their activities. This is demonstrated by variations in community decision making that can be correlated with frequency of flooding (Kates, 1962).

Magnitude (depth)

Magnitude of a flood may be expressed in physical, or probabilistic terms. The physical measures are rate of flow measured in cusecs, m^3/sec (cubic meters per second), or river stage in meters (or feet) above some datum (reference) point. Both of these require carefully established measuring in-

stallations to obtain reliable data. Flow can be graphically plotted versus stage to give a stage-discharge relationship, which can then be useful in predicting damages. This relationship can be quite complex, and considerable care should be exercised in its use.

The probabilistic measure is a statistical method of ordering various magnitudes of flow and stating the probability that a given flow will be exceeded. Under this procedure, a 5-year flood is a flow or stage which will be equaled or exceeded 20 times, on the average, in a 100-year time span. The statistical method has validity only in areas where good flow records over a long period of time (at least 30 years) are available, although some simulation can be done. The ability to determine probability of flood magnitudes is not the same as the ability to state when such floods will occur. Some of the parameters upon which magnitude depends and which are useful in classifying the flood characteristics of a given area include the type, intensity, duration, areal extent, and distribution of precipitation; basin size and shape; floodplain topography; surface conditions of soil; and land use. There is an effort in Japan, for example, to elaborate a system of flood hazard classification through the use of landform analysis. It is suggested that such analysis will provide for predictability of flood current, ranges of submersion, depth of stagnant water, and length of period of stagnation (Oya, 1969).

Depth can be important in terms of both the kinds of damage and possible adjustments, for example, floodproofing of structures.

Rate of Rise

Rate of rise is the time from flood stage (or zero-damage stage) to flood peak and is a measure of the intensity of the flood. As Sheaffer (1961) notes, this time between flood stage and flood peak represents an adjustment time during which persons affected by flooding can engage in activities to lessen the damage. Generally, people will not respond to the danger of flooding until at least flood stage, so this time period is critical. It is clear that there is a relationship between the nature of the drainage system and the rate of rise—upstream areas will have a more rapid rise and shorter duration of flooding than will downstream areas.

Seasonality

Seasonality is one of the more significant factors for agricultural damage and probably the main basis for the adjustments made by preindustrial riverine societies. Clearly, the hazard increases where the growing season is limited and coincides with the season of flooding. Winter floods might also account for increased loss and disruption in urban areas where heating and sanitary facilities are needed to guard against increase in disease and discomfort.

Duration

The time of inundation for flood flows can vary from a few minutes to more than a month. The duration is highly correlated to the rate of rise and fall of flood crests except where drainage of land area is impeded by obstructions. Flood duration is dependent upon such parameters as source of runoff; runoff characteristics including slope and surface conditions; nature of obstructions impeding recession of waters; and man-created controls such as reservoirs, levees, and channelization.

Nature of Floodplain Occupance

Floodplain occupance includes the density of settlement; types of facilities; extent of fixed facilities, buildings, and equipment; and value of facilities. Obviously, every increase in such occupance will increase the potential damage and call for some kind of adjustment, whether protective works, warning systems, and public relief capabilities, or a willingness to accept the losses.

Efficacy of Forecasting and Warning Systems

The ability to forecast the occurrence of overbank flooding is limited to a time span in which the hydrologic conditions necessary for flooding to occur have begun to develop. The formulation of a forecast for flood conditions requires information on current hydrologic conditions such as precipitation river stage, water equivalent of snowpack, temperature, and soil conditions over the entire drainage basin as well as weather reports and forecasts.

In small headwater regions a forecast of crest height and time of occurrence is all the information required to initiate effective adjustments since the relatively rapid rate of rise and fall makes the period of time above flood stage relatively short. In lower reaches of large river systems where rates of rise and fall are slower, it is important to forecast the time when various critical stages of flow will be reached over the rise and fall. Reliability of forecasts for large downstream river systems is generally higher than for headwater systems.

Warning time for peak or overbank conditions can range from a few minutes in cloudburst conditions to a few hours in small headwater drainages to several days in the lower reaches of large river systems. As with forecasting, the time and reliability of the warning increase with distance downstream where adequate knowledge of upstream conditions exists.

Clearly, the amount of information required, the data collection network necessary for collecting the information, the technical expertise required for interpretation, and the communication system needed to present the information in time to potential victims are such as to preclude many poor and developing nations from having an adequate service. The World Meterological Organization of the United Nations, through its World Weather Watch and Global Data Processing System, hopes to coordinate efforts to improve forecasting. A recent report (Miljukov, 1969) notes that quantitative precipitation forecasts for 24 to 49 hr in advance are provided in parts of Australia, Byelorussia, Cambodia, Canada, Czechoslovakia, France, Federal Republic of Germany, Hong Kong, India, Iraq, Japan, Mauretania, Norway, Pakistan, Philippines, Rhodesia, Romania, Sweden, Ukraine, the Soviet Union, and the United States. Precipitation forecasts for hydrological purposes are provided in Australia, Canada, Czechoslovakia, France, Federal Republic of Germany, India, Iraq, Japan, the Soviet Union, and the United States. The report also notes: "Precipitation forecasts are not accurate and reliable enough, in the present state of meteorological science, for use in the preparation of quantitative forecasts of river discharges" (Miljukov, 1969, p. 10). Most developing nations will have to rely on much less data than are ideally needed for forecasting and warning, which in turn will lessen the effectiveness of this factor with respect to flood losses.

Efficacy of Emergency Services

The helplessness of many small and poor societies is exemplified by the situation in Bangladesh during the tropical cyclone of 1970. Where resources are not available for planning, for the physical effort of relief and evacuation, and for coordination with other activities, little will be done outside of contributions of international aid agencies. Even in more developed areas local conditions of transportation and public attitudes will lessen the usefulness of emergency aid, for instance, the disaster at Buffalo Creek, West Virginia, in 1972. The more elaborate and dependable such services are, however, the more there is a tendency to rely on such aid as a major adjustment and to reject consideration of less costly and more effective adjustments. There clearly must be provisions for first-level emergency aid where settlement already exists and

Table 27-2. Adjustments to the Flood Hazard

Modify the Flood	Modify the Damage Susceptibility	Modify the Loss Burden	Do Nothing
Flood protection	Land-use regulation and changes	Flood insurance	Bear the
(channel phase)	Statutes	Tax write-offs	loss
Dikes	Zoning ordinances	Disaster relief	
Floodwalls	Building codes	volunteer	
Channel	Urban renewal	private	
improvement	Subdivision regulations	activities	
Reservoirs	Government purchase of lands and	government aid	
River diversions	property	Emergency measures	
Watershed treatment	Subsidized relocation	Removal of persons	
(land phase)	Floodproofing	and property	
Modification of	Permanent closure of low-level	Flood fighting	
cropping practices	windows and other openings	Rescheduling of op-	
Terracing	Waterproofing interiors	erations	
Gully control	Mounting store counters on wheels		
Bank stabilization	Installation of removable covers		
Forest-fire control	Closing of sewer valves		
Revegetation	Covering machinery with plastic		
Weather	Structural change		
modification	Use of impervious material for		
	basements and walls		
	Seepage control		
	Sewer adjustment		
	Anchoring machinery		
	Underpinning buildings		
	Land elevation and fill		

Source: Adapted from Sewell (1964), pp. 40–48; and Scheaffer, Davis, and Richmond (1970).

where alternative moderations of the hazard are difficult to implement or costly, but these should normally not supplant other measures to reduce losses.

THE RANGE OF ADJUSTMENTS AND THEIR ADOPTION

The accompanying table (Table 27-2) suggests that the cumulative experience of centuries provides for any society or group wishing to alleviate the social and economic costs of flood losses a choice of methods, to be used singly or in strategic combinations. Much of the wisdom with respect to the need for such strategies, adapted to local conditions of basin hydrography, settlement characteristics, and economic capabilities, has been gained in industrial nations after painful and costly trial and error. It has been suggested (Goddard, 1969) that developing nations need not repeat the errors of the past, that they have models for actions and policies that would provide a much more coherent and suitable response to the flood hazard. It is

clear, even with such models, that adaptation to local circumstances in any society will not be a simple matter. It is increasingly evident that various combinations of individual psychology, institutional inertia, costs, governmental policies and philosophies, and historical precedents help to condition the choice of adjustments.

Modify the Flood

This category includes engineering works affecting the channel that represent the most widely accepted feasible adjustment with the possible exception of bearing the loss. Such protective works are justified where benefits exceed the costs of implementation and especially where high damage potential exists for relatively intensive settlement in urban and industrial situations. In such cases the high value of fixed facilities will justify levees, dams, and channelization even when 100 percent protection cannot be guaranteed. While benefits accrue to both private and public sectors, the costs are necessarily largely public. Partly for this reason this adjustment strongly tends to encourage persistent settlement and even attracts, through a false sense of security, further floodplain encroachment. On the other hand, such engineering works are important components of multiple-purpose projects that are directed toward comprehensive land and water planning, and they can be complemented, for floodplain management, with other measures. A major problem is to encourage engineers and officials to think in terms of nonstructural alternatives or supplements to protective works.

Watershed treatment practices have more subtle implications with respect to flood control and are frequently more significant for their contribution to improved *in situ* land management. All such measures have their limitations with respect to major flood events. There may be some contribution to lowering the depth in small floods and to lengthening the flood-to-peak interval, but essentially the appeal of such practices is lower costs. About 90 percent of the costs for such practices are public, while benefits accrue largely—about 85 percent—to private land users. Land treatment measures often complement and make more effective protective measures but will also tend to encourage continued settlement for flood-prone areas.

Weather modification is a fairly recent technique with respect to flood control, and too little is known about its effectiveness. One major problem, even given scientific certainty about effectiveness, will be the necessity to allay public fears that tampering with weather processes will increase rather than lessen floods. Recent news stories of the use of weather modification techniques in the Indochina War will not make this task easier (Shapley, 1972). The immediate postflood news reports from the 1972 Rapid City, South Dakota, flood suggest that this has already become an issue. Costs for weather modification are entirely public, while benefits are about equally divided between private users and the public.

Modify the Damage Susceptibility

Given that there may be a need to encroach on floodplains or to accept present settlement patterns, certain measures are possible that are either less costly than protective works or bearing the loss, or that will lessen the actual damages even more. There is also a greater shift of cost bearing to private interests, especially in the case of floodproofing, with resultant increased awareness of the need for flood adjustments.

Land use regulation, including changes in occupance, is especially suitable where there is competition for floodplain land for uses other than agricultural or recreational. The legislative and policy powers of the state can be used to control

and guide development of floodplains. According to Goddard, in the United States "about 35 states have adopted regulations and 500 additional places in 41 states have them in adoption process" (Goddard, 1971). Encouraging this is the 1969 Federal Flood Insurance Act, which provides for governmental flood insurance subsidy to individuals in communities that agree to adopt floodplain regulation guidelines. These measures tend to encourage more efficient and less costly use of floodplains, and there is a greater shared responsibility between floodplain users and authorities. Strong leadership and a commitment to long-range planning and rational allocation of land uses are also prerequisites to widespread adoption of such measures.

Structural changes and floodproofing (including land elevation) provide for even larger shifts of costs, as individual users may bear all the costs and share benefits with the public on an equal basis. Such adjustments are most appropriate where flooding is not intense either in velocity or depth and where some warning time is possible. Floodproofing especially requires a network of forecasting and warning facilities along with a flood hazard information program that will encourage preflood adjustments. Structural modifications are possible for existing structures as well as for new structures, although this will increase costs. In many cases it would be too expensive to modify old buildings. Some types of buildings are better suited to modification than others, but clearly damage reduction is related to size of structures and costs of modification. These adjustments tend to encourage persistent occupance and lose effectiveness where flood frequency is low. At the same time, they place more responsibility on the user and thus heighten sensitivity to and knowledge about the flood hazard.

Modify the Loss Burden

There is much more emphasis in this category on humanitarian responses rather than calculated economic rationale, based on the inevitability of flooding and the unlikely possibility of preventing all damage by eliminating floodplain occupance. Losses will thus occur even in the face of widespread use of appropriate adjustments. When people suffer trauma and loss, there can be little question of a social obligation to provide assistance. The dilemma for rational flood damage reduction, however, is that relief measures and emergency assistance unless properly designed tend to encourage persistent occupance and reluctance to accept more rational adjustments.

Insurance and tax write-offs will not decrease

flood losses, but there will be a spreading of loss over time and a shift of some costs to the general public. As the flood insurance program has been worked out in the United States, the insurance subsidy by the government to private carriers must be coupled with community planning for land use regulation and other adjustments to lessen potential damage. In Hungary, where levees and flood fighting are the principal adjustments, agricultural insurance was extended in 1968 to cover flood damages (Bogardi, 1972). Whether this works as an incentive for private adjustments is not clear. There is obviously a sensitive line between encouraging further encroachment or private irresponsibility and alleviating the damages to those who must occupy floodplains. Purchases of insurance, according to recent reports after the June 1972 floods in the eastern United States, have not been commensurate with the danger nor with the benefits to eligible individuals. Problems resulting from a hazard insurance program that was not thoroughly planned have been noted.

Disaster relief is a necessary adjustment in order to lessen the immediate impact of a flood event and to ease the implementation of rehabilitation efforts. Whether government or private, the major disadvantage is that such measures, necessary though they may seem when disaster strikes, strongly encourage the belief that nothing else need be done.

The effectiveness of emergency measures depends largely upon the nature of the flood hazard (ideal combination of high flood frequency, low velocity and depth, long flood-to-peak interval, and short duration) and the quality of forecasting. The immediate governmental obligation is generally seen to be the removal of persons and property from flood-threatened areas.

Do Nothing

Bearing the loss is still the major adjustment for large numbers of floodplain occupants in developing countries and is frequently modified in developed nations only by the widespread expectation of relief and emergency measures. In all cases, however, it is clear that an increasing effort to clarify public interest in floodplain situations will restrict the choice of doing nothing, and management strategies will become more common (Sheaffer, Davis, and Richmond, 1970).

REDUCTION OF LOSS

One element in the acceptance by any group of decision makers of a particular mix of components in a flood damage reduction program is the assessment of the comparative return from each possible choice of adjustment or combination of adjustments. If damage assessment after the fact of flooding is extremely difficult, it is even more difficult to predict what the damage will be under a set of assumptions about responses of various kinds. White and Burton have suggested methods whereby maximum damage reduction and minimum cost can be calculated for particular situations (White, 1964; Burton, 1969). Such methods may hopefully provide an additional planning tool in those situations where encroachment onto the floodplain is neither as intensive as in some industrial countries nor necessary. Some such tool is essential also to ensure the most efficient allocation of scarce resources, whether of materials, manpower, or money.

The relative contributions of each possible adjustment to reduction of potential damage can only be crudely measured at present. Such measurement is further complicated by the fact that only infrequently is a single adjustment adopted. Clearly, any protective works that provide for 100 percent security under any feasible flood condition will provide 100 percent loss reduction, although costs of providing such protection are likely to be unacceptable. Such security is a highly improbable, both because of costs and imperfect knowledge of potential floods. The damage reduction to structures may range from 40 to 100 percent, dependent upon the size of the flood experienced and the nature of structures (White, 1964).

Watershed treatment data are inconclusive with respect to damage reduction, and there are no data available for weather modification. Land use regulation and change can provide for up to 90 percent damage reduction dependent upon the effectiveness of the regulations and the speed of application.

Data from one United States town (White, 1964) suggest that even minimal floodproofing of present structures under conditions of frequent but shallow flooding can be very effective, reducing damages by 60 to 85 percent. Great depths and/or high velocities would call for consideration of floodproofing as part of building design.

Emergency action increases in effectiveness where there is a long flood-to-peak interval, high flood frequency, low depths, short duration, and low velocity. Where such conditions prevail, and assuming adequate warning facilities plus personnel and equipment, emergency action can reduce damages by 15 to 25 percent. A lower range of 5 to 10 percent is more probable.

Adequate warning would seem to be a minimal requirement for communities subject to flood hazard, but it is not simple nor inexpensive to provide a good system. Meteorological services and communications are part of the costs. Even given an excellent network of knowledge about the physical event, it may be difficult to convey that information to persons who will have to make adjustment decisions. Factors involved in a less-than-optimal warning system include the following:

1. Reluctance of officials to give false alarms.
2. Lack of complete coverage of media used to transmit warnings (radios, telephones, etc.)—communities and individuals may not be able to afford facilities.
3. Reluctance of people to see themselves affected by distant events (storms, runoff).
4. Individual interpretation of warning messages, especially where several messages may be contradictory or the messages may be incomplete.
5. Failure to provide exact information about what recipients of warnings are to do.
6. Impossibility of warning in time for much else than rapid evacuation.
7. Dramatic warning signals triggering an influx of the curious, which negates warning advantages.

Flood insurance and tax subsidies spread the burden through time and shift much of the loss to the general public but do not reduce damage.

Another indication of the relative efficacy of various adjustments in reducing damage is the importance placed on them in national and regional plans. A recent report on Hungary (Bogardi, 1972), for example, suggests that reduction of damages is to be achieved largely through levee construction and maintenance, flood fighting, and, to some extent, insurance. Recommendations for Malaysia (*Flood control*, 1968) are for a flood control program involving better data collection and improved organization for relief and evacuation, combined with structural controls, land use regulations, and flood-resistant crops. It is estimated that these measures would reduce anticipated damage from presently known levels of flooding by 50 percent. This report does not consider dams in catchment areas justifiable for flood control alone. Engineering works are still considered primary tools for India, although some attention is being given to catchment area management and weather modification. The Japanese have extended their management approach to include regulatory measures (Oya, 1969). A comprehensive summary of national efforts to cope with floods as one of many natural hazards would probably justify a comment in a report from the United Nations (1969b): "Although there is still a considerable gap in many countries between the needs for governmental action and the actual institutional framework, new administrative patterns have evolved in others which responded to

the need for a more coordinated and system oriented approach to resource administration."

PERCEPTION OF HAZARD AND ADJUSTMENTS

The global nature of the flood hazard is suggested not only by maps of large floods and by tables of deaths but also by reference to international interest in the problem. The special agencies of the United Nations are involved in a wide spectrum of activities, including hydrological and meteorological data collection, flood-forecasting methods, world catalogue of large floods, problems of health due to floods, and relief and aid to victims. Agencies involved include the Economic Commission for Africa, Economic Commission for Asia and the Far East, World Meteorological Organization, World Health Organization, and UNESCO. In many cases small nations will have to rely on technical help and assistance through United Nations channels. There is a discernible diffusion of efforts to plan and implement comprehensive programs including Canada, Japan, United Kingdom, and the United States. Because river basin management is so popular as an economic development tool, this opens the door for widespread consideration of comprehensive flood control as a component of such programs. From the global and national institutional viewpoints there is probably adequate sensitivity to the nature of the problem, if not to the possible range of adjustments. At the individual level it is more difficult to judge whether the knowledge gained in recent years about perception of the hazard in the United States (Kates, 1962; Burton and Kates, 1964; James, Laurent, and Hill, 1971) is applicable to individual perception in developing or industrializing nations—or even industrial nations with different social and political conditions. A summary of some of the findings of these hazard perception studies, especially of floodplain occupants in Georgia, may be listed in the form of planning guidelines (adapted from James, Laurent, and Hill, 1971):

1. It cannot be assumed that accurate knowledge of the flood hazard will inhibit all persons from moving onto the floodplain.
2. The flood hazard itself will process people over time in terms of perception of the hazard and willingness to make adjustments. Management programs can short-circuit the unhappy experiences of those who remain unaware of the hazard and reluctant to adopt adjustments by preventing their settlement (e.g., through insurance programs).

3. Prospective floodplain occupants who are initially unaware cannot be swayed by large amounts of technical information; they also tend to be people who avoid contact with public officials and are not observant with respect to natural features.
4. In contrast, people who are knowledgeable about the flood hazard and settle anyway on floodplains will be responsive to more sophisticated information than is usually presented.
5. Delineation of flood hazard areas on a map is ineffective as a form of communication.
6. Officials who disapprove of settlement on floodplains or who think in technical terms about risk will not be effective with those who are unaware of the risk.
7. Those who know about the flood hazard will be sensitive to depths, if not to frequency, and will therefore be open to floodproofing and possibly insurance as adjustments.
8. Time reduces awareness of the hazard, especially for those moving into a hazard area where indications of past flood events are not evident.
9. The wave of concern for environmental issues has brought with it evidence that those who are unaware of the flood hazard, but who have a concern for environmental damage, may respond more to appeals that land use regulations are ecologically sound than to information about potential property damage.
10. Flood damage sufferers who contact, or who are contacted by, officials are a biased sample

in terms of response to flood hazard. Frequently, this bias is associated with speculation as a motive for owning floodplain property.
11. Upstream development frequently becomes the scapegoat for downstream floodplain users threatened by floods.
12. Floodplain users who are alienated from government or authority because of other contacts are poor candidates for participation in floodplain management programs.
13. Extended delays in programs to reduce flood losses will increase alienation and make user participation more unlikely.
14. Encouragement of particular users should be part of policy; for example, it should be made easy for those who are unaware of the hazard and/or reject adjustments to leave and be replaced by those who know something about the hazard and will be willing to adopt reasonable adjustments, including insurance and floodproofing or structural change. Where even these adjustments are too costly in the light of potential damage, the policy should be to consider purchase and reversion to open space and recreational use.

It is hard to believe that persons would vary much with respect to a number of factors involved in determining the degree of knowledge about flood events, anticipation of future events, and willingness to consider various possible adjustments. Confirmation of this belief awaits further investigations of human response in diverse societies.

REFERENCES

Baroyan, O. V. 1969. Problems of health due to floods. Tbilisi, U.S.S.R., United Nations Interregional Seminar on Flood Damage Prevention Measures and Management.

Bogardi, I. 1972. Floodplain control under conditions particular to Hungary. International Commission on Irrigation and Drainage, 8th Congress.

Burton, Ian. 1969. Methods of measuring urban and rural flood losses. Tbilisi, U.S.S.R., United Nations Interregional Seminar on Flood Damage Prevention Measures and Management.

Burton, Ian and Robert W. Kates. 1964. The perception of natural hazards in resource management. *Natural Resources Journal*, 3 (2):412–441.

Flood control. 1968. Report of the Technical Subcommittee for Government of Malaysia, Director of Drainage and Irrigation.

Goddard, James E. 1969. Comprehensive flood damage prevention management. Tbilisi, U.S.S.R., United Nations Interregional Seminar on Flood Damage Prevention Measures and Management.

——. 1971. Flood-plain management must be ecologically and economically sound. *Civil Engineering—ASCE*, 000:81–85.

James, L. Douglas, Eugene A. Laurent, and Duane W. Hill. 1971. *The flood plain as a residential choice: Resident attitudes and perceptions and their implication to flood plain management*. Atlanta: Georgia Institute of Technology, Environmental Resources Center.

Kates, Robert W. 1962. *Hazard and choice perception in flood plain management* (Chicago: University of Chicago), Department of Geography, Research Paper No. 78.

Miljukov, P. I. 1969. Review of research and development of flood forecasting methods. Tbilisi, U.S.S.R., United Nations Interregional Seminar on Flood Damage Prevention Measures and Management.

Oya, Masahiko. 1969. Flood plain adjustments, restricted agricultural uses, zoning, and building codes as damage prevention measures. Tbilisi, U.S.S.R., United Nations Interregional Seminar on Flood Damage Prevention Measures and Management.

Sewell, W. D. F. 1964. *Water management and floods in the Fraser River Basin* (Chicago: University of Chicago), Department of Geography, Research Paper No. 100.

Shapley, Deborah. 1972. News and comment. *Science,* 176: 1216–1220.

Sheaffer, John R. 1961. Flood-to-peak interval. In Gilbert F. White, ed., *Papers on flood problems,* ed. Gilbert F. White (Chicago: University of Chicago), Department of Geography, Research Paper No. 70.

Sheaffer, John R., George W. Davis, and Alan P. Richmond. 1970. *Community goals—Management opportunities: An approach to flood plain management* (Chicago: University of Chicago), Center for Urban Studies. Report by Institute for Water Resources, Department of the Army, Corps of Engineers.

Sheehan, Lesley and Kenneth Hewitt. 1969. *A pilot survey of global natural disasters of the past twenty years* (Toronto: University of Toronto), Natural Hazards Research Working Paper No. 11.

United Nations. 1969a. *Integrated river basin development.* New York: rev. reprinting.

———. 1969b. Some institutional aspects of adjustments to floods. Tbilisi, U.S.S.R., Resources and Transport Division, Department of Economic and Social Affairs, United Nations Interregional Seminar on Flood Damage Prevention Measures and Management.

White, Gilbert F. 1945. *Human adjustment to floods: A geographical approach to the flood problem in the United States* (Chicago: University of Chicago), Department of Georgraphy, Research Paper No. 29.

———. 1964. *Choice of adjustment to floods* (Chicago: University of Chicago), Department of Geography, Research Paper No. 93.

Wittfogel, Karl A. 1957. *Oriented despotism: A comparative study of total power* (New Haven, Conn.: Yale University Press).

28

DETERIORATION OF WATER SUPPLY SYSTEMS

HARRISON J. GOLDIN

Oh, Danny boy, the pipes, the pipes. . . .

Do you know where New York City's oldest water main and sewer are?

Or how much it would cost to replace the city's sewage collection system?

The answers to these—and countless other—questions can be found in Comptroller Goldin's new inventory of the city's water and sewage systems. In fact, virtually everything you ever wanted to know about New York's water and sewer pipes is contained in the recently completed inventory, the first such compilation of data in the city's history.

And this information has far more important uses than providing material for trivia buffs or grist for barroom bets. The comptroller explained that the inventory is not only another key milestone in the city's quest for complete financial disclosure but is also a valuable tool for capital and budgetary planning and management.

In fact, in order to make the inventory even more useful to city officials and engineers, the Comptroller's Office, which compiled the data with the assistance of the Department of Environmental Protection and the accounting firm of Ernst & Whinney, went beyond the requirements of generally accepted accounting principles (GAAP) in recording statistics on the water distribution and sewage collection systems (Figure 28-1). Data on the systems, which are considered part of the city's infrastructure, are optional under GAAP inventories of fixed assets, Mr. Goldin explained.

As a result of the additional analysis, however, the city now knows the age, original cost, and replacement cost of every facet of its water and sewage systems and is better equipped to plan for their repair and replacement, he said.

Among the key findings of the inventory:

- More than 23 percent of the city's sewers and more than 17 percent of its water mains were built prior to 1906 (see Figure 28-1) and, assuming an estimated life of 75 years, must be considered for replacement.
- Of the $263 million in capital expenditures on the two systems in fiscal year 1981, $194 million was spent on expansion and only $69 million on replacement, an amount Mr. Goldin considers $161 million less than was needed.
- Shortfalls in replacement expenditures for the city's sewage treatment system are "not significant at the present time," because the system is relatively new.

The Department of Environmental Protection (DEP) is already using the inventory for planning and other purposes. For example, DEP has arranged with Con Edison to be notified whenever the utility plans to break ground to install pipes or cables, so the agency, using the inventory, will be able to identify any replacement needs in the affected area.

It took more than three years to compile the inventory. Thousands of records and maps from DEP offices throughout the city and upstate were analyzed by engineering contractors and other professionals. Less technical data was gathered by 150 college students who recorded water mains and sewer pipes on a block-by-block basis and made entries of hydrants, manholes, valves, and other elements of the systems on more than 57,000 special worksheets. Approximately 1 million "units," each comprising one block of water main and sewer pipe, are recorded in the inventory.

Oh, yes. The oldest water main in the city still in service is located under Chambers Street, between Broadway and Centre Street, across from the Comptroller's Office in the Municipal Building.

Mr. Goldin is comptroller for the City of New York. This material is taken from *City of New York comptroller's report,* vol. 7 (11) (May 4, 1982).

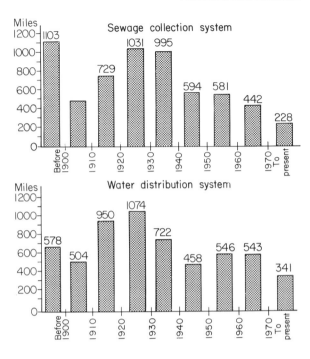

Figure 28-1. Mileage of in-service pipes by decade of installation.

It is made of concrete and was laid in 1873. The oldest sewer here still being used is even older: It was built of brick in 1851 under Bleecker Street between Elizabeth and Mott Street and there it still is, working.

There are 5816 mi of water main and 6197 mi of sewers in the city, and it would cost $3.1 billion to replace the water mains (they cost only $472 million to build) and $13 billion to replace the sewers, which cost $1.9 billion to build.

And if this information is not enough for you, wait: Still to be inventoried are the city's roads, bridges, curbs, gutters, streets, sidewalks, parks, subway tracks, and tunnels.

PART V

LAW, ECONOMICS, AND MANAGEMENT OF WATER

Development of the nation's water resources is vital to every aspect of our daily lives. As the U.S. Bureau of Reclamation points out, dams built across rivers control and regulate stream flow to store water for municipal, industrial, and irrigation purposes; to generate hydroelectric power; to protect people and land from floods; to dilute and freshen polluted water; to improve navigation, and to create other important benefits. Reservoirs, created by these dams, along with surrounding land areas, provide outstanding recreational opportunities for people and create habitat and refuge for wildlife.

Besides surface water, other sources being developed include aquifers, from which water is pumped for irrigation and other purposes; geothermal water, from which potable water will be obtained; the sea, from which water is being desalted and converted to fresh water; and the atmosphere, from which, by means of weather modification, we seek to increase the fall of available moisture onto the arid land.

Part IV discussed many hazards and problems created when humans utilize water resources. What can be done about limiting or alleviating them? Part V deals with the legal, economic, and management problems associated with ensuring a high-quality water resource when, where, and in the amount we need. For example, in Chapter 29, ''The Consequences of Mismanagement,'' Sandra Postel uses international examples to broaden the discussion of problems caused by misuse of water, looks at policies and technologies necessary for extending the use of available water, and examines present economic problems of water use.

Table V-1 indicates the interrelationship between water resource problems and resource management. The table was developed to describe the situation that exists in many Asian countries, but we think that it is applicable to the United States as well, because so many water rights lie with the individual states.

Chapter 30, ''Federal protection of ground water,'' gives a brief introduction to water rights in the states and the federal laws affecting ground water. Wendy Gordon points out that the federal system is a patchwork system of laws that were designed primarily to deal with some other environmental problems. Indeed, we can see in Figure V-1 that there is no common set of laws among the states for dealing with ownership of ground water. Four different basic doctrines apply. Riparian or common-law doctrine holds that the owner of the surface of the land has absolute ownership of the underlying water, with no limit on the quantity that can be removed. Reasonable-use doctrine restricts the riparian right to limit the amount of water that can be withdrawn when there is insufficient supply or when such withdrawal is directly harmful to other landowners. Appropriation doctrine is based on the order of claims for use of the

Table V-1. Linkage of Water Resources Problems, Water Resources Management,
and Organization and Administration

Water Resources Problem Expressed in Output Terms	Direct and Indirect Physical Manifestations	Implications for Water Resources Management	Implications for Organization and Administration
Erosion and sedimentation: economic losses to agricultural and fishery production and hydroelectric energy generation because of premature loss of reservoir storage capacity	Increased rates of reservoir sedimentation, caused by erosion of watershed lands, in turn caused by improper agricultural and grazing practices, timber cutting, and road construction	Implies the absence of or deficiencies in watershed management that includes (1) effective controls on both private and public use of public lands, (2) protection and restoration programs on forest, grazing and agricultural lands, and (3) technical and financial aid to private landowners	The primary water resources management agency rarely has responsibility for watershed management; in fact, no single agency typically has this responsibility. For some types of land use, e.g., agriculture, no agency may be able to exert effective control. With a number of agencies nominally responsible, there may be no effective leadership and no coordinating mechanism
Flooding: Economic losses to agriculture, households, industry, and infrastructure because of inundation and force of floodwaters	Increased flood peaks caused by loss of forest cover, diminished infiltration capacity of land, aggradation of streambeds due to increased sediment loads, and increased occupancy of floodplains	Implies the absence of or deficiencies in watershed management, as above, and in floodplain management programs, including controls over land use, floodproofing, and flood-warning systems.	The primary water resources management agency, which has responsibility for building and operating flood damage reduction facilities, rarely has responsibility for watershed management or for floodplain management measures such as land use controls and floodproofing

Source: Blair T. Bower and Maynard M. Hufschmidt, 1984, *A Conceptual Framework for Analysis of Water Resources Management in Asia,* Table III, Natural Resources Forum, 8 (4): 343–356. Copyright by the United Nations and used with permission of the publisher, Graham & Trotman Ltd.

water: "First in time, first in use." Thus in times of shortage those with later claims receive no water, while those with prior appropriation rights receive their full allotment. Correlative rights holds that all landowners have a proportional right to the underlying ground water. Note in Figure V-2 that riparian doctrine is most common in the eastern states, where there is usually an abundance of rainfall and stream runoff, whereas the appropriation law is applied in the arid and semiarid western states.

Even states with similar legal doctrines concerning ground water ownership can vary drastically when it comes to specific regulations. For instance, Table V-2 illustrates the variations among states using the High Plains aquifer as a ground water source.

What is ground water? What is surface water? Can science make distinctions that have legal bearing? For example, surface water is divided into overland flow and water within well-defined channels. Ground water often has been differentiated into percolating ground water, water in underground streams, and springs. Scientific investigation has shown that these water categories are not static, that ground water seeps through the sides of stream channels and nourishes the

Water Resources Problem Expressed in Output Terms	Direct and Indirect Physical Manifestations	Implications for Water Resources Management	Implications for Organization and Administration
Salinity: Economic losses to agriculture, forestry, and domestic and industrial water users because of limitations on uses of highly saline water	Excessive use of water for irrigation and/or excessive leakage from unlined canals. Inadequate drainage facilities for irrigation projects. Excessive pumping of ground water or reduction of surface water flows in coastal areas can cause saline water infiltration from the sea	Implies the absence of or deficiencies in management of irrigation water by farmers, or excessive ground water pumping in coastal aquifers	Irrigation agencies often have no responsibility for controlling waterlogging and salinity buildup on farm lands. No single water resources management agency may have full responsibility for controlling excessive ground water pumping
Water demand-supply imbalances: Economic development opportunities can be limited by inadequate and uncertain supplies of precipitation and runoff	Variability of precipitation and runoff from year to year causes uncertainty in supplies. Drought years may bring extreme reduction in agricultural production. The wet-dry season phenomenon limits agricultural production without external supplies	Implies difficult management problems at the river basin, irrigation district, and farm levels in attempts to adjust to variability	Responsibility for water resources development may be unassigned or may be scattered among a number of different agencies
Water pollution: Economic losses to agriculture, fisheries, industry, and households, plus threats to public health, from biological and chemical pollution of streams, lakes, and estuaries	Biological pollution caused by inadequate disposal of human and animal wastes in both rural and urban settings. Chemical pollution from agricultural fertilizers and pesticides, and chemical wastes, including toxics, from industries and urban areas	Implies the absence or inadequacy of programs of basic sanitation in rural areas and waste water collection and treatment in urban areas. Improper use of fertilizers and pesticides in agriculture	No agency has full responsibility to handle water pollution problems. Authority to control some types of pollution may be unassigned. Water resources management agencies may have no responsibility for pollution control

stream flow. In many instances stream water recharges ground water. What should have been asked earlier in the paragraph is: Can the legal profession make any scientific distinctions about water?

This issue of ownership of water is in conflict with our discussion in Part I that water is a common property. We also noted that The *Global 2000 Report* stated that water was very inexpensive. But Chapter 31, "Studies of the Economic Value of Water" by the Office of Technology Assessment, discusses both the direct and the implied costs of water for its various uses. Indeed, the same water can vary in value depending on what it is used for—and who is using it. For example, the federal government reserves the right to claim water if needed for national security and to supply the water needs of federal lands and facilities. Some judicial rulings claim that the Federal Wilderness Act of 1964 protects wilderness water resources and places limits on upstream water uses that override any state ownership rights. All Indian tribes have the same right to claim water for their reservations, and the Pueblo tribes have the additional right to communal water supplies in former Spanish territories.

Table V-2. Summary of Regulations Controlling Ground Water Development from the High Plains Aquifer, by State

		Control of Ground Water			Regulations Affecting Ground Water Development						
		Administering Agencies				Minimum		Limited			
State	Owner-ship	State	Local	Well Permits	Flow Metering	Well Spacing (ft)	Drilling Restrictions	Allocation (acre-ft/acre/year)	Planned Rate of Depletion	Conservation Measures	Remarks
Colorado	State	Ground water Commission, State Engineer	Ground water management districts	Required statewide	Can be required	2640 in ground water basins	In areas of planned rate of depletion	2.5 in N. High Plains, 3.5 in S. High Plains	40 percent in 25 years in N. High Plains, none in S. High Plains	Can be required	Most regulations only apply to designated ground water basins
Kansas	State	State Board of Agriculture, Division of Water Resources	Ground water management districts	Required	Can be required	1320 to 2640 depending on area and well yield	Restricted in some areas	1.5 to 2.0 depending on location	Varies by management district	Required	Rules vary by management district
Nebraska	State	Department of Water Resources	Ground water conservation and natural resources districts	Registration required	In some areas	600 to 1000 in control areas	Permit required in control areas	Varies by control area	None	Statewide irrigation runoff control	Rules vary by conservation and natural resources district
New Mexico	State	State Engineer	None	Required	In some areas	None	Restricted in declared ground water basins	3.0	Not sooner than 40 years starting in 1952 or 1956	Waste not allowed	Regulations only apply to declared ground water basins

State	Ownership	Agency								Comments
Oklahoma	Land-owner	Water Resources Board	None, but can be organized	Required	None, but can be required	None, but can be required	None	Temporary permits for 2, permanent permits based on supply	Not sooner than 1993	Waste not allowed	Regulations began in 1973; only well permits and allocations controlled in 1982
South Dakota	State	Department of Water and Natural Resources	None	Required	None	None	None	2 maximum	Allocation limited to rate of recharge; depletion not allowed	Waste not allowed	Regulations do not apply to Indian reservations
Texas	Land-owner	Department of Water Resources	Ground water management districts	Required	None	600 to 1200 depending on well size and yield	Restricted in certain areas	None, but can be limited	None	Waste not allowed	Rules only apply in management districts
Wyoming	State	Board of Control, State Engineer	Advisory board	Required statewide	None	None	Permit required	Restricted in control areas	None	Waste not allowed	Restrictions apply to control areas

Source: E. D. Gutentag, F. J. Heimes, N. C. Krothe, R. R. Luckey, and J. B. Weeks, 1984, *Geohydrology of the High Plains aquifer in parts of Colorado, Kansas, Nebraska, New Mexico, Oklahoma, South Dakota, Texas, and Wyoming,* U.S. Geological Survey Professional Paper 1400–B.

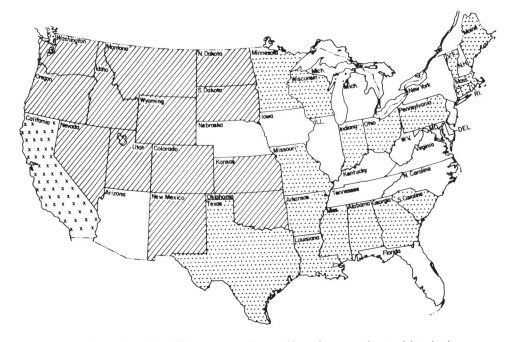

Figure V-1. Ground water laws. Four different systems of ownership apply to ground water rights: riparian or common law (shaded), reasonable use (unmarked), appropriation (lined), and correlative rights (checked). (*Source:* 1983, *Ground water: Issues and answers.*)

Some of these problems are discussed by Kenneth D. Frederick in Chapter 32, "Water Policies and Institutions." He also stresses that laws were developed for regions with plenty of water and applied to regions with little water; population growth and the associated increase in water demand are placing greater stress on what was once considered a free good. Frederick calls for modification of ownership and the establishment of state and/or federal "water banks" where water transfers can be facilitated.

In Chapter 33, "Instream Water Use: Public and Private Alternatives," James L. Huffman focuses his discussion on water in streams and the problems of use, flow maintenance, regulation, and acquisition. He feels that treating water as a commodity will sort out competing uses and help solve the problem of often-conflicting laws and regulations. Disagreeing with Frederick, Huffman prefers as little governmental regulation as possible and sees the privatization of water allocation as the only sensible long-term solution to water supply.

The very real political and social problems involved with the allocation of water and the transfer of those allocations are discussed by F. Lee Brown, Brian McDonald, John Tysseling, and Charles DuMars in Chapter 34, "Water Reallocation: Conflicting Social Values." The excerpts we have included clearly show that there are strong reasons to favor transfer of water rights, that such transfers place stress on existing institutions, that market situations of a sort already exist, and that social conflict is to be expected.

The augmentation of existing water supplies is another major water management problem. This topic is briefly discussed by Frederick at the conclusion of Chapter 32 and in more detail by Postel in Chapter 29. In an extensive study of the problem the Office of Technology Assessment (Chapter 31) lists technologies for increasing water availability based on combinations

Table V-3. Technologies for Increasing Water Availability

Effect	Technology	Comments
Increase surface water runoff	1. Weather modification	Increase snowpack of mountain watersheds (see Chapter 4)
	2. Watershed management	Vegetation removal or change (see Chapter 35)
	3. Stream flow forecasts	Allows better timing of use; models still a problem
Store and augment surface water	4. Storage increase	United States already has over 2000 dams
	5. Desalination	High costs and brine production give limits
	6. Interbasin transfer	Major political, economic, and social problems (see Chapter 34)
Conservation of surface water	7. Flexible delivery	Delivery based on crop management rather than water management
	8. Seepage control	Limits loss through sides and bottoms of reservoirs and canals
	9. Evaporation control	Limit surface, reflect sunlight, cover reservoir chemically or mechanically
	10. Vegetation management	Remove competing vegetation and aquatic plants (expensive mechanically and pollutes if done chemically)
Supplement soil water	11. Surface management	Shape soil surface, control cover, change properties of soil to conserve precipitation
	12. Irrigation	Many techniques; soil salinization is a major threat
Improve efficiency of water user	13. Use of adapted plants and animals	Limits water stress; generally not major agricultural products
	14. Biotechnology	Introduce genetic variations in major products to limit water stress
	15. Breeding innovation	Identification of desired characteristics to be modified remains a problem with both 14 and 15
Conservation of ground water	16. Ground water recharge	Depends on surplus of surface water sometime during year
	17. Pollution control	Only feasible way to ensure quality of ground water; limited technology to clean up once polluted
Management techniques	18. Information management	Computer revolution here, too
	19. Land management	Alternative agriculture; multiple land use; crop and animal mixtures must be matched to local conditions
	20. Water management	UNDERSTAND THE OPERATION AND LIMITATIONS OF THE NATURAL HYDROLOGIC CYCLE

Source: Adapted from Office of Technology Assessment, 1983, *Water-related technologies for sustainable agriculture in U.S. arid/semiarid lands.*

of hydrology, plant and animal science, engineering, and land use management. These technologies are indicated in Table V-3, a summary table.

Chapter 35, "Water Yield Augmentation" by Stanley L. Ponce and James R. Meiman, demonstrates that water yield can be increased through forest and range management practices. The degree to which this increase will be achieved—if it can be—depends on the local demand for water, the costs of alternative methods of increasing the supply, and the ownership of the watershed. Again, legal and economic issues affect what technologies can be effectively applied. B. R. Beattie, in "Irrigated Agriculture and the Great Plains: Problems and Policy Alternatives" (see complete reference in Appendix C), indicates that, at least for the Ogallala aquifer, the main challenge is to effectively use what water is there, rather than waste resources in attempts to transfer water from other regions—an action that he thinks is sure to fail.

The last chapter (36), "The Future of Water" by Peter P. Rogers, indicates that there is indeed a future for water—and even for our use of it. We have, however, the legacy of institutions, laws, and management techniques initiated to freely use that inexpensive, common, ubiquitous commodity. We now discover that these techniques have been so successful that the use is pushing the amount available, that pollution problems associated with the use of any natural resource are overwhelming some areas, that social values conflict with existing economics, and that humans have to face the fact that water is a fragile resource after all.

29

THE CONSEQUENCES OF MISMANAGEMENT

SANDRA POSTEL

When a resource begins to show physical signs of abuse, economic and ecological consequences are usually not far behind. Water's seeming ubiquity has blinded society to the need to manage it sustainably and to adapt to the limits of a fixed supply. Mounting pressures are currently manifest in pervasive pollution, depletion of ground water supplies, falling water tables, and damage to ecological systems. Failure to heed these signs of stress, and to place water use on a sustainable footing, threatens the viability of both the resource base itself and the economic systems that depend on it.

Each liter of polluted water discharged untreated contaminates many additional liters of fresh water in the receiving stream. The disposal of synthetic chemicals and heavy metals, which pose dangers in extremely low concentrations, is an especially grave threat to the quality of water supplies. Without adequate treatment the growing volume and toxicity of wastes could render as much as a fourth of the world's reliable supply unsafe for use by the year 2000.[1]

Many industrial countries now require that waste waters meet specified standards of quality before they are discharged. Yet in most Third World countries pollution controls are either nonexistent or unable to keep pace with urbanization and industrialization. In China, for example, only about 2 percent of the 28 billion m³ of waste water discharged each year is treated. Already, a third of the water in its major rivers is polluted beyond safe health levels, and fish and shrimp have disappeared from 5 percent. China's first large waste water treatment plant began operating in Beijing in the fall of 1980, but the volume of sewage far outpaces the facility's capacity to treat it. Waste water flows

in Beijing have increased twenty-seven-fold over the last three decades, and volumes for the country as a whole are projected to triple or quadruple by the end of the century. Vaclav Smil, a specialist on China's environment, writes that the country's water pollution problem "will require very heavy and sustained investment—not to achieve zero discharges but merely to bring the appalling situation within reasonable limits after decades of no control.[2]

In virtually all of Latin America municipal sewage and industrial effluents are discharged into the nearest rivers and streams without treatment. The pulp and paper and the iron and steel industries—two of the region's biggest polluters—have been growing twice as fast as the economy as a whole. Yet cleanup efforts have typically been postponed because of their high cost. Purifying Colombia's Bogota River, for example—one of the continent's most contaminated waterways—would cost an estimated $1.4 billion, a high price for a debt-ridden country to pay. Unless governments begin attacking urban and industrial pollution soon, however, they will inevitably face the prospect of a water supply too polluted for their people to drink.[3]

A similar situation exists in the Soviet Union. Industrial waste waters comprise 10 percent of the Volga River's average flow at Volgograd, and three-fourths of the wastes are untreated. A major effort was begun in the midseventies to cleanse the river, but apparently enforcement has been too slack to encourage industries to install the costly technologies. Under these conditions the Volga simply cannot sustain the existing high level of withdrawals and also remain of acceptably quality. According to Thane Gustafson, a U.S. specialist on Soviet affairs: "Footdragging by industry on pollution control will make it necessary to use more water for dilution. All these effects add up to a greater demand for water by the end of the century than the available supplies can satisfy.[4]

Vast quantities of the earth's water move slowly underground through the pores and fractures of geologic formations called aquifers. Some hold water thousands of years old and receive little annual replenishment from rainfall. Like oil reserves, water in these "fossil aquifers" is essentially nonrenewable; if tapped, it will in time be depleted. Even where recharge does occur, ground water is often pumped at rates that exceed the replenishment, causing water tables to fall and depleting future water reserves. Such overpumping—which geologists call *water mining*—supports only a fragile and short-term prosperity at best, for eventually the water becomes too salty to use, too expensive to pump to the surface, or runs out altogether.

One-fifth of the irrigated cropland in the United States is supported by water mined from a vast underground reserve called the Ogallala aquifer. Stretching from southern South Dakota to northwest Texas, the aquifer underlies portions of eight states and spans an area roughly three times as big as the state of New York. Natural recharge is minimal in this semiarid region, and farmers have profitably irrigated corn, sorghum, and cotton only by drawing on water stored for thousands of years. Irrigation with Ogallala water began to expand rapidly in Texas in the forties, and when powerful pumping and irrigation systems were introduced, it spread northward into Oklahoma, Kansas, and Nebraska during subsequent decades. By 1978 over 8 million hectares (ha) were under irrigation in the states most heavily dependent on the Ogallala, compared with just 2.1 million in 1944. (See Figure 29-1.) Over the last four decades 500 cubic kilometers (km^3) of ground water have been withdrawn. Hydrologists estimate that the aquifer is now half depleted under 900,000 ha of Kansas, New Mexico, and Texas.[5]

Faced with rising pumping costs, diminishing well yields, and low commodity prices, farmers are taking land out of irrigation. After several decades of steady growth, the total irrigated area in the High Plains is now declining. In just four years, 1978 to 1982, irrigated land in Texas dropped by 20 percent, in Oklahoma by 18 percent, and in New Mexico by 9 percent. Collectively, in these and the other three states that draw most heavily on the Ogallala (Colorado, Kansas, and Nebraska), the total area under irrigation declined by 592,000 ha, or 7 percent. In Nebraska, where a smaller portion of the Ogallala has been depleted irrigation

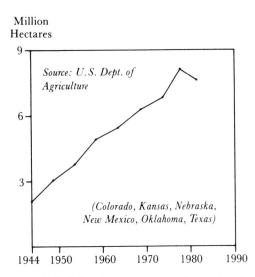

Million
Hectares

Source: U.S. Dept. of Agriculture

(Colorado, Kansas, Nebraska, New Mexico, Oklahoma, Texas)

Figure 29-1. Irrigated area in six states that rely heavily on the Ogallala aquifer, 1944–1982.

is still expanding. Yet in 1982 net returns from Northern Plains production of corn—the dominant irrigated crop in Nebraska—were less than half the national average, and it appears that eventually farmers there will begin switching crops, converting to dryland farming, or leaving agriculture altogether.[6]

Economists and government leaders are concerned about the potential collapse of a lucrative regional farming economy. The U.S. Army Corps of Engineers has even looked at the feasibility of massive river diversions to supply water to farmers now dependent on the diminishing Ogallala. But few have asked the more fundamental question of whether it makes sense to deplete this resource at a time when the nation can afford to preserve it. The U.S. government is paying farmers to idle rain-fed cropland in order to lessen a price-depressing surplus of crops; at the same time it is allowing the wholesale exhaustion of a unique water reserve to grow those same crops. Moreover, among the consequences predicted for much of the central and western United States from the rising level of atmospheric carbon dioxide is a reduction in the renewable water supply and an increase in the frequency and severity of droughts.[7] By exploiting the Ogallala today, farmers are foreclosing options to draw on it in the future when it may really be needed to meet vital food needs domestically and abroad. Failure to preserve this resource is shortsighted, and an error future generations will rightfully find hard to forgive.

Many other U.S. aquifers are suffering from

overuse. Among the severest cases is one under-lying Tucson, Arizona—the largest American city completely dependent on ground water. Only about 35 percent of the water withdrawn to supply Tucson's residents, farms, and copper mines is replaced each year by recharge, and water tables in some areas have fallen over 50 m. The Santa Cruz River is no longer sufficiently fed by underground water to keep it flowing during dry spells. Water levels have also dropped precipitously around El Paso in Texas and Ciudad Juarez in Mexico from the mining of the aquifer they share. In portions of the Dallas-Fort Worth metropolitan area, water tables have fallen more than 120 m over the last 25 years.[8]

Though rarely as well-documented as cases in the United States, excessive ground water pumping and subsequent lowering of the water table appears to be increasingly common worldwide. (See Table 29-1.) Over the seventies, water levels dropped 25 to 30 m in areas of Tamil Nadu in southern India, a consequence of uncontrolled pumping for irrigation. Overpumping is epidemic in China's northern provinces, where some ten major cities rely heavily on ground water for their basic supply. In Beijing annual ground water withdrawals exceed the sustainable supply by 25 percent, and water tables in some parts of the city have been dropping over 1 m each year. In one district of Tianjin, a major manufacturing and commercial city, water tables are falling an astonishing 4.4 m annually.[9]

Large withdrawals of ground water may have other costly effects besides the depletion of future supplies. If water pumped from an aquifer susceptible to compaction is not replaced by recharge, the aquifer may compress, resulting in subsidence of the overlying land. Subsidence in Mexico City has damaged buildings and streets and disrupted the sewage system. In China portions of Beijing have been sinking 20 to 30 centimeters (cm) an-

Table 29-1. Selected Cases of Excessive Water Withdrawals

Region	Status
Colorado River Basin, United States	Yearly consumption exceeds renewable supply by 5 percent, creating a water deficit; Colorado River is increasingly salty; water tables have fallen precipitously in areas of Phoenix and Tucson
High Plains, United States	The Ogallala, a fossil aquifer that supplies most of the region's irrigation water, is diminishing; over a large area of the southern plains the aquifer is already half-depleted
Northern China	Ground water overdrafts are epidemic in northern provinces; annual pumping in Beijing exceeds the sustainable supply by 25 percent; water tables in some areas are dropping up to 1 to 4 m per year
Tamil Nadu, India	Heavy pumping for irrigation has caused drops in water level of 25 to 30 m in a decade
Israel, Arabian Gulf, and coastal United States	Intrusion of seawater from heavy pumping of coastal aquifers threatens to contaminate drinking water supplies with salt
Mexico City; Beijing, China; Central Valley, California; Houston-Galveston, Texas	Ground water pumping has caused compaction of aquifers and subsidence of land surface, damaging buildings, streets, pipes, and wells; hundreds of homes in a waterfront Texas community have been flooded
California, United States	Water from Owens Valley and Mono Basin have been diverted to supply southern water users; Owens Lake has dried up, and Mono Lake's surface area has shrunk by a third
Southwestern Soviet Union	Large river withdrawals have reduced inflow to the Caspian and Aral seas; the Caspian sturgeon fishery is threatened; the Aral's fisheries are virtually gone, and the sea's volume may be halved by the turn of the century

Source: Worldwatch Institute, based on various sources.

nually since 1950, and rates of 10 cm/year have been measured in Tianjin. In the Houston-Galveston area of Texas, where water levels have declined 60 m during the last half century, portions of the land surface have sunk over 2 m. High tides in the gulf have flooded residential developments that, because of subsidence, are now closer to sea level.[10]

In coastal areas heavy pumping may alter the volume and flow of ground water discharging to the ocean and thereby allow sea water to invade the aquifer. Saltwater intrusion threatens to contaminate the drinking water supplies of many cities and towns along the U.S. Atlantic and Gulf coasts; it is especially severe in several Florida cities where pumping has pulled the water table below sea level. Israel, Syria, and the Arabian Gulf states are also battling threats of saltwater intrusion. Once it occurs, such contamination is difficult, if not impossible, to reverse.[11]

Excessive demands also take a toll on lakes, estuaries, and inland seas that are sustained by fresh water inflow from nearby rivers and streams. The Aral Sea in the southern Soviet Union is shrinking because of large withdrawals from its two major tributaries, the Amu Darya and Syr Darya. These two rivers help support Soviet central Asia's lucrative agricultural economy, which includes roughly half the nation's irrigated cropland. The population of several central Asian republics has grown by 30 percent over the last decade, adding to pressures on the available water supply and to the importance of maintaining a thriving economy to secure more jobs in the region. The Aral's level had remained fairly stable between 1900 and 1960 but has since dropped 9 m. Fisheries that once figured prominently in the regional economy have virtually disappeared. Although officials are taking some measures to save portions of the Aral, they appear resigned to it shrinking further. Some scientists have projected that before the end of the century the sea may drop another 8 to 10 m and its volume may be reduced by half.[12]

A similar scenario threatens to unfold further west, in the Caspian Sea. The Volga River is the Caspian's main source of inflow, helping to replenish the large quantities of water evaporated from the sea each year. Construction of huge dams on the river during the fifties and subsequent large irrigation withdrawals dramatically reduced the river's discharge into the Caspian. The sea reached its lowest level in centuries in 1977, having dropped more than 3 m over the preceding half century. The level has risen somewhat in recent years because of unusually heavy rains that increased the Volga's flow. But Soviet scientists do not expect this fortuitous occurrence to continue. According to U.S. geographer Philip Micklin, who discussed the situation with scientists during a five-month stay in the Soviet Union in 1984, additional diversions for irrigation are planned for the Volga, and the Caspian's level is expected to drop further over the next decade. The sea supports bountiful fisheries, including 90 percent of the world's catch of sturgeon. Salmon and migratory herring spawn in the Volga and feed in the North Caspian. Substantial damage to these fisheries is likely to occur if the sea's level declines much further.[13]

Shrinking inland seas are a dramatic consequence of heavy water withdrawals to meet irrigation and other water demands. But an equally grave threat is the quiet loss of fish and other aquatic life from rivers and streams whose altered flow patterns can no longer sustain them. As long as water withdrawals remain well below a region's average sustainable supply, stream flows will be sufficient to safeguard most ecological values. Yet where a large share of surface water is diverted from its natural channels, these benefits may be lost.

Over the last decade many nations have begun to realize this danger, but they are not prepared to avert it. Setting minimum flow levels to protect wildlife requires large quantities of data and the expertise of hydrologists, fishery biologists, and aquatic ecologists. The quick and inexpensive methodologies are simply not accurate enough to be reliable. A common one, for example, sets minimum flow requirements as a fixed percentage (such as 10 percent) of the average annual flow. But this makes no allowance for the large flow variability that typifies many river basins nor for the long-term, cumulative effects on fish of low flows for extended periods of time. More sophisticated methods usually involve a computer model that quantifies, for each particular species, the amount of habitat available in a given stretch of the stream at each stage of its life cycle and under varying stream flow conditions. Though more accurate, such methods are time-consuming and costly, requiring much field data and scientific expertise to interpret them.[14] A paper issued in 1984 by the Canadian Inquiry on Federal Water Policy acknowledges that "in Canada, we are only beginning to appreciate the magnitude of water needs for the support of the ecosystem. We do not have very reliable estimates of instream requirements."[15]

Among the least affordable consequences of irrational water use is the degradation of valuable cropland from poor irrigation practices. Irrigation water is typically brought to crops through unlined canals and ditches that allow vast quantities of water to seep down to the water table. Where drainage is inadequate, the water level gradually rises, eventually entering the crops' root zone and waterlogging the soil. In the Indian state of Madhya Pradesh, for example, a large irrigation project that originally was expected to increase crop production tenfold led to extensive waterlogging and, consequently, a reduction in corn and wheat yields. Farmers there now refer to their once fertile fields as "wet deserts." [16]

In dry climates waterlogging may be accompanied by salinization as water near the surface evaporates and leaves behind a damaging residue of salt. According to some estimates, waterlogging and salinization are sterilizing some 1 million to 1.5 million ha of fertile soil annually. The problem is especially severe in India and Pakistan (where an estimated 12 million ha have been degraded), the Valley of Mexico, the Helmud Valley in Afghanistan, the Tigris and Euphrates basins in Syria and Iraq, the San Joaquin Valley in California, the North Plain of China, and Soviet central Asia. [17] In these areas waterlogging and/or salinization threaten to diminish the very gains in food production that costly new irrigation projects are intended to yield.

AUGMENTING DEPENDABLE SUPPLIES

When natural water supplies become inadequate to meet a region's demands, water planners and engineers historically have responded by building dams to capture and store runoff that would otherwise flow through the water cycle "unused" and by diverting rivers to redistribute water from areas of lesser to greater need. As the demand for water has increased, so have the number and scale of these engineering endeavors to augment available supplies. Tens of thousands of dams now span the world's rivers. Collectively, their reservoirs store roughly 2000 km^3 of runoff, increasing by 17 percent the 12,000 km^3 of naturally stable runoff derived from ground water and lakes. Most of this capacity has been added since midcentury, when the pace of large dam construction abruptly quickened. All but 7 of the 100 largest dams in the world were completed after World War II. [18]

Many industrial countries are now finding, however, that the list of possible dam sites is growing shorter and that the cost of adding new storage facilities is rising rapidly. In the United States, for example, reservoir capacity grew on average 80 percent per decade between the twenties and the sixties. As the narrow valley sites were gradually exploited, any new capacity required broader, earth-filled dams. By the sixties, 36 times more dam material was needed to create a given reservoir capacity than in the twenties. With a corresponding escalation in construction costs, reservoir development markedly declined. [19]

In most of Europe a favorable climate and geography for securing water supplies has lessened the need to build large storage reservoirs, compared with, for example, the western United States. Yet to meet rising demands, many European nations plan large increases in reservoir capacity over the next decade. (See Table 29-2.) A 1981 report prepared by the U.N. Economic Commission for Europe (ECE) raises doubts, however, about the ambitious plans of several countries materializing. Both high costs and growing opposition to the flooding of farmlands and valleys are becoming major barriers to dam construction. Notwithstanding government forecasts that "optimistically predict" a doubling or tripling in reservoir capacity, the ECE assessment concludes that some countries have already reached the practical limits of their reservoir development. [20]

Lagging the industrial world's big dam era by two decades, dam construction in the developing world is now in its heyday. Two-thirds of the dams over 150 m high slated for completion this decade are in the Third World. [21] Designed mainly for generating hydroelectric power and supplying water for irrigation, large dams and reservoirs offer promises of greater energy independence and food self-sufficiency. Their lure is understandable as

Table 29-2. Reservoir Capacity in Selected Countries, 1970, with Projections for 1990

Country	Total Capacity (km^3)	Projected Increase in Capacity, 1970–1990 (%)
Belgium	0.1	79
Bulgaria	2.7	296
Canada	518.0	—
Czechoslovakia	3.3	76
East Germany	0.9	156
France	2.0	—
Greece	8.7	78
Poland	26.0	127
Portugal	5.3	119
Romania	2.6	746
Sweden	27.1	0
Soviet Union	830.0	60
United Kingdom	1.5	47
United States	670.0	15
West Germany	2.3	—

Source: United Nations Economic Commission for Europe, 1981, Long-term perspectives for water use and supply in the ECE region (New York: United Nations).

large-scale solutions to a set of large development dilemmas. Unfortunately, high costs, poor planning, and environmental disruption are leaving a legacy of failed expectations that suggest they are not the panacea once envisioned.

Sri Lanka's Mahaweli Development Programme encompasses construction of four large dams across the Mahaweli River to help achieve goals of tripling the nations' electric generating capacity and irrigating an additional 130,000 ha of cropland. Yet with only two dams completed, the project has already been plagued with problems. Capital costs nearly doubled in just four years, severely straining the government's finances. Inspections by agencies donating to the project—including the Agency for International Development and the World Bank—uncovered serious design and construction problems that in 1982 led to the conclusion that without major corrective efforts the irrigation canals would not function as planned. Studies had warned that unless deforested hillsides were replanted, runoff would wash large amounts of soil downstream, threatening a buildup of silt in reservoirs and irrigation canals and a lowering of soil fertility. Yet reforestation did not begin until more than a decade after initiation of the project, and by the end of 1982 replanting had taken place on less than 1 percent of the area targeted for it. Writer John Madeley notes, "The homes of 45,000 people are being flooded by the Victoria Dam, and, when they move into the new resettlement zone, their hopes of making a new living will not have been helped by the lack of attention to replanting."[22]

The experience Sri Lanka has had with the Mahaweli project is by no means unique. Though undertaken with good intentions of raising food production and living standards, large dam schemes are often so costly and complex that other critical tasks—often essential to the project's success—are neglected. As described earlier, vast areas of valuable cropland are becoming waterlogged and salt-laden because of excessive seepage from reservoirs and canals and poor drainage from fields. Deforestation and overgrazing are disrupting water's flow through the landscape. Natural forests and grasslands absorb runoff and allow it to move slowly through the subsurface. As hillsides are denuded, rainfall and soil run rapidly off in floods, filling expensive reservoirs with silt and causing dry weather stream flows to disappear.

Especially in the Third World, managing watersheds to stabilize runoff is critical to reversing a vicious cycle of flooding, soil loss, declining crop production, and perennial drought. In Malaysia conversion of natural forest to rubber and palm oil plantations has doubled peak runoff and cut dry season flows in half. Deforestation on the small island of Dominica has contributed to a 50 percent reduction in dry weather flows there.[23] Though virtually impossible to quantify, it may well be that deforestation—now estimated at 11.3 million ha/year—is diminishing the Third World's stable runoff by as much as expensive new dams and reservoirs are augmenting it. Unless the threats posed by deforestation, waterlogging, and soil salinization are countered, large dam schemes may end up wasting capital and degrading land while bringing few lasting benefits to those they are intended to serve.[24]

As with dams and reservoirs, projects to divert water from one river basin to another have grown in number and scale in response to rising demand. Proposals to import water from some distant source have been made for virtually every major region facing a shortage. Most were developed during an era of cheap energy, relatively cheap capital, and when environmental values rarely entered the debate over project costs and benefits. The collective history of these large diversion schemes is marked by long study times, periodic abandonment, multibillion-dollar cost estimates, and growing concern over their ecological effects. (See Table 29-3). Some of these projects will probably never leave the drawing boards. Those that do, and that are actually completed, may be more a product of political expediency than an objective analysis of alternative ways to achieve a given end.

In China, officials and scientists began in the early fifties to study the possibility of diverting water from the Chang Jiang (Yangtze) River Basin in central China to the water-poor regions of the north. After years of lying dormant, the project was given a boost in February 1983 when the government approved the first stage of work on what is known as the East Route. This mainly involves reconstructing the old Grand Canal, which will offer navigation benefits regardless of whether other phases of the project are completed. The long-term plans call for pumping water 660 km north to the Huang He, the Yellow River, from which it would flow an additional 490 km by gravity into the vicinity of Tianjin. Chinese water planners estimate that the diversion will require several dozen pumping stations with a total installed capacity of about 1000 megawatts (MW)—equal to one very large nuclear or coal plant. The system would transfer about 15 km^3 of water in an average year and up to double that volume in a dry year. Most of the water would be used to expand or improve irrigation on 4.3 million ha; the remainder would enhance Tianjin's municipal and industrial water supply.[25]

With an estimated price tag of $5.2 billion, which analysts say could easily double, Chinese

Table 29-3. Current Status of Selected Major River Diversion Projects

Project	Distance (km)	Planned Annual Volume (km³)	Estimated Capital Cost (Billion Dollars)	Current Status
Chan Jiang River–North China Plain, China	1150	15.0	5.2[a]	Decision in 1983 to begin construction
Northern European Rivers–Caspian Sea Basin, Soviet Union	3500	20.0	3.1	Construction to begin 1986
Siberian Rivers–Central Asia, Soviet Union	2500	25.0	41.0	Preparing engineering designs; decision pending
Central Arizona Project, United States	536	1.5	3.5	Deliveries to Phoenix to begin December 1985; to Tucson, 1991.
California State Water Project, United States	715	5.2	3.8[b]	Operating at 60 percent of planned capacity
Midwest Rivers–High Plains, United States[c]	600–1600	2.0–7.4	5.5–35.0	No action

Source: Worldwatch Institute, based on various sources.

[a] A published estimate considered low by project analysts; cost could easily double. [b] Includes only costs incurred and projected through 1995; state has yet to develop new proposals (and cost estimates) to significantly increase the project's capacity over existing levels. [c] Five different diversions were studied. Lower figure of each range is for diversion of Missouri River into western Kansas, the least costly alternative; higher figure is for diversion of several south central rivers into Oklahoma and Texas panhandles, the most costly alternative.

officials are understandably proceeding cautiously. Bruce Stone, one of a team of experts studying the Chinese diversion proposals, makes a convincing case that the water transfer may be an unnecessarily costly and risky way to raise grain production from the North China Plain. He notes that most of the irrigated cropland near Tianjin now yields only 1.8 tons/ha, while a smaller portion yields 2.3 tons. The production increase gained by expanding irrigation to 1 average-yielding hectare could therefore be obtained equally by upgrading 3 or 4 ha already under irrigation to produce the higher yields. Moreover, without better management and drainage of irrigated lands, the diverted water may worsen the salinization of North Plains' farmland. Salinization is already reducing yields on 2.7 million ha, and another 4.7 million are threatened.[26]

Officials in the Soviet Union have in recent years revived century-old ideas of diverting north-flowing rivers to the more populous southern European and central Asian regions. One project aims to transfer water from northern European lakes and rivers to the Volga drainage basin, the primary purpose being to stabilize the level of the Caspian Sea.

Even more ambitious is the proposed diversion of Siberian rivers south to the central Asian republics, where water deficits of 100 km³ are projected by the turn of the century. The region's burgeoning population and intensifying political clout have increased pressure to find some solution to its pending water shortage and unemployment problems. Thane Gustafson observed in 1980 that apparently "the latitude enjoyed by technical specialists to criticize or oppose the diversion projects has become hostage to the projects' political priority." The greatest single obstacle to proceeding with the diversions, he noted, was "the tightness of investment capital, which makes a full-scale commitment by the leadership unlikely in the near term."[27]

In January 1984, nevertheless, the USSR Council of Ministers called for a detailed engineering design for the entire 2500 km route from the Ob' River to the Amu Darya. Construction could begin by 1988 if the designs are accepted, and water that now drains into the Arctic may be heading to the cotton lands and industries of central Asia by the end of the century. Cost estimates for the initial transfer capacity of 25 km³ are $18 billion for the main diversion canal and $23 billion for the facilities to distribute the water once it reaches its destination. Meanwhile, some Soviet scientists still maintain there is considerable potential to increase the efficiency of water use in the destination region. According to one estimate, conservation in agriculture and industry could save up to half the initial volume of the proposed transfer. Moreover, as with China's project, the diverted water could spread the already severe salinization of irrigated land.[28]

In the United States no new federal water projects have been authorized since 1976, though since the turn of the century authorization bills have been introduced into the U.S. Congress about every two years. More importantly, actual funding for water project construction (excluding wastewater treatment) has declined steadily over the past eight years; appropriations in 1984 where about 70 percent less in real terms than in 1976.[29] Tight capital and $200-billion federal deficits are forcing to an end a long era of massive water subsidies. Historically, few of these projects have returned sufficient benefits to justify their high costs. Long before the first drops of Central Arizona Project water were destined for Phoenix and Tucson, for example, economist Thomas Power of the University of Montana stated that not only was the project's benefit-cost ratio less than one, "it may well only return a few cents of each dollar invested in it."[30]

Public opposition is adding another large hurdle to water project construction in the United States—in some cases, perhaps an insurmountable one. The California State Water Project (SWP) is a case in point. One of the most complex water schemes ever designed, SWP is now operating at 60 percent of its planned annual capacity. Capital costs to date total about $3.4 billion, and the need to lift much of the water 590 m over the Tehachapi Mountains guarantees high energy bills: Pumping costs in 1983 totaled over $100 million.[31]

Two successive state administrations in California have failed to win sufficient support for additional SWP facilities that would allow more northern water to be transferred to Los Angeles and the agricultural valleys in the south. The voters rejected one proposal, called the Peripheral Canal, in a 1982 referendum. This defeat reflected concern about the canal's ecological effects around the Sacramento–San Joaquin Delta and, more fundamentally, about the merits of costly water exports versus stronger conservation efforts by southern water users. Another proposal, known as the "through-delta" plan, died in the California assembly in August 1984 when it appeared to proponents that another public referendum could not be avoided. Approval of any plan within the next

few years that would substantially increase the volume of water shipped south appears increasingly doubtful.[32]

As the prospects for dams and diversions to augment dependable water supplies become less promising, the potential to store surplus runoff underground is receiving more attention. Artificially recharging underground aquifers—either by spreading water over land that allows it to percolate downward or by injecting it through a well—is one way to both stabilize water tables and increase the amount of runoff stored for later use. Underground storage also avoids damming a free-flowing river, minimizes competition for valuable land, and prevents large losses of water through evaporation, which are among the principal objections to surface reservoirs.

More than twenty countries now have active projects to artificially recharge ground water. Yet in just a few cases has the practice been adopted on a large scale. Israel transports 300 million m³ of water from north to south every year through its National Water Carrier System and stores two-thirds of it underground. The water is used to meet high summer demands and offers a reliable source of supply during dry years. In the United States local water agencies in California, which have been recharging ground water since the twenties, now place nearly 2.5 billion m³ in underground basins each year. The state's Department of Water Resources also began to seriously investigate ground water storage as the options for damming more surface streams became increasingly limited. By 1980 the department had 34.5 million m³ stored in two separate State Water Project demonstration areas. Preliminary estimates for seven ground water basins indicate a potential for augmenting the SWP's annual yield by about 500 million m³, at unit costs at least 35 to 40 percent lower than the median cost of water from new surface reservoirs. Also, the U.S. Congress enacted legislation in the fall of 1984 authorizing demonstration projects in 17 western states to recharge aquifers, including the diminishing Ogallala.[33]

Underground storage may hold special potential for Third World countries subjected to the destructive flooding and perennial dry spells of a monsoon climate. Capturing excessive runoff and storing it underground can convert damaging flood waters into a stable source of supply while avoiding the large evaporation losses that occur with surface reservoirs. In India subsurface storage has sparked interest as a way of providing a reliable source of irrigation water for the productive soils of the Gangetic Plain. According to some estimates, a fully irrigated plain could grow crops sufficient for three-fourths of India's population. On the North

Plain of China, also prone to chronic drought, water from nearby surface streams is diverted into an underground storage area with a capacity of 480 million m^3. When fully recharged, the aquifer will supply irrigation water for 30,000 ha of farmland. Several countries in Hebei Province are also artificially recharging aquifers to combat sinking water tables.[34]

Many aquifers are also recharged unintentionally by seepage from irrigation canals. In such cases, managing ground water in conjunction with the surface irrigation water can help prevent waterlogging and salinization and may allow for an expansion of irrigated area without developing additional surface water sources. Such a strategy has been tried in the Indus Valley of Pakistan where a 60,000-km network of canals sits atop a vast ground water reservoir. By the midsixties leakage from the canals had tripled the volume of recharge to the aquifer, and the resulting rise in the water table caused extensive waterlogging. Following a World Bank–sponsored study of the area, the Pakistan government began to subsidize the installation of tube wells to tap the vast amount of water that had collected underground over the decades. About 11,000 public wells have been installed under the government program, and individual farmers have constructed over 100,000 private wells, which, though built to supply them with irrigation water, also help control waterlogging. Unfortunately, much of the water pumped is too saline for use unless mixed with purer surface water, and poor operation and maintenance have apparently made the public wells a burden to the government. Yet the strategy of jointly managing ground water and surface water may offer substantial benefits where the physical setting is right and the needed technical and institutional coordination can be developed.[35]

Artificial recharge on a small scale has helped augment local water supplies for decades. The North Dakota town of Minot, for example, opted for this approach when faced with chronic water shortages and rapidly declining ground water levels. Its complete recharge system cost only 1 percent as much as building a pipeline to the Missouri River, another of the town's supply alternatives. After six months of operation, water levels in portions of the aquifer had risen more than 6 m.[36] Despite a host of similar local-level success stories, however, the practice is far from realizing its potential. According to Jay H. Lehr, executive director of the National Water Well Association in the United States, the efficiency of storing surplus runoff underground "has been proven the world over. The costs, while by no means negligible, are reasonable in the face of other sound alternatives and a steal when compared to the grandiose water

schemes of the mega minds of the Army Corps of Engineers and the Bureau of Reclamation."[37] Soviet scientist M. I. L'vovich has predicted that "the 21st century will undoubtedly be the century of underground reservoirs."[38]

Of the less conventional ways to augment a region's fresh water supplies—such as seeding clouds to induce precipitation, towing icebergs, and desalting sea water—desalination appears to hold the greatest near-term potential. Indeed, with the oceans holding 97 percent of all the water on earth, desalted sea water seems to offer the ultimate solution to a limited renewable fresh water supply. Several technologies have proved effective, but their large energy requirements make them too expensive for widespread use. Desalting sea water is typically 10 times more costly than supplying water from conventional sources, and applying the process to brackish (slightly salty) water is 2.5 times more costly. Total desalination capacity worldwide is now 2.7 km^3/year, less than one-tenth of 1 percent of global water use. Sixty percent of the world's capacity is in the Arabian Peninsula and Iran, where surface water is virtually nonexistent and even ground water is often too salty to drink. Yet even in these energy-rich countries, producing and transporting the desalted water inland is in some cases prohibitively expensive. Though perhaps the ultimate source, desalination is unlikely to deliver its promise of a limitless supply of fresh water any time soon.[39]

CONSERVING WATER

As affordable options to augment dependable water supplies diminish, the key to feeding the world's growing population, sustaining economic progress, and improving living standards will be learning to use existing supplies more efficiently. Using less water to grow grain, make steel, and flush toilets increases the water available for other uses as surely as building a dam or diverting a river does. The outlines of a strategy to curb water demand are clear, though no single blueprint can apply to every region. The challenge is to combine the technologies, economic policies, laws, and institutions that work best in each water setting.

Since agriculture claims the bulk of most nations' water budgets and is by far the largest consumer, saving even a small fraction of this water frees a large amount to meet other needs. Raising irrigation efficiencies worldwide by just 10 percent, for example, would save enough water to supply all global residential water uses. As discussed previously, vast quantities of water seep through unlined canals while in transit to the field, and much more water is applied to crops than is necessary for them to grow. The rising cost of new irrigation projects, the limited supplies available to expand watering in many areas, and the high cost of pumping are forcing governments, international lending agencies, and farmers alike to find ways of making agricultural water use more efficient.

Most farmers in developing as well as industrial countries use gravity flow systems to irrigate their fields. The oldest method, and generally the least expensive to install, these systems distribute water from a ground water well or surface canal through unlined field ditches or siphons. Typically, only a small portion reaches the crop's root zone; a large share runs off the field. Sprinkler systems, which come in many varieties, apply water to the field in a spray. They use more energy than gravity systems and require a larger capital investment to install, but they have brought irrigation to rolling and steep lands otherwise suited only for dryland farming. One design—the center-pivot system—was largely responsible for the rapid expansion of irrigation on the U.S. High Plains in recent decades.[40]

Drip or trickle irrigation systems, developed in Israel in the sixties, supply water and fertilizer directly onto or below the soil. An extensive network of perforated piping releases water close to the plants' roots, minimizing evaporation and seepage losses. These costly systems thus far have been used mainly for high-value orchard crops in water-short areas. Today drip irrigation is used on about 10 percent of Israel's irrigated land, where experiments in the Negev Desert have shown per-hectare yield increases of 80 percent over sprinkler systems. Introduced into the United States in the early seventies, these systems now water nearly 200,000 ha and are slowly being used on row crops too. In Brazil's drought-plagued northeast a project sponsored by the Inter-American Development Bank is experimenting with one design to irrigate crops where farm incomes are low and water supplies are scarce.[41]

Most irrigation experts agree that the actual efficiency of water use obtained in the field depends as much on the way the irrigation system is managed as on the type used. Although drip irrigation may be inherently more efficient by design, the wide average range of efficiency for each system— 40 to 80 percent for gravity flow, 75 to 85 percent for a center-pivot sprinkler, and 60 to 92 percent for a drip system—shows that management is a key determinant. Farmers using conventional gravity flow systems, for example, can cut their water demands by 30 percent by capturing and recycling the water that would otherwise run off the field. Some U.S. jurisdictions now require these tailwater reuse systems. Farmers are also finding, however,

that they often make good economic sense because pumping tailwaters back to the main irrigation ditch generally requires less energy than pumping new water from the source, especially from a deep well.[42]

Farmers can also reduce water withdrawals by scheduling their irrigation according to actual weather conditions, evapotranspiration rates, soil moisture, and their crops' water requirements. Although this may seem like fine tuning, careful scheduling can cut water needs by 20 to 30 percent. At the University of Nebraska's Institute of Agriculture and Natural Resources, a computer program called IRRIGATE uses data gathered from small weather stations across the state to calculate evapotranspiration from the different crops grown in each area. Farmers can call a telephone hot-line to find out the amount of water used by their crops the preceding week and then adjust their scheduled irrigation date accordingly. The California Department of Water Resources is launching a similar management system with a goal of saving 740 million m^3 of water annually by the year 2010. The department is also demonstrating irrigation management techniques through mobile laboratories equipped to evaluate the efficiencies of all types of irrigation systems—gravity, sprinkler, and drip—and to recommend ways that farmers can use their water more efficiently.[43]

Israel has pioneered the development of automated irrigation, in which the timing and amount of water applied is controlled by computers. The computer not only sets the water flow, it also detects leaks, adjusts water application for wind speed and soil moisture, and optimizes fertilizer use. The systems typically pay for themselves within three to five years through water and energy savings and higher crop yields. Motorola Israel Ltd., the main local marketer of automated systems, has begun exporting its product to other countries; by 1982 over a hundred units had been sold in the United States. Israel's overall gains in agricultural water use efficiency, through widespread adoption of sprinkler and drip systems and optimum management practices, have been impressive: The average volume of water applied per hectare declined by nearly 20 percent between 1967 and 1981, allowing the nation's irrigated area to expand by 39 percent while irrigation water withdrawals rose by only 13 percent.[44]

In the Third World, where capital for construction of new projects is increasingly scarce, better management of existing irrigation systems may be the best near-term prospect for increasing crop production and conserving water supplies. Lining irrigation canals, for example, can help reduce water waste, prevent waterlogging, and eliminate the erosion and weed growth that makes irrigation ditches deteriorate.[45] Yet canal lining is expensive, and other options may prove more cost-effective. Seepage from canals is not necessarily water wasted, since it increases the potential ground water supply. By coordinating the use and management of ground water and surface water, as in the case of the Indus Valley described earlier, the total efficiency of water use in an agricultural region can be increased.

Farmers also need control of their irrigation water in order to make good use of fertilizer and other inputs that increase crop yields. Concrete turnouts that allow farmers to better dictate the timing and flow of water to their fields, for example, are being built in India, Pakistan, and elsewhere. At a pilot project in Egypt, funded by the U.S. Agency for International Development, improved management of irrigation systems is largely credited with boosting rice yields 35 percent. Water savings alone will often justify such investments: By some estimates, better irrigation management in Pakistan could annually save over 50 km³—four times the storage capacity of the nation's Tarbela Dam—at one-fourth the cost of developing new water supplies.[46]

Curbing industrial demand for water, the second major draw on world supplies, tackles problems in two ways: It frees a large volume of fresh water to meet other competing demands, and it can greatly reduce the volume of polluted water discharged to local rivers and streams. In most developing countries industry's demand for water is growing faster than that of either agriculture or municipalities. A slowdown is thus essential for sustained economic growth in water-short regions and for battling pollution problems that are fast making available supplies unfit for use.

In many industries much of the water used is for cooling and other processes that do not require that it be of drinking water quality. A large share of the water initially withdrawn can thus be recycled several times before disposing of it. Thermal power plants can cut their requirements by 98 percent or more by using recycled water in cooling towers rather than the typical once-through cooling methods. Palo Verde, a nuclear power plant built in the desert outside Phoenix, Arizona, for example, is near no body of water; it will draw on nearby communities' treated waste water, which the plant will reuse 15 times. The water needs of other industries also vary greatly, depending on the degree of recycling: Manufacturing a ton of steel may take as much as 200,000 liters (L) or as little as 5,000, and a ton of paper may take 350,000 L or only 60,000. Moreover, recycling the materials themselves can also greatly cut industrial water use and waste water discharges. Manufacturing a ton

of aluminum from scrap rather than virgin ore, for instance, can reduce the volume of water discharged by 97 percent.[47]

For the manufacturing industries that use a great deal of water—primary metals, chemicals, food products, pulp and paper, and petroleum—the cost of water is rarely more than 3 percent of total manufacturing expenses. Incentives to use water more efficiently have come either from strict water allocations or stringent pollution control requirements. In Israel, where virtually all available fresh water supplies are being tapped, the government has set quotas on the amount of any industrial plant may receive. A water use standard per unit of production is established for each industry, and a particular plant's allocation is then calculated by multiplying the standard by the anticipated level of production. As new technologies are developed, the standards are made more stringent. Consequently, average water use per unit value of industrial production has declined in Israel by 70 percent over the last two decades.[48]

In Sweden industrial water use quintupled between 1930 and the midsixties but has since shown a marked decline. Strict environmental protection requirements for the pulp and paper industry, which accounts for about 80 percent of the country's industrial withdrawals, fostered widespread adoption of recycling technologies. Despite more than a doubling of production between the early sixties and late seventies, the industry cut its total water use by half—a fourfold increase in water efficiency. Indeed largely because of these savings, Sweden's total water withdrawals in the midseventies were only half the level projected a decade earlier.[49]

Pollution controls spawned by federal and state laws are also helping to curb manufacturing water use in many areas of the United States. Surveys of Californian industries show, for example, that total water use in manufacturing declined during the seventies despite a 14 percent increase in the number of plants. Echoing Sweden's experience, the pulp and paper industry led in water reductions, with a 45 percent decline in withdrawals between 1970 and 1979. Nationwide, industrial withdrawals have not yet turned the corner, probably because of long delays in passing the pollution control requirements authorized by the Clean Water Act. Yet declines should occur when and where strict standards are enforced.[50]

Developing countries are in a prime position to take advantage of these new recycling technologies. Building water efficiency and pollution control into new plants is vastly cheaper than retrofitting old ones. Experience in the West shows that industries will have little incentive to adopt these measures without either sufficiently high water and waste water fees or stringent pollution control requirements. Many of the technologies available are able to reduce water use and waste water flows at least 90 percent and thus can contribute greatly to alleviating water supply and pollution problems in growing industrial areas. A recent study of an integrated iron and steel plant near Sao Paulo in Brazil, for example, showed that the plant was withdrawing 12,000 m³ of water per hour—and that it was discharging 22,000 tons of iron oxide and 2600 tons of grease annually into the nearby Santos estuary. For an estimated $15 million, or less than $1 per ton of annual production, the plant could install a recirculating water system that would cut water use by 94 percent and pollutant discharges by 99 percent.[51]

Household and other municipal water demands rarely account for more than 15 percent of a nation's water budget, and worldwide they claim only about 7 percent of total withdrawals. Yet storing, treating, and distributing this water, as well as collecting and treating the resulting waste water, is increasingly costly. Large capital investments are required, making water and waste water utilities especially sensitive to scarce capital and high interest rates. In the United States, for example, water and waste water utilities require an average of $8.5 billion in new investment each year. Capital needs for 1982–90 are expected to total about $100 billion, and some estimates go much higher.[52] Reducing municipal water use can ease these financial burdens by allowing water and waste water utilities to scale down the capacity of new plants, water mains, and sewer pipes and to cut the energy and chemical costs associated with pumping and treating the water.

Many household fixtures and appliances use much more water than necessary to perform their varied functions. Most toilets in the United States, for example, use 18 to 22 L per flush, while water-conserving varieties recommended by the Plumbing Manufacturers Institute average about 13. A typical West German toilet requires only 9 L per flush, and a new model that meets government standards uses about 7.5 L, just a third as much as conventional U.S. models. Showerheads often spray forth 20 L or more per minute; water-conserving designs can cut this at least in half. Water-efficient dishwashers and washing machines can reduce water use 25 to 30 percent over conventional models. With simple conservation measures such as these, indoor water use can easily be reduced by a third.[53] (See Table 29-4.)

Consumers installing these devices and appliances will almost always save money, since they will reduce not only water use but the energy used in heating water. A typical household in the United States, for example, could expect investments in common water-saving fixtures and appliances to pay for themselves through lower water, sewer, and energy costs in just a few months, or within four years at most. Israel, Italy, and the states of California, Florida, Michigan, and New York now have laws requiring the installation of various water-efficient appliances in new homes, apartments, and offices.[54]

Despite its potential financial benefits to consumers and utilities, municipal conservation is still typically viewed only as a means of combating drought, rarely as a long-range water strategy. Programs developed by water-short communities to foster lasting reductions in water use, however, have yielded fruitful results. In Tucson, Arizona, a combination of price increases and public education efforts to encourage installation of household

Table 29-4. United States: Annual Household Water Use and Potential Savings with Simple Conservation Measures

Activity	Share of Total Indoor Water Use (%)	Without Conservation (1000 L per Capita)	With Conservation	Savings (%)
Toilet flushing	38	34.5	16.4	52
Bathing	31	27.6	21.8	21
Laundry and dishes	20	18.0	13.1	27
Drinking and cooking	6	5.5	5.5	0
Brushing teeth, miscellaneous	5	4.1	3.7	10
Total	100	89.7	60.5	33

Source: Adapted from U.S. Environment Protection Agency, Office of Water Program Operations, 1981, Flow reduction: Methods, analysis procedures, examples (Washington, D.C.: U.S. EPA).

Note: Estimates based on water use patterns for a typical U.S. household. European toilets, for example, often use less water than the figures given here would imply.

water-saving devices and replacement of watered lawns with desert landscaping led to a 24 percent drop in per capita water use. As a result, the Tucson utility's pumping costs were reduced and the drilling of new water supply wells was deferred. Planners thus expected customer water bills to be lower over the long term than they would have been without the conservation efforts.[55]

In El Paso, Texas, one of the most water-short cities in the United States, pricing and education efforts are also credited with a substantial reduction in water use. Long-term water supply projections show conservation meeting about 15 to 17 percent of the city's future water needs. Besides slowing the rate of depletion of El Paso's underground water supplies, the conservation measures are saving water for an average cost of about $135 per 1000 m^3—8 percent less than the average cost of existing water supplies.[56]

Many other options are available to reduce the demand for fresh water. Some areas are finding, for example, that brackish water and treated waste water can meet many of their water needs. In Saudi Arabia brackish water irrigates salt-tolerant crops such as sugar beets, barley, cotton, spinach, and date palms, thereby saving the best-quality water for drinking and other household uses. Treated municipal waste water is also reused there to irrigate crops and gardens, to recharge aquifers, and as a supply for certain industries. Power plants in Finland, Sweden, the United Kingdom, and the United States are beginning to use brackish water or salt water for cooling.[57]

In perennially dry South Africa, water policy specifically calls upon users to "make use of the minimum quantity of water of the lowest acceptable quality for any process." Over the next several decades cities and industries are projected to recycle between 60 and 70 percent of the water they withdraw. Engineers estimate that the cost of treating raw sewage to a quality suitable for drinking is very likely competitive with that of developing the next surface water source. In Israel 30 percent of municipal waste water was already being reused in 1981, most of it for irrigation. With completion of the Dan Region Waste Water Reuse Project serving the Tel Aviv metropolitan area, projections are that the proportion of municipal waste water reused will climb to 80 percent by the turn of the century.[58]

PRIORITIES FOR A NEW WATER ECONOMY

Much of the profligate waste and inefficiency in today's use of water results from policies that promote an antiquated illusion of abundance. People rarely pay the true cost of the water they use. Economists often suggest pricing water at its marginal cost—the cost of supplying the next increment from the best available source. Consumers would thus pay more as supplies become scarcer. Market forces would foster conservation and a reallocation of water supplies to their highest-valued uses. In California, for example, the value added per cubic kilometer of water is 65 times greater in industry than in agriculture.[59] Increasing competition for water and rising prices thus dictate a shift in water use from farming to manufacturing. The extent to which a market-driven reallocation should take place is partially a political decision, since it would alter a region's basic character and social fabric; but by economic criteria, it is efficient.

In reality, water is rarely priced at marginal cost; charges often bear little relation to the real cost and quantity of water supplied. Many homeowners in Great Britain, for instance, are charged for water according to the value of their property, a practice that dates to Victorian times. In Indonesia, Malaysia, Saudi Arabia, South Africa, Tanzania, most East European countries, and many others, the government pays all or most of the capital costs for major irrigation projects. Farmers in the United States supplied with irrigation water from federal projects pay, on average, less than a fifth of the real cost of supplying it.[60] Taxpayers are burdened with the remainder, and farmers use more water than they would if asked to pay its full cost.

When water users supply themselves rather than relying on a public project, they typically pay only the cost of getting the water to their farm, factory, or home. But if their withdrawals are diminishing a water source or harming an ecosystem, they should bear the costs that their private actions impose on society. American farmers pumping water from the Ogallala aquifer, for example, pay nothing extra for the right to earn their profits by depleting an irreplaceable resource. On the contrary, many get a tax break by claiming a depletion allowance based on the drop in water level beneath their land that year. The greater the depletion, the greater the allowance—hardly an incentive to conserve.[61] A more appropriate policy would be to tax ground water pumping in all areas where aquifers are being depleted. That way the public gets some compensation for the loss of its resource, and farmers are encouraged to curb their withdrawals.

In much of the Third World, where the cost per hectare of building new irrigation systems often exceeds per capita gross national product, pricing water at its full cost may not always be feasible. Water is often supplied for free or is heavily subsidized because it is so vital to food production. Yet most experts agree that the inefficient operation and poor maintenance of irrigation systems is largely due to farmers' perceptions that they have no responsibility for them. International lending agencies are now investing handsome sums to rehabilitate irrigation systems that sound operation and maintenance could have kept in good working order. Having farmers pay some share of water costs gives them a stake in the system, besides generating revenue to improve operations.[62]

A combined strategy of charging Third World farmers for some share of system costs and organizing them into "water user associations" to coordinate management tasks and the collection of fees appears a promising way of improving irrigation management. Arguing for more attention to pricing and water user organizations in Thailand, economist Ruangdej Srivardhanaat of Kasetsart University in Bangkok says that in order for Thai farmers to improve their practices "the feeling that the irrigation facilities belong to and are useful to them is crucial."[63] Charging a modest price for an initial allotment and higher fees for water used above this amount would encourage farmers to conserve without overburdening them. Moreover, where ground water supplies are available, farmers may be able to profitably construct irrigation wells with minimal public support. In India over 1.7 million private tube wells have been installed by the late seventies, aided by the availability of credit with very reasonable interest and repayment terms. For many farmers on the Indo-Gangetic Plain, installation of these wells has yielded rates of return greater than 50 percent.[64]

Water users must also begin to pay for treating the water they pollute. Especially in many areas of the Third World, water bodies cannot long be expected to provide a source of high-quality drinking and irrigation water *and* to dilute the increasing tonnage of waste dumped into them each year. Dilution alone simply cannot maintain adequate water quality in a society undergoing rapid industrialization and urbanization. Industries should pay the full cost of using water in their production, which includes the cost of discharging most of it in a form suitable for reuse. Controlling pollution is costly: Funds for protecting quality now account for over half the U.S. budget for water resource development and amount to $25 billion annually.[65] Developing countries may not have the financial resources to subsidize costly pollution controls while at the same time continuing to improve irrigation systems and install drinking water services. Industrialization should proceed in tandem with industries' ability to pay for controlling the pollution they generate. Sacrificing water quality for industrial growth cannot be a winning proposition in the long run.

Existing laws and methods for allocating water supplies are often heavily biased toward those wanting to withdraw water and against those desiring that it remain in place. The old English common law, which required that a riparian landowner not diminish the quantity or quality of water remaining for downstream users, inherently protected stream ecology and habitats. Yet this rule was changed early in the American experience to give riparians the right to "reasonable use" of the water, thus allowing for alterations in stream flows. In the drier states of the American West an appropriative system was adopted that is even more biased toward withdrawals: Water rights are allocated successively to those who put water to "beneficial use." Establishing such a use, and thus a water right, often required an actual diversion from the stream. As legal expert James Huffman notes, this was not a problem "until the combination of changing values and diminishing water supplies brought the issue of instream flow maintenance to the public attention."[66]

A number of options exist for governments seeking to preserve an ecological balance in their rivers and streams. In the United States, for example, Montana passed a law in 1973 that allows government agencies to acquire prospective water rights. Much of the state's water has not been appropriated, so under this legislation a large share of it can be reserved to protect stream ecology. Because of these reserved rights, much of the Yellowstone River will never be withdrawn for use. Many rivers and streams in the United States, however, are already fully appropriated during the dry season of the year. Preserving water quality and fish and wildlife habitats thus requires some form of regulation that limits withdrawals during periods of diminished flow. One of the most powerful tools available, though as yet little used, is what legal experts call the "public trust" doctrine. Dating back to Roman times, it asserts that governments hold certain rights in trust for the public and can take action to protect them from private interests. Its application has potentially sweeping effects since even existing water permits or rights could be revoked in order to prevent violation of the public trust.[67]

In a landmark decision handed down in February 1983, the California Supreme Court declared that the water rights of the City of Los Angeles, which allow diversions from the Mono Lake Basin, are subject to the public trust doctrine. Mono Lake, a hauntingly beautiful water body on the eastern side of the Sierra, has diminished in surface area by a third, largely because Los Angeles is diverting water from its major tributaries. The lake is also becoming more saline, threatening its brine shrimp population, which in turn feeds millions of local and migratory birds. By invoking the public trust doctrine, the California Court paved the way for a state agency or the courts to decide that Los Angeles must reduce its diversions from the Mono Lake Basin. California law professor Harrison C. Dunning writes: "Although ramifications of the ruling may not be apparent for years, there can be no doubt that it will raise new obstacles for those who would divert California's natural stream flows to farm and city use. . . . From now on, the state must protect what the court calls 'the people's common heritage of streams, lakes, marshlands and tidelands'.[68]

Where demands are already at the limits of the available supply, regulations may be necessary to put water use on a sustainable footing. Strategies geared toward balancing the water budget are lacking in most areas of falling water tables or shrinking surface supplies. Despite pleas by hydrologists, for example, no Indian states have passed laws to regulate the installation of tube wells or to limit ground water withdrawals. In the southern state of Tamil Nadu, authorities are doing little to curb overpumping that in some areas has caused ground water levels to drop 30 m in just a decade. Hydrologists note that the "long-term effects are probably understood, but until the water disappears, it is hardly likely that anyone is going to do anything about the situation."[69]

At least one example worth emulating has emerged in the United States: the 1980 Arizona Groundwater Management Act. Facing a rapidly dwindling water supply, the state is requiring its most overpumped areas to achieve "safe yield" by the year 2025. At this level no more ground water is withdrawn than is recharged; the resource is thus in balance. Achieving this goal will by no means be painless. Conservation measures will be required of all water users and all ground water withdrawals will be taxed. No subdivided land can be developed without proof of an assured water supply. If by the year 2006 it appears that conservation alone will not achieve the state's goal, the government can begin buying and retiring farmland. Shifts in Arizona's economy have already begun: Between 1978 and 1982 the state's irrigated area declined 8 percent. Other water-short regions should recognize that such shifts are bound to occur, and that they will be less traumatic if, as Arizona is doing, they are eased by thoughtful planning. Many governments will be watching as the real test of Arizona's law begins in the nineties.[70]

Finally, planners and educators must dispel the myth that conservation is exclusively a short-term strategy to alleviate droughts and other immediate crises. Only in such dry nations as South Africa and Israel is conservation made an integral part of planning future water supplies. In these countries, which are already tapping most of their available sources, continually striving to increase the efficiency of water use is imperative if growth is to continue. But even in nations with untapped rivers and aquifers, measures to conserve, recycle, and reuse fresh water may in many cases make the resource available at a lower cost and with less environmental disruption than developing these new supplies. Conservation's potential will never be realized until it is analyzed as a viable long-term option comparable to drilling a new well or building a new reservoir.

Steps toward this end were taken in the United States during the late seventies. In a June 1978 water policy message to the nation, President Carter resolved to make conservation a national priority. Government agencies began to make federal grants and loans for water projects conditional upon inclusion of cost-effective conservation measures. Numerous analyses suggested that substantial savings would accrue both to the government and to communities and their residents from measures to curb water demand. Unfortunately, the Reagan administration took several steps backward when it demoted these conservation requirements to voluntary guidelines and disbanded the Water Resources Council, which had been pushing for a more economically efficient and environmentally sound water policy. California has taken the lead where the federal government has faltered: A 1983 law requires every major urban water supplier in the state to submit by the end of 1985 a management plan that explicitly evaluates efficiency measures as an alternative to developing new supplies.[71]

Most governments continue to expect traditional dam and diversion projects to relieve regional water stresses. Yet the engineering complexities of these projects, along with their threats of ecological disruption, multibillion-dollar price tags, and 20-year lead times leave little hope that they will deliver water in time to avert projected shortages—if, indeed they are completed at all. In the Third World, unless deforestation and erosion are curbed and irrigation systems are better managed and maintained, large projects may waste scarce capital and diminish the productivity of cropland. Moreover, even the most grandiose schemes will not be ultimate solutions to regional water problems. The Soviet Union's planned diversion of the Siberian rivers, for example, may meet only one-fourth of the deficit expected in central Asia. Water delivered to Arizona through the Central Arizona Project will make up for only half of the state's annual ground water depletions and thus will not alone balance the water budget. Against an insatiable demand, the best any dam or diversion can do is to slow the depletion of supplies or delay the day when they fall short.

In an area of growing competition for limited water sources, heightened environmental awareness, and scarce and costly capital, new water strategies are needed. Continuing to bank on new large water projects, and failing to take steps toward a water-efficient economy, is risky: Vital increases in food production may never materialize, industrial activity may stagnate, and the rationing of drinking water supplies may become more commonplace.

Alternatives to large dam and diversion projects exist. Water crises need not occur. Securing more dependable supplies in the Third World can and should continue, but it may better be done with smaller projects more amenable to coordinated land and water management, with incremental development of ground water, and especially with joint management of surface and underground supplies. In water-short areas of industrial countries people and economic activity must begin adapting to water's limited availability. Supplies in Soviet central Asia, for example, simply cannot support a booming population and an expanding farming economy for long. Oasis cities such as Phoenix and Los Angeles can no longer expect to grow and thrive by draining the water supplies of other regions. Conservation and better management can free a large volume of water—and capital—for competing uses. Thus far, we have seen only hints of their potential.

NOTES

1. Water rendered unusable by pollution by year 2000 estimated at 3000 km^3 in Robert P. Ambroggi, 1980, Water, *Scientific American* (September).

2. Zheng Guanglin, 1984, *Research program on China 2000* Institute of Scientific and Technical Information of China, Beijing, draft, February; Smil, *The bad earth*.

3. U.N. Economic Commission on Latin America, 1977, *The water resources of Latin America: Regional report* (Santiago de Chile); Bogota River cited in Peter Nares, 1984, Colombian towns threatened by polluted Bogota River, *World Environment Report,* May 30.

4. Thane Gustafson, 1977, Transforming Soviet agriculture: Brezhnev's gamble on land improvement, *Public Policy* (Summer).

5. Share of U.S. irrigated land from Resource Analysis Section, Colorado Department of Agriculture, 1983, *Colorado High Plains study: Summary report* (Denver: November); for background on the Ogallala's development, see Kenneth D. Frederick and James C. Hanson, 1982, *Water for western agriculture* (Washington, D.C.: Resources for the Future) and Morton W. Bittinger and Elizabeth B. Green, 1980, *You never miss the water till . . .* (Littleton, Colo.: Water Resources Publications); data on water depletion from U.S. Geological Survey, 1984, *National water summary 1983—Hydrologic events and issues* (Washington, D.C.: U.S. government Printing Office).

6. U.S. Department of Agriculture, 1983, *Agricultural statistics 1983* (Washington, D.C.: U.S. Government Printing Office) and Bureau of the Census, 1984, *1982 census of agriculture* (Washington, D.C.: U.S. Department of Commerce); Nebraska corn production statistics from U.S. Department of Agriculture, Economic Research Service, 1983, *Economic indicators of the farm sector: Costs of production 1982* (Washington, D.C.: U.S. Government Printing Office).

7. See Dean Abrahamson and Peter Ciborowski, 1983, *North American agriculture and the greenhouse problem,* Report of the Humphrey Institute Symposium on the Response of the North American Granary to Greenhouse Climate Changes, Minneapolis, Minn., April, and Kellogg and Schware, Society, science and climate change.

8. Tucson information from Tony Davis, 1984, Trouble in a thirsty city, *Technology Review* (August/September); Texas and Mexico references in Tommy Knowles and Frank Rayner, 1978, Depletion allowance for groundwater mining: Pros and cons, *Journal of the American Water Works Association* (March).

9. Reference to Tamil Nadu in Widstrand, *Water conflicts and research priorities;* China references from Smil, *The bad earth.*

10. Thomas G. Sanders, 1977, Population growth and resource management: Planning Mexico's water supply, *Common Ground* (October); China references from Sinking city under control, *Beijing Review,* February 23, 1981; Smil, *The bad earth;* Texas situation from Knowles and Rayner, Depletion allowance: Pros and cons.

11. For U.S. citations, see U.S. Geological Survey, *National water summary 1983;* other countries cited in Tony Samstag, 1984, Too much of a good thing, *Development Forum* (April).

12. Cropland figure from M. I. L'vovich and I. D. Tsigel'naya, 1981, The potential for long-term regulation of runoff in the mountains of the Aral Sea drainage basin, *Soviet Geography* (October); population and employment issues discussed in Thane Gustafson, 1980, Technology assessment, Soviet Style, *Science* (June 20); information on the Aral from Philip P. Micklin, Department of Geography, Western Michigan University, Kalamazoo, Mich., private communication, September 5, 1984; scientists' projections from G. V. Voropaev et al., 1983, The problem of redistribution of water resources in the midlands region of the USSR, *Soviet Geography* (December).

13. Estimate of water level decline from O. K. Leont'yev, 1984, Why did the forecasts of water-level changes in the Caspian Sea turn out to be wrong?, *Soviet Geography* (May); background information from Grigorii Voropaev and

Aleksei Kosarev, 1982, The fall and rise of the Caspian Sea, *New Scientist* (April 8); Micklin, private communication; Caspian fisheries discussed in Philip P. Micklin, 1977, International environmental implications of Soviet development of the Volga River, *Human Ecology,* (2).

14. Philip C. Metzger and Jennifer A. Haverkamp, *Instream flow protection: Adaptation to intensifying demands* (Washington, D.C.: The Conservation Foundation, June).

15. Inquiry on Federal Water Policy, *Water is a mainstream issue.*

16. Anupam Mishra, 1981, *An irrigation project that has reduced farm production,* (New Delhi: Centre for Science and Environment).

17. Global estimate from V. A. Kovda, 1983, Loss of productive land due to salinization, *Ambio,* 12 (2); India and Pakistan estimates from Gilbert Levine et al., 1979, *Water,* prepared for Conference on Agricultural Production: Research and Development Strategies for the 1980's, Bonn, West Germany, October 8–12; other areas from Biswas, Major water problems, and various other sources.

18. Estimate of reservoir capacity based on M. I. L'vovich, 1979, *World water resources and their future,* translation ed. Raymond L. Nace (Washington, D.C.: American Geophysical Union); figures on large dam construction from van der Leeden, *Water resources of the world,* and from Philip Williams, 1983, Damming the world, Philip Williams & Associates, San Francisco, Calif., unpublished, April.

19. U.S. Geological Survey, *National water summary 1983.*

20. UNECE, *Long-term perspectives for water use and supply.*

21. Williams, Damming the world.

22. Donor agency inspections from U.S. General Accounting Office, 1983, Irrigation Assistance to developing countries should require stronger commitments to operation and maintenance (Washington, D.C.); John Madeley, 1983, Big dam schemes—Value for money or non-sustainable development?, *Mazingira,* 7 (4).

23. For an excellent discussion of these water and land interactions, see Malin Falkenmark, 1984, New ecological approach to the water cycle: Ticket to the future, *Ambio,* 13 (3); Malaysia example from Eneas Salati and Peter B. Vose, 1984, Amazon Basin: A system in equilibrium, *Science* (July 13); Dominica example from Robert S. Goodwin, 1984, Water resources development in small islands: Perspectives and needs, *Natural Resources Forum* (January).

24. For an assessment of selected large dam projects, see Environmental Policy Institute, 1984, *Fact sheets on international water development projects* (Washington, D.C.).

25. History and 1983 government support from Asit K. Biswas, 1983, Water where it's wanted, *Development Forum* (August/September); Yao Bangyi and Chen Qinglian, 1983, South-north water transfer project plans, in *Long-distance water transfer,* ed. Asit K. Biswas et al. (Dublin: Tycooly International).

26. Bruce Stone, 1983, The Chang Jiang diversion project: An overview of economic and environmental issues," in Biswas et al., *Long-distance water transfer;* cost estimates also given in Asit K. Biswas, 1982, US $12 billion plan to redistribute China's water wealth, *South* (April).

27. For discussion of Siberian diversion plans, see O. A. Kibal'chich and N. I. Koronkevich, 1983, Some of the results and tasks of geographic investigations on the water-transfer project, *Soviet Geography* (December); quote is from Gustafson, Technology assessment, Soviet style.

28. Philip P. Micklin, 1984, Recent developments in large-scale water transfers in the USSR, *Soviet Geography* (April); cost estimates from Micklin, private communication, October 16, 1984; water-saving potentials and salinization risks from summary of remarks made by O. A. Kibal'chich at conference in Irkutsk, Soviet Union, August 1983, and published in *Soviet Geography* (December 1983).

29. U.S. Congressional Budget Office, 1983, *Efficient investments in water resources: Issues and options* (Washington, D.C.: U.S. Government Printing Office).

30. Thomas M. Power, *An economic analysis of the Central Arizona Project, U.S. Bureau of Reclamation,* (Missoula: Economics Department, University of Montana).

31. Costs to date from Steve Macauley, Supervisory Engineer for the State Water Project Analysis Office, California Department of Water Resources, private communication, October 23, 1984; pumping costs from California Department of Water Resources, 1983, *Management of the California State Water Project* (Sacramento: California Resources Agency).

32. Jack Foley, 1984, Governor's water bill dead for this year, *San Jose Mercury News* (August 7); private communications with California Department of Water Resources personnel, June and August 1984.

33. Projects worldwide from Jay H. Lehr, 1982, Artificial ground water recharge: A solution to many U.S. water-supply problems, *Ground Water* (May/June); Israel project cited in Robert P. Ambroggi, 1977, Underground reservoirs to control the water cycle, *Scientific American* (May); California Department of Water Resources, 1980, *The California State Water Project—Current activities and future management plans* (Sacramento: California Resources Agency); California Dept. of Water Resources, *Management of the water project;* Helen Peters, Ground Water Staff Specialist, California Department of Water Resources, private communication, October 1984; median cost estimate for new surface reservoirs from Ronald B. Robie, 1981, Irrigation development in California—Construction or water management?, in *Irrigation challenges of the 80's* (St. Joseph, Mich.: American Society of Agricultural Engineers); Environmental and Energy Study Institute, *Weekly Bulletin,* March 26, 1984; Russell Brown, Subcommittee on Water and Power, Senate Committee on Energy and Natural Resources, private communication, October 1984.

34. Gangetic Plain estimates from Widstrand, *Water conflicts and research priorities;* information on China from Smil, *The bad earth.*

35. Ambroggi, Underground reservoirs; William R. Gasser, 1981, *Survey of irrigation in eight Asian nations* (Washington, D.C.: U.S. Department of Agriculture; problems with the project cited in Ian Carruthers and Roy Stoner, 1981, *Economic aspects and policy issues in ground water development* (Washington, D.C.: The World Bank) and by Douglas Merrey, Agency for International Development, private communication, October 1984.

36. Wayne A. Pettyjohn, 1981, *Introduction to artificial ground water recharge* (Columbus, Ohio: National Water Well Association).

37. Lehr, Artificial ground-water recharge.

38. L'vovich, *World water resources and their future.*

39. U.S. Office of Technology Assessment, 1983, *Water-related technologies for sustainable agriculture in U.S. arid/semiarid lands* (Washington, D.C.: U.S. Government Printing Office); desalination cost estimates from U.S. Comptroller General, 1979, *Desalting water probably will not solve the nation's water problems, but can help* (Washington, D.C.: U.S. General Accounting Office); use in Arabian Peninsula from M. A. Khan et al., 1984, Development of supplies & sanitation in Saudi Arabia, *African Technical Review* (June).

40. A concise description of irrigation systems is contained in Office of Technology Assessment, *Water-related technologies for sustainable agriculture in U.S. lands.*

41. Negev experiments from Israel's water policy: A national commitment, in U.S. Office of Technology Assessment, 1983, *Water-related technologies for sustainable agriculture in arid/semiarid lands: Selected foreign experience* (Washington, D.C.: U.S. Government Printing Office); Office of Technology Assessment, *Water-related technologies for sustainable agriculture in U.S. lands;* Jay H. Lehr, 1983, Increased irrigation efficiency will ultimately silence the water-short blues of the wasteful West, *Ground Water* (March/April); Brazilian information from *IDB News,* Inter-American Development Bank, Washington, D.C., vol. 10, no. 4.

42. Efficiency ranges from J. Keller et al., 1981, Evaluation of irrigation systems, in *Irrigation challenges of the 80's;* tailwater reuse discussed in Gordon Sloggett, 1982, *Energy and U.S. Agriculture: Irrigation pumping 1974–1980* (Washington, D.C.: U.S. Government Printing Office); E. G. Kruse et al., 1981, Advances in surface irrigation, in *Irrigation challenges of the 80's.*

43. Nebraska program from Paul E. Fischbach, 1981, Irrigation management (scheduling) application, in *Irrigation challenges of the 80's;* California system described by Edward Craddock, California Department of Water Resources, Office of Water Conservation, private communication, June 21, 1984; California Department of Water Resources, undated, *The mobile agricultural water conservation laboratory,* information pamphlet prepared by the Office of Water Conservation, Sacramento, Calif.

44. Israel's water policy: A national commitment.

45. General Accounting Office, *Irrigation Assistance to developing countries;* D. B. Kraatz, 1977, *Irrigation canal lining* (Rome: U.N. Food and Agriculture Organization).

46. Worth Fitzgerald, U.S. Agency for International Development, private communication, April 25, 1984; Egyptian pilot project cited in Mark Svendsen et al., 1983, Meeting the challenge for better irrigation management, *Horizons* (March).

47. Harte and El-Gesseir, Water and energy; Thirsty desert plant has unique water system, *The Phoenix Gazette,* June 27, 1984; ranges of water use for steel and paper from United Nations, *Resources and needs;* reductions with aluminum recycling from R. C. Ziegler, 1976, *Environmental impacts of virgin and recycled steel and aluminum,* Calspan Corporation, Buffalo, N.Y.

48. Saul Arlosoroff, 1984, *Water management policies under scarce conditions: A case study—Israel,* presented at Conference on Water for the 21st Century: Will It Be There?, Dallas, Tex., April.

49. Swedish Preparatory Committee, *Water in Sweden.*

50. Reductions by California pulp and paper industry from California Department of Water Resources, 1982, *Water use by manufacturing industries in California, 1979* (Sacramento: California Resources Agency).

51. See 3M Company, 1982, *Low- or non-pollution technology through pollution prevention,* prepared for United Nations Environment Programme, St. Paul, Minn., June; Brazil study from Division for Industrial Studies, 1981, *Water use and treatment practices and other environmental considerations in the iron and steel industry* (Vienna: United Nations Industrial Development Organization, December).

52. John J. Boland, 1983, *Water/waste water pricing and financial practices in the United States* (Washington, D.C., Metametrics, Inc., August).

53. For estimates of water and energy savings and costs of various water-conserving measures, see U.S. Environmental Protection Agency, Office of Water Program Operations, 1981, *Flow reduction: Methods, analysis procedures, examples* (Washington, D.C.); reference to West German toilets from *World environment report,* April 4, 1984.

54. Typical U.S. household savings from Environmental Protection Agency, *Flow reduction;* Stefano Burchi, 1983, Regulatory approaches to the use of water for domestic purposes, *Natural Resources Forum* (July); Barbara Yeaman, Consultant to Facilities Requirements Division, U.S. Environmental Protection Agency, private communication, August 10, 1984.

55. Adrian H. Griffin et al., 1979, Changes in water rates and water consumption in Tucson, 1974 to 1978, *Hydrology and Water Resources in Arizona and the Southwest,* 10; Stephen E. Davis, 1979, Tucson's tools for demand management, *Hydrology and Water Resources in Arizona and the Southwest,* 8.

56. Lee Wilson and Associates, Inc., 1981, *Water supply alternatives for El Paso,* prepared for El Paso Water Utilities Public Service Board, Santa Fe, N. Mex., November, 1981.

57. Kahn et al., Development of supplies & sanitation in Saudi Arabia; Dennis J. Parker and Edmund C. Penning-Rowsell, 1980, *Water planning in Britain* (London: George Allen & Unwin; Swedish Preparatory Committee, *Water in Sweden.*

58. Quote and estimates of advanced treatment costs from Axel F. Zunckel and Maria P. Oliveira, 1981, South African water reuse policy and its practical implications, in *Proceedings of the Water Reuse Symposium II,* vol. 1 (Denver, Colo.: AWWA Research Foundation); reuse projections from Mike Nicol, 1984, South Africa will require wastewater recycling before year 2000, experts say, *World Environment Report,* (June 27); Hillel I. Shuval, 1981, The development of the waste water reuse program in Israel, in *Proceedings of the Water Reuse Symposium II.*

59. Value-added figures cited in Ambroggi, Water.

60. Great Britain practice from Burchi, Regulatory approaches to the use of water for domestic purposes; subsidies in selected countries from J. A. Sagardoy et al., 1982, *Organization, operation and maintenance of irrigation schemes* (Rome: U.N. Food and Agriculture Organization; costs to U.S. farmers from Congressional Budget Office, *Efficient investments in water resources.*

61. Knowles and Rayner, Depletion allowance: Pros and cons; Institute of Agriculture and Natural Resources, Cooperative Extension Service 1983, *IRS extends ground water depletion deduction to Nebraska irrigators* (Lincoln: University of Nebraska, March 18).

62. Importance of improved operation and maintenance discussed by Guy Le Moigne, Irrigation Advisor, World Bank, private communication, April 1984, and by Fitzgerald, private communication.

63. See Frances F. Korten, 1982, *Building national capacity to develop water users' associations: Experience from the Philippines* (Washington, D.C.: The World Bank; quote from Ruangdej Srivardhanaat, 1984, No easy management: Irrigation Development in the Chao Phya Basin, Thailand, *Natural Resources Forum* (April).

64. See James A. Seagraves and K. William Easter, 1983, Pricing irrigation water in developing countries *Water Resources Bulletin* (August); B. D. Dhawan, 1983, *Development of tubewell irrigation in India* (New Delhi: Agricole Publishing Academy).

65. Peter Rogers, 1984, *Fresh Water,* prepared for The Global Possible Conference, World Resources Institute, Wye, Md., May 2–5.

66. James Huffman, 1983, Instream water use: Public and private alternatives, in *Water rights: Scarce resource allocation, bureaucracy, and the environment,* ed. Terry L. Anderson (Cambridge, Mass.: Ballinger).

67. Ibid.; Metzger and Haverkamp, *Instream flow protection.*

68. For an excellent review of this decision, see Ellen Sullivan Casey, 1984, Water law—Public trust doctrine, *Natural Resources Journal* (July); Harrison C. Dunning, 1983, A new front in the water wars: Introducing the 'public trust' factor, *California Journal* (May).

69. Dhawan, *Development of tubewell irrigation in India;* Tamil Nadu observation from Widstrand, *Water conflicts and research priorities.*

70. Arizona Groundwater Management Study Commission Staff, 1980, Summary: Arizona Groundwater Management Act, briefing presented to the Arizona Groundwater Management Study Commission and the Arizona State Legislature, June 5; Scott Hanson and Floyd Marsh, 1982, Arizona ground-water reform: Innovations in state water policy, *Ground Water* (January/February); decline in irrigated area from U.S. Department of Agriculture, *Agricultural statistics 1983,* and Bureau of the Census, Census of agriculture.

71. See Environmental Protection Agency, *Flow reduction,* and Institute for Water Resources, 1979, *The role of water conservation in water supply planning* (Fort Belvoir, Va.: U.S. Army Corps of Engineers); policy changes under Reagan administration from Water resources, in The Conservation Foundation, 1984, *State of the environment: An assessment at mid-decade* (Washington, D.C.), and from Yeaman, private communication; California law from Reprint of Assembly Bill No. 797, *Legislative council's digest,* California Assembly, October 1983.

30

FEDERAL PROTECTION OF GROUND WATER

WENDY GORDON

Having suffered irreparable damage from wide-spread contamination, a significant portion of the nation's ground water may be seriously limited as a future drinking water source. Those who have recognized the ground water contamination problem doubt that our existing legal system provides either adequate protection for underground supplies or prevention from further contamination. The present system, composed of a patchwork of federal and state authorities, lacks comprehensiveness and coordination and has resulted in confusing and complicated policies and programs. At present there is no federal program aimed specifically at the problem of ground water contamination. Instead, at least eight federal statutes, directed primarily at other environmental problems, serve to provide some protection to ground water:

· Safe Drinking Water Act (SDWA).
· Resource Conservation and Recovery Act (RCRA).
· Comprehensive Environmental Response, Compensation and Liability Act (CERCLA).
· Clean Water Act (CWA).
· Toxic Substances Control Act (TSCA).
· Federal Insecticide, Fungicide and Rodenticide Act (FIFRA).
· Surface Mining Control and Reclamation Act (SMCRA).
· National Environmental Policy Act (NEPA).

These statutes grant regulatory authority to several federal agencies, the most important being the Environmental Protection Agency (EPA) and the Department of Interior (DOI). Effective ground water protection has been limited by the absence of a comprehensive federal statute on ground water protection in combination with only partial implementation of existing authorities.

The Safe Drinking Water Act (SDWA) of 1974 is the only statute that is intended to ensure the quality of water at the tap, including ground water. It requires the EPA to establish drinking water quality standards that must be met by water supplies nationwide. It also contains an underground injection control (UIC) program, which is designed to prevent ground water contamination by the underground disposal of wastes via wells. Injection wells must be granted permits in order to operate and must meet the permit specifications. A unique feature of the SDWA is the sole-source aquifer program, which enables EPA to protect those areas of the country that have only one aquifer as the principal source of drinking water.

The Resource Conservation and Recovery Act (RCRA) of 1976 established guidelines for the treatment, storage, and disposal of municipal solid wastes and hazardous wastes. All such facilities must be granted permits and comply with the requirements of the permit. Ironically, prior to the enactment of RCRA, pollution control programs, such as the Clean Air Act and the Clean Water Act, significantly increased waste management activities in and on the land. These unregulated land disposal practices created many situations where leachate could migrate into ground water, causing serious contamination. RCRA was enacted to pro-

Wendy Gordon received her M.A. in environmental health sciences from Harvard Graduate School of Public Health and is a scientist with the Natural Resources Defense Council. Her relevant issues at NRDC include handling of hazardous and toxic substances and the protection of drinking water sources. This material is taken from Groundwater protection—An overview of the laws, regulations, and institutions, in *Handbook on Groundwater Protection* (New York: NRDC, 1984), Chapter VII. Used by permission of the author and NRDC. Copyright © 1984 by National Resources Defense Council.

vide for careful planning and management of solid and hazardous wastes disposal practices.

The Comprehensive Environmental Response and Liability Act of 1980 (CERCLA, or Superfund) authorizes the federal government to clean up contamination caused by inactive waste disposal sites or spills. Many of the thousands of sites identified as requiring cleanup represent immediate threats to the quality of ground water. Cleanup activities are to be carried out under the terms of a National Contingency Plan and financed by a $1.6 billion fund, or Superfund, that is derived in part from a tax on the chemical and petroleum industries and in part from general revenues. Where parties responsible for a site can be identified, the burden of financing cleanup is placed on them.

The Clean Water Act of 1977 (CWA) requires that EPA establish a program to equip and maintain a water quality surveillance system for surface water and ground water, but little has been done to implement the ground water provision. In large measure this is because many states own the ground water and challenge the jurisdiction of the federal government to act to protect it. One of the few ground water-related federal programs, the area-wide water resources planning program funded by Section 208 of the Clean Water Act, has proved of limited value because implementation was not mandatory. Although the program is no longer funded, it embodied a potentially effective means of controlling ground water pollution. Plans developed under this program could still be used as part of new efforts to control activities in sensitive aquifer recharge areas.

The Toxic Substances Control Act (TSCA) of 1976 enables EPA to regulate toxic pollutants throughout their manufacture, use, and disposal cycles. While the act does not provide directly for ground water protection, it can indirectly do so by controlling specific toxic substances. In fact, TSCA gives EPA the authority to control and even prohibit manufacture, use, storage, distribution or disposal of a substance if it may present an unreasonable risk. The Federal Insecticide, Fungicide and Rodenticide Act (FIFRA), which was enacted to manage pesticide use and disposal, is much like TSCA in that it does not specifically address the impact of pesticides on ground water. However, it does give EPA broad powers to regulate pesticides and authorizes the agency to review the environmental effects associated with a pesticide as part of the registration process.

Under the Surface Mining Control and Reclamation Act of 1977 (SMCRA), the DOI is authorized to prevent the contamination of ground water that could result from strip mining. Permits must be obtained for mining operations, which must be in compliance with all permit requirements. Areas especially susceptible to drinking water contamination from a mining operation can be designated unsuitable for mining.

Finally, the National Environmental Policy Act (NEPA) may be used effectively to identify and help prevent ground water contamination. This act requires that all major federal actions must be evaluated and studied for their potential adverse effects on the environment, including ground water.

Table 30-1 outlines the opportunities provided by each of the eight federal statutes for citizen participation in its implementation.

Each state has one or more laws pertaining to ground water contamination, as well as basic common-law doctrines of negligence and nuisance. Most, however, lack comprehensive statutes that are designed specifically to protect ground water and are implemented in well-coordinated programs. Frequently, state laws bearing on ground water are administered by several different state agencies, creating a disorganized regulatory program. Moreover, these efforts typically receive less attention and fewer resources than do corresponding surface water programs.

The considerable variation in the natural quality of ground water and the regional characteristics of some of the sources of ground water contamination may account, in part, for the diversity in state regulatory mechanisms and organizational structures. The absence of a federal model also contributes to the variability. However, most state regulations that affect the quality of ground water fall into three broad categories:

- Regulation of contaminant sources.
- Classification of aquifers and establishment of ground water quality standards.
- Control of land use in areas overlying aquifer recharge zones.

Traditionally, local governments have had jurisdiction over land use. Significant reduction in the potential for ground water contamination could be achieved if local authorities applied their power over zoning to the development of protective land use plans.

Regulatory disorganization at the local level, however, has made ground water protection particularly difficult. As with the state government, the structure of local government splits responsibility among several departments (the engineering department, the health department, the water commission, etc.). The existence of so many unrelated and often competitive entities raises serious questions about the political feasibility of regulations and protection at the local level. Development

Table 30-1. How Citizens Can Use the Federal Laws to Protect Ground Water

Statutes	Citizens Are Invited to Participate in State Program Authorizations	Citizens Are Invited to Participate in the Permitting Process	Citizens Are Empowered to Bring Citizen Suits Against Violators of Any Enforceable Regulation or Permit Condition Established Under the Act	Administrator is Empowered with Emergency Powers to Halt Any Action That May Present an Imminent Hazard	Citizen May Petition for the Issuance, Amendment, or Repeal of Any Regulation Under the Act	Other Aspects of the Laws That Bear on Ground Water Protection
SDWA	Enforcement responsibility of public water systems. Underground Injection Control (UIC) Program	UIC permits	Yes	Yes	Yes	Public notification requirements Sole-source aquifer protection
RCRA	Hazardous Waste Management Program Solid Waste Management Program	Treatment, Storage and Disposal Facilities (TSDFs) permits	Yes	Yes		
CERCLA			Citizens may sue polluters for response expenses incurred in reducing risk (e.g., relocation, new water supply)	Yes	Yes	Releases of reportable quantities of hazardous waste
CWA	Discharge of Dredged or Fill Materials Program National Pollution Discharge Elimination System (NPDES)	NPDES permits	Yes	Yes	Yes	Areawide waste treatment management/planning. (Section 208 planning)
TSCA			Yes	Yes	Yes	Citizens may petition EPA to control uses of toxic products that result in contamination or to require testing substances found in ground water
FIFRA				Yes	Yes	
SMCRA	Regulation of surface mines	Surface mine permits	Yes	Yes	Yes	Mine inspections Designation of areas unsuitable for mining
NEPA			Yes		Yes	

pressures on local governments increase the difficulties. Moreover, the extent of an aquifer is not controlled by political boundaries. The recharge zone, which is highly vulnerable to contamination, may be beyond the jurisdiction of the town that draws its water from the aquifer. It may very well be in another state. This only augments the difficulties associated with program coordination and development.

Many people have begun to consider the ways in which strategies for ground water protection might be improved. Some find that existing federal and state legislation provide sufficient resources for the task but need clarification of federal and state responsibilities and a plan for ensuring that these responsibilities are met. Others—skeptical of what can be achieved without specific legislative direction—believe that the promulgation of new laws dealing specifically with ground water protection is imperative. In either case, what is needed is a coordinated control policy. Such a policy would apply an understanding of the regional nature of the resource, the multiplicity of interests and responsibilities of existing ground water management entities, and the need for regional management systems to deal with the problem of protecting and utilizing the ground water system.

31

STUDIES OF THE ECONOMIC VALUES OF WATER

OFFICE OF TECHNOLOGY ASSESSMENT

THE ECONOMIC VALUE OF WATER FOR IRRIGATION

The direct value of water in irrigation is measured in terms of the increment of profit to the producer *with* irrigation as compared with profits *without* irrigation. Several methods may be employed to make this calculation. One is an *ex ante* (before-the-fact) approach, which computes the change in net income from assumptions about crop prices, yields, production technology, and production costs. An alternative technique may be labeled *ex post* (after the fact), which relies on statistical analysis of actual production data. The *ex ante* method is often most convenient for planning in specific cases and is generally used by the Bureau of Reclamation and other government agencies that deal with water. Various statistical approaches serve to validate the analytic measures and are regarded by many analysts as more reliable owing to their base in "real, observable data." Any analytic measures—*ex ante* or *ex post*—can be abused by improper assumptions about prices, yields, and/or input requirements or some cost items that may be ignored. Experience has shown, however, that when properly performed, the methods yield similar results.

What is the value of irrigation water? The value of the marginal unit of water may reflect water scarcity as well as the cost of supplying the marginal unit. Local production conditions such as rainfall, temperature, length of growing season, and market situations will also have an impact, so considerable variation in water value across the West can be expected. Highly productive areas

such as the Imperial Valley or the San Joaquin Valley in California will have high values for water. Marginal production areas such as the high meadows of Wyoming will show low values.

Beattie and Frank (2) used 1974 census data as the basis for a statistical analysis of agricultural output. One of their purposes was to learn how agricultural output is influenced by resource inputs, including land, labor, machinery, chemicals, and irrigation water. The results yielded water values (expressed in current 1982 dollars) of $10 to $15 per acre-foot in the intermountain valleys of the Upper Colorado and Snake River basins; $20 to $25 per acre-foot in the desert Southwest and central California; and $40 to $45 per acre-foot in the Ogallala ground water region of the High Plains.

Howitt et al. (9) reported similar results using a much different technique. Their interregional supply-demand model for California yielded prices at the margin of $23 to $35 per acre-foot in the Central Valley and southern California and $7 in the Imperial Valley. Gollehon et al. (6) show estimated prices for irrigation water in 11 Rocky Mountain subregions. This study is somewhat atypical since it studies the value of water that might be lost to the region or transferred to other uses. When the water supply is reduced by 20 percent, two regions showed water valued in excess of $20 per acre-foot, four were between $10 and $20 per acre-foot, and six were below $10 per acre-foot.

The Department of Commerce recently sponsored a study of water value in the Ogallala region of the High Plains. The study showed a value of $60 to $80 per acre-foot for water used in irriga-

This material is excerpted from Institutions, in *Water-related technologies for substantial agriculture in U.S. arid and semiarid lands*, (1983), Appendix C, 388–392. This material was originally published as R. Young, Allocating the water resource: Market systems and the economic value of water (OTA commissioned paper, 1982).

tion. These values move upward with the passage of time, reflecting (assumed) increases in crop prices and yields through the year 2000.

The estimate of the value of water used to produce certain specialty crops (e.g., flowers, spices, berries) may be somewhat higher than the figures cited above. However, such cases will account for less than 10 percent of total irrigation water use in the foreseeable future. These crops are not, and probably will not be, of much significance for the formation of national water policy. This being the case, a rough estimate suggests that 90 percent of all irrigation demand is probably for water that costs no more than $40 per acre-foot.

THE VALUE OF WATER IN INDUSTRY

Energy production is the major consumer of water used for industrial purposes in the arid West. Most of this water is used for cooling thermal-electric power plants. Several processes can be used for cooling, depending on water scarcity and price.

Young and Gray (15) use an alternative cost approach to show that it is economical to convert existing plants from a pass-through cooling system to an evaporative cooling tower when water costs rise above $5 per acre-foot (1982 price levels). Methods designed to conserve cooling water are much more expensive. Gold et al. (5), in a study for the U.S. Environmental Protection Agency, report that the break-even points for combination wet-dry cooling systems run around $600 per acre-foot, while the shift to a completely dry, water-free cooling system would be economical only if water were extremely expensive—perhaps as much as $1400 per acre-foot. Abbey's (1) comprehensive analysis of water and energy problems in the Colorado River Basin provides similar estimates. Hence the large-scale stem plants proposed for several areas in the West could, if necessary, be willing to pay an amount many times the value of water in neighboring and competing agricultural uses.

Recent experience suggests, however, that even the large water requirements of huge power plants can be met with relatively little loss of water to agriculture in the surrounding area. Much of the 45,000 acre·ft required by the Intermountain Power Project (IPP) in Utah will be met by using conveyance losses or water used on saline soils that have little or no present agricultural value.

Leigh (10) has studied the value of water for coal slurry pipelines. His values are based on cost savings that accrue from not having to rely on rail transportation to move the coal (the alternative cost method of measurement). The value of water in a Colorado-to-Texas pipeline system is estimated to exceed $1600 per acre-foot. The estimate of value is, however, extremely sensitive to changes in the level of railroad freight rates. Reductions in freight rates could reduce the imputed value of water, although it is not likely to drive the value below willingness to pay for irrigation water. That is, agriculture cannot expect to compete with this use of water.

The need for water in recovery of hydrocarbons from oil shale has received considerable attention. Valuing water in this use could be accomplished by using the alternative cost method or by estimating the change in net income accruing to oil-producing firms. The alternative cost approach suggests that water could substitute for considerable capital and labor in the refining process and hence be very valuable. The change-in-net income approach requires that the production process be profitable before positive residual income can be imputed to water. Under current and anticipated petroleum prices, shale oil extraction is not economically feasible; therefore water has a zero or negative value in this use.

VALUE OF WATER IN HOUSEHOLDS

While willingness to pay for water delivered to households is readily observed and has been studied by many analysts, deriving a marginal value of water to households that is comparable and commensurate with estimates of raw water values in streams is, however, quite difficult. Household water that is treated (filtered and chlorinated), stored, and delivered to the user on demand is a much different economic commodity than the raw and untreated river water that is used in irrigation or industry. Hence a deduction for treatment, storage, and delivery costs must be made to make the prices and values comparable. An estimate may be derived by using a method suggested by Young and Gray (15) and based on data developed by Howe and Lineaweaver (8). This approach finds that lawn sprinkling is valued at about $150 per acre-foot and in-house uses at $250 per acre-foot (in 1982 dollars). A weighted average of water in the two uses would be about $220 per acre-foot. In another study Howitt et al. (9) do not distinguish between industrial and household demand. Their municipal and household sector estimates for 1980 (in 1982 prices) are about $160 to $200 per acre-foot.

An alternative estimate can be derived from market values of water in the Colorado–Big Thompson project (in northeastern Colorado) that can be transferred to urban uses. Gardner and

Miller (4) report that the price of water rights—that is, the price of exclusive rights to water—averaged $2450 per acre-foot in 1981. Converting this figure to an annual acre-foot value requires assumptions regarding the capitalization rate and expectations about future inflation. However, if the interest rate is about 8 to 9 percent (which seems plausible), and the planning horizon is long, the value of water is nearly equivalent to the $240 determined by Young and Gray (15) and Howitt et al. (9).

HYDROELECTRIC POWER GENERATION

Because evaluation of hydroelectric projects has usually proceeded on the assumption that falling water is a free good, recorded efforts to value water in this use are rare. In recent years competition for water—even falling water—has intensified, so evaluation methods have had to be developed. The procedure that has emerged centers on the cost of generating electricity by using some alternative method of generation (alternative cost method). When this method is used, the value of water is derived by deducting capital and operating costs of the generation and transmission system from the revenue earned by selling the power. The residual, if any, is attributed to the water resource (change-in-net-income method). Specific-value estimates vary, depending on the differences in head (the distance the water falls before turning the turbines), distances to load centers, costs of the steam-generating alternative, and the construction costs of the dam and storage facilities behind it. Even given these variables, values are also expressed for one site only or for several sites on a given reach of a river. Young and Gray (15) report single-site values ranging from $3.30 to $10 per acre-foot in 1982 prices in the western states. The higher values are associated with sites that have relatively high heads and can thus turn larger turbines. Most of these sites are found on the Colorado River. Whittlesey and Gibbs (14) report values for power generation in the Columbia Basin of over $30 per acre-foot (1982 prices) for water that goes through all of the dams below Franklin Roosevelt Reservoir, including Grand Coulee. This figure is higher than that reported by Young and Gray because of continued reuse at several generating stations and because of the higher costs associated with alternative energy sources in or near the Columbia River Basin. While single-site values for hydropower are not large relative to the values found in diversionary uses, diversions that are made high in a basin can lead to loss of large cumulative benefits stemming from reuse as the water passes through a number of facilities.

VALUING WATER IN WATER LOAD DILUTION

Water released for dilution of pollutants has value to the extent it reduces damage that the pollutants may impose on subsequent users. Precise estimates are difficult to derive since the detrimental effects depend on the particular pollutant, distance downstream, water temperature, rate of flow, and the quality of the waste-receiving water used for dilution. Most analysts have estimated values by assuming that the value of a unit of "clean" water is equivalent to the cost of treating effluent so that it does not reduce the quality of the water.

The results of these studies generally imply that dilution values are generally quite low. Merritt and Mar (12) showed dilution water in the Willamette Basin (Oregon) to have a value of about $1.30 per acre-foot (1982 price levels). Gray and Young (7) applied the aforementioned technique to several regions in the West. Their estimates of value in dilution ranged from $0.08 per acre-foot (Colorado Basin) to $3.25 per acre-foot in the Lower Missouri. Employing data from the Colorado River Board of California, Young and Gray (15), however, derived a value of water for dilution or reduction of salinity in the Colorado Basin at about $15 per acre-foot.

THE VALUE OF WATER IN WATER-BASED RECREATION

Water-based recreational services, by tradition and policy, are not often priced by market processes. Indeed, recreation and recreation services are so varied and so abstract that many people scoff at the notion that any reasonable value can be attributed to the resources used to produce them. The normal problems of valuing water are compounded since the value of water for recreation must be derived from a prior, synthetic, and sometimes arbitrary imputation of the value of the recreational services themselves. The problem is further complicated because the recreational uses of water are often complementary to other water uses rather than competitive with them. Water stored for irrigation, hydropower production, or flood control can be enjoyed by swimmers or fishing enthusiasts without diminishing its usefulness in its other uses. In such cases it is difficult to value the water and only slightly less difficult to ascertain the value of the recreational experience.

However, the growing demand for recreation is creating situations in which recreational uses are beginning to compete with other classes of instream or offstream use. At this time few analysts are working on measuring water values that are suitable for comparing allocations among alternative uses that include recreation.

Daubert, Young, and Gray (3) formulated a direct-interview procedure to elicit hypothetical bids from recreationists on the value of water in flowing streams. Applied to a sample of visitors to the Poudre Canyon in northeastern Colorado, this approach yielded estimates of economic value related to river flow used for fishing, white-water kayaking, and noncontact streamside recreation such as picnicking. The resulting marginal bid values for typical summer stream-flow were converted to dollars per acre-foot and were $9 per acre-foot for fishing, $5 per acre-foot for white-water sports, and $7 per acre-foot for the noncontact recreational experiences. Walsh et al. (13) performed similar analysis on western Colorado streams, reporting $13 per acre-foot for fishing, $4 per acre-foot for kayaking, and $2 per acre-foot for rafting when flows were maintained at 35 percent of maximum.

These findings lend support to the notion that nonconsumptive uses, even though they are non-marketed, have economic value to users. While many are skeptical of the validity of benefit estimates based on responses to questions regarding hypothetical situations, a preferable alternative technique to generate quantitative estimates of in-stream flow values has not been developed. While recognizing that estimates using this technique are subject to more than the usual error, they appear to be reasonable reflections of user preferences. Since these estimates are for values in a public, nonexclusive use, they must be used with great care, especially when incorporated into water management policy decisions.

FISH AND WILDLIFE HABITAT

Efforts to value habitat directly in economic terms are relatively recent. Many suffer from one or more of the potential difficulties noted earlier, particularly in valuing total product rather than incremental units of water. *

The problem remains of relating physical water requirements to habitat productivity, an issue that appears not to have been addressed in literature that is readily accessible. The estimates made by Lynn et al. (11) indicate a marshland value of less than $1 per acre. Water supplies per acre for the habitat of one crab species would not be highly valued in strict economic terms.

NAVIGATION

Provisions of facilities for inland waterways navigation has always been an important part of federal water policy. Estimates of the value of water for this purpose are almost nonexistent, since the usual approach to benefit-cost analysis of navigation projects implicitly assumes water to be a free good (as with hydropower). A sample approach (15) credited water with the savings from transporting commodities by water rather than by rail, pipeline, or truck. They reported positive values for water used for navigation only on the Mississippi, Ohio, and Tennessee river systems. Elsewhere, such as on the Missouri and the Columbia rivers, the total cost of building and maintaining a navigation system exceeded the savings: No benefit was creditable to navigation.

NOTES

1. David Abbey, 1979, Energy production and water resources in the Colorado River Basin, *Natural Resources Journal*, 19(2):275–314.

2. B. R. Beattie, 1980, Department of Agricultural Economics and Economics, Montana State University, Bozeman, Mont., personal communication to R. Young in note (ref. 15).

3. J. T. Daubert and R. A. Young, with S. L. Gray, 1979, *Economic benefits from instream flow in a Colorado Mountain stream*, Completion Report 91 (Fort Collins, Colo.: Water Research Center, Colorado State University, June).

4. Richard Gardner and T. R. Miller, 1982, *An explanation of price behavior in the water rights of northeastern Colorado*, Department of Economics, Colorado State University, Fort Collins, Colo., proceedings of a paper prepared for delivery to the Annual Conference of the American Agricultural Economics Association, Logan, Utah, August.

*See Lynn et al. (11) for an analysis of the conceptual issues and some empirical estimates relating to blue crab production on the Florida Gulf Coast.

5. G. Gold et al., 1977, *Water requirements for steam electric power generation and synthetic plants in the Western United States,* Environmental Protection Agency Report No. 600/7–77–037, February.

6. N. R. Gollehon, R. R. Lansford, et al., 1981, Impacts on irrigated agriculture from energy development in the Rocky Mountain region, *Southwestern Review of Management and Economics,* 1 (Spring): 61–88.

7. S. L. Gray and R. A. Young, 1974, The economic value of water for waste dilution: Regional forecasts to 1980, *Journal of the Water Pollution Control Federation,* 46(4)(July):1653–1662.

8. C. W. Howe and F. P. Lineaweaver, 1967, The impact of price on residential water demand, *Water Resources Research* 3(2).

9. R. E. Howitt, D. E. Mann, and H. J. Vaux, Jr., 1982, The economics of water allocation, 1982, in *Competition for California water: Alternative resolutions,* ed. Ernest A. Englebert (Berkeley, Calif.: University of California Press), 136–162.

10. Marie Leigh, 1982, *Competition for water: Energy v. agriculture,* paper prepared for ASCE Conference on Water and Energy: Technical and Policy Issues, Fort Collins, Colo., June.

11. G. D. Lynn et al., 1981, Economic valuation of marsh areas for marine production processes, *Journal of Environmental Economics and Management,* 8(2)(June):175–186.

12. L. B. Merritt and B. W. Mar, 1969, Marginal value of dilution waters, *Water Resources Research,* 5(6)(December).

13. R. C. Walsh et al. 1980, *An empirical application of a model for estimating the recreation value of instream flow,* Completion Report 101 (Fort Collins, Colo.: Water Resources Research Institute, Colorado State University, October.

14. Norman Whittlesey and Richard Gibbs, 1978, Energy and irrigation in Washington, *Western Journal of Agricultural Economics,* 3(1):1–11.

15. R. A. Young and S. L. Gray, 1972, *The economic value of water: Concepts and empirical estimates,* Technical Report, U. S. National Water Commission (Springfield, Va.: National Technical Information Service, PB 21–356).

32

WATER POLICIES AND INSTITUTIONS

KENNETH D. FREDERICK

The very nature of water resources and the problems inherent in valuing the benefits and costs of many possible uses create special problems for the management of these resources. While the hydrological cycle makes water renewable, it also makes it fugitive in time and space. As water flows from one property to another, supplies are accessible to many but belong to no one until they are withdrawn for use. Thus water supplies are common-property resources.

Water management problems are accented because a user seldom bears the full costs of use. For instance, virtually all the costs of using a stream for waste disposal are borne by those downstream rather than by the polluter. Even when a farmer spends sizable sums to pump ground water, these costs do not include the impacts on neighboring or potential future users associated with use of a common aquifer. When the capacity of streams to assimilate wastes, or the rates of natural recharge to ground water stocks are exceeded, the costs not borne by the users can be substantial. Consequently, when left to private decisions, water use tends to exceed socially desirable levels. On the other hand, it is extremely difficult to determine what the socially desirable uses of water should be. This requires some determination of the benefits and costs of alternative uses, as well as a knowledge of the distribution of these good and bad effects.

In part because of the limitations of the market mechanism for allocating common-property resources, water seldom is allocated through markets in the manner of most resources. Instead, government laws, institutions, and regulations control the distribution and use of water. There has been no effort to impose a single regulatory system over

the nation's waters; the federal role is limited largely to ensuring sufficient water for federal and Indian lands, federal water projects, environmental requirements, and the negotiation and enforcement of international and interstate water agreements. The states establish the principal controls over water use. A variety of state institutions, laws, and regulations have emerged, reflecting variations among states in the supply and demand for water and in their views of property rights and the role of government.

THE STATE ROLE

Surface Water

Although each state has adopted a somewhat unique approach, one of two basic doctrines—riparian rights or prior appropriation—underlies state surface water law. The doctrine of riparian rights holds that the owner of land adjacent to a water body has the right to make use of that water. Under this doctrine the right is inseparable from the land, and the owners can make any "reasonable" use of the water on the riparian land that does not unduly inconvenience other riparian owners. There is no priority in right among riparian users, and all such users share in reducing consumption in time of shortage. The riparian doctrine is used in the relatively water-abundant eastern states, and it provided the legal basis for the earliest water diversions in the western states.

Since western streams are not as numerous or reliably watered, and since development has been more dependent on large water diversions in the West than in the East, western development soon

Dr. Frederick is director of the Renewable Resources Division of Resources for the Future. See Chapter 11 for additional information about him. This material is excerpted from Water supplies, in *Current issues in natural resource policy*, ed. Paul R. Portney (Washington, D.C.: Resources for the Future, 1982), Chapter 7. Used by permission of the author and Resources for the Future. Copyright © 1982 by Resources for the Future, Inc.

required diverting water beyond the riparian lands and increasing the assurance of supply. These needs led to development and adoption of the doctrine of prior appropriation, which establishes the basic principle of "first in time, first in right." In contrast to the riparian doctrine, appropriative water rights are not tied to use on land bordering a stream or pond nor are shortages shared equally. Appropriation rights are acquired by diverting water from its natural channel and putting it to some "beneficial" use. The diverter can then apply for a permit authorizing similar annual diversions for as long as the use remains beneficial. The right has priority over (is senior to) all rights acquired afterward and is junior to all previously acquired rights. Thus in time of shortage the full burden of the shortfall is borne by the holders of the junior water rights. All 17 western states have adopted the appropriation doctrine as the basis of their water law, although several of these states also have retained elements of the riparian doctrine.

While state water laws and institutions have evolved in response to changing needs, further adjustments are needed if the nation's waters are to be efficiently utilized and not become a major obstacle to further regional development. Current and anticipated conditions in many states now differ from those prevailing when most existing water laws were written, the water rights acquired, and the water distribution organizations established. Increasing demands on eastern water supplies have started to focus on deficiencies of the riparian doctrine for allocating water when demand starts pressing upon supply. Even though the West started adjusting to similar pressures a century ago, the most pressing needs for reform of water law and institutions still are found in the West where continued development is becoming increasingly dependent on improving the efficiency with which its waters are utilized.

Several features of western water law limit the incentives to conserve water or transfer it to more highly valued uses. The beneficial-use provisions, although intended to prevent wasteful water use, actually have the opposite effect in some states. For example, farmers may be discouraged from temporarily transferring water rights to other parties for fear that the transfer will be construed as evidence the water no longer can be put to beneficial use in farming and thereby serve as the basis for loss of the water rights. Similarly, farmers may not initiate measures to conserve water for fear that the water savings might be declared in surplus of the amount that is beneficially used.

Irrigation water rights are appurtenant (or tied) to the land specified in the permit and cannot be changed without approval of the agency granting the permit. The original intent of these provisions was to help prevent fraudulent land and water sales common to early settlement schemes. However, to the extent that appurtenancy provisions impair water transfers to alternative lands and uses, they may deter efficient water uses. A few states continue to use a relatively strict application of their appurtenancy provisions, but most permit water transfers so long as the rights of third parties (i.e., those who are not a direct part of the transfer) are not impaired.

Protecting the rights of third parties is perhaps the most important obstacle to water transfers. In general, transfers of appropriative rights are prohibited if the rights of third parties are expected to be impaired. But in view of the common-property nature of the resource, transfers that have no impact on the quantity or quality of either the surface water or ground water flows of some nonparticipating party are rare. Efforts to resolve potential conflicts can be time-consuming and expensive, especially if litigation is involved, and these costs often exceed the benefits of the transfer.[1]

Ground Water

Initially, ground water pumping did not lead to obvious conflicts requiring government intervention. Consequently, the earliest legal doctrine governing ground water use was that of absolute ownership in which economic factors imposed the only constraint on a landowner's use of the underlying waters. Uncontrolled pumping, however, contributed to a variety of problems. Where ground water withdrawal exceeds recharge, higher pumping lifts coupled with lower well yields push up water costs; sea water may intrude into and contaminate an aquifer; or overlying and neighboring lands may subside. Since many aquifers are interconnected with surface flows, ground water pumping may jeopardize the rights of surface water users. The emergence of such problems, along with improved knowledge of ground water hydrology, has brought greater appreciation of the need for better ground water management.

With the exception of Texas, the western states have abandoned the doctrine of absolute ownership of ground water. Most western states now employ the doctrine of prior appropriation to ground as well as to surface waters. The basic principle is to grant ground water permits only where use does not adversely affect prior ground water appropriations and where the water will be beneficially used. In practice, however, there are wide variations among the states in the application of the doctrine. It is seldom applied to prevent ground water depletion, and the negative impacts ground water mining may impose on all users of an aquifer. However, Arizona's new Ground Water Management Act, which establishes active management areas and imposes strict limitations on ground water use in these areas, may be the start of a move toward more comprehensive management of the ground waters that are being depleted.

The costs of installing and operating a well and pump have been the principal regulators of ground water use. In recent years high energy costs have encouraged irrigators to adopt a variety of water-saving techniques. These costs, together with the greater control ground water users have over the timing and quantity of withdrawals, explain why ground water generally is used more efficiently than surface water in western irrigation. Nevertheless, in view of the common-property nature of most aquifers, private pumping costs generally are less than social costs. The difference between private and social ground water costs depends significantly on the characteristics of the aquifer. If lateral water movement is very slow or negligible, as with the Ogallala aquifer underlying the High Plains, the impacts on others are small. But where lateral flow is significant, the effects of an individual's pumping are spread over a wide area, and the distortions between private and social costs may be important. In these cases the "use it or lose it" view toward water resources prevails, and current use exceeds socially efficient levels. Efforts to curb such misuses of ground water have been limited to restrictions on pumping and drilling imposed by some states. While such restrictions tend to protect existing users from the additional damage that might be imposed by new users, they do not provide for an efficient long-term use of scarce ground water resources.

Further inefficiencies in ground water use result from restrictions some states have imposed on transfers of pumping permits from irrigation to nonagricultural uses. In particular, such restrictions have added to the problem some energy companies have had in securing needed water rights.[2]

FEDERAL WATER PROJECTS

Two federal agencies—The Corps of Engineers and the Bureau of Reclamation—have provided much of the planning, financing, and construction of the nation's major water development projects. Since political considerations often preempt economic criteria in the selection of projects undertaken by these agencies, enormous sums of public funds have been spent on projects of questionable merit. Further, some of these projects, especially irrigation projects, contribute to the inefficient use of the nation's water resources. Support for these statements is readily found by examining the activities of the Bureau of Reclamation.

The bureau was established in 1902 to encourage settlement of the arid West through irrigation. By 1980 federal projects supplied either full or supplementary irrigation water to over 11 million acres, a considerable testament to the success of the 1902 legislation. This achievement has not come cheaply, however. Although the initial legislation stipulated that farmers be charged enough to recover all construction costs except interest, the intent bears little relation to what has transpired.[3]

The costs and problems of establishing irrigation on previously unfarmed arid lands were much greater than anticipated, and many farmers were unable to meet repayment requirements. Relief came with the 1914 Reclamation Extension Act, which increased the repayment period from ten to twenty years and provided a 5-year grace period. New irrigators paid 5 percent of the costs up front and the balance in 15 annual installments beginning in the sixth year. But even these more lenient repayment schedules were not met by many farmers, and the 1914 legislation proved to be just the first of a series of legislative and administrative adjustments providing enormous subsidies for federal irrigation projects. In 1926 about $17.3 million, or 13 percent, of the costs on 21 projects were written off, and the repayment period was extended to 40 years on all projects. The policy of not charging interest remained intact.

Further relief came when the Reclamation Project Act of 1939 specified that irrigators were responsible only for that portion of the debt they were able to repay. While implementation of the ability-to-pay criterion has been very beneficial to the farmers, an even more important concession has been fixing the rates charged farmers so that there is no adjustment for inflation throughout the repayment period, which extends as long as 50 years, with a 10-year grace period. Thus almost regardless of the initial rates charged farmers, enormous subsidies are assured.

Estimates of the level of subsidy provided to those using federal irrigation water vary depending in part on the projects considered and the interest rate used to discount future payments by irrigators. In all cases, however, dispassionate analysis indicates the subsidies have been extraordinarily generous. Analysis of 18 irrigation districts by the U.S. Department of Interior shows an average subsidy of $792 per acre in 1978 dollars. The range in the subsidy varies widely among projects, varying from $58 per acre for Moon Lake, a small project receiving only supplemental water, to $1787 per acre for the Wellton-Mohawk district. The subsidy ranges from 57 percent of the total project costs for Moon Lake to 97 percent for the East Columbia Basin.[4] An alternative estimate of irrigation subsidies suggests that at current collection rates and costs, farmers will repay only 3.3 percent of the $3.62 billion the bureau has spent for irrigation construction.[5] In some cases the effects of inflation on long-term fixed charges have reduced the rates paid by irrigators below the point where they even cover the project's operating and maintenance costs.

A 1981 General Accounting Office (GAO) report to the Congress suggests that high subsidy levels will continue with new water projects.[6] The GAO report summarizes their assessment of six Bureau of Reclamation projects under construction. The total cost of these projects exceeds $2.1 billion, nearly half of which is attributed to irrigation facilities. With a 7.5 percent discount rate the fees established by the bureau imply irrigation subsidies ranging from 92.2 to 97.8 percent of the construction costs. The estimated costs of water delivered from these federal projects range from $54 an acre-foot for the Fryingpan–Arkansas project, which distributes supplementary water through existing facilities, to $130 an acre-foot for the Pollock–Herreid project. Yet the charges to recover the construction costs of this water range from $0.27 (only $0.07 according to GAO calculations) to $9.82 per acre-foot.

As noted earlier, the damage of these projects extends beyond the waste of federal funds at a time when the budget is very tight indeed. The farmers fortunate enough to be serviced with this highly subsidized water must use the water on their farms or lose it. Thus there are no incentives to conserve this water, and no opportunities to put it to nonagricultural uses. Consequently, federal irrigation projects have created, at enormous public cost, isolated areas within the water-scarce West where water is viewed and treated as virtually a free resource. Of course, this extravagance adds to the overall problems of water scarcity in the West.

One of the more controversial aspects of the federal water projects has been the Bureau of Reclamation's failure to enforce provisions of the 1902 Reclamation Act limiting an individual to 160 acres and a farm couple to 320 acres of land receiving federal water. Congress's initial intent to provide water solely for small family farms has been violated in practice through lax enforcement of the law and loose administrative practices, which permit unlimited leasing and multiple-ownership arrangements. While the great majority of farm operations comply with a 160- or 320-acre limitation, much of the irrigable land is operated in larger units. Of the 126,000 owners of 8.8 million irrigated acres supposedly subject to acreage limitations, nearly 91 percent own 160 acres or less and 98 percent own 320 acres or less. The remaining 2 percent, however, own 27 percent of the land. The 340 largest owners (comprising less than one-third of 1 percent of the farmers receiving water from federal projects) own 11 percent of the land receiving subsidized water.[7]

The Bureau of Reclamation's procedures allowing such large holdings to receive subsidized water were challenged in 1977 when National Land for People (a small public interest organization centered in Fresno, California, and consisting largely of small farmers and their sympathizers) brought suit against the federal government over violations of the acreage limitation provisions of the 1902 legislation. The court ruled in favor of the plaintiffs and issued an injunction against sales of excess lands until the Congress changed the law or the Department of Interior altered its rules and regulations. Interior proposed new rules in 1978, but actual changes in the rules have been stalled until at least mid 1982 pending final comments on a court-mandated environmental impact statement. In the meantime, Congress has considered several proposed changes in the 1902 law that would increase the acreage limitations.

Resolution of the acreage limitation issue is likely to eventually reduce the subsidized water provided the very large landowners. (Former Secretary of Interior Watt supported charging farmers the full cost for acreages in excess of those stipulated by Congress.) Nevertheless, the changes are not likely to have any significant impact on the farm output grown with this water or the efficiency of its use. Proponents of eliminating or relaxing the acreage provisions have claimed there are significant economies of scale to be gained by allowing large farmers to use the water. A recent U.S. Department of Agriculture study, however, suggests that 98 percent of the cost advantages achieved by larger operations are captured by farms of 320 to 640 acres.[8] All the legislation under consideration by the Congress as well as the Department of Interior's proposed changes in the rules and regulations would permit the use of federally subsidized water on operations of this size.

Although the acreage limitation issue is essentially a question of who receives the subsidized water, the debate over this issue has focused national attention on the appropriateness of federal financing of irrigation projects and the manner in which they are administered. As noted earlier, an important part of the subsidy results from setting water rates for long periods, with no provision for interim adjustment. In response to the pressures of public scrutiny, new contracts negotiated by the

Bureau of Reclamation call for adjusting water prices every five years to allow for rising operation and maintenance costs. These changes are not retroactive and thus only affect new contracts or old contracts as they expire. In California over 80 percent of the water delivered under current contracts will not be renegotiated until the 1992–1996 period. Even then, federally supplied water will remain grossly underpriced unless much more sweeping policy changes are made.

WATER FOR FEDERAL AND INDIAN LANDS

Water rights for federal lands were not specified when, in the nineteenth century, the states were granted jurisdiction over all nonnavigable waters in the public domain. Nor were the water rights of Indian lands specified when the reservations were established. These oversights became the origin of much anxiety since the states proceeded to grant water rights without any special regard to the rights of federal and Indian lands. The 1908 U.S. Supreme Court ruling in *Winters* v. *United States* provides a legal basis for Indian water rights. The Winters doctrine, which has been supported in subsequent judicial rulings, holds that when the federal government withdrew lands for any purpose, it at the same time implicitly withdrew unappropriated waters from the public domain to accomplish this purpose. Accordingly, Indian water rights have a senior claim to western waters dating from the time a reservation was created. Furthermore, several Supreme Court decisions since 1955 suggested that all federal lands have reserved water rights.[9] Since these rights would be senior to those of most other users, Indian and federal claims to western waters threatened the existing allocation systems established by the western states. Although the legitimacy of Indian and federal claims has been established in principle, these rights have not been quantified. Even though current use for these purposes is small, the Indian claims at least are potentially large.[10]

In recognition that uncertainties created by unresolved Indian and federal water claims are detrimental to western development, the Carter administration announced in 1977 its intent to resolve quickly the quantity of water claimed for these lands. The Reagan administration has taken the position that there are no federal "nonreserved" water rights and that "federal agencies must acquire water as would any other private claimant within the various states."[11] Nonetheless, the uncertainties at least concerning Indian claims are likely to remain for many years to come. Currently, federal policy is to allow concerned parties to negotiate a settlement on a case-by-case basis and let the courts resolve any differences. In the meantime, a cloud of uncertainty envelops western water use and adds to the risks borne by investors dependent on its use.

CONCLUSIONS AND RECOMMENDATIONS

The United States is not running out of water. Supplies appear adequate to meet both existing needs as well as possible additional ones arising from energy development, the expansion of irrigated agriculture, growing household and industrial demands, and even an increasing national appetite for outdoor recreation and amenity resources.

This optimistic forecast depends upon a number of important qualifications. The first is this: The growing and increasingly more competitive demands for water cannot be met successfully until and unless it is recognized that water is a scarce resource that will not be used efficiently under existing institutional arrangements. Water simply must be priced more rationally and exchanged more freely among alternative users. Unless users pay a price for water that is more nearly in accord with its opportunity cost, they will have little or no incentive to take the many available conservation measures that would free up existing water supplies for the new uses mentioned previously. And unless water is exchanged more freely among users in marketlike settings, it will not be put to its most socially valuable uses.

Although marketlike exchange has many desirable features, the common-property nature of both surface waters and ground waters limits the utility of an unfettered market system for allocating scarce water resources. Nevertheless, the federal and especially state governments can create and oversee the operation of pseudomarkets where normal markets would not arise naturally or where they would not result in wise resource use. One example of a marketlike mechanism is "water banking," a scheme to facilitate water transfers in areas where it is scarce.[12] The bank would be a state-sanctioned agency established in a particular water district to serve as a broker in arranging transfers between users and to determine the effects of such transfers on third parties. However, water banking would have to be classified as a beneficial use to avoid forfeiture under the appropriation doctrine. Under such a scheme water prices would be determined by supply and demand, with the buyer's and seller's prices differing by an assessment to cover legitimate third-party effects as well as transfer costs. As in cost markets, participation would be voluntary, but the opportunity to sell water without fear of losing the rights to it would provide a great incentive to conservation that is now absent.

As indicated earlier, ground water generally is used more efficiently than surface water. Nevertheless, social costs often exceed private costs because of the external effects on others. In theory, a tax on pumping could adjust for such differences. In practice, however, such a tax would encounter strong resistance, and it would be very difficult to determine the correct tax rate. Well or pumping quotas are a more acceptable means of limiting overpumping since they have the political advantage of protecting the early users. Many states permit the formation of special local districts to regulate ground water; such districts are common in areas with critical ground water problems— California, Texas, Nebraska, Kansas, and Colorado, for instance. Although the controls established by these districts are unlikely to result in a socially optimal ground water use over time, there is evidence that the regulated outcome better approximates the social optimum than does a policy of unrestricted pumping.[13]

In some situations major improvements in water management are possible through conjunctive management of ground water and surface water resources. This generally involves utilizing surface water supplies whenever they are available, either for direct delivery to users or for recharging ground water stocks. Ground water serves as a reserve and becomes the main source of supply when the less expensive but fugitive surface waters are insufficient to meet demand. Conjunctive management has been employed successfully by some local water agencies in California for some years, but it is the exception rather than the rule.

Our dictum about rational water pricing applies not only to natural surface water and ground water supplies but also to the water the federal government makes available through reclamation projects. Massive subsidies to water use (95 percent, or more in some cases, as pointed out earlier) are hard to justify even in a rapidly growing economy that is generating considerable tax revenues. They are nearly scandalous at a time when badly needed social programs are being cut in an all-out effort to reduce the size of government spending. Once again, users should pay the full social costs of the water they receive. This includes both principal *and* interest, where the latter should reflect the true cost of money to the government rather than some artificially low rate. The sooner the Corps of Engineers and the Bureau of Reclamation begin more sensible water pricing, the sooner will water conservation measures begin to be adopted.

There are other possibilities for augmenting or stretching supplies in water-scarce areas. Dams and reservoirs can increase water availability for specific purposes such as irrigation, and interbasin transfers can relocate water to more closely approximate demand. Structural measures, however, no longer offer a source of inexpensive water for the West. The best dam sites already have been exploited, ecological and aesthetic costs at certain sites are significant, and the costs of increasing usable supplies through impoundment are high. For example, the dams and reservoirs under consideration in California will cost $200 to $300 per acre-foot of water added to effective supply. Such costs are about an order of magnitude above the prices currently paid by California farmers receiving water through state projects and nearly two orders of magnitude above the highly subsidized rates charged those fortunate enough to receive water from federal projects. Interbasin transfers tend to be even more costly. For instance, it would cost $360 to $880 per acre-foot to bring water into the High Plains, where current use is depleting ground water stocks.[14] Furthermore, these costs assume water has no value in its current use and location. The increasing value of instream uses makes such an assumption increasingly difficult to justify, however.

Nontraditional methods of increasing water supplies include weather modification, icebergs, and desalinization. Weather modification is the only method with potential for providing relatively inexpensive water in selected areas. Studies suggest winter cloud seeding within the mountain ranges of the Upper Colorado Basin might add 1.4 to 2.3 million acre-feet of water a year at a cost of about $5 to $10 per acre-foot.[15] However, major institutional obstacles may prevent adoption of any sizable cloud-seeding program even if scientists and economists can agree it makes sense. Compensating the losers (whether real or imagined), dealing with the concerns of people downwind from the seeding, and allocating the additional water pose major challenges to use of weather modification for enhancing water supplies.

The enormous quantities of fresh water trapped as polar ice have attracted the interest of some arid areas. Precise cost estimates for the transportation and use of icebergs await resolution of some outstanding technical problems. Even if iceberg harvesting appears promising after examination of technical, economic, and environmental factors, international political and legal issues concerning resource rights could prevent its use.

Costs impose the principal constraint to desalinization as a means of increasing fresh water supplies. There are several processes for desalting, but the least expensive methods cost $250 to $300 per acre-foot even when the process starts with waters with salt levels well below those of sea water and ends with less than pure water.[16] These are plant-site costs, and delivery costs would be additional. Although such costs can be justified in isolated instances, they far exceed the marginal value of water even in the arid West.

Another important qualification to our optimistic conclusion concerns water quality. If discharges of conventional or toxic pollutants increase—because of either relaxations in current standards or an unwillingness or inability to enforce them—at least certain kinds of water availability problems could reoccur. The most likely problems would involve forgone instream uses, probably recreational and commercial fishing losses. However, a failure to control certain toxic substances could have even wider public health effects if sources of municipal drinking water are affected.

The Clean Water Act—which directs the Environmental Protection Agency (EPA) to establish discharge standards for tens of thousands of individual water polluters—is the most important water quality statute. Unfortunately, a number of analyses have suggested serious shortcomings with the act—shortcomings related to the controls imposed on both industrial and municipal polluters.[17] Moreover, these analyses have suggested that much more effort should be going into the control of non–point sources. Almost everyone who has reviewed the act agrees that much more water quality can be ''produced'' by the existing funds. The Clean Water Act is up for reauthorization, and this may provide the occasion to review these and other issues that affect water quality.

As suggested earlier, the quality of ground water supplies also has an important bearing on overall water availability. Moreover, there is evidence of serious local ground water contamination problems. Since monitoring has been very limited, there is cause for concern that these problems may be much more extensive and serious than currently realized.

One difficulty with management of ground water quality is that authority for it exists under at least three, and perhaps four, statutes administered by EPA. Under the Resource Conservation and Recovery Act (RCRA) of 1976, EPA is to promulgate either design or performance standards that must be met by hazardous and other waste disposal sites. In part, these regulations are intended to prevent the future contamination of underground aquifers by pollutants leaching from such sites. Under the so-called Superfund act passed by Congress in 1980, EPA is to begin identifying and cleaning up abandoned waste disposal sites that may be contributing to current ground water contamination and other problems. Finally, under the underground injection control provisions of the Safe Drinking Water Act of 1974, EPA is to regulate the practice by which the muds and brines that are by-products of energy exploration are disposed of in deep shafts drilled in the earth. These latter regulations are clearly related to drinking water protection but have the obvious effect of preserving aquifers used for other purposes as well.

Because of these conflicting and overlapping authorities, the Carter administration tried to develop an overall ground water strategy. Apparently wishing to begin afresh, the Reagan administration is making its own examination of such a coordinated approach. This is being made difficult, however, by the pressure EPA is facing from both Congress and the courts to issue regulations under RCRA, a law that is now more than five years old but that has yet to be translated into meaningful controls on waste disposal sites. Several of the key issues to be determined in any ground water protection strategy are, first, whether zero degradation of ground water will be the goal (as opposed, say, to establishing a limit on the amount of degradation permitted); and, second, whether aquifers that are not used as drinking water sources might be used as underground waste disposal sites. Clearly, the final decision will have considerable bearing on both the quality and quantity of ground water available, not only for current use but for future generations as well.

NOTES

1. Some illustrations of the kinds of inefficiencies that result from existing interpretation of water law in many states are presented in The John Muir Institute, 1980, *Institutional contraints on alternative water for energy: A guidebook for regional assessments,* prepared for the U.S. Department of Energy, DOE/EV/10180–01 (November), 53–63.

2. Ibid.

3. Exempting interest is a tremendous subsidy in itself. Imagine, for example, how small a homeowner's monthly mortgage payments would be if it were only necessary to repay the principal of a loan over its lifetime. Typically, total interest payments will be two or three times the amount of the principal.

4. U.S. Department of Interior, Water and Power Resources Service, 1980, *Acreage limitation,* Interim Report (March), 37–42.

5. E. Phillip LeVeen, 1978, Reclamation policy at crossroads, in *Public affairs report,* Bulletin of the Institute of Governmental Studies, University of California, Berkeley, vol. 19, no. 5 (October), 2–3.

6. U.S. General Accounting Office, 1981, *Federal charges for irrigation projects reviewed do not cover costs,* PAD–81–97 (March 3).

7. U.S. Department of Agriculture, 1980, *Farmline,* vol. 1, no. 6 (September), 4.

8. U.S. Department of Agriculture, 1980, *The U.S. Department of the Interior's proposed rules for enforcement of the Reclamation Act of 1902: An economic impact analysis,* ESCS–04 (Washington, D.C.).

9. Heidi Topp Brooks, 1979, Reserved water rights and our national lands, *Natural Resources Journal,* 19 (April): 433–435.

10. See Allen V. Kneese and F. Lee Brown, 1981, *The Southwest under stress: National resources issue in a regional setting* (Baltimore: Johns Hopkins University Press, for Resources for the Future), 70–94.

11. Secretary of the Interior James Watt told the western governors on September 11, 1981, that "federal land managers must follow state water laws and procedures except where Congress has specifically established a water right or where Congress has explicitly set aside a federal land area with a reserved water right." See *Federal lands,* September 21, 1981 (New York: McGraw-Hill), 9. In June 1982 U.S. Attorney General William French Smith confirmed that the Department of Justice supported the view that there are no federal nonreserved water rights.

12. A water-banking scheme is proposed and described in Sotirios Angelides and Eugene Bardach, 1978, *Water Banking: How to stop wasting agricultural water* (San Francisco: Institute for Contemporary Studies).

13. Jay E. Noel, D. Delworth Gardner, and Charles V. Moore, 1980, Optimal regional conjunctive water management, *American Journal of Agricultural Economics,* 62(3)(August):489–498.

14. Based on the Corps of Engineers estimates for the Six-state High Plains–Ogallala aquifer regional resources study, Congressional briefing, February 25, 1981, p. 25.

15. Personal communication with Bernie Silverman, Chief Atmospheric Water Resources Management, Water and Power Resources Service, Denver, February 1980.

16. U.S. Bureau of Reclamation, California, Department of Water Resources, and California State Water Resources Control Board, 1979, *Agricultural drainage and salt management in the San Joaquin Valley* (June 1979), 8–4 and 8–5.

17. For instance, see A. M. Freeman, 1978, Air and water pollution policy, in *Current issues in U.S. environmental policy,* ed. Paul Portney (Baltimore: Johns Hopkins University Press, for Resources for the Future); Charles Schultze and Allen Kneese, 1975, *Pollution, prices and public policy* (Washington, D.C.: Brookings Institution); and David Harrison and Robert Leone, 1981, Federal water pollution control policy, draft manuscript prepared for the American Enterprise Institute, Washington, D.C.

33

INSTREAM WATER USE: PUBLIC AND PRIVATE ALTERNATIVES

JAMES L. HUFFMAN

For the early European settlers of the American West, the fact that water occurred for much of the year only in the often widely dispersed streams and rivers was a serious problem. They needed water to pursue the mining, and later farming, that was so critical to the success of their settlements. They constructed viaducts and ditches to get the water out of the streams, and they devised a system of water rights to make their efforts legal.[1] Although the nineteenth-century settlers were very successful in their effort to put the limited water resources of the West to productive use, they would never have imagined the extent to which those who followed them in the twentieth century would devise methods for getting the water out of the stream. Transmountain diversions, using pipes of such enormous size that the early settlers could have driven their wagons through rather than over the mountains, are only the extreme example of an impressive array of modern technological solutions to the problem of getting the water out of the stream. The law has generally followed apace with the technology, with only the inevitable jealousies of states to establish legal barriers—barriers not to getting the water out of the stream but to taking the water out of the state.

In the midst of this often massive effort to get the water out of the stream, there always have been a few who have sought to keep water in the stream. The federal government since its formation has been concerned with maintaining the navigability of the nation's waterways, but hydrology, topography, and the technology of modernizing transport have fortuitously kept the navigators and the out-of-stream water users from significant conflict. The navigability of the nation's streams and rivers was threatened not by those who would remove the water from the streams but by those who would place obstructions in the streams. As a result, there was little reason for anyone to be concerned about a legal system designed to facilitate the task of getting the water out of the stream. By the 1970s, however, the out-of-stream water users were faced by competition from those who wanted to keep water in the streams. The language and soon a technology of minimum flows and instream flows rapidly became as commonplace as those of diversion, consumption, and return flow in the nation's legislatures, agencies, and tribunals. The demand for the maintenance of instream flows became an important part of a revolution in American water law. The harsh realities of life on a finite planet ensured that efforts to get even more water out of the streams would coincide with a new demand to keep at least some of the water in the stream.

In a country that has relied increasingly upon government to resolve the problems of resource allocation and wealth distribution, it is not surprising that both those seeking to divert more water and those seeking to curtail diversions would look to the government for assistance. Both sides have experienced successes, in the sense that both the state and federal governments have constructed or subsidized often massive water development projects, setting aside specified flows of water to be left in the streams and rivers. In the midst of this

James Huffman is Director of the Natural Resources Law Institute and Professor of Law at Lewis and Clark Law School, Portland, Oregon. He has taught water law for several years and has published a comparative study of instream water allocation in Colorado, Idaho, Montana, and Washington.

Excerpted from a chapter of the same title by James L. Huffman in WATER RIGHTS: SCARCE RESOURCE ALLOCATION, BUREAUCRACY, AND THE ENVIRONMENT, Terry L. Anderson, editor, Pacific Institute for Public Policy Research, San Francisco, California. Used with permission. Copyright 1983 by Pacific Institute for Public Policy Research.

rush to compete in the politics of water management, an occasional voice has suggested that there might be a better way. Frank Trelease, one of the few lawyers to point out that the "goddamned bureaucrats" may not have all of the answers, has argued that the system of private rights in water should not be so willingly abandoned in favor of various schemes of public water management.[2] Writing in the 1950s and 1960s, he focused on the allocation of water to agriculture and industry, demonstrating that Americans will derive more benefit from the available water supplies if the allocation of water to consumptive uses is left to the private market. His persuasive arguments for the restoration of private rights in water have brought many to his side. Although no doubt there is much to be gained by a privatization of rights to use water for agricultural and industrial purposes, the efficiencies of the marketplace would still be limited by the fact that most state water laws give the states the primary role in the allocation of water to instream uses. There seems to be a general assumption that water will not be allocated to most instream uses unless the state undertakes to make the allocation. In the water resource field the allocation of water to minimum stream flow maintenance has become the archetypal case for government intervention, as Ralph Johnson observes: "In recent years it has become increasingly clear that the appropriation system, if allowed to continue unrestrained, will adversely affect and in some cases destroy valuable in-place commercial and recreational water uses."[3] This chapter will examine the validity of this presumption in favor of state intervention based upon in-depth studies of four state approaches to the allocation of water to instream flows.[4] The first section briefly examines the history and nature of people's interests in minimum stream flow maintenance. The following section examines the constraints that the private rights system has placed upon the private and public provision of instream uses and describes the basic approaches that states have used to overcome these constraints. Case studies in Idaho, Washington, and Montana then are discussed, followed by a description of Colorado's sharply contrasting approach to instream flow allocation. Finally, the prospects for the allocation of water to instream flows through a purely private system of water rights are considered.

INSTREAM USES OF WATER

Before the nineteenth-century development of mining and agriculture in the American West, most uses of water were instream uses. In England and colonial America there was normally ample precipitation for the growing of crops, and industrial uses were limited to the use of falling water to power various types of mills. Although it was often necessary to divert water into millponds in order to achieve an adequate head, the diversions were ordinarily small in relation to the size of the streams, and almost all of the water was returned to the stream within a short distance of the point of diversion. Occasionally, large mills were located in small streams, with a resultant disruption in the stream flow, but as a general rule, industrial uses of water had very little impact on the natural flow of the stream.[5] Indeed, as Chancellor Kent observed, the English common law required that mill operators not alter the flow of the stream with respect to either the quantity or the quality of the water. "Every proprietor of lands on the banks of a river, has naturally an equal right to the use of the water which flows in the steam adjacent to his lands, as it was wont to run (currere solebat) without diminution or alteration."[6] The English rule that every riparian landowner has a right to the natural flow of the stream, undiminished in quantity and unaltered in quality, which proponents of instream flow protection could not improve upon, was modified very early in American jurisdictions by the qualification that the rights of riparians extend to the reasonable use of the water.[7] This modification of the English rule opened the door to consumptive uses by riparians even though they altered the flow of the stream.

Consistent with the general development of the law in the western states, the riparian doctrine of water law was imported along with the rest of the common law. In some areas, particularly western Washington and Oregon, the riparian system was suited to the climatic conditions. But most of the West was not blessed with abundant supplies of water, and settlers in these arid regions were quick to abandon the riparian doctrine in favor of the appropriation system of water rights, which recognizes water rights in those individuals who have put the water to beneficial use.[8] The appropriation system allows individuals to acquire water rights without also having title to riparian lands and permits the use of water on any lands without regard to their location in relation to the stream from which the water is diverted. The limits on water use under the appropriation system are largely technological. Subject to the requirement that the water be put to beneficial use, water rights owners can use the water wherever they are able to transport it. The costs of water development virtually assure that any use to which water is put will be beneficial.

Unlike the English riparian rule, which assured

the continued flow of water courses, the appropriation system made it very difficult for any individual to assert a right to the flow of the stream except to deliver the water to a particular point of diversion. This situation arose in response to the need for appropriative rights holders to be able to demonstrate the existence of their rights. Under the appropriative system water rights have priority in order of time of first use, which means that one must be able to establish the date on which the water was first applied to beneficial use. In a dispute between water rights claimants, each claimant must offer evidence of the date of first use. Given that most states did not have a system of recordation until fairly recently, the best evidence of first use is the date on which the water was physically diverted from the stream. As a result, diversion has become an essential element of any claim of an appropriative water right.[9] The water had to be taken out of the stream, but it was not a problem until the combination of changing values and diminishing water supplies brought the issue of instream flow maintenance to the public attention.

Historically, the most important reason for the protection of stream flows was navigation. In 1824 the U.S. Supreme Court determined that navigation is a form of interstate commerce, which is subject to federal control under the Constitution.[10] But the obstacles to navigation normally were physical obstructions, not inadequate flows.[11] On those streams where flow maintenance was likely to be a problem, navigation was generally of little importance. Thus although the issue of navigability has been important to recent legal disputes relating to stream flow maintenance, it is generally in the context of jurisdictional disputes between states and the federal government, not because commercial navigation is threatened by inadequate stream flows. An exception is the recent growth of commercial services for recreational river travel, but this seldom raises constitutionally relevant issues of navigability.

The instream uses of water that have contributed to the recent pressures for instream flow protection are of three broad categories: wildlife protection, recreation, and pollution control. Perhaps the earliest legal action to ensure stream flows for wildlife, specifically fish, was in the context of federal water development projects that affected commercial fisheries. The hydroelectric dams of the Columbia River, in particular, had an enormous impact on the anadromous fishery that was important to the economy of the Northwest.[12] The solution, the effects of which are still being debated, was to provide "ladders" for the fish to use in their upstream migration. Of course, these concerns for

fish habitat protection did not stem from a scarcity of water and thus were more analogous to the earlier concerns over instream obstructions to navigation. Not until the 1950s were any significant legal actions taken to protect fish habitat from the reduction of stream flow to a level at which the fish could not survive.[13] More recently, the concern over the impact of stream flow levels on wildlife habitat has extended beyond fish to other forms of plant and animal life dependent upon the aquatic ecology.

Although wildlife habitat protection has been the primary source of concern for stream flow maintenance, recreation has been implicit in the arguments of those concerned for the fish. The biggest and most influential lobby for legal changes to protect instream flows has been promoted by organizations that represent recreational fishing. Their efforts generally have been supported by state fish and wildlife agencies, which normally depend heavily on sport fishing for political and financial support. Participants in other forms of outdoor recreation also have argued for instream flow protections, although at least as many forms of outdoor recreation are dependent upon the manipulation of natural stream flows as upon their maintenance.

In recent years pollution control has emerged as yet another use for instream flows, a use that may prove to be at least as important as wildlife and recreation. Consistent with the purist orientation of the first years of the environmental movement, reducing the concentration of water pollutants through dilution was not considered a viable alternative in the early efforts to control water pollution. However, with the growing realization that zero pollution is seldom if ever a viable or defensible option,[14] agencies charged with the control of water pollution levels have increasingly looked to the effect of stream flow levels on the concentration of pollutants. Although dilution may not be the long-run solution to pollution, in the short run there can be little doubt that fresh water streams will be used to convey various types of industrial and agricultural waste, the concentration of which will be affected by stream flow levels.

APPROACHES TO INSTREAM FLOW MAINTENANCE

Wildlife habitat protection, outdoor recreation, and pollution control commonly have been viewed as public goods that will only be provided through some type of public effort. In the context of stream flow maintenance the presumption for state or federal action is fortified by the long-standing view that water is owned by the public in general and committed to private use only in the form of usufructuary rights.[15] Lawmakers and legal interpreters have been at pains to point out that under both riparian and appropriation doctrines of water law, the right holder possesses the right to use the water but does not own the water itself. The water belongs to the state, which has chosen to allocate it through a system of private rights in its use. In this context water uses such as wildlife habitat protection, recreation, and pollution control represent the almost ironclad case for public action. To the extent that there is a justification for the provision of water to these uses, it has been generally assumed that it will only be achieved through state or federal action.

Because other water users had been successful for many years in inducing the state and federal governments to subsidize consumptive water uses, those advocating government action to keep water in the streams were faced by opponents experienced in the politics of public water policy formation. The existing law of private water rights posed some unexpected obstacles to the implementation of instream flow legislation. The most common approach of the new legislation was to authorize the state, through one of its agencies, to appropriate water to maintain minimum stream flows. Some states proceeded to appropriate water for instream flows only to be told by the courts that the absence of a central feature of the appropriative doctrine—diversion—prevented such appropriations.[16] Furthermore, it was not always clear that the instream water uses were beneficial under existing law, nor was it easy to determine when the state had ceased using the water for instream flows and thus abandoned its right. To the extent that these constraints were statutory, it was an easy matter for the state legislatures to change the law, assuming a favorable political climate. But in some states the elements of an appropriative water right were claimed to be a part of the constitutional law, which made change more difficult.

That these factors impeded government appropriations of water for instream flows to a large extent was simply a product of inadequate legislation and the lawyer's eye for legal loopholes. If the states really wanted to get involved in protecting stream flows, there was no doubt that the law would eventually be adapted to the achievement of that end. The more significant aspect of these constraints was that they were, and still are in most states, part of the system of private rights in water. Until it was proposed that the state appropriate water for stream flow protection, the appropriation system was widely believed to exist exclusively for the acquisition of private rights. Because it is generally assumed that instream flows will only be

protected through state action, those aspects of water law that initially prohibited the states from appropriating water for instream flows still operate to prohibit individuals from making such appropriations or from acquiring existing water rights for instream flow purposes.

Regulation

Perhaps the most obvious form for government action to protect instream flows is regulation. The states' expansive police powers presumably can be used for regulating the use of water as easily as for regulating the use of land. For complex historical reasons political constraints on state regulation of water use in the West are more severe than on the regulation of land use. There is no reason to assume, however, that the legal powers of the states are significantly different in the case of water than they are for land. Pursuant to the states' police powers, instream flows might be preserved by actions analogous to zoning regulations, which would limit private water rights holders to actions that would not be detrimental to the public interest in maintaining stream flows to reduce pollutant concentrations. Even without the justification of protecting the citizens' health and welfare, however, there is ample precedent for state actions designed to preserve aesthetic and other noneconomic values.[17] Barring significant change in the judicial interpretation of constitutional limits on the states' ability to promulgate regulations that limit private property rights, it would seem that the only serious constraint on state regulation to require instream flows is the political climate in which water rights owners have substantial influence.

Conditional Rights

A second alternative for state action to preserve instream flows is the imposition of conditions on newly acquired or transferred water rights. Like regulation, the state's power to impose conditions arguably derives from the police power; but unlike regulation, many states already possess statutory authorization that can be interpreted to give the state the power to issue conditional rights. Most state laws provide that the state can apply a public interest standard in the recognition of new appropriative rights or in the transfer of existing rights. Pursuant to this power, the relevant state agency could refuse to approve a new right or a transfer of right because it is contrary to the public interest in instream flows or could recognize the right on the condition that it be revoked if stream flows are negatively affected. As in the case of regulation, the political climate no doubt operates as a constraint on such state action, but under existing law it appears that the state has the power to recognize rights made conditional on minimum stream flow maintenance. The principal disadvantage of the approach from the point of view of the state is that it can only be applied to new appropriations or to water rights transfers, whereas regulation presumably could apply to all existing rights.

Reservation

A third approach the state might employ is to reserve unappropriated water from future appropriation. This approach resembles the reservation process employed by the federal government to exclude portions of the public domain from availability for private acquisition under the homestead and other land disposition laws. Because the state is the owner of the water and has chosen to make that water available for private rights in its use, there is no legal reason why the states cannot decide to reserve all or some of the unappropriated water from future appropriation. Although it has been argued in some states that there is a private constitutional right to appropriate unappropriated water, there is no indication that the claim will be upheld.[18]

State Acquisition of Water Rights

As was indicated previously, the most common approach to the state protection of instream flows has been state acquisition of water rights. This can be accomplished in three basic ways: appropriation, purchase, and eminent domain. Legislative authorization for a state agency to appropriate unappropriated waters for minimum flows has the political advantage of avoiding conflicts between the state and holders of existing water rights. However, the approach severely limits the opportunity for the state to protect stream flows because it can act only on those streams that are not fully appropriated and normally will result in the acquisition of very junior rights, which may be ineffective in maintaining adequate minimum flows. Authorization for the state to purchase existing rights will permit the state to acquire more senior rights and thus more adequately protect stream flows. The same end can be accomplished through eminent domain if the method is a legal form of state action to protect stream flows.

Public Trust

A final approach to the states' protection of instream flows is reliance on the public trust doctrine. Pursuant to this approach, the state might claim or be assigned a responsibility to protect instream flows pursuant to its role as trustee of a public right in the maintenance of stream flows. The public trust theory, which has historical roots in common law, has been variously revived in recent years for the purpose of asserting rights or duties of state action in the public interest.[19] The basic concept holds that the public, or all individuals in common, have a right to certain natural conditions—in this case, minimum stream flows—that supersedes any private rights in the use of the natural resource, and that the state has a responsibility to assure that those private rights do not infringe upon the public right.[20]

CASE STUDIES OF STATE INSTREAM FLOW PROGRAMS

Idaho—Setting Flows by Statute

In 1977 a group of Idaho citizens sought by initiative petition to get the Hydro-Power Protection and Water Conservation Act on the November 1978 ballot. The proposed law would have set minimum stream flows on all Idaho rivers and streams with unappropriated waters while recognizing the validity of all existing water rights. Prior to that time the only Idaho law relating to minimum stream flows was an authorization for state officials to appropriate water for the maintenance of water levels in designated lakes and flows in designated springs.[21] The constitutionality of this law had been upheld in 1974 in the face of claims that the state constitution prohibited the state from appropriating water, excluded minimum flows from the definition of beneficial use, and required an actual diversion of water for an appropriation to be valid.[22] The 1977 initiative drive was abandoned when the 1978 Idaho legislature adopted a statute that established three specific base flows on the Snake River and authorized the Idaho Water Resources Board to appropriate waters for the purpose of maintaining instream flows, subject to the approval of the state legislature.[23]

Under the 1978 statute the Water Resources Board may apply to the director of the Department of Water Resources for a permit to appropriate water for instream flows on its own initiative or in response to a request for such action by any private party. For an application to be approved by the director, it must (1) not interfere with any vested water rights; (2) be in the public interest; (3) be necessary for the preservation of fish and wildlife habitat, aquatic life, recreation, aesthetic beauty, navigation, transportation, or water quality; (4) be a minimum rather than an ideal or desirable flow;

(5) be capable of being maintained as indicated by past flow records. In addition, the application, like all other applications to appropriate water, must not conflict with the local public interest.

Although a decision on the application is dependent upon technical information relative to stream flows, existing rights, and minimum flows necessary for various purposes, the director has considerable discretion in the application of the two public interest standards. It was presumably in recognition of that discretion that the legislature required that all approved applications for permits to appropriate water for instream flows be submitted to the legislature within five days of the beginning of the legislative session. The legislature then may either approve or disapprove the director's determinations; a failure of the legislature to take any action is understood to be an approval of the application.

Because the system is relatively new and because the Water Resources Board has proceeded cautiously, it is too early to determine the impacts of the law. Some minimum flows have been established in addition to the base flows set by the legislature in the original legislation. However, these have generally been in noncontroversial areas where competing demands for water are minimal. Because the system is so closely linked to the political process by the requirement for legislative approval, it is extremely unlikely that the Water Resources Board will take any actions that are contrary to the interests of any politically influential group in the state. Agricultural and mineral interests, both heavy users of water in Idaho, are very influential in the state legislature. Hence it is unlikely that the Idaho approach to instream flow preservation will result in the protection of flows in the more populous and developed areas of the state. If such streams are protected, it will be a result of the ability of those valuing instream water uses to garner the necessary political power in the state legislature.

Washington—Setting Flows by Bureaucratic Expertise

Washington had one of the earliest state laws designed to give the state a role in maintaining instream flows. The power to deny or make conditional appropriation permit applications was granted explicitly in a 1949 amendment to the Washington appropriative permit law.[24] The director of the Department of Ecology (DOE) is required to notify the Departments of Fisheries and Game of pending applications for appropriation permits. From the recommendations of those two departments and the DOE's assessment, the director can either deny permits or make their granting conditional on the maintenance of minimum stream flows. Pursuant to this power, about two hundred fifty Washington streams have been effectively closed to any future consumptive appropriations.

In 1969 the Washington legislature authorized the DOE on its own initiative or at the request of the Departments of Fisheries or Game to establish minimum stream flows for wildlife habitat maintenance, recreation, aquatic life protection, other environmental values, and water quality control.[25] As of 1980, the DOE had received 26 requests for the setting of minimum flows, one of which had been established.[26] The lack of action under the 1969 law was largely due to the 1971 adoption of another statute that authorized the DOE to set base flows defined as the flow sustained in a stream during extended periods without precipitation. Pursuant to that statute, the department launched two major programs for setting base flows, one for western Washington and one for the Columbia River and its tributaries. The principal difference between the two statutes is that the DOE is required to establish base flows for all perennial rivers and streams, while the setting of minimum flows is entirely discretionary. Thus the task of setting base flows, given the department's interpretation of the law, is highly technical.

The DOE has established a five-step process for setting base flows.[27] The first step is a stream system analysis, which involves the collection and analysis of historical stream flow data. Then the various streams are rated on the basis of their value for the various instream uses outlined in the statute. Hydrographs are developed to determine the percentage of time flow duration and the discharge duration for each of the streams in question. The stream-rating system then is used to determine what the base flow for each stream should be. Except for the task of rating the streams, the process is strictly technical. Although no doubt there are professional disputes about appropriate methodology, the task is simply to chart the volume and nature of the flow on an annual, seasonal, and daily basis. Once this is accomplished, a formula is applied to each stream depending upon its rating, yielding a base flow.

The Washington system is heavily dependent upon the work of technical experts in the fields of hydrology and fish and wildlife biology. Although public participation is encouraged at various stages in the process, the technical nature of the DOE approach makes it difficult for people other than technical experts to become involved. Because the stream-rating system, the heart of the policy, comes early in the process, it is easily lost in the debate over technical questions. The stream ratings result from consultation among various state officials, who are subject to political pressures, in contrast to the Idaho approach, in which the state legislature has the final say. The Washington system allows the DOE to waive base flow requirements if hydropower water use is threatened in low-flow years and requires a reassessment every five years. It would appear, however, that the reassessment is likely to take the form of refinement of technical data rather than reevaluation of the water allocations resulting from the base flow program.

Montana—Setting Flows by Reservation

Montana has assumed by far the most aggressive state role in the setting of minimum stream flows. Although a reservation statute exists in the state of Washington, it has been of no importance to the maintenance of instream flows.[28] However, Montana's reservation law has made the state the dominant allocator of the state's water resources. Prior to 1973 when the reservation law was adopted, both the Department of Fish and Game and the predecessors of the Department of Natural Resources and Conservation (DNRC) had authority to appropriate water for public purposes.[29] Although it is not clear whether DNRC can appropriate for minimum stream flows, the Fish and Game statute was enacted specifically for that purpose, and the department made 12 appropriations under the law prior to its repeal in 1973.[30] Both departments also have the power to acquire water rights by purchase, condemnation, exchange, and lease, although neither has been in a financial or political position to use that authority for the acquisition of instream flow rights.[31]

The 1973 Water Use Act was adopted pursuant to the legislature's 1972 constitutional mandate to "provide for the administration, control, and regulation of water rights."[32] The act authorizes the United States, the state, and its political subdivisions to apply for water reservations for existing or future beneficial uses or to maintain minimum flows, levels, or quality of water. Reservation applications, like ordinary applications to appropriate water, are to be made to the DNRC with approval to come from the Board of Natural Resources and Conservation, consisting of seven members appointed by the governor for staggered terms. In the context of western water law, the Montana reservation law is a striking change of direction in that it permits the acquisition by government agencies of prospective water rights. There was some precedent for government appropriation of water rights, but the concept of rights in future use is contrary to the history of a water law system that granted rights on the basis of use and took them away on the basis of nonuse.[33]

Although the statute does not require that the reservations on any particular stream be considered in a single proceeding, the circumstance of pending applications for significant appropriative rights on the Yellowstone River led the DNRC to organize a single massive proceeding to consider all reservation applications on that river and its tributaries.[34] The result was a protracted process of application, hearing, and debate leading to allocation of all of the unappropriated water in the Yellowstone River. Because private water users could not apply for reservations, the board sought to ensure that the reservations that it granted did not tie up all of the water and thus prohibit any future private development of water. However, the variable nature of the stream flow and the inadequacy of much of the data available to the board raises some doubt about the prospects for future private water development.

Proponents of minimum stream flows fared extremely well in the Yellowstone reservation proceeding. The Department of Fish and Game applied for a reservation of 8.2 million acre·ft/year for fish habitat maintenance, an amount that in some years would exceed the total annual flow of the river. The Department of Health and Environmental Sciences applied for an instream reservation of 6.4 million acre·ft. The board eventually granted a reservation of 5.4 million acre·ft to Fish and Game near the point where the river leaves the state and numerous other flows at various points on the river and its tributaries.[35] The Department of Health was granted the same reservation at the downstream boundary of the state. Of course, the board granted numerous other reservations for consumptive water uses by municipalities and agriculturalists, but by far the lion's share of the water was reserved for instream flows.

It is difficult to determine which factors played an important role in the board's decision. There is little doubt that the board was heavily dependent upon the information supplied by the various state departments in their applications, particularly the DNRC's environmental impact assessment and recommendations. The enormous volume of data that the board had to assess was technically complex. Although the members of the board are political appointees, they have been relatively free of political pressure, at least in the Yellowstone proceeding, and thus have been able to exercise their personal judgment about what decision is in the best interests of the people of the state. Under the provisions of the reservation statute, the board is required to review its reservation decisions at least once every ten years, but there is not likely to be significant change with respect to the instream reservations since the basis for reassessment is whether the amount of water reserved is necessary for the intended use. In the case of instream water use that issue will be very difficult to evaluate, except on the basis of information supplied by the departments of state government that hold the reservation right.

Implications of the Three State Programs

The instream flow programs of Idaho, Washington, and Montana involve two general approaches. The more dominant approach in all three states is to change the fundamental basis for the initial assignment of rights in water to permit the state to exclude certain waters from availability for private appropriation. Although the Idaho law speaks in terms of the state's acquisition of appropriative rights, the fact that the legislature must agree to every state appropriation for minimum flows differentiates the state's appropriative rights from private water rights. For example, it is most unlikely that these minimum flow rights can be abandoned or transferred to private parties, at least not without specific authorization from the legislature. The terminology of appropriative water rights is applied to the state's regulatory actions as a convenience to facilitate their integration with the existing law, rather than to establish the state on a par with private water rights holders. The Montana legislature repealed a statute that authorized the state to acquire appropriative rights, and Washington has never employed the terminology of appropriation to describe the state's role in allocating water to instream flows. Thus although only the Montana approach is self-described as a reservation system, the practical effect of both the Idaho and Washington approaches is to reserve water from availability for private acquisition.

The second general approach in the three states has been the direct regulation of existing private rights. In fact, this approach exists far more in theory than it does in practice. In all three states the relevant administrative agency may refuse to approve water rights transfers that are found to be contrary to the public interest, which presumably includes the maintenance of minimum stream flows. All three states also have authority to deny appropriation permits on the basis of the public interest, but the exercise of this power will have the same effect as the reservation of water from private appropriation. The regulatory approach has been little used in any of the three states, in part because of the widespread political power of vested private water rights holders. However, the increasing urbanization in all three states and the resultant disassociation of individual economic welfare from the legal mechanisms for water allocation are likely to improve the political viability of the regulatory approach. Certainly, all three states are well imbued in the regulatory philosophy and legal mechanics of land use regulation, which could be readily applied to water use regulation.

Some variation on reservation has been the dominant approach to date, the impact of which on the allocation of water in a particular state depends largely upon the existing conditions in that state. In Montana, where there is an abundance of unappropriated water, the reservation approach to instream flow protection will give the state a dominant role in water use decision making. In states where most water has been appropriated by private users prior to the implementation of a state reservation program, the state's role in water allocation will be far less significant. Those states will not achieve a dominant role in water allocation without resorting to regulation similar to that common in land use decisions.

One implication of the reservation approach to instream flow maintenance is that there is no logically valid public trust principle under which such flows are to be maintained. A public trust duty by the states to maintain stream flows theoretically predates any private rights and can be implemented by the simple denial, suspension, or cancellation of appropriation permits that would result in the violation of the public trust. The affirmative reservation of water from future appropriation implies that the water would otherwise be available for appropriation. Of course, the existence of a public trust duty on the part of the state may be little different from the power that the state may assert pursuant to its police powers, except that the state has discretion in implementing the latter. If the states recognized a trustee duty, they could easily base their stream flow protection actions on the existence of the public trust and would not be constrained to protecting flows on streams that are not yet fully appropriated.

Given the theoretical foundation of western water law, which recognizes that the states have authority to grant private rights of use in publicly owned water supplies, there is little doubt that the states have the authority to pursue the reservation approach to instream flow protection. The reluctance of the states to resort to regulatory approaches must be attributed to political rather than legal constraints, particularly considering past experience with land use regulation. Thus it is important to consider the policy arguments that might be invoked to justify future resort to regulation.

The assertion that minimum stream flows are a public good that will not be provided without state intervention is based on the assumption that no private individual or corporate entity will provide for instream flows because it will be unable to profit from the necessary investment. The inability of investors in instream flows to gain an adequate return on their investment is attributed to the fact that many individuals experience external benefits for which they cannot be forced to pay because of an assortment of transactions costs and the existence of free riders. In other words, it is argued that water resources will be inefficiently allocated unless the state intervenes to ensure that a sufficient amount of water is allocated to instream flows.

Whether state intervention can actually improve upon efficiency is an empirical issue that cannot be resolved on the basis of existing data. There is reason, nevertheless, to resist this theory. It is possible that the costs of state transactions will be as high as or higher than the costs of private transactions. But more importantly, it may well be found that existing inefficiencies in water allocation result from deficiencies in the private rights system rather than from alleged market failures. The existing water law seriously limits private acquisition of instream flow rights, so we cannot be sure from experience that the initial public good assumption is accurate. And even if we accept the public good assumption with respect to stream flow maintenance, there are alternative approaches to state involvement that are far more likely to approach the goal of efficient water allocation than the systems being employed in Idaho, Washington, and Montana. The state of Colorado has implemented one such alternative.

Before we consider the Colorado approach, it will be useful to briefly detail the reasons for urging caution in the further implementation of the existing instream flow laws. The question of what water should be reserved from future private appropriations, whether addressed in the context of a particular situation, an entire water basin, or political jurisdiction, turns on whether the people of the state will benefit more from the allocation of particular waters to instream flow maintenance or to private acquisition of some other use. To the extent that water reservations are generally stated, the allocational problem will be delegated to an agency bureaucrat, legislator, or board member, depending on the context. An examination of the instream flow programs in Idaho, Washington, and Montana suggests that the designated public officials are in no position to make such allocational decisions with respect to the objective of allocational efficiency. Whether in the context of stream rating under Washington law or water reservation under Montana law, the decision makers have very little information about the relative values of the water for the competing uses. If one could equate a democratic result with an efficient one, the Idaho

approach might be defensible on an efficiency basis, but the hard truth of the matter is that the delegation of any issue such as water use to a state agency will result in a decision based upon distributional rather than allocational considerations. Particularly in a democracy, public officials will decide on the basis of who benefits from water use rather than on which water uses will produce the most net benefits.

In Idaho, where the popularly elected representatives have an opportunity for a direct say on every instream flow allocation, the process of political compromise and logrolling will be most obvious. In the case of local streams, local interests are likely to dominate. In the case of major rivers like the Snake, the instream flow issue will be subject to the same political factors as any other decision of statewide importance. In Montana the political process will determine the outcome in the long run, since those board members whose views lead them to unpopular decisions will be replaced by others whose views conform to those of the current governor. Even if the governor and the board members free themselves from any political considerations, the board will be faced with the incomprehensible task of allocating the waters of an entire river basin with only their own sense of what is in the interest of the people of the state. As a result, they will tend to look to state agencies for counsel, and the decision may not differ from that which the agencies would have arrived at on their own. In Washington the difficulty of coping with the value questions leads the decision-making agencies to focus on the technical issues. The stream-rating system employed by Washington's Department of Ecology, although not arbitrary, does not reflect the efficient allocation of water in the particular streams. The fact that a stream is a good habitat for trout in no way is determinative of whether the stream should be maintained as trout habitat. That issue can only be resolved in the context of the possible alternative uses of the water in the particular streams at a particular time.

The impossibility of deciding the optimal use of a given quantity or flow of water leads public officials to establish water use priorities or standards, which are as difficult to challenge as they are to justify, and then focus on the technical problems of meeting the established standards. It is clear in both Montana and Washington that these technical problems play the dominant role in the instream flow deliberations. When a resolution is finally reached, there is a tendency to believe that the allocational issue has somehow been resolved. In reality, however, the allocational efficiency of the resultant instream flow reservation depends entirely upon the efficiency of the initial standards that defined the technical problem. The technical expert can provide the evidence on how much water is required to maintain a trout habitat but has only personal opinion to offer on the issue of whether the state should preserve that trout habitat. The reservation processes do not begin to provide a solution to the allocation problem; rather, they provide a mechanism for particular interests to guarantee that their preferences will be controlling.

COLORADO—STATE PARTICIPATION IN THE MARKET

In 1963 the Colorado legislature authorized the Colorado River Conservation District to appropriate water of any natural stream sufficient to maintain the fish habitat for public use.[36] The district's implementation of the instream appropriation power was challenged, and the Colorado Supreme Court found "no support in the law of this state for the proposition that a minimum flow of water may be 'appropriated' in a natural stream for piscatorial purposes without diversion of any portion of the water 'appropriated' from the natural course of the stream."[37] A 1973 amendment to the 1969 Water Rights and Administration Act eliminated the statutory diversion requirement, which had been the basis of the Supreme Court's decision on the 1963 law, and broadened the definition of beneficial use to include minimum stream flows "to preserve the natural environment to a reasonable degree."[38] Again, the power of the state to allocate water to instream flows was challenged, this time on constitutional grounds. In 1979 the Colorado Supreme Court held that the 1973 amendment did not deprive Colorado citizens of their constitutional right to appropriate water and that the constitution did not require a diversion of water for a valid appropriation.[39]

On the surface, the Colorado law appears to be very similar to the authority of other states to appropriate unappropriated waters, an approach that effectively reserves waters from future private appropriation. However, two characteristics of the Colorado law make it significantly different. The Colorado Water Conservation Board has authority to acquire instream flow rights by means of other than appropriation. The law is not specific on how the board may acquire instream flow rights, but certainly, purchase of existing rights is the most obvious approach the board might take. The statute also specifically forbids the acquisition of instream flow rights by eminent domain, the clear intent of which is that the state, acting through the Water Conservation Board, is to participate in the water rights acquisition system on the same terms as any other prospective water rights owner. Obviously,

the state's potentially deep pocket can give it a significant competitive advantage against most private parties, but the prohibition on resort to eminent domain ensures that state instream flow rights will only be acquired from willing sellers.

A committee consisting of officials from several state and federal agencies was formed in 1973 to identify priorities for the appropriation of instream flows.[40] Although faced with the same difficulties in setting priorities that Washington officials face in classifying streams under their rating system, Colorado officials encounter serious constraints on which instream rights they can actually acquire, while Washington's DOE is required to set base flows on every perennial stream in the state. The two constraints faced by the Colorado board both have characteristics of market influences. Because the board must satisfy the same requirements for an appropriation as anyone acquiring a private water right, the resources of the implementing agencies limit the number of appropriation permit applications the board can file. More importantly, any rights that the state acquires by purchase will be acquired at the going market rate, thus forcing the board to carefully order its priorities. In Washington and Montana the state officials may have access to water rights market information, but they are required to reserve water for instream purposes, and there are no direct incentives to take into account the relative market values of various water uses. The Colorado system removes the inevitability of instream flow action by the state while forcing the state to deal with the realities of the water rights market.

Of course, the Colorado system is not neutral with respect to the market allocation of water rights. The state does have enormous potential resources to draw upon, but taxpayers are likely to be far more cautious about public expenditures when they are able to identify the specific returns on those expenditures, as they are with the public school systems. The Colorado requirement that the board obtain specific legislative appropriations for the acquisition of instream rights gives taxpayers a similar sense of where their money is going. Under the reservation systems the cost of instream flow maintenance are impossibly obscured because the states do not compete in the private market and the reserved waters have values in alternative uses that will be known only in the context of future markets. The appropriative part of the Colorado system suffers from the same problem in theory, but the realities of Colorado's water situation greatly diminish its significance in that state. In contrast to the situation in Montana, where nearly 4 million acre·ft of water per year are reserved to instream flows where the Yellowstone flows through Billings, the state's largest city, the appropriated instream flow rights in Colorado are all on very small segments of high alpine streams. Thus for significant minimum flows to be established in the populated areas of the state, it will be necessary for the board to acquire instream flow rights in the water rights market rather than through appropriation.

The Colorado system offers no magical solution to the problem of determining what level of streamflows will be in the public interest. Although the Colorado system restricts the state to what it can afford, instream flow preservation is only one of a multitude of budget items for the state. There is probably a tendency to rely on technical experts within the bureaucracy, as is the case in other states where financial constraints are largely absent. The Water Conservation Board relies on the Division of Wildlife in the determination of where instream flow rights should be required and how much flow should be maintained.[41] Because that agency's personnel are only competent to estimate how much flow will be necessary for particular purpose, their recommendations are relevant only after the state has determined the priority of purposes it will pursue. The difficulty of that task will no doubt lead the priority question to be submerged in the data of the technical issues of fish habitat maintenance.

These problems notwithstanding, the Colorado approach offers promise as a means for the state to participate in water allocation decisions without totally distorting or ignoring the value information that the water rights market provides. In the context of land use regulation constitutional provisions requiring compensation by the state to parties whose property rights are taken, if enforced, can help to preserve some market influences on government action. However, water presents a different problem in states such as Montana where a significant amount of water is unowned and the state can reserve water as if it had no value or a different value from water privately controlled. It is not surprising that water-short Colorado would turn to the market approach while water-rich Montana would turn to the reservation approach.[42] It is relatively easy to tie up water that thus far has been insufficiently valuable to justify its acquisition. However, the value of water will only rise with increasing demands, and without political changes the reserved waters of Montana and Washington will be unavailable even if the instream uses are worth far less than alternative uses.

It is not clear whether the Water Conservation Board has the authority to dispose of as well as acquire instream flow rights. The law does not specify that the state may exchange its appropriated rights for other rights that might have greater instream value, but if the purpose of the Colorado approach is to take advantage of the allocative advantages of the market, it is essential that such transfers of state-owned water rights be permitted. Indeed, most of the market advantages that justify requiring the state to purchase instream rights without resorting to eminent domain would be lost if the state cannot in turn sell those rights if subsequent market values warrant such an action. The state's holding instream rights in perpetuity, even if those rights were purchased, will have the same long-run allocative consequences as if the state initially had acquired the rights by eminent domain, reservation, or regulation.

PRIVATE RIGHTS TO INSTREAM FLOWS

It is a virtual certainty that the possibility of private provision of instream flows was not even considered by state legislators when instream flow programs were adopted. Yet it may well be possible to provide for instream water uses through the existing systems of private rights in water. No doubt the presumption in favor of public action is in part a consequence of the relative paucity of private rights in instream flows, but it does not follow that the only way to provide for minimum stream flows is by state action. Three factors have contributed to the paucity of private rights in instream flows. It has been widely believed that water is a common-pool resource and therefore not amenable to allocation to many uses through a system of private rights. As a consequence, private rights systems have not been designed to accommodate instream flow rights, and governments have undertaken often massive programs to provide uses that supposedly cannot be provided privately. The following discussion assesses the validity of the common-pool argument as applied to water resources and examines the existing law of private water rights and the extensive involvement of government in the provision of consumptive water uses. It should be clear from the discussion that instream water uses can be privately supplied if private rights in water are clearly defined, enforced, and transferable through appropriate institutional changes.

The Common-Pool Problem

The notion that water is a common-pool resource that cannot be privately allocated may have had some historic validity. The traditional argument has taken the following form: Almost all surface water contributes to the flow of water in drainage basins, which increase in size from the peaks of the mountains to the shores of the continent. Indeed, much of the ground water is tributary to these surface flows, which themselves are part of the earth's complex hydrologic cycle. Because of the migratory and integrated nature of the water resource, it is generally assumed that the problem of defining private rights in water is far more complicated than the definition of private rights in land. More closely analogous is the problem of assigning property rights in oil and gas resources, the history of which is replete with inefficiencies and wasted resources. The solution in oil and gas law has been voluntary or state-imposed pooling and unitization. The solution in water law has been the appropriation doctrine, which assigns rights with temporal rather than geographic points of reference. When dealing with migratory resources such as water, we cannot put up a fence to mark the boundaries of private property. When the use of migratory resources is consumptive, rights can have volume as a parameter; but when the resource is in its natural migratory environment, it is not possible to define with adequate precision the content of any particular right. The best solution is a rule of capture, which, though adequate for a society of hunters and gatherers, provides none of the certainty necessary to resource development in a modern society. The temporal priority scheme of appropriation doctrine works well when the water is diverted and can be measured, but when the water is left in the stream, it is impossible for private rights holders to enforce their rights. The owner of a private right to a minimum stream flow will not know the right has been violated until the flow is too low and the fish are dead.

Although the definition of private rights in water is generally more difficult than the definition of private rights in land, and instream rights definition may not have been possible during the nineteenth century when appropriation doctrine was developed, the argument summarized in the preceding paragraph is not convincing in the context of instream flow protection in the 1980s. Sophisticated technologies of stream flow monitoring can serve the law of instream flow rights just as the technology of barbed wire served the nineteenth-century law of private rights in grazing land. Defining the parameters of a right to instream flows is no more difficult than defining the parameters of a right to divert water for agriculture or industry. Indeed, if the states have any expectation of endorsing minimum stream flows, they must surely recognize that the common-pool arguments are no longer a justification for precluding private instream rights, at least in terms of rights definition and enforcement.

A few private rights to instream flows exist despite the institutional obstacles. The Nature Conservancy owns an instream flow right on Boulder Creek near the city of Boulder, Colorado. Individuals in Montana own rights to spring creeks of the Yellowstone Valley, and people willingly pay for the opportunity to fish in those creeks. Private investors have proposed the purchase of lands bordering the lower Deschutes River in Oregon, access rights to which they would market on a membership or temporary-use basis in order to protect the stream for sport fishing. Arizona appears to authorize private acquisition of instream rights.[43] The most persuasive evidence of the private willingness to invest in instream flow water uses is the enormous support of the efforts of the many organizations that seek to influence the public allocation of water resources. Through lobbying and litigation, organizations such as the Sierra Club and the National Wildlife Federation have spent millions of dollars to influence the enactment and ensure the implementation of legislation allocating water to instream flows. That people clearly value instream water uses, even in the face of enormous institutional obstacles, occasionally has been demonstrated through the acquisition and retention of private rights.

The valuation of water for instream uses does not have to be a mysterious process. Certainly instream uses are not of infinite value, although the reservation approach to state provision of instream flows seems to assume that no other uses will ever be sufficiently valuable to justify state abandonment of instream flows. Those who would claim that minimum stream flows should be provided no matter what other demands for water may exist are preservationists in the tradition of the national park and wilderness advocates of generations past. Few would not feel the loss if Old Faithful were plugged with a power generator or the cliffs of Yosemite were dynamited for their minerals. But we deceive ourselves if we contend that humans will choose to perish from a lack of resources in preference to destroying the aesthetic wonders of our national parks. Fortunately, we seldom, if ever, will face those extreme choices because of the expanding availability of substitutes for scarce resources. Instream water uses also have their value, which may be sufficiently high that we are depriving ourselves of certain benefits by relying on the government to provide the desired flows. An inevitable consequence of government actions that overproduce certain benefits is that other benefits must be underproduced. To the extent that the government may be overproducing certain water uses through subsidization, society will experience the underproduction of other water uses.

As a general rule, the value of a particular volume or flow of water is a function of the temporal priority of the associated right and of the location of the diversion or instream use. Normally, greater seniority will give the right greater value, and downstream rights will be more valuable than upstream rights because of the numerous geographic advantages of locating most productive activities downstream. The value of an instream water use, like any other water use, will depend upon the values of alternative uses of the water. As a general rule, then, the more alternative uses forgone because of a particular instream use, the higher the value of that instream use must be to justify the lost opportunities.

Instream water uses, like all other water uses, have values that can be translated into private investment in water rights. Whether or not a private entity will choose to invest in an instream water use depends upon the value of that use in relation to other water uses and upon the legal possibilities of owning instream water rights. Once an institutional framework that permits and does not discriminate against private ownership of instream water rights is implemented, the existence of instream water rights will depend upon the values that people place on instream uses. Aside from the few examples of private instream rights that exist despite the extant institutional obstacles, there is no conclusive evidence that private instream rights will in fact be acquired when the institutional obstacles are removed. However, through a hypothetical example it can be demonstrated that such rights can and will be acquired if they are sufficiently valued.

To make an appropriation of water for an instream flow, assuming no institutional obstacles, private party X needs only to value that use in excess of the costs of making the appropriation. Those costs will be relatively low since no diversion is required, and X simply needs to comply with the existing state procedures for an appropriation. Once X has the instream right, assuming no obstacles to rights transfers, other parties may propose to purchase the right so that the water can be used in some other way. Although X may place a very high value on the instream use, X will be willing to sell the right to Y if Y offers to pay an amount that will allow X to acquire even larger instream values at some other location. If X values the instream use more highly than Y or any other prospective purchaser, X will retain the right, and the water to which X is entitled will remain instream.

In many areas of the United States the preceding hypothetical scenario is not a real possibility because many streams are fully appropriated during the low-flow seasons of the year. That X will not be able to appropriate an instream flow during the dry season when instream flow maintenance is a problem does not alter X's ability to acquire and retain instream rights if X values the instream use of particular waters more than the party presently owning the relevant water right. If Y has a right to use 20 cubic feet per second (ft^3/s) of water during the months of June, July, and August and produces $2000 of additional net income by using that water for irrigation, Y will be willing to sell the right to use that water during the summer to X for some price in excess of $2000. Party X might make optimal use of its resources by thus leasing water in years of unusually low flow, or X might purchase the water right if the value of the permanent maintenance of that 20-ft^3/s flow exceeds the value of the right to Y. As with other water uses, the value placed on an instream use will have to be higher in some situations than in others if an instream right is to be acquired and retained. But there is nothing about instream uses that prevents their being provided privately. If the instream use is valued more highly than uses that would be forgone, the instream use will be provided.

Legal Obstacles

As the earlier discussion of state approaches to instream flow maintenance indicated, the private water rights system, designed to facilitate consumptive water use, has erected numerous obstacles to the acquisition of instream water rights. Although many state laws have been changed to permit state acquisition of instream water rights, most do not permit private acquisition of instream rights. In some states it is not yet clear whether the changes will be found to apply to private acquisition of instream flow rights. Five specific aspects of appropriation doctrine pose obstacles to the private acquisition of instream rights.

The diversion requirement, which prohibited Colorado's first instream flow appropriations, still may be a major obstacle to private rights in instream flows. As was indicated previously, the central purpose of the diversion requirement is to facilitate determining the temporal order of appropriations and measuring the amount of water being used. The earliest unit of water measurement commonly used in the West, the miner's inch, was based upon the diversion mechanisms employed by the early miners. Obviously, an instream water user has no need or desire to divert water, and the requirement is no longer of any value with respect to consumptive rights, since records establish evidence of priority and modern measuring methods have long made the miner's inch an historical curiosity. The diversion requirement must be repealed if private instream rights are to exist.

Two legal obstacles to private instream flow rights, the beneficial-use requirement and the rules of abandonment or nonuse, have similar historical roots. The beneficial-use requirement is an outgrowth of the same judicial reasoning used to justify the reasonable-use requirement of American riparian doctrine. However, where the reasonable-use rule was essential to the consumptive use of water under a riparian system, the beneficial-use rule had no similar purpose under the appropriation system, the whole purpose of which was to allow consumptive water use. While the beneficial-use rule made it clear that the state would permit only socially meritorious water uses, a reading of the case law will reveal that the rule has seldom been invoked as a restraint on the appropriation of water. Beneficial uses have always been designated in general terms such as industry and agriculture, with the implicit recognition that any water use worth investing in is beneficial to the public. Occasionally, the law has required specific amendment, as in the case of recreation, and excluded specific activities such as coal slurry from the definition of beneficial use.[44] In general, however, the economics of water use ensure that water will not be applied to nonbeneficial uses.

The beneficial-use standard is linked with the law relating to abandonment and nonuse, by the nineteenth-century bias against speculation in the future values of resources. By requiring rights claimants to demonstrate that they were putting the water to a beneficial use, the state was able to preclude individuals from diverting water and then not using it in anticipation that they or others might be able to use it in the future. In this respect the water law system was far more effective in discouraging speculation than the public land disposal laws, which were consistently violated for speculative purposes. Although the law of abandonment generally was justified on the logical premise that a right acquired by use is lost by nonuse, its purpose was clearly to prevent speculation. It was argued that not only were idle resources unproductive, an anathema to growth-conscious westerners, but permitting one individual to hold water rights without using the water allowed that individual to profit from the future increased demand and higher value of water that would inevitably come with western development. Although speculation has distributive consequences that may have seemed undesirable to nineteenth-century judges and legislators, it has no negative impact on the efficiency of water resource allocation.

Both the beneficial-use standard and the law of abandonment are current obstacles to private acquisition of rights in instream flows. The beneficial-use standard can either be amended to include instream uses or repealed without any harm to efficient water allocation. The rules of abandonment can be easily adjusted to recognize that the fact that water remains in the stream does not mean that the water is not being used. The legitimate purpose of abandonment law, to ensure that water

is not excluded from productive use by uncertainties about title claims, will decrease in importance as recording of water rights becomes the rule rather than the exception.

Another legal obstacle, restraints on the transfer of existing water rights, often makes it difficult for water rights to be shifted from one use to another. Assuming the legal recognition of private instream rights, the efficient level of instream water use will not be achieved unless instream water users are able to purchase or otherwise acquire water rights now being applied to other uses. In many areas of the West streams are fully appropriated, and instream benefits cannot be realized without the acquisition of existing rights. In addition, it must also be possible to change the point of diversion of a right on a particular stream or basin. Often the instream user's water rights will not be strategically located on the stream, and there should be no arbitrary restriction on changing the place of use of the water. Of course, such changes or rights transfers must not negatively affect other water rights, but many legal restrictions are not for the purpose of protecting vested rights. Often it will be possible to transfer a water right from a consumptive user to an instream user without affecting the total consumptive use of water on the stream. If the stream right extends over only a segment of a stream, additional water will be made available for consumptive use downstream. The law should be structured so that instream right holders can transfer the consumptive components of their acquired rights to downstream water users and thus finance a part of the costs of the acquired rights, provided the sale does not negatively affect third-party rights.

If instream uses are put on the same legal footing as all other water uses, and if the water rights system is generally improved to allow for a free market in water rights, it will be possible for instream values to be realized through strictly private actions. However, even if the legal obstacles are eliminated, a major obstacle to private rights will remain. Ironically, that obstacle, employed to justify public action to protect instream flows, is a function of public actions in support of competing water uses.

Government Subsidization of Competing Water Uses

One need only travel the major river basins of the West to appreciate the dominant role of federal and state government in the allocation of water. Rivers and many small streams are "regulated" with dams, the vast majority of which have been built at public expense. The water from these dams is used to irrigate hundreds of thousands of acres, generally at far below market cost to farmers. Electricity is generated and sold at less than market prices. The state of California is a spider's web of publicly funded canals and pipelines, all of which subsidize certain water uses. The state of Montana has implemented a water reservation system that excludes the reservation of water for industrial use but allows state agencies to tie up water for agricultural, municipal, and instream uses. Sometimes, as in Montana, the instream uses receive a far bigger slice of the pie than they would get in a private market.

The vast bulk of government subsidies, however, benefit consumptive water users. So long as the existing system of government subsidization of consumptive water use continues, it is unlikely that instream water users will be able to afford to compete in a private rights market. Their only alternative is to seek government subsidies themselves, which is precisely what they have been doing. It is very likely that many of the critically low flows that are now the rule on most western rivers and streams during the late summer months would not exist but for government subsidization of competing water uses. Of course, the late summer flows of some streams are made possible by those government projects, but that fact only underscores the disruptive impact of existing government water programs on the water market.

It has long been argued that government subsidization is necessary to water development because of the enormous costs of large dams and massive water transfer projects. But private capital has been aggregated in amounts far exceeding the cost of any existing or proposed public water project. If the large water development promises an adequate rate of return, private resources can bear the cost, unless, of course, the government is willing to do it. As long as competing uses continue to be subsidized, repairs to the legal system clearly are not sufficient for the privitization of rights in instream water uses.

CONCLUSION

For most of America's history, instream water uses have been supplied because of abundant water sources, the proximity of human activities to those water sources, and relatively little demand for consumptive water uses. People now live in the once-foreboding deserts and use water for an array of activities never contemplated a century ago. Instream water uses are no longer free goods. The uniform response in the western Untied States, where water problems are most serious, has been to resort to various types of state action designed to protect and maintain minimum stream flow levels. The effects of these programs largely reflect the scarcity of water in particular streams and rivers. The reservation approach, sometimes achieved in the name of appropriation, places the entire burden of determining the proper allocation of water to instream uses on state officials, whose decisions are apt to reflect the relative power of the various water user lobbies. The valuation issue is often lost in technical argument over how much water is necessary for a particular purpose, which may have been chosen without any consideration of allocative efficiency. In states with abundant water, instream users will probably do well, at least until water scarcity forces a reassessment. In water-scarce states instream users must compete with other water user groups with considerable experience in water politics.

Assuming that public officials are concerned about the efficiency of their water allocation decisions, the unavoidable problem, whether water is abundant or scarce, is to determine what allocation will serve the best interests of the people of the state. Where regulations and subsidies distort market information on the relative values of competing water uses, the prospects for an efficient allocation are dim indeed. The only solution is the privatization of water allocation decisions through the existing or altered systems of private rights in water. With specific changes in the existing law of property rights in water and extensive alteration of the role of government in water allocation, significant instream water use should result from private action. The achievement of that objective could take years. In the interim states should consider the Colorado approach, which potentially involves the state in the water rights market and thus promises that state actions will be at least marginally informed on the issues that are critical to efficient allocation of water.

NOTES

1. For a summary account of western American water law generally assumed to have been the product of local mining camp customs influenced by common and civil law, see Wells A. Hutchins, 1971, *Water rights laws in the nineteen western states,* vol. 1 (Washington, D.C.: U.S. Department of Agriculture), 159–175.

2. Frank J. Trelease, 1974, The model water code, the wise administrator and the god-dam bureaucrat, *Natural Resources Journal*, 14:207.

3. R. Johnson, 1980, Public trust protection for stream flows and lake levels, *U. C. Davis Law Review*, 14:256–257.

4. The information on the four state programs is drawn from a study funded by the Office of Water Research and Technology, United States Department of Interior. James L. Huffman et al., 1980, *The allocation of water to instream flows: A comparative study of policy making and technical information in the states of Colorado, Idaho, Montana and Washington*, vols. I–V (Portland, Ore.: Natural Resources Law Institute).

5. See *Mason et al.* v. *Hoyle*, 56 Conn. 255, 14 A 786 (1888).

6. Quoted in Charles J. Meyers and A. Dan Tarlock, 1971, *Water resources management* (Mincola, N.Y.: Foundation Press), 53.

7. See *Red River Roller Mills* v. *Wright*, 30 Minn. 29, 15 N.W. 167, 169 (1883).

8. See *Irwin* v. *Phillips*, 5 Cal. 140 (1855).

9. The standard elements of a valid appropriation generally include the following: intent to appropriate, notice of appropriation, compliance with state laws, diversion of water, and application to a beneficial use.

10. See *Gibbons* v. *Ogden*, 9 Wheat 1, 6 L. Ed. 23 (1824).

11. See *Wilson* v. *The Black Bird Creek Marsh Co.*, 2 Pet. 245, 7 L. Ed. 412 (1829).

12. For a detailed historical account see Michael C. Blumm, 1981, Hydropower v. salmon: The struggle of the Pacific Northwest's anadromous fish resources for a peaceful coexistence with the federal Columbia River power system, *Environmental Law*, 11:211.

13. A 1949 Washington statute authorized state action to protect stream flows (Revised Code of Washington, 72.20.050). Oregon adopted a similar law in 1955 [Oregon Revised Statutes 536.310(7), 1977]. As early as 1915, the Oregon legislature acted to preserve the flows of streams tributary to the scenic aterfalls of the Columbia River Gorge (Oregon Revised Statutes 538.200).

14. See William Baxter, 1974, *People or penguins: The case for optimal pollution* (New York: Columbia University Press).

15. Water rights owners are held to possess a right to use the water, but they do not own the water itself.

16. See *Colorado River Water Conservation District* v. *Rocky Mountain Power Company*, 158 Colo. 331, 406 P.2d 798 (1965).

17. See *Berman* v. *Parker*, 348 U.S. 26 (1954).

18. In Colorado the claim has been specifically rejected. In The matter of the application for water rights of the Colorado Water Conservation Board in the Roaring Fork River and its tributaries, District Court for Water Division No. 5, Findings of Fact, Conclusions of Law, and Order for Entry of Judgment (26 June 1978), pp. 13–14.

19. See J. Sax, 1970, The public trust doctrine in natural resource law: Effective judicial intervention, *Michigan Law Review*, 68:471; J. Sax, 1980 Liberating the public trust doctrine from its historical shackles, *U. C. Davis Law Review*, 14:185.

20. See Johnson, Public trust protection.

21. Idaho Code, 67–4031 to 67–4311 (1971).

22. *State Department of Parks* v. *Idaho Department of Water Administration*, 96 Idaho 440, 530 P. 2d 924 (1974).

23. Idaho Code, 41–1503 et seq.

24. Revised Code of Washington, 75.20.050.

25. Revised Code of Washington, 90.22.010 to 90.22. 040.

26. Huffman, et al., *Allocation of water to instream flows*, vol. V, 17.

27. Ibid., 21.

28. Revised Code of Washington, 90.54.050.

29. Montana Code Annotated, 89–801 (1947) and 85–1–209 (1979).

30. Huffman et al., *Allocation of water to instream flows*, vol. IV, 17–18.

31. Montana Code Annotated, 87–1–209 and 85–1–204(1) (1979).

32. Montana Code Annotated, Constitution, Article IX, Section 3 (1979). Pursuant to that provision, the 1973 act was adopted. Montana Laws, Chapter 452, Section 2 (1973).

33. See *Utt* v. *Frey*, 106 Cal. 392, 39 P. 807 (1895).

34. In 1974 the Montana legislature adopted a moratorium on water appropriations on the Yellowstone River for a three-year period pending completion by the DNRC of the reservation proceedings. Montana Code Annotated 85–2–601 to 608 (1979). In 1977 the legislature extended the moratorium, which was further extended by action of the Montana Supreme Court. Montana Laws, Chapter 26, Section 1 (1977). See Huffman et al., *Allocation of water to instream flows*, vol. IV, 25–29.

35. Huffman et al., *Allocation of water to instream flows*, vol. IV, 72–74.

36. Colorado Revised Statutes Annotated, 37–46–107 (j).

37. *Colorado River Water Conservation District* v. *Rocky Mountain Power Company*, 158 Colo. 331, 406 P. 2d 798,800 (1965).

38. Colorado Revised Statutes Annotated, 37–92–103(3).

39. *Colorado River Water Conservation District* v. *Colorado Water Conservation Board, Colo.*, 594 P. 2d 570 (1979).

40. Huffman et al., *Allocation of water to instream flows,* vol. II, 53.

41. Ibid., 54.

42. James L. Huffman, 1980, Water and public control, in *New directions in water policy* (Corvallis, Ore.: Oregon State University Press).

43. Arizona Revised Statutes 45–141 (A) (Supp. 1978). For a discussion of the prospects for private instream rights in Arizona, see Tom Scribner, 1979, Arizona water law: The problem of instream appropriation for environmental use by private appropriators, *Arizona Law Review,* 21:1095.

44. See Hutchins, *Water rights laws,* pp. 542–544: Montana Code Annotated 85–2–102(2) (1979).

34

WATER REALLOCATION: CONFLICTING SOCIAL VALUES

F. LEE BROWN, BRIAN MCDONALD, JOHN TYSSELING, and CHARLES DUMARS

In recent years water experts and the public at large have increasingly recognized that much of the native water resources of the western United States is reaching a condition of full appropriation. In part, this recognition has been fostered by the academic literature on water, which has increasingly focused on problems of water *re*allocation (1) as opposed to the dominant theme of "new appropriations" that occupied the literature for the first two-thirds of the twentieth century. In part, this recognition has been stimulated by various water demand projections of federal agencies and others (2)—particularly those associated with energy development—which have forecast a condition of full appropriation for most basins in the West by the early years of the next century. Fundamentally, however, this recognition has been forced on the West by a very concrete circumstance—namely, in basin after basin whenever a new water use is contemplated, it is soon discovered that the new use can be accommodated only through transfer of water or water rights from previously established appropriators. As one result, more and more water right transfers are occurring throughout the region through many singular arrangements agreed to by purchaser and seller (3). These transfers place new demands on the region's water institutions, which were originally evolved to support *new* appropriations of water and to protect those appropriations once achieved. Now these same institutions are being asked to *facilitate* transfers of water and water rights from these previously established uses to new ones while simultaneously protecting the equity interests of those established appropriators.

It comes as no surprise to discover that the new task of water reallocation places considerable strain upon institutions whose original objectives included protection of a static pattern of water use once the original pattern was achieved. Nor is it surprising to find the transfers continuing at the pace of recent years (4). A condition of full appropriation of water is a new phenomenon for the people of the West. Learning to cope with the situation and mitigating its negative consequences are not easy tasks. In this context it is important to examine the existing water institutions that govern the reallocation function for society and consider their evolutionary future.

A PROBLEM IN POLITICAL ECONOMY

Water in the West is a scarce resource subject to a number of competing demands that taken together would exceed the available supply if an apportioning method were not available for determining which demands will be met and which not. In this circumstance water fulfills the classic definition of a scarce commodity, whose allocation has been studied by economists for centuries. "The rules, the patterns of behavior, the values and purposes, and organizational structures" (5) that society evolves for performing this allocation task jointly comprise the *institutions,* as the term is used here.

As an initial observation, it is clear that society may choose from a large variety of institutional forms in conducting its water business. The eastern United States has codified riparian doctrine in which

Lee Brown is professor of economics; Brian McDonald is associated director for research, Bureau of Business and Economic Research; John Tysseling is a graduate student with the bureau; and Charles DuMars is professor, School of Law; all at the University of New Mexico. This material is excerpted from Water reallocation, market proficiency, and conflicting social values, in *Water and agriculture in the western U.S.: Conservation, reallocation, and markets,* ed. Gary D. Weatherford (project of the John Muir Institute, published by Westview Press, 1982), 191–255. Used with permission of the authors.

those uses physically more proximate to the water sources are given higher claim on the water, while western states have given a higher claim to those who have put the water to beneficial use at the earlier chronological time even if that use is physically more remote from the water source than a later use. Yet even in the West there have been significant exceptions to the prior-appropriation doctrine. Perhaps the most notable exceptions are the two Colorado River compacts (6) themselves, which have given congressionally ratified claims to Colorado River water to those Upper Basin states that are nearest the physical source of most of the river's flow even though at the dates of the negotiated agreements the Upper Basin states had not put their shares to beneficial use.

There are numerous other examples of institutional forms that at least a portion of society has adopted for the water allocation task, including irrigation districts that may "share shortages" among individual farmers even though a prior-appropriation system does not require a shortage sharing between the irrigation district taken as a unit and other water right holders (7). No attempt will be made to catalogue these forms. The important point here, instead, is the diversity of the forms that society has adopted. Although in the West the essential themes are "priority in time" and "beneficial use," it is clear that these themes have been modified or even waived when other important considerations intervened. These "other important considerations" make the examination of the societal choice of institutional form a problem in *political* economy rather than simply a search for an institutional form that will yield an economically efficient allocation.

THE IMPORTANT SOCIAL VALUES

Economic Improvement

As just indicated, in making its choice of the institutional forms that will handle the water allocation task, society does not have the luxury of simply measuring the various forms against a single standard of social value. If economic improvement were the sole standard, then the choice would be relatively easy for economic theorists who have long since established the superior virtues of the market institution in moving a scarce commodity to its highest-value use. By pushing a scarce resource or other commodity toward those uses in which it is economically most productive, the market system helps ensure that the maximum economic reward will be achieved for any given combination of resources. Certainly, for a subeconomy (the West) that still contains some of the

economically poorest people in the nation, economic improvement remains a social goal. As a general rule, if a power plant can bid for more water rights than a farming enterprise, it is because the power plant is an *economically* more productive (8) use of water as measured by the dollars and cents that reflect society's comparative valuation of electricity and agricultural production.

At the same time that the market bidding process determines the allocation of the scarce resource among the different, competing uses, a market price for that resource is also established, thereby signaling all users of the resource of its relative worth. Consider the irrigation farmer who, although he has no interest in selling his water right and leaving the business, may indeed be willing to invest in water-conserving technologies such as sprinkler or drip irrigation. With an increasing market value for his water rights, he may find it possible to afford these conserving technologies (and even increase his profit) by selling off a portion of his water rights and "spreading" the remaining rights over the same amount of land or even more land. As a caveat to this last remark, there are legal problems with this "spreading" possibility, which will be discussed below. Nevertheless, this brief review of "market economics" shows that the conservation-signaling information conveyed by market prices can be a substantial by-product of its allocative role.

Environmental Preservation

A second standard against which society measures the acceptability of a particular institutional form for allocating water is the degree to which that form preserves the natural character of the region. For many westerners it is the beauty of the region's natural environment that is the compelling social value (9). Faced with a stark choice between preservation of the natural environment and economic improvement, they would choose environmental preservation every time. For these individuals any prospect of market-induced shifts of water to energy or other uses that might dry up mountain streams or otherwise diminish the region's natural heritage is anathema.

This social value has found its clearest expression in the state of Colorado, which has enacted legislation that establishes the instream uses such as maintenance of surface flow as legitimate beneficial uses of water (10). Other Colorado events such as the decision in 1972 against hosting the Winter Olympics in that state (11), while not directly tied to water, nevertheless help reveal the relative values of citizens of that state. In keeping with these values, it is noteworthy that Colorado's government has been active in investigating alternative institutional forms for accomplishing the water allocation task that do not rely so heavily on the market institution and instead constrain it with devices such as zoning (12) and other nonmarket values. This investigation will be reviewed in more detail below.

Agricultural Preservation

A third standard is the degree to which an institutional form preserves the social character of the region, specifically, the agricultural and ranching industries and accompanying rural and small-town life-styles that historically provided the genesis for much of the region's economic development. Up until modern times, water and agriculture in the West had been so intertwined as to be virtually synonymous. Even now in a state such as New Mexico, approximately 90 percent of the water consumption occurs in irrigated agriculture (13). The symbolic refrain that supported the massive water development projects of the early and middle twentieth century was "to make the desert bloom" (14) through irrigated agriculture. The value structures as well as the economic well-being of entire families and communities rest on the maintenance of agriculture, and that maintenance in the arid West depends heavily upon the availability of water. Agriculture is already threatened in many areas by declining water tables (15) and rapidly rising costs of pumping (16). Couple that condition with a prospective large-scale shift of water rights from agriculture to other uses, and a potential foundation for resistance to further development of a market institution is laid.

Access to Water

Under the water laws of the western states, a water right is a property right with a legal footing similar to that of land titles and other property rights (17). Yet despite this solid statutory affirmation that prior capture and application to beneficial use conveys a private equity interest in the water supply, western legislatures have historically exhibited a certain ambivalence toward the private ownership of water. The water right itself is a *usufructuary* right (18), granting the owner the use of the water while the water itself remains in the public domain. Even more to the point, some legislatures have guaranteed individuals the right to obtain, at any time, water for domestic purposes even in basins that are already fully appropriated (19). By this action, however, meritorious on its face, legislatures have simultaneously attempted to defy hydrological realities and diluted the equity interests of previous appropriators.

At the heart of this ambivalence is the social norm that asserts that all individuals must be guaranteed access to water even if previous appropriations and private property rights would otherwise militate against that access. Institutional forms for performing the water allocation task must take account of this norm also.

CONFLICTING VALUES

The existing water institutions were shaped by the array of social values through cooperation, conflict, and compromise into acceptable social arrangements for conducting the region's water business (20). Actions stemming from these and other values will also pound on these same institutions until they are molded into acceptable instruments for performing the reallocation task. The challenge is to develop sufficient ingenuity of thought and action that some of the more severe strains upon the social fabric during this process can be avoided and others resolved as amicably and in as timely a manner as possible. Otherwise, society may be faced with debilitating oscillations in which first one value and interest is dominant—thereby making the rules to its liking—and then another. Such a process creates much social friction as well as prevents the stability and continuity of public purpose and rule making that is essential for sound private decision making.

SUMMARY AND CONCLUSIONS

This summary section begins with a restatement of themes. The principal function to be performed by western water institutions in the coming decades is *re*allocation of a fully appropriated scarce resource. Yet the existing institutions were not originally shaped by societal pressures for this reallocative purpose. Consequently, the current period in the evolution of these institutions is an era of stress as the new demands are increasingly loaded on top of old habits, customs, and routines. The business of these institutions is changing, and the institutions themselves must also change if they are to successfully perform this new task assigned to them by society. In this context this study has examined two questions concerning the water reallocation function and the evolving form of the institutions that handle that function.

Market Proficiency

First, there are the *market* institutions as they have developed in the various water basins. In those basins where freely negotiated water transactions and transfers are permitted, the water rights markets are indeed becoming increasingly proficient in strong correlation with increasing prices for water rights. Stated in other terms, there is strong evidence to support a conclusion that the more proficient water rights markets will be found in those precise locations where the increasing demand for water is greater relative to the available supply.

From an economic perspective this is a solidly positive conclusion, for economic theory has well confirmed the social value of markets in allocating and reallocating scarce resources in a manner that will achieve the highest economic value. With regard to the social goal of economic improvement, then, the evidence supports the general conclusion that a healthy evolution of water institutions is indeed occurring in at least the basins studied.

However, there are caveats to this sanguine conclusion. First, in at least one basin—the Gila in New Mexico—there is strong evidence that the ownership of water rights has been consolidated into the hands of a few large corporate enterprises. Although the consolidation has simply been a by-product of the development of large copper mining and milling operations and no socially detrimental results were apparently intended, the consolidation has produced an oligopolistic pattern of ownership in which a few water right holders could strongly influence future water prices if they so choose.

A second caveat concerns legal restrictions on the transfer of conserved water. Ideally, one of the principal beneficial effects of market measurement of the relative scarcity of water is price-induced water conservation. As water becomes more valuable in the marketplace, a right holder is rewarded if he can conserve in his use of water and sell his economic surplus to others. However, if there are legal restrictions that prevent this sale, then the price information provided by the market loses its social usefulness, and conservation may not occur.

There are complex reasons underlying the existing legal restrictions on transfer, some of which are designed to protect third parties who might otherwise be injured by a transfer of conserved water. These must be dealt with constructively in the future so that institutional procedures can be formulated that both protect third parties and yet allow conserving practices to be rewarded. Other arguments supporting legal restrictions on the transfer of conserved water simply reflect outmoded ideas that will eventually give way as the importance of conserved surpluses becomes apparent.

Conflicting Values

In turning to the second question examined, more ambiguity was found, and no definite conclusion can be drawn. It is quite clear, however, that the pressures associated with water reallocation are creating considerable social tension, leading to extensive litigation, and demanding increasing political attention in the West. It is equally clear that no single path has yet been found that will result in the codification of a standard body of law and procedure for performing the task of water reallocation, analogous to the existing doctrine of beneficial use and prior appropriation that has served as a common institutional structure to most of the water development task throughout the region.

It is basic social values that are, indeed, in conflict in much of the region. The conflict is not an abstract construction but is concretely expressed in the opinions and actions of farmers, housewives, mining officials, and others throughout the region. It is found in the opinion of environmentalists who passionately assert that market forces cannot be allowed to eliminate stream flows. Or it is found in the opinion of a farmer who stoutly defends beneficial use as requiring a "taking and using," thereby placing instream uses outside the law. On the question of whether society is succeeding in the development of institutions that can settle these conflicts, the evidence is mixed.

There are examples such as Colorado in which instream values have been legitimatized through constitutional change. Yet many Coloradans do not feel that limitations on market forces are strong enough. There is Arizona in which pressures against the historical dominance of agricultural interests in water matters have accumulated to the point that strong, even extraordinary, legislative action has been taken in delegating authority to a special commission. Yet it appears that the coalition of nonagricultural interests may have redressed an imbalance in favor of agriculture with an imbalance in opposition to it. In Utah and New Mexico institutional problems still persist, but there is reason to believe the basic institutional structure may be adequate to the task.

Some Lessons

While a clear, general conclusion may not be possible on the basis of the evidence that has been compiled, certain observations that do emerge may be instructive in winnowing the institutional paths from which a state may successfully choose.

1. In Arizona the institutional structure has for decades strongly favored agricultural interests through the riparian ground water doctrine established by the court system. Having closed the door on reasonable access to water by other parties and interests, the pressures accumulated until the outcome may be an imbalance in the opposite direction. This observation is instructive in suggesting

that *all societal values should be provided an equal opportunity to obtain access to water*. In particular, it strongly suggests that other states should emulate Colorado in legalizing the holding of water rights for instream purposes. Otherwise, by closing the door to straightforward access on an equal footing with other interests, such a policy invites the development of indirect, obstructing barriers in the institutional process with little resulting gain for society as a whole. Moreover, parties that close the door on others invite retribution at some future time.

Having cast Colorado as a model to be followed, the same lesson from a slightly different perspective may be instructive to that state also. Namely, because Colorado has achieved an equal footing under the law for instream uses compared with other conventional uses, prudence may dictate the avoidance of additional institutional protection for instream values that might ultimately lead to a backlash.

2. The discussion of the legal restrictions inhibiting the transfer of conserved surplus revealed a facet of the water reallocation issue that reappears frequently in many contexts. Third parties, who may not enter directly into the bargain struck by a buyer and a seller of water rights, may nevertheless have a stake in the outcome. That stake may range from the case of another right holder whose right is impaired to a farm implement dealer whose business is declining as a result of transfers away from agriculture. While the former will generally have legal standing under the law to intervene in the institutional proceedings governing the transaction, the latter will not, even if the financial consequences were greater. There are questions of equity and fairness here that the water institutions have not yet explicitly addressed. Even the formulation of these questions is beyond the scope of this study. However, it is very clear that one source of opposition to improve markets for water rights arises from those interests that have no standing in market transactions or the institutional procedures associated with the transfer. The equity interest of third parties needs to be examined and clarified.

3. The increasing proficiency of the market institutions in the water basins of the West suggests that society may be well served if it can find means for accommodating the noneconomic values of society *through* the marketplace, as has occurred with instream values in Colorado, rather than seeking to preserve these values by restricting the market institution itself.

Whatever the institutional path that will ultimately be chosen for the reallocation task, the value of advance thought and conscious action based upon conclusions thoughtfully obtained is strongly supported by the evidence accumulated in this study. To borrow a term from engineering jargon, *retrofitting* of social institutions is extremely costly.

NOTES

1. Charles J. Meyers and Richard A. Posner, 1971, *Market transfers of water rights: Toward an improved market in water resources,* National Water Commission (Arlington, Va.: National Technical Information Service); Willis H. Ellis and Charles T. DuMars, 1978, The two-tiered market in western water, *Nebraska Law Review,* 57:333–367 and F. Lee Brown, James W. Sawyer, and Rahman Khoshakhlagh, 1977, Some remarks on energy related water issues in the Upper Colorado River Basin, *Natural Resources Journal* 17 (1).

2. Allen V. Kneese and F. Lee Brown, 1975, Water demands for energy development, *Natural Resources Lawyer,* 8.

3. Rahman Khoshakhlagh, F. Lee Brown, and Charles T. DuMars, 1977, *Forecasting future market values of water rights in New Mexico* (Las Cruces, N.M.: Water Resources Research Institute, Report no. 092).

4. Office of the Executive Director, Colorado Department of Natural Resources, 1978, *The Colorado water study—Directions for the future; Introduction* (Denver: Colorado Department of Natural Resources) (hereinafter cited as *Colorado water*).

5. Dean E. Mann, 1979, The impact of institutional arrangements on water conservation practices in the West: A policy analysis (mimeographed, Santa Barbara, Calif., University of California, Department of Political Science, October).

6. Norris Hundley, 1975, *Water and the West: The Colorado River Compact and the politics of water in the American West* (Berkeley: University of California Press).

7. Arthur Maass and Raymond L. Anderson, 1978, . . . *And the desert shall rejoice—Conflict, growth and justice in arid environments* (Cambridge: The MIT Press).

8. David Abbey, 1979, Energy production and water resources in the Colorado River Basin, *Natural Resources Journal,* 19 (2).

9. Edward Abbey, 1975, *The monkey wrench gang* (Philadelphia: Lippincott).

10. In 1973 the Colorado legislature passed S.B. 97, which authorized the Colorado Water Conservation Board to

appropriate or purchase water rights for "such minimal flows . . . as are required to preserve the natural environment to a reasonable degree. . . ."

11. A referendum on hosting the 1976 Olympic Games in and around Denver was overwhelmingly rejected in a 1972 vote. One of the factors in that rejection was publicity regarding the destruction of sensitive mountain environments by rapid development and population growth.

12. Office of the Executive Director, Colorado Department of Natural Resources, *Colorado water.*

13. U.S. Bureau of Reclamation and New Mexico Interstate Stream Commission, 1972, *New Mexico State water plan, situation assessment report* (Santa Fe: New Mexico State Engineer's Office).

14. This phrase has been repeated several times by authors but seems to originate with an article written in 1929 by the commissioner of the Federal Reclamation Bureau. Elwood Mead, 1929, Making the American desert bloom, *Current History,* 31:123–132.

15. Two frequently studied aquifers with declining water tables are the Ogallala in eastern New Mexico, west Texas, and Oklahoma [see *Impact of the declining water supply on the economy of Curry and Roosevelt counties* (Las Cruces: New Mexico State University, Agricultural Experiment Station Bulletin 588, 1971)] and the Tucson Basin in central Arizona [see Adrian H. Griffin, James C. Wade, and William E. Martin, 1979, The economic effects of changes in water use in the Tucson Basin, Arizona (mimeographed, University of Arizona, Department of Agricultural Economics)].

16. Hugh Harman, 1970, *Prospective costs of adjusting to declining water supplies in the Texas High Plains* (College Station, Texas: Department of Agricultural Economics, Texas A & M University, Technical Report 71–3).

17. For a complete description of the property nature of water rights, see Wells A. Hutchins, 1971, *Water rights law in the nineteen western states,* vol. 1 (Washington, D.C.: U.S. Government Printing Office), Chapter 5.

18. *Salt Lake City* v. *Salt Lake City Water & Electric Power Company,* 24 Utah 249, 266, 67 Pac. 672 (1902); *Murphy* v. *Kerr,* 296 Fed. 536, 541 (D.N. Mex. 1923); *Garner* v. *Anderson,* 67 Utah 553, 565, 248 Pac. 496 (1926).

19. According to New Mexican law, the state engineer has no control over a domestic well that provides no more than 3 acre·ft annually [N.M. Stat. Ann. Sec. 74–11–1 (Repl. 1968)].

20. Hundley, *Water and the West;* and Maass and Anderson, *Desert.*

35

WATER YIELD AUGMENTATION

STANLEY L. PONCE and JAMES R. MEIMAN

Recent papers on water yield management offer a good opportunity to assess progress during the period since 1965 as well as to once again look ahead to the future.

Predictions made in 1965 (Meiman and Dils, 1965) were reasonably accurate. The current set of papers (Douglass, 1983; Harr, 1983; Hibbert, 1983; Kattelmann et al. 1983; Krutilla et al., 1983; Troendle, 1983), along with the literature cited in them, and the paper by Anderson et al. (1976) attest to the considerable gain in knowledge in water yield management. The moderate forecasts for U.S. population growth have held; however, the large redistribution to the Sun Belt was not foreseen.

The cybernetic revolution is now very much here, although one significant development—the use of computers to model hydrologic processes and resource management decisions—was not predicted in the earlier paper. The multiple-use concept of resource management has developed to an extent far beyond that anticipated in 1965. Although water has become a key concern in such management, management for water yield augmentation has not progressed very far.

The demand for water continues to grow. Current knowledge of watershed processes indicates that vegetation removal increases water yields from upland watersheds. The current set of papers suggests that the reasons why water yield augmentation has been slow in adoption include the problems associated with large-scale applications, the failure to deal realistically with the economics and especially the value of increased yields, and legal problems. Each of these issues are discussed with ref-

erence to the current set of papers. Finally, some suggestions for future consideration are given.

ISSUES AND CONCERNS FOR THE FUTURE

Biophysical Processes

Research results show that water yields can be increased substantially through vegetative manipulation. Troendle (1983) reports, "Through prudent management, we could realize a 2 to 6 cm increase in flow from each hectare of land managed to optimize water yield in the subalpine." Douglass (1983), summarizing work of others, states, "Water yield from well-stocked northeastern forests could be increased by 4 to 12 in. the first year after complete cutting." Hibbert (1983) reports, "Chaparral watershed experiments in Arizona and California demonstrate the mean annual stream flow can be increased by as much as 150 mm (6 in.) by converting brush to grass;" and "stream flow increases of 25 to 75 mm may be expected from clear-cutting mature ponderosa pine." Other work (Anderson et al., 1976; Harr, 1983; Kattelmann et al., 1983) also support this concept.

However, given the current status of the land, what is the actual potential for increasing water yield on a management unit through forest and range management? Several of the papers in the preceding series (Douglass, 1983; Harr, 1983; Hibbert, 1983; Kattelmann et al., 1983; Troendle, 1983) pointed out that most of the wildland watersheds that are available for timber harvesting or

Mr. Ponce is a hydrologist specializing in upland watershed systems. He has been director of the Watershed Systems Development Group for the U.S. Forest Service and chief of the Water Rights Branch for the National Park Service. Dr. Meiman is professor of watershed science and associate vice president for research at Colorado State University. This material is taken from Water yield augmentation through forest and range management. Issues for the future, *Water Resources Bulletin*, 19 (3) (1983): 415–419.

The authors greatly appreciate the review of this manuscript by Robert Winokur, David Herrick, Freeman Smith, Charles Troendle, Gordon Warrington, and Owen Williams. Used with permission. Copyright © 1983 by American Water Resources Association.

species type-conversion have already been treated to some degree. As a result, water yield increases from large areas are normally not as great as those obtained from research watersheds that were in an undisturbed condition before vegetative manipulation.

For example, Harr (1983) reports that water yield from small experimental watersheds in western Oregon increased 360 to 540 mm after 100 percent clear-cut logging and 100 to 300 mm after partial cutting. These increases occurred immediately after harvesting and were observed to have diminished as revegetation occurred. Applying this information to forested lands covering western Oregon and western Washington, and assuming that no timber harvest has occurred in the past and that average annual water yields of the watersheds are unaugmented, he predicts a 4 to 5 percent increase in annual water yield from most large watersheds in the region. Harr points out that the omission of past land use changes and logging from the analysis most likely has resulted in an overestimate of potential increases.

Kattelmann et al. (1983) support this argument. They state that if national forestlands in the Sierra Nevada Mountains were managed almost exclusively for water production, while meeting the minimum legal standards, stream flow would increase by an estimated 2 to 6 percent. Upon imposing additional constraints to meet other resource demands under multiple-use and sustained-yield guidelines, estimates of attainable increased water yield drop to 0.5 to 2 percent of current yield. Practically speaking, they feel that water yield from these lands probably can be increased by only about 1 percent under intensive forest watershed management because most of the potential has been achieved by current practices.

In some areas large watersheds do not have the limitations discussed previously. These watersheds have prime potential for water yield increases and could be expected to yield larger volumes of water relative to the previous estimates.

When one is evaluating the potential for water yield increases, the relationship between timber rotation length and recovery period also should be considered. Harr (1983) reports vegetation recovery in western Oregon and western Washington to be about 27 years. If the rotation length is 100 years, under typical harvesting conditions approximately 27 percent of the management unit would be contributing yield above baseline flow conditions. If the rotation age was lowered to 70 years, the area contributing additional yield could be raised to nearly 40 percent.

In the Colorado Rockies Troendle (1983) reports recovery to be between 60 and 80 years. Given a rotation age of 120 years, conservatively speaking, one-half of the watershed could be contributing additional yield. Lowering the rotation age to 100 years would increase the area contributing additional yield to 60 percent. Consequently, the longer the rotation age, the smaller is the area contributing additional water yield. Because rotation ages are gradually being shortened, this fact alone will tend to increase water yields from forestlands.

Water yield increases, in general, strongly depend on the amount of annual precipitation. Increases in water yield will be larger during wet years relative to dry years. Table 35-1 illustrates that the increase in yield from the North Fork of Deadhorse Creek is three times higher during wet years (1979 and 1980) than during a drier year (1981).

Harr (1983) reports for the period 1965–1981, 75 percent of the total variance in water yield increases (Y) on the experimental watershed HJA–1 in western Oregon is accounted for by Equation (1):

$$Y = 513.2 - 19.1X_1 \qquad (1)$$

where X_1 is the number of years after logging. However, adding annual precipitation (X_2) to the model (Equation 2),

$$Y = 308.4 - 18.1X_1 + 0.87X_2 \qquad (2)$$

results in accounting for 89 percent of the total variance in size of yield increases at HJA–1. Consequently, it is important that the stochastic nature of total annual precipitation, as well as seasonal distribution, be considered when one is designing a water yield augmentation program.

Timing is of critical concern. When will the augmented yield be realized? In the intermountain

Table 35-1. The Observed Increase in Flow Following Timber Harvesting of 36 Percent of the Forest by Clearcutting 12 Circular Units Five Tree Heights (5H) in Diameter During 1976 and 1977 on the North Fork on Deadhorse Creek

Year	Annual Precipitation (cm)	Annual Flow (cm)	Flow Increase (cm)
1978	62.7	23.1	3.6
1979	70.9	21.1	5.8
1980	71.9	23.1	6.6
1981	51.8	9.4	2.0
Mean	64.8	19.3	4.6

Source: Troendle (1982).

West, where water yield can be augmented by snowpack management, the increased yield is generally obtained on the rising limb of the snowmelt hydrograph, and little, if any, increase is available as summer base flow (Troendle, 1983). Demand for the additional water generally occurs during the summer. Consequently, storage facilities are required to hold this additional water if it is to be available at a later date. Such facilities require land and money for construction and may adversely affect the other beneficial values of the water resource.

Another major issue and concern is the effect of water yield augmentation on the environment. Kattelmann et al. (1983) note that in some areas of California steep slopes are too unstable for timber harvesting. The risk of mass wasting is high in these areas, and the likelihood of increased sediment loading to the stream courses is greatly increased. Similar concerns also exist for other regions of the country.

In addition, the expected impact of increased yields on stream channel morphology should be evaluated. Most stream channels are in a state of dynamic equilibrium (Schumm, 1974). Flow increases change the energy regimen of the channel system, which may change the sediment transport characteristics significantly and adversely affect the aquatic ecosystem.

Economic Issues

Krutilla et al. (1983) discuss in detail the assumptions that must be made in arriving at an economic evaluation of watershed management for joint production of water and timber. They observe that there are often two different markets for water, each with its own supply and demand characteristics. The choice of the agricultural irrigation market or the expanding urban-industrial market can make an order-of-magnitude difference in the value placed on water yield increases. Traditionally, the lower agricultural values have been used.

Although all the authors point out the restrictions on areas treatable for water yield increases, the need for water so produced is identified as very high in the Rocky Mountains (Troendle, 1983) and western rangelands (Hibbert, 1983); limited but developing in the eastern United States (Douglass, 1983); and relatively low in the Sierra Nevada (Kattelmann et al., 1983) and Pacific Northwest (Harr, 1983). Kattelmann et al. (1983) suggest that delaying stream flow instead of increasing yield may be the most valuable water yield practice in the Sierra Nevada.

Most of the authors recognize the potential for water yield management for carefully targeted local areas. Hibbert (1983) provides an extreme example in this regard in his discussion of water harvesting on microwatersheds. Kattelmann et al. (1983) mention the benefits to expansion of mountain communities and small-scale hydroelectric facilities in the Sierra Nevada. Douglass (1983) suggests the greatest opportunities in the eastern United States are on the 4 to 5 percent of lands in the Northeast and South in municipal watersheds. Krutilla et al. (1983), using the results from the Fraser Experimental Forest, estimate that the value of that increment in yield used by the Denver Water Board for municipal industrial applications is about $112 per acre-foot per year.

The issue of the economic values resulting from large-scale but diffuse application of water yield management over entire basins is less clear. Most of the authors tend to discount the basinwide effects because they are such small percentage increases (1 to 5 percent) as contrasted to the 25 to 30 percent increases on small watersheds. Nonetheless, there is little evidence to suggest that in most cases these increases are no present even though unmeasurable, and although they are small percentages in a large basin, they may represent very large volumes of water. The issue is further clouded by the fact that varying degrees of timber harvest have been underway for many years on most watersheds.

The Use of Models

From a management perspective, system modeling may help solve several problems. The first is extrapolation of research results from selected watersheds to other basins. Several watershed models currently exist for prediction, at the planning level, of hydrologic response to vegetative manipulation on basins of fourth order or less in size. The most notable of these models is WET (Williams, 1983), which is the computerized version of the hydrology chapter of WRENSS (U.S. EPA, 1980), which was developed by using results from experimental watersheds nationwide. The WET model incorporates simulations produced by the Subalpine Water Balance Model (WATBAL) (Leaf and Brink, 1973a, 1973b) for snow-dominated regions and PROSPER (Goldstein et al. 1974) for rain-dominated regions. The WET model has been designed as a planning tool intended to predict changes in water yield relative to an "undisturbed" condition. Water yield and timing for the existing condition as well as for proposed treatments can be readily assessed, relative to the undisturbed condition. The model requires readily available input data and will use regional functions if local hydrograph/flow duration information is not available.

A problem with the use of system models for predicting water yield in upland watersheds is that data are seldom available for site-specific calibration or validation. This leads to the question of what level of confidence the user (manager) can place on the results. When one lacks an analytical assessment, professional judgment is needed to assess the credibility of the predicted increase in water yield. To a large degree, confidence is dependent on the degree to which the model is based on scientific principles underlying the hydrologic process, the degree to which it has been validated in other areas, and how credible its predictions are for the watershed of interest.

Another application for water resource modeling is to simulate routing of additional water to downstream users. Both Harr (1983) and Krutilla et al. (1983) point out that incremental yields produced on upland watersheds are difficult to measure downstream because the magnitude of increase is generally less than the measurement error. Furthermore, it is even more difficult to accurately account for the use of the water by downstream users and to determine the amount of consumptive use and transmission loss throughout the distribution system. Although some models estimate evaporation and transmission losses and return flows, considerable uncertainty is associated with their estimates. Clearly, this is an area where further study is needed.

Given the ability to estimate increased yield from a specific land area and to estimate a realistic value of the water to downstream users, the manager is faced with the problem of optimizing various combinations of outputs of natural resource products at different cost levels. These models are generally extremely complex linear programs, such as FOR-PLAN (Johnson et al., 1980). Because of their complexity, it is often difficult to assess if the output for the allocated products is reasonable. By contrast, in the area of water resource development analytical tools are available to assess trade-offs when one is developing the water resource.

Legal Issues

Of all the problems involved in water yield augmentation perhaps none are so vexing as the question of legal rights. Douglass (1983) succinctly sums the problem up when he says, "there is no incentive to expend money to increase production of a product (water) that the landowner cannot sell." Until these legal problems are resolved, water yield augmentation by vegetation management will be confined to public lands, to be used for the general public good.

The legal problems are formidable, particularly in that the areas of greatest need and potential are often under the prior-appropriation doctrine. In these situations water rights are defined in terms of season, place, point of diversion, type of use, and return flow characteristics. Furthermore, water yield increases from vegetation management vary from year to year, as noted previously. Another problem is that legal descriptions are often difficult, if not impossible, to interpret in hydrologic terms (e.g., such terms as *new water* and *salvage water*).

Finally, there is the question of what constitutes adequate proof of increased yield. It is not possible to measure such increases in large basins, nor is it practical to install a stream gage on each subwatershed and calibrate water yield for many years before treatment for augmenting yield. The advances in hydrologic modeling offer a promising alternative if such models would be accepted in the courts.

SUMMARY

The greatest potential for water yield augmentation appears to be on carefully selected watersheds that have the biophysical potential, produce water that is used for high-value purposes, and can be managed under sound multiresource management. The first such watersheds to be managed in all regions are those that supply municipal-industrial water directly or contribute directly and significantly to hydropower generation. However, management of these types of watersheds on any large scale will require changes in approaches to economic analysis and in water laws. We predict that such changes and the resulting management will come in the near future.

In general, the demands for timber and water are expected to increase in the future. While current land management practices do enhance water yield, opportunity still exists to achieve an optimum production of water subject to the constraints of competing land uses. However, full realization of potential requires a more accurate prediction of the quantities of water produced and a clearer definition of the associated values.

Although research has shown that water yield can be increased by forest and range management, the means to transfer this information to larger areas is limited. We see system models as effective tools for performing this task. Current models need to be verified and validated so that their reliability can be assessed over a broad range of environmental conditions. Improved models also will be needed. In general, additional local data are needed to calibrate system models so that they can perform at an acceptable level. Improved methods are necessary to determine the value of additional water to downstream users so that realistic resource optimizations, consistent with sound multiresource management, can be determined. In addition, the issue of court acceptance of system models as proof of water yield augmentation needs to be resolved.

The opportunity to augment water yield as a large-scale program may not be as great as has been demonstrated on small experimental watersheds. This is because of many physical and administrative constraints. The implementation is further restricted because there is little incentive for private landowners, who own a significant portion of the eastern forests as well as large tracts in the West, to manage for the production of a commodity they cannot market. Until incentives to participate in a water yield program become available to the private landowner, their management strategies will not emphasize water production.

Land managers and resource management agencies have clearly recognized that in the water-short areas of the West the conservation of existing supplies and increasing yield through prudent forest and range management are issues of great concern. They have also realized that in some areas the potential to increase water yield over existing flows is substantial.

Based on the experience of the past 17 years, it is apparent that the extent of water yield augmentation using vegetation management will depend on the need for additional water as well as the alternative means available to meet this need. The high cost of energy is forcing a critical look at agricultural use of water and is stimulating conservation in the most consumptive uses of water. Conversion of water use from highly consumptive agricultural applications to much less consumptive municipal/industrial uses, more efficient use on farms, and water yield augmentation are alternatives. As better techniques are developed to predict water yield from vegetation management and assuming that pressures develop to solve the legal and economic issues, both private and federal land managers are likely to employ water yield augmentation in their management strategies in response to these pressures.

REFERENCES

Anderson, H. W., M. D. Hoover, and K. G. Reinhart. 1976. *Forests and water: Effects of forest management on floods, sedimentation, and water supply.* Pacific Southwest Forest and Range Experiment Station, USDA–Forest Service General Technical Report PSW–18/1976, Berkeley, California.

Douglass, J. E. 1983. The potential for water yield augmentation from forest management in the eastern United States. *Water Resources Bulletin,* 19(3):351–358.

Goldstein, R. A., J. B. Mankin, and R. J. Luxmore. 1974. *Documentation of PROSPER: A model of atmosphere-soil-plant water flow.* Environment Science Division Publishing no. 579, Oak Ridge National Laboratory, Tennessee.

Harr, R. D. 1983. Potential for augmenting water yield through forest practices in western Washington and western Oregon. *Water Resources Bulletin,* 19(3):383–393.

Hibbert, A. R. 1983. Water yield improvement potential by vegetation management on western rangelands. *Water Resources Bulletin,* 19(3):375–381.

Johnson, K. N., D. B. Jones, and B. Kent. 1980. *Forest planning model (FORPLAN). A usersguide and operations manual.* Land Management Planning, USDA–Forest Service, Fort Collins, Colorado.

Kattelmann, R. C., N. H. Berg, and J. Rector. 1983. The potential for increasing streamflow from Sierra Nevada watersheds. *Water Resources Bulletin,* 19(3):395–402.

Krutilla, J. V., M. D. Bowes, and P. Sherman. 1983. Watershed management for joint production of water and timber: A provisional assessment. *Water Resources Bulletin,* 19(3):403–414.

Leaf, C. F., and G. E. Brink. 1973a. *Computer simulation of snowmelt within a Colorado subalpine watershed.* USDA Forest Service Research Paper RM–99, Rocky Mountain Forest and Range Experiment Station, Fort Collins, Colorado.

——. 1973b. *Hydrologic simulation model of Colorado subalpine forest.* USDA Forest Service Research Paper RM–107, Rocky Mountain Forest and Range Experiment Station, Fort Collins, Colorado.

Meiman, J. R., and R. E. Dils. 1965. *Watershed management—Practice and trends in the United States.* Proceedings of the First Annual Meeting of the American Water Resources Association held December 1–3 at the University of Chicago, 215–221.

Schumm, S. A. 1974. Geomorphic thresholds and complex response of drainage systems. In. *Fluvial geomorphology,* ed. M. Morisawa, 299–309. Proceedings of the Fourth Annual Geomorphology Symposium. Publications in Geomorphology. New York: SUNY-Binghamton.

Troendle, C. A. 1982. The effects of small clearcuts on water yield from the Deadhorse watershed, Fraser, Colorado. In. *Proceedings of the 50th Annual Western Snow Conference* (Conference in Reno, Nevada, April 19–23, 1982). Fort Collins, Colo.: Colorado State University.

————. 1983. The potential for water yield augmentation from forest management in the Rocky Mountain region. *Water Resources Bulletin,* 19(3):359–373.

U.S. EPA 1980. *An approach to water resources evaluation of non-point silvicultural sources (a procedural handbook).* EPA–600/8–80–012. Athens, Ga.: III–1—III–173.

Williams, O. R. 1983. *WET: The computerized version of Chapter III—Hydrology—of the WRENSS handbook.* WSDG–AD–00007. Fort Collins, Colo.: Watershed Systems Development Group, USDA–Forest Service.

36

THE FUTURE OF WATER

PETER P. ROGERS

Is there a "water crisis" in the United States? The impression has certainly been created in recent years that there soon won't be enough water to go around, and that much of what is left is likely to be poisoned. The 1982 report on the state of the environment by The Conservation Foundation, a nonprofit group devoted to the study of environmental policy, describes how excessive consumption of surface water has led to a reliance on ground water, which has itself been depleted and also stressed by contamination. By last fall the Environmental Protection Agency (EPA) had found that the ground water supplies of 29 percent of the 954 cities in its sample were contaminated, chiefly by toxic wastes that had leached out of landfills.

Most of what one reads about the pollution of water in every region, and the short supply in some parts of the country, is true. It is not true, however, that nothing can be done. Water's depletion and contamination are not irreversible. We have the technology and management skills to rise to the challenge; only the will to do so has been lacking.

Unfortunately, for many people the response to news of contamination and shortage—or to the way in which the news is presented—is fear rather than determination. The anxiety arises in part from the fact that since the turn of the century, we have been able to take unlimited amounts of clean water for granted. But as recently as the end of the nineteenth century, the mortality and morbidity rates from waterborne diseases were very high. (Because they remain high for the majority of the world's population, the United Nations has designated the 1980s the International Drinking Water Supply and Sanitation Decade.) The engineering achievements of the nineteenth century and the early years of the twentieth were, in effect, too

successful: As water in short supply and of poor quality became a distant memory, purity and plentifulness began to be assumed rather than systematically safeguarded.

Although the results of our poor stewardship are not beyond repair, they should not be minimized. Of these, contamination is by far the most notorious. In 1974 the Safe Drinking Water Act, which Congress passed largely in response to the efforts of environmentalist groups, forced many communities, often for the first time, to search for chemical contaminants in their water supplies, as well as for the traditionally monitored biological sources of pollution. People have been quite surprised by the exotic toxic chemicals that were found—and are still being found. Disclosures of the reckless disposal of toxic wastes, which, by leaching into the earth, can contaminate ground water as well as soil, have not inspired confidence in either government or industry, and the addition of Times Beach, Missouri, to a long roster of contaminated sites has led many people to question the safety of their own surroundings.

Some contaminants are showing up in water not because they have been recently introduced but because scientists have only recently become able to measure them in small concentrations. Moreover, the science of assessing the amount of contaminants in water is a good deal more precise than the science of assessing the effects of contaminants on human health. Here we find that definitive judgments are extremely difficult to make. Although much work remains to be done to improve these assessments, scientists may never know more about the toxic effects of many chemicals than they do now. Too many variables—occupation, per-

Dr. Rogers, a professor at Harvard University, teaches city planning at the Kennedy School of Government and environmental engineering in the Division of Applied Sciences. He is a consultant to the Ford Foundation and the Government of Bangladesh. This material is excerpted from an article of the same title in *The Atlantic Monthly* (July 1983): 880–892. Used by permission. Copyright © 1983 by *The Atlantic Monthly*.

sonal habits, diet, the presence of environmental insults other than the chemical in question, and so forth—exist in any given geographic area to allow clear linkages of diseases to specific contaminants. This frustration is not likely to diminish significantly with time; it represents an inexorable deficit in scientific understanding.

The public has little tolerance for the ambiguities that complicate statements about contaminated water's effects on health. Confronted with the lack of sure knowledge, many people assume that they are being manipulated for devious reasons. Often, in fact, the suspicion is well founded. Although the uncertainties about toxicity ought to be a warrant for extreme caution in any effort to purify water, they have served also as a refuge for manufacturers and the overseers of landfills, who can cite the uncertainties as a way of understanding their liability once contamination is discovered. Under these circumstances rational debate on how best to protect water is hard to sustain.

Safeguarding water's quantity and quality are two parts of a single enterprise—management— and most of the threats to water in this country stem directly from the fact that it is a common resource that has been managed chiefly in the service of private profit. Believing that water is limitless, we have made it available to industry and agriculture essentially at zero cost and with few stipulations. Therefore we should not be surprised or offended by the profligacy with which these users have exploited it. Few corporate officers who look to the bottom line of a financial balance sheet would insist on investing large sums to conserve or clean up a resource for which their companies did not have to pay much in the first place.

To worsen the consequences of this fundamental error, the country's shortsighted and rudimentary control of the environment makes it inevitable that the regulators will trail behind those regulated. Even as we restrict the permissible levels of some contaminants, we discover others and must set regulations on those as well, in a never-ending game of catch-up. Similarly, as we undertake large and costly engineering projects to transfer water to farms and factories in regions where it is dwindling, we ensure that the practices that have led to a supply's depletion will persist. Both these obstacles are chiefly matters of policy, however, and policy—unlike the laws of nature—can be changed.

Poor planning is particularly obvious in the country's efforts to deal with chemical wastes, which are a chief source of water's contamination. Although a variety of treatment methods have been developed to make hazardous wastes safe, most wastes have been either stored untreated or indiscriminately dumped. Unfortunately, storage is at best a temporary solution and an unreliable policy for the future.

The annual tonnage of hazardous wastes produced in the United States and other data equally fundamental to any reckoning of the threat to the water supply are by no means firm. The EPA and other government agencies have not been collecting data for long, and the release of interim numbers to the scientific community and the press has given rise to inconsistent but often-cited figures. For example, a lengthy report on hazardous wastes, which Congress's Office of Technology Assessment (OTA) published in March, cited an estimate by the EPA that between 28 million and 54 million tons of federally regulated hazardous wastes are released in the United States each year. The OTA estimates that the total is instead somewhere between 255 million and 275 million tons. The enormous difference between these estimates, which is explained by the OTA's more liberal definition of hazardous wastes, can only complicate the government's efforts to control them. Moreover, within both the OTA's estimate and the EPA's, the margins for error are so wide—at least 20 million tons—that either one alone is a poor basis for policy.

Since no one can say precisely how many tons of hazardous wastes are produced in the United States each year, the fact that no sure figures are available on the extent to which chemicals have infiltrated the nation's water should come as no surprise. The United States Geological Survey (USGS) oversees more than five hundred stations around the country, which supply regular information on the quality of surface water. These stations do not keep track of all pollutants, however. They do not measure most of the potentially toxic chemicals, and those they do test for elude detection at the low concentrations that would be expected in large volumes of water—concentrations at which the pollutants may nevertheless have significant adverse effects. Thus, although the data the USGS does provide indicate that the country has been able to hold its own with respect to conventional pollutants, such as municipal sewage, the extent of contamination of surface water by unconventional pollutants is barely known.

As for ground water, *no* systematic monitoring has been undertaken. Instead, we have only incidental evidence, brought to light, by and large, by concerned individuals. In the San Gabriel Valley of California, for example, 39 wells that supplied water to 13 cities had to be closed in 1980 when they were found to be polluted by high concentrations of trichloroethylene (TCE), an industrial solvent and degreaser. Seven municipal wells and 35 private wells near Atlantic City, New Jersey, had been closed by March 1982, after wastes from a chemical dump site were found to have seeped through the sandy soil into ground water. In Bedford, Massachusetts, 4 wells providing 80 percent of the town's drinking water were closed in 1978 when they proved to have been contaminated by high concentrations of dioxane and TCE, among other toxic chemicals. By 1979 drinking water in a third of the communities in Massachusetts had been found to be contaminated. In the spring of 1980 the federal Council on Environmental Quality compiled all the reports of ground water contamination that had been filed at the EPA's regional offices. The results showed that contamination had occurred in at least thirty-four states: in nearly every state east of the Mississippi River and in the less industrial western states as well.

Although the EPA is conducting tests of drinking water wells scattered nationwide, it has concentrated chiefly on locating potential sources of contamination. By the end of 1982 the agency had found 80,263 pits, ponds, and lagoons in which industrial, municipal, agricultural, and mining wastes are impounded. Fewer than 10 percent of these sites have been investigated, and of these only half are being checked on a regular basis to ensure that cleanup is proceeding. In addition to these impoundments, approximately 16.6 million septic tanks routinely disperse their effluent—about 800 billion gallons (gal) a year—into the ground. Septic tanks can be a source of chemical as well as biological water pollution, because of the popularity of cleaning fluids sold to flush out sludge. The fluids contain TCE and benzene, which are both known carcinogens. Finally, the 14,000 active municipal landfills, many of which receive industrial wastes as well as the solids from sewage treatment plants, and the 75,000 active industrial landfills generally are unlined and leach their contents into ground and surface waters.

Given the volume of contaminants released into ground water, the fact that they have taken so long to show up may seem puzzling. But most of the chemicals are tasteless and colorless, and therefore hard to detect, especially at low—but nevertheless, potentially toxic—concentrations. Moreover, water travels through the ground quite slowly; a contaminant may emerge in a well years after it was released from a source even a short distance away.

The insidiousness of the poisons, their wide geographic distribution, and the frequency with which they appear when tests are conducted understandably provoke reactions of panic or despair. Remedies do exist, however. We can insist that manufacturers reduce the amount of waste they produce; we can treat the wastes before disposing of them; we can treat water after it has been polluted; finally, we can restore an entire area after wastes have been discharged. Of course, some combination of these strategies is also possible. It is generally recognized, however, that the cost of cleaning a site after untreated wastes have been dumped is somewhere between ten and a hundred times the cost of treating wastes before disposing of them. The first two options are therefore clearly the best policy.

Any landfill, no matter how well engineered, is likely to leak eventually. Thus to start with, landfills might be designed in such a way that the leaks would be directed into collectors; even then, they will have to be checked over a much longer period than the 30 years now required by the EPA, because of the extreme corrosiveness and long lives of many of the chemicals contained. Beyond this, technologies for treating wastes are already in place at many manufacturing plants, and the advantages and disadvantages of potential improvements on these should be explored. Segregating the harshest wastes before they are released to a treatment facility (rather than shipping them in mixed form) and altering manufacturing processes to make their by-products more benign and to allow for recycling appear to be relatively cheap but effective ways to control potential contaminants at their source. Wastes can also be burned. Incineration is not appropriate for all chemicals, however. The most toxic ones must be subjected to extremely high temperatures—an expensive and not yet perfected technology—to keep the gases given off from polluting the air. Dumping hazardous wastes in canisters in the ocean and injecting them in the earth at a depth of 5000 ft or more—that is, far below easily accessible water tables—are two other possibilities. (Even now, in many coastal cities the effluent of sewage treatment plants is freely discharged into the ocean.) Both plans are risky, however, because they bring untreated wastes into direct contact with the environment. Eventually, canisters will leak or break up in the ocean, but according to proponents of this strategy, the wastes released would be so diluted that their toxicity would be nil. Shifts in ocean currents could defy this theory, though, and wash the wastes in close enough to shore to pose a threat after all. The chemicals could also be ingested by fish and find their way into human diets. As for deep-injection wells, wastes could migrate upward along unsuspected fault lines; earthquakes and other unforeseen geophysical phenomena could disturb them too. Chemically breaking down batches of wastes into less toxic forms or decomposing them with bacteria are perhaps the best tactics tried so far, because they ensure that all hazards have been eliminated before a substance is released from a plant.

There is no doubt that, at a price, the country can safely handle its wastes. Prices quoted by nine large firms that treat chemical wastes to bring them up to acceptable safety standards range between $55 and $240 per metric ton. Generously assuming a price of $250 a ton to treat 54 million tons of hazardous wastes a year (the maximum volume estimated by the EPA), the annual cost would be about $13.5 billion. This cost could rise, of course, as more chemicals are analyzed and found dangerous.

Is $13.5 billion, or more, a year a high or a low price to pay for the safe disposal of chemical wastes? There is no easy answer to this question. Consider, however, that this sum represents 2 percent of the annual contribution of manufacturing to the GNP and 0.4 percent of the GNP as a whole. It should also be kept in mind that this $13.5 billion would not be lost, as payments for imported oil are. Rather, it would create and stimulate a substantial industry—and jobs—in its own right.

A task more arduous than controlling wastes before they assault the environment is cleaning them up after the fact. How a site is cleaned up varies with its proximity to houses and drinking water supplies, the type of soil, the disposition of ground water, the specific chemicals dumped, and how they were dumped (in liquid or solid form; in containers or not). Generally, though, problem dump sites can be handled in one of three basic ways: isolating the site by surrounding it with trenches lined with clay or some other nonporous material, and lining and capping the site itself; diverting the flow of any water through the site, underground or on the surface; removing the wastes and the soil and water they have polluted, decontaminating them by one of the methods cited earlier, and storing them at a controlled landfill.

These options are not uniform in cost. Clearing a contaminated site and treating everything removed from it entails a good deal of money and labor, but once the task is accomplished, little else is required. In contrast, steering water away from a contaminated site is less expensive initially, but purification plants nearly always must be built to ensure that the water is truly safe, and so maintenance costs can be high. Moreover, regular tests for toxicity are essential, no matter how a contaminated site is handled. The EPA's crisis management strategy has been simply to transfer the contents of a dump site, without treating them, to other landfills, which are themselves prone to leaks.

Although the EPA's response has the advantage of being fast and relatively cheap in the short term, its ultimate wisdom and cost-effectiveness are dubious.

The obstacles to cleaning up hazardous wastes, like the obstacles to treating them, are economic, and therefore political, rather than technical. The OTA reports that some 15,000 uncontrolled dump sites are in need of attention. The total cost is estimated at between $10 billion and $40 billion—many times the $1.6 billion that the federal government expects to collect from chemical manufacturers for its environmental Superfund, established in 1980. If the country chooses to accept the expense, however, the goal is within reach.

Aside from the steps the country can take to protect water from contamination, there are ways to protect the public from the toxic effects of contaminants already in water. No contaminants appear to have an irreversible effect. What we put in water, we can usually extract—at some cost. Indeed, only the cost and long-term efficacy of the available technologies remain to be established.

The most reliable method that has been developed so far is the granular-activated-carbon (GAC) filter—essentially ground-up charcoal. (Other filter mediums, such as synthetic resins, are also being studied.) The filter works by a physical process called *adsorption*. Molecular attraction causes the chemicals flushed through a carbon filter to adsorb, or stick, to the carbon's surface. Once the grains of carbon are covered completely, they are either restored by being subjected to a high temperature, which drives off the chemicals adsorbed (if the filter is a large one, at a municipal plant), or discarded (if it is small, attached to a household tap). There is little risk involved in the disposal of household filters, because they are small and because, in any case, the molecular bond is extremely strong.

A special advantage of carbon filters for municipal drinking water is their proven ability to trap not only chemical wastes that have infiltrated the system from outside but also unwanted chemical by-products of chlorination. The great reduction in waterborne diseases in the United States has been sustained by chlorine, which rids water of harmful bacteria. It turns out, however, that chlorine reacts with natural as well as man-made chemicals in water to create a family of chemicals known as trihalomethanes, one of whose members is chloroform, which has been shown to cause cancer in animals. Flushing water that has been disinfected with chlorine through a carbon filter at the end of the treatment cycle can efficiently remove trihalomethanes and any other hazardous organic chemicals present, according to a report released in 1980 by the National Academy of Sciences. The report cautions that further study is needed to make sure that potentially toxic chemical reactions do not occur on the carbon and that no microorganisms build up in the filters, but it concludes that there is already "ample evidence for the effectiveness of GAC" and advocates the filter for municipal supplies threatened by contamination.

In 1978 the EPA proposed carbon filters as the option of choice for the control of hazardous chemicals in water. Studies sponsored by the EPA have shown that for cities with a million or more residents, the cost of water would rise between 17 and 21 percent. In relative terms one could argue that the increased cost would be large. But in absolute terms, and considering how cheap water is everywhere in the United States at present, the increase would be modest; based on the EPA's figures, for a family of four it would range between $11.81 and $60.25 a year.

Many water utilities have been using carbon filters to control taste and odor and to treat wastewater; some have installed them specifically to control toxic chemicals. People who are concerned that their water might not be safe could ask their utilities for the results. Local health departments may also be able to provide this information, and some may even be willing to accept a sample and test it themselves. Independent tests of domestic water supplies are expensive, but they are the only recourse for the 33 million Americans whose water comes from their own wells.

Fortunately, some of the household-sized carbon filters on the market do a very good job of ridding water of chloroform, trihalomethanes, and other toxic chemicals. In Rockaway Township, New Jersey, a dozen houses relying on contaminated well water have been fitted with such small carbon filters, and tests show that the water is up to acceptable safety standards.

An investment in the treatment of contaminated water need not be an endorsement of the uncontrolled release of hazardous substances into the environment. But treating contaminated water can protect the public health for relatively low additional costs while the country goes about cleaning up the mess it has already made.

A good engineering approach is to be very conservative and to set the highest standards affordable. It is legitimate to ask whether or not the cost of improving the quality of water is too high, but it is also legitimate to ask, Compared with what? If the cost is compared with what industry, agriculture, and households now pay for water, it does

seem large. But if we compare the cost with the value of water that won't make us sick, it is small.

Although by now it is a commonplace to say that we live in an era of limits, water is still managed in accord with presumptions, laws, and institutions that do not reflect a recognition of these limits. Abel Wolman, professor emeritus of environmental engineering at Johns Hopkins University and one of the world's foremost experts on water supply, was quoted in *The New York Times* on August 9, 1981, as saying:

> Water is cheaper than dirt. That means there is no orderly design as to when and where to use it. In a vast country such as ours we have never been able to organize a thoughtful, logical national plan, and I am very doubtful we ever will.

One does not have to be an economist to see the problem of underpricing. Common sense tells us that people waste commodities that they do not value highly or commodities that, though valued highly, are supplied cheap.

Both in the Southwest, where water shortages are permanent and likely to worsen if present consumption rates continue, and in the Northeast, where shortages are periodic and rarely dire, water prices are extremely low. Indeed, they are lower by half here than they are in Europe. This disparity in prices is reflected in the sharp contrast in rates of consumption between the United States and Europe. The average amount of publicly supplied water for all nonagricultural purposes in the United States is about 180 gal per capita per day (gpcd), of which 120 gal go to households and for municipal uses, such as street washing and fire fighting; the rest goes to manufacturing and businesses. Yet in countries with comparable levels of social and economic development, consumption is a fraction of the U.S. rate.

Price differentials alone do not explain the large differences in water consumption in these societies. Consider that in New York City, where water prices are on a par with those elsewhere in the East, the rate of consumption is reported to be 180 gpcd, and supplies to most residential users are unmetered. (As it turns out, the rate of consumption in cities where water *is* metered is roughly half the rate in cities without meters.) In Boston, water prices are also in the range of prices elsewhere in the East, yet the rate of consumption is 233 gpcd—well above the national average. As much as 40 percent of that amount is lost to leaks in the system. But guesses as to the effect of doubling Boston's 1978 price of $1.18 (now *down,* to $1.00) per thousand gallons suggest that consumption would be reduced at most to 160 gpcd.

Something else in addition to price seems to influence water consumption in the United States. Americans may be able to recognize, in a theoretical way, the pressures that serve to diminish supply, but they seem unable to keep conservation in mind when devising distribution schemes that borrow from reserves necessary for the future, when confronted with the need to renovate old and wasteful utilities, and when watering lawns and taking showers. Thus even where water is least plentiful in the United States, prices rarely reflect supply. Indeed, prices are lowest where water supplies are lowest: $0.53 in El Paso and $0.59 in Albuquerque, compared with $1.78 in Philadelphia or $1.00 in Boston. Moreover, a rise in price does not seem to provoke a commensurate decline in demand. Perhaps American prices are so low that even when they are raised, they fail to make an impression on consumers. Can the average homeowner even remember the amount of his last water bill? And how does that expense compare with the cost of an evening's entertainment or a tank of gas?

However low the price of municipal water in the United States appears in comparison with prices in other countries, it is high in comparison with what American farmers pay. The price of agricultural irrigation water in the United States ranges between about $0.009 and $0.09 per thousand gallons, and federal subsidies make up the difference between price and cost. The amount of subsidy in the more arid parts of the country is staggering, in fact. In 18 western federal irrigation projects that the U.S. Department of the Interior studied in 1980, the government was providing between 57 and 97 percent of the costs over the lifetimes of the projects. Until recently, a 160-acre farm was the largest eligible for federally subsidized water under the 1902 Reclamation Act. An amendment to the act was passed during 1982, however, which raised the ceiling to 960 acres for individuals and to 640 acres for corporations. The amendment merely acknowledged the fact that the Bureau of Reclamation had been providing subsidies to farms larger than 160 acres for some time. Senator Daniel Patrick Moynihan of New York, who has been persistent in calling for more rational water policies, argued on the Senate floor that the bill represented "all that is wrong with federal involvement in water resources: chaos, arbitrariness, inequity, and waste."

Water is not limitless, but contrary to the views of alarmists, there is no absolute danger that it will run out. The United States as a whole can count on at least fifty years without serious shortages, even at the present wasteful rates of consumption. Extremely difficult choices will have to be made soon in some regions, however, and almost everywhere methods of allocating present supplies and developing new ones need to be more conservative. The nation has the power to abandon various water-intensive activities in favor of others less demanding on supplies, and the opportunities are numerous enough to make the transition relatively painless.

Annual precipitation in the continental United States is about 4200 billion gallons per day (bgd), on average. Of this amount, 1400 bgd should be available for all our uses; the rest evaporates. But yearly fluctuations in precipitation further diminish the supply, in 95 out of 100 years, to 675 bgd. The variability of precipitation should be obvious to everyone. Notwithstanding the costs that result from flooding and soil erosion, rains could bring important benefits: increased supplies in surface reservoirs, an increased volume of ground water, and increased soil moisture to ease the demand for irrigation. Unfortunately, though, the amount of water that can be carried over from wet years to drier years is quite small, because the capacity of American reservoirs to store it is limited.

According to the U.S. Water Resources Council, the nation's total average daily consumption in 1975—the most recent year for which the council made an estimate—was 106 bgd. Of this amount fully 83 percent was used in agriculture. Improved irrigation efficiencies could reduce this share substantially. A tremendous asymmetry exists between agriculture, which consumes 83 percent of the nation's water but contributes only 3 percent to the GNP, and manufacturing, which consumes less than 6 percent of the water but contributes 27 percent to the GNP.

Nearly a quarter of the total amount of water consumed in the United States is drawn from the ground. Between 1950 and 1975 the demand for ground water increased 140 percent. Ground water accounts for half of the irrigation water used in the West, and some appealing features make it a prime target for development. First, a tremendous quantity is stored in American aquifers—underground streams and basins. The Water Resources Council estimated in 1978 that more than a thousand times the amount of water consumed in 1975 lay within 2500 ft of the surface. Second, ground water can be used to even out the natural fluctuations in precipitation from year to year. And third, in many areas ground water is the only source that remains to be appropriated.

Unfortunately, in addition to those advantages, there are some drawbacks. Extracting water from the ground demands a good deal of energy. Also, the rate of replenishment, or "recharge," is slow—typically less than 10 percent a year, and only slightly faster when precipitation is above average. These drawbacks are now becoming critically evident in the West, where, since 1950, development has depended largely on ground water sources. As a result, annual withdrawals of water in excess of the amount recharged to an aquifer—"overdrafts"—range between 4 and 95 percent in those regions in which farm acreage is increasing most rapidly. In Texas and the Oklahoma High Plains (the site of the vast Ogallala aquifer), the annual overdraft is equivalent to the unregulated flow of the Colorado River—approximately 14 million acre·ft a year. In all of the western states the overdrafts exceed 20 bgd, or 22.4 million acre·ft a year. In the remaining states overdrafts have been less substantial, rising to 1 billion gal. These data show that enough water to irrigate 10.7 million acres is being withdrawn in excess of replenishment from American aquifers. Although overdrafts for short periods do no harm, for the long term they deplete aquifers, and pumping costs rise as a result.

Consumption has meaning only in terms of supply, and supply is not always easy to predict, because it is subject to technology, politics, and the economy, as well as to natural constraints. For example, a serious effort has been made during the past 15 years to modify weather in ways likely to increase precipitation. These techniques—seeding clouds with silver iodide crystals, for example—are still experimental, but they may prove workable eventually. A problem that engineers have yet to solve is that efforts to increase rainfall in one area might decrease it in another. Then there is desalinization—a technology that has attracted a good deal of publicity and that has been employed successfully in harshly arid parts of the world, such as the Middle East and some islands in the Caribbean. However, desalinization is well enough established now to support the prediction that in the United States it will always be too expensive for farming purposes—the only source of demand that could seriously deplete supply here.

Just as human intervention can enhance the water supply, human intervention can diminish it. For example, if the federal government were to impose a ban on overdrafts of ground water, the supply available to consumers could drop by 25 percent in the country as a whole and by an even greater percentage in the West. The burning of fossil fuels may have provoked a buildup of carbon dioxide in the atmosphere, and evidence is accumulating that if such a buildup has indeed occurred, the small changes in temperature that will result will lead

gradually to large changes in rainfall in the country's midlatitudes. This could create major agricultural dislocations in Colorado, Utah, and Arizona. So far, though, the change in the earth's atmosphere and its potential effect on weather are only hypotheses.

The effect of politics and the economy on the country's water supply is easy to demonstrate. A huge federal pork barrel is based in part on the construction of dams and the transfer of water from one basin to another, both of which serve to increase a water supply when it runs short. Between 1965 and 1980, $52 billion was spent on all water resource projects—inland waterways, ports and harbors, multipurpose dams, and so forth—by four federal agencies: the U.S. Army Corps of Engineers, the Bureau of Reclamation, the Tennessee Valley Authority, and the Soil Conservation Service.

The Bureau of Reclamation's Central Arizona Project is a good example of how the money is used. This project has been under construction since 1973 and is 38 percent complete. Ultimately, it will divert 1.2 million acre·ft of water a year from the Colorado River Basin to the cities of Phoenix and Tucson and to the agricultural land in between. The water has to be lifted 1200 ft over a granite reef to a 300-mi-long aqueduct. Most of the water is intended to compensate for the depletion of aquifers by farmers for irrigation; the project will not provide enough water to allow for any expansion in irrigated farmland or for any further development of cities and industries.

The bureau estimates the project's final cost at between $1.4 billion and $2 billion; estimates made by others are substantially higher. When federal agencies undertake projects like this one, they pay most of the cost; the share borne by the beneficiaries is very small. Moreover, only 6 percent of the $52 billion budget for water projects between 1965 and 1980 was allocated to the Northeast, where 24 percent of American taxpayers reside and where water systems are generally old—in need of renovation and expansion. The West received 35 percent of the expenditures, though its share of the population is only 17 percent. As a result of this inequity, many congressional representatives of states and districts outside the West have opposed western water reclamation projects in recent years; indeed, no major package of water resource projects has been authorized since 1976. A good deal of money remains to be spent, however. According to the General Accounting Office, the cost of the 289 authorized projects still under construction by the Bureau of Reclamation and the Army Corps of Engineers totals $57.4 billion; of this sum $22 billion had been spent by the end of fiscal year 1981. Another 645 projects have been authorized, with funds yet to be appropriated.

One can object to expensive water projects on the narrow grounds that if they are truly needed, then the beneficiaries should pay for them. But the question of their financing aside, many water projects entail no riders to ensure that the increased supplies will not merely perpetuate a wasteful status quo, in which users who have consumed water indiscriminately can persist in doing so. Federal water projects attract development in regions that have already been developed to a point that strains existing supplies, and they encourage farmers to cling to thirsty crops and inefficient irrigation practices that ought to be abandoned. For example, water supplied by the Bureau of Reclamation to the Grand Valley of Colorado is so highly subsidized that the farmers rely on flood irrigation—they simply cover entire fields with water. Flood irrigation uses up to six times more water than "big wheel" and fixed-sprinkler systems, which are themselves by no means the most advanced and efficient techniques available.

If water and water rights were freely bought and sold, the demand for water would be easier to control; the market would help to establish the most efficient use of the resource. But most water supplies and rights are protected under legal and institutional arrangements that have been worked out over time in accord with objectives other than equity and economic efficiency. The two major legal doctrines—riparian rights and prior appropriation—were designed to secure the land titles of settlers. Riparian rights, which are observed chiefly in water-abundant eastern states, allow the owner of land adjacent to a body of water to make use of the water so long as no inconvenience is caused to his neighbors. Prior appropriation, which is observed in water-short western states, ranks the claims of settlers according to how early their homesteads were established. When a property owner fails to exercise his rights, they lapse. Thus prior appropriation, far from being an incentive for farmers to use less water, is an incentive for them to use as much as they can. In many cases water rights are attached to parcels of land or specify particular uses. As a result, late arrivals—industries, municipalities, or other farmers—often have trouble acquiring water at any price.

If Indian water rights in the western states are ever enforced, they will reduce substantially the amount of water available to everyone else in those states. The Winters Doctrine, based on the 1908 Supreme Court ruling in *Winters* v. *United States,* holds that when the federal government claims land for any purpose (specifically, to create an Indian reservation), it at the same time implicitly claims sufficient water to accomplish the purpose for which the land was set aside. At present Indian water rights remain unquantified, but the Navajos, for example, could legitimately claim as much as a third of the flow of the Colorado. In New Mexico pending claims by several tribes add up to many times the present allocations of water to all the irrigated farmland in the state.

Water is largely in the hands of the states, and it was only in the nineteenth century—with the passage of the Rivers and Harbors Act of 1826, which established federal control over interstate navigation; the Refuse Act of 1899, which controlled dumping in navigable waters; and the Reclamation Act of 1902, which established a federal role in enhancing water supplies in the West—that the federal government involved itself at all. The state laws are complex, having evolved, much like federal tax laws, over a long period, and there is little chance that a single national code will ever be devised, much less approved, to replace them. In fact, such a code would not even be desirable in this country, where there is so much variation in supply and demand from one region to another. But it is clearly in the interest of the states, perhaps with the guidance of the federal government, to plan equitably in ways that exploit the legal structure rather than conflict with it.

Water rights may be unbreakable, but they can be bent. For example, a state might offer grants to farmers to offset the cost of adapting farms so that they consume water more efficiently. In exchange, the state would be entitled to the surpluses achieved—water that could then be diverted for municipal and industrial uses and to increase reserves. Alternatively, since the economic return on water for agriculture is significantly lower than the return on water for manufacturing, industries in arid parts of the country might be enabled, through state intervention, to meet their needs by a mechanism as simple as competitive bidding.

Arizona's Groundwater Management Act—the first comprehensive state law to limit the pumping of groundwater—is a worthy precedent for the thoughtful redistribution of water. It was the severity of ground water shortages in that state that prompted construction of the Central Arizona Project, but the project will supply only two-thirds of the water needed by the year 2000, at current rates of consumption. Thus the act calls for wells to be registered and levies fees on the water withdrawn. It calls on farmers to improve irrigation efficiencies, and if these efficiencies are not adequate by the year 2005, it allows the state to buy and retire as much farmland as necessary to restore the natural balance. The act also requires housing developers to secure state certification that a 100-year supply of water is available before construction can proceed.

Harvests need not suffer if farmers use less water than their water rights entitle them to; the agricultural and social optimums rarely coincide, and given the subsidies, there is no reason why they should. On average, irrigated crops consume only 53 percent of all the water that a farmer withdraws. At farms dependent on surface water, irrigation efficiencies are typically less than 45 percent. At farms dependent on ground water, the efficiencies are better than 60 percent. The difference is a matter of economics. Irrigation water from surface supplies, which reaches farms through federal or privately managed canal systems, costs a good deal less than ground water, which farmers must pump themselves. With improved irrigation systems these efficiencies could rise to between 65 and 75 percent or higher.

Many farmers—not just those in Colorado's Grand Valley—irrigate by flooding. Alternatives to this practice are "permanent furrow systems," which collect rainwater and surplus irrigation water and channel it to storage ponds for future use; concrete or plastic pipes that carry water underground from the water source to the furrows, reducing the losses to evaporation and seepage incurred by open channels; and drip irrigation, in which plastic tubes feed small increments of water directly to the roots of each plant. Drip irrigation both inhibits evaporation and allows plants to be fed only as much water as they need; farms in Israel using this method have been reported to achieve efficiencies as high as 95 percent.

Besides reducing total water consumption, farmers might also tap supplies of brackish water, for which there is no competing demand. A rule of thumb in agriculture holds that "if you can drink the water, it's all right to use it for irrigation." Yet some crops—barley and cotton, for example—can flourish with brackish water. Crops less tolerant of salt, such as corn and beans, can also be grown in brackish water if it is supplied in such quantity that the salt is washed past the roots. Running tiled drains under the root zones of the plants also helps to keep salt concentrations low. Brackish water occurs inland as well as near the sea. Thus adapting drainage systems to accommodate brackish water and growing crops that can thrive on it—leaving sweet water for other uses—the conservation options for farmers everywhere in the United States, not just those in coastal regions. Brackish water could also be substituted for sweet in air conditioners, washing machines, and industrial washing and rinsing processes.

In many arid regions of the United States, farmers have been able to cultivate crops that are more water-intensive than the resource ultimately can support because water is cheap. But when a water supply is exploited to the limit of its capacity, nature asserts itself in ways that should force farmers to conserve. As a water table drops, pumping becomes more expensive. For a while farmers may persist in their withdrawals, choosing to pay for more fuel to run the pumps or to drill new wells. Eventually, however, the cost will rise higher than farmers can afford, and at that point they will abandon their farms or turn to the crops that can thrive with little or no irrigation, such as sorghum, wheat, and soybeans.

Society can be content with this self-regulating system so long as the irrigated crops lost are not essential to health or economic survival. The system can be hard, however, on the farmers at its mercy, and some form of government subsidy ought to be made available to compensate them at least partially for the cost of converting to different crops. Such compensation is sure to cost much less than the federal projects conceived to increase water supplies artificially.

Does the nation depend on crops irrigated with water from the Ogallala aquifer? Do we need Arizona's cucumbers? The most recent comprehensive study of the question, conducted last year by Kenneth Frederick, shows that we do not. Frederick found that out of a national total of 547 million acres of cropland and pastureland, only about 61 million acres are irrigated. The Center for Agricultural and Rural Development, at Iowa State University, has predicted that only 32 million irrigated acres will be needed to meet the demand for food and fiber in the United States in the year 2000. The center asserts that even this acreage can be irrigated at lower cost, in both water and dollars, using existing technologies. Clearly, if irrigated acreage can safely decline by half over the next 20 years, the country does not desperately need to water the desert, nor does it need to devote so much of the water it consumes to the task. With a decline in irrigated farmland and the adoption of more efficient irrigation methods, an enormous amount of water can be saved for drinking and other uses.

Industry has already markedly reduced its demand for water, as a result (unintended) of the Water Pollution Control Act, which Congress passed in 1972. Predictably, economic constraints have proved a powerful motive for change. The act requires industries to treat their own effluents or to pay others to do so; to cut the cost, firms have found it in their interest to produce less waste. The Water Resources Council estimated in 1978 that by the year 2000 industry will withdraw 62 percent less fresh water than it did in 1975, although its consumption of water will increase by as much as 150 percent. Industries will make up the difference by recycling the water they do withdraw from the nation's supply.

Industries can reduce their demand for fresh water substantially merely by keeping a cleaner house, resorting to such simple practices as shutting off taps when no water was needed and using some washwaters more than once.

The nation's demand for energy is another source of stress on the water supply. However, although this demand is sure to rise substantially in the future, the amount of water consumed to generate electricity from stream is not likely to rise commensurately. In fact, steam power plants will consume about 11 percent *less* water in the year 2000 than they did in 1975, according to the Water Resources Council's last forecast, even though the amount of water they require could be seven times greater. The huge difference between these two figures will be accounted for by improved technologies for cooling water, so that more of it can be recovered.

The synthetic-fuels industry, which converts coal and oil shale into more efficient sources of energy, is based chiefly in the West. Although it has been stalled by the buyer's market in oil, its revival will place it in fierce competition for water that is already running short. A coal gasification plant that can produce 250 million ft^3 of gas a day will require between 3000 and 15,000 acre·ft of water a year. A coal liquification plant that can produce 50,000 barrels of fuel a day will require between 5000 and 10,000 acre·ft of water a year. An oil shale plant that produces 50,000 barrels of oil a day will require between 2000 and 9000 acre·ft of water a year. These are small amounts of water, however, in comparison with agriculture's requirements. A coal liquification or gasification plant would use only enough water to irrigate between 1000 and 5000 acres, with harvests worth about $250,000. If the same amount of water were used to produce synthetic fuels, it would be worth many times that sum.

Although reasonable pricing is not the only solution to the country's poor stewardship of water, it would have powerful ramifications. No state can be expected to refuse a dam or waterway as long as the federal government promises to foot the bill, even though such schemes may delay its coming to grips with the true sources of a shortage. No consumer in any category can be expected to use less water as long a water bills do not reflect the need. Industry cannot be expected to pay for adequate treatment as long as its failure to do so is met at most by nominal fines and more often by silence. Finally, water utilities cannot be counted on to be careful of chemical contamination and to take steps to prevent it, as long as the EPA is too timid to insist.

The consumers of water must be made aware of its cost for any effort at conservation to succeed. Consider, for example, the aftermath of the federal Water Pollution Control Act, which provided substantial subsidies for the construction or renovation of municipal sewage systems. By 1982 grants to municipalities totaled about $36 billion. The EPA estimates that by the year 2000 an additional $118 billion will be needed to bring the country's sewerage systems into compliance with the act's standards. Because initially the government agreed to pay 75 percent of all construction costs, but nothing for land or maintenance, it ensured that communities would turn to conventional plants, which are expensive to build but cheap to maintain and which require little land. By the same token, it ensured that communities would avoid innovative systems, which typically cost little to build and relatively more to maintain and which require large tracts of land. (Innovative sewage treatments tend to rely on nature itself; they employ liquid sewage to irrigate and fertilize crops or they decompose sewage biologically in oxidation ponds or meadow/marsh systems.) The terms of the grants have since been changed in a way that partially corrects this defect and compels communities to pay close attention to a project's true cost. In 1985 the government's share will drop to only 55 percent for conventional treatment plants, but it will remain at 75 percent for innovative systems. The grants will still cover only the cost of construction, however, and as a result, a community's incentive to innovate is still not obvious.

The Reagan administration has effectively eliminated the three federal bodies that were in a position to correct the nation's faulty planning: the Council on Environmental Quality, the Water Resources Council, and the Office of Water Resources Technology. Even though the Council on Environmental Quality was composed of too many lawyers and too few scientists and engineers, it played an important role in educating the public about the threats to water and the progress toward solutions. The Water Resources Council studied on a continuous basis the adequacy of water supplies in all parts of the country. The Office of Water Resources Technology, in the Department of the Interior, was the major funding agent for studies of water's proper management. By consistently pointing out the flaws in how water is being managed, it made an enemy of the EPA, whose administrators did not appreciate negative evaluations of their performance. It also alienated the four agencies holding the purse strings of the government's water projects and their chief constituencies—farmers, real estate developers, mining companies, and so on.

Water policy is thoroughly fragmented: the Department of the Interior and the USGS have chief jurisdiction over problems of quantity, and the EPA over problems of quality, but no agency is trying to look at the nation's water in all its aspects. Even when it was functioning, the Water Resources Council, which might have studied the interrelationships between supply and purity and the competing demands of the states, did not do so effectively. There ought to be a commission on *water,* to seek ways to resolve the battles between East and West and to integrate the activities of the various agencies. And it must do so without engaging in the partisan politics in which these agencies are so often embroiled. The American people are owed a careful explanation of the choices they face, both as taxpayers and as consumers of water.

Perhaps, after all, this country can devise a "thoughtful, logical national plan" for water. First on the agenda is to compel users to appreciate water's economic value. Higher prices could influence water consumption radically, and with little trauma.

Higher prices by themselves will not untangle the snarl of competitive demands for water in regions where it is running short in absolute terms. But higher prices can reveal the true economic value of water and make the political choices among these demands clearer. Similarly, the public health is not merely in the domain of cost-benefit analyses. Standards of water purity are ideally studied with no consciousness of the financial burdens they would impose. These burdens must then be confronted, however, and borne as part of our commitment to ourselves and to future generations. Compared with the nation's other expenses, these burdens will not be so very large. Insofar as a water crisis exists, it is a crisis of political will. If the country acts wisely and is willing to pay the price, it can prevent politics from transforming the water crisis of the headlines into a true crisis in the natural world.

APPENDIX A
Glossary

Acre-foot (acre·ft) The volume of water required to cover 1 acre of land [43,560 square feet (ft^2)] to a depth of 1 foot; equivalent to 325,851 gallons (gal).

Alluvium A general term for unconsolidated material deposited by a stream or other body of running water.

Aquifer A geologic formation, group of formations, or part of a formation that contains sufficient saturated permeable material to yield significant quantities of water to wells and springs. See also *Confined aquifer* and *Unconfined aquifer*.

Basal ground water or **basal lens** A term that originated in Hawaii and refers to a major body of fresh ground water in contact with saline water and occurring in the lower part of the freshwater flow system.

Base flow Sustained or fair-weather flow of a stream. In most places base flow is derived from ground water inflow to the stream channel.

Benthic organism A form of aquatic life that lives on the bottom or near bottom of streams, lakes, or the oceans.

Billion gallons per day (bgd) A rate of flow of water.

Brackish water Water that generally contains 1000 to 10,000 milligrams per liter (mg/L) of dissolved solids. See *Saline water*.

Brine Water that generally contains more than 100,000 milligrams per liter (mg/L) of dissolved solids. See *Saline water*.

Capillary fringe The zone above the water table in which water is held by surface tension. Water in the capillary fringe is under a pressure less than atmospheric.

Commercial water use Water used by motels, hotels, restaurants, office buildings, commercial facilities, and institutions, both civilian and military. The water may be obtained from a public supply or be self-supplied. See also *Public supply* and *Self-supplied water*.

Cone of depression A depression in the potentiometric surface around a well from which water is being withdrawn.

Confined aquifer An aquifer in which ground water is confined under a pressure that is significantly greater than atmospheric pressure. Synonym: artesian aquifer. See also *Aquifer* and *Unconfined aquifer*.

Connate water Water entrapped in the interstices of a rock at the time of its deposition.

Consumptive use The quantity of water that is no longer available because it has been evaporated, transpired, or incorporated into products, plant tissue, or animal tissue. Also referred to as *water consumption* and *water consumed*.

Conveyance loss Water that is lost in transit from a pipe, canal, conduit, or ditch by leakage or evaporation. Generally, the water is not available for further use; however, leakage from an irrigation ditch, for example, may percolate to a ground water source and be available for further use.

Domestic water use Water used for normal household purposes, such as drinking, food preparation, bathing, washing clothes and dishes, flushing toilets, and watering lawns and gardens. Also called *residential water use*. The water may be obtained from a public supply or may be self-supplied. See also *Public supply* and *Self-supplied water*.

Drawdown The difference between the water level in a well before pumping and the water level in the well during pumping. Also, for flowing wells it is the reduction of the pressure head as a result of the withdrawal of water. See also *Pressure head*.

Eutrophication The process by which waters become enriched with plant nutrients, especially phosphorus and nitrogen.

Eutrophic lake A standing body of water containing an excessive concentration of plant nutrients, especially phosphorus and nitrogen, which results in excessive algal production, especially bluegreen algae.

Evapotranspiration A collective term that includes water lost through evaporation from the soil and surface water bodies and by plant transpiration.

Fresh water Water that generally contains 0 to 1000 milligrams per liter (mg/L) of dissolved solids.

Ground water In the broadest sense, all subsurface water, as distinct from surface water; as more commonly used, that part of the subsurface water in the saturated zone.

Hydraulic conductivity The capacity of a rock to transmit water; expressed as the volume of water that will move in unit time under a unit hydraulic gradient through a unit area measured at right angles to the direction of flow.

Hydraulic gradient Change in head per unit of distance measured in the direction of the steepest change.

Hydraulic head In ground water, the height above a datum plane (such as sea level) of the column of water that can be supported by the hydraulic pressure at a given point in a ground water system. For a well the hydraulic head is equal to the distance between the water level in the well and the datum plane.

Industrial water use Water used for thermoelectric power (electric utility generation) and other industrial uses such as steel, chemical and allied products, paper and allied products, mining, and petroleum refining. The water may be obtained from a public supply or may be self-supplied. See also *Public supply* and *Self-supplied water*.

Instream use Water use taking place within the stream channel for purposes such as hydroelectric power generation, navigation, water quality improvement, fish propagation, and recreation. Sometimes called *nonwithdrawal use* or *in-channel use*.

Irrigation water use Artificial application of water on lands to assist in the growing of crops and pastures or maintaining recreational lands such as parks and golf courses.

Livestock water use Water used by livestock. Livestock, as the term is used here, includes cattle, sheep, goats, hogs, and poultry. Also included are animal specialties such as horses, rabbits, bees, pets, fur-bearing animals in captivity, and fish in captivity. See also *Rural water use*.

Mineralized water Water containing dissolved minerals in concentrations large enough to affect use of the water for some purposes. A concentration of 1000 milligrams per liter (mg/L) of dissolved solids is commonly used as the lower limit for mineralized water.

Mining of ground water Ground water withdrawals in excess of recharge. See also *Overdraft*.

Non–point source of pollution Pollution from sources that cannot be defined as originating from discrete points, such as areas of fertilizer and pesticide application and leaking sewer systems.

Offstream use Water withdrawn or diverted from a ground water or a surface water source for public supply, industry, irrigation, and rural uses. sometimes called *off-channel use* or *withdrawal use*.

Overdraft Withdrawals of ground water at rates perceived to be excessive. See also *Mining of ground water*.

Perched ground water Unconfined ground water separated from an underlying main body of ground water by an unsaturated zone.

Permeability The capacity of a porous rock, sediment, or soil for transmitting a fluid without altering its physical structure; a measure of the relative ease of fluid flow under pressure.

pH (hydrogen–ion activity) A number used by chemists to express the acidity of solutions, including water. A pH value lower than 7 indicates an acidic solution, a value of 7 is neutral, and a value higher than 7 indicates an alkaline solution. Most ground waters in the United States have pH values ranging from about 6.0 to 8.5.

Phreatophyte A deep-rooted plant that obtains its water supply from the zone of saturation.

Point source of pollution Pollution originating from any confined or discrete source, such as the out-

flow from a pipe, ditch, tunnel, well container, concentrated animal-feeding operation, or floating craft.

Porosity The volume of openings in a rock. When expressed as a fraction, porosity is the ratio of the volume of openings in the rock to the total volume of the rock.

Potentiometric surface An imaginary surface representing the level to which water will rise in wells.

Pressure head Hydrostatic pressure or force per unit area expressed as the height of a column of water that the pressure can support, relative to a specific datum such as land surface or sea level.

Primary porosity The openings or pores in a rock at the time it was formed.

Prior appropriation A concept in water law under which users who demonstrate earlier use of water from a particular source are said to have prior-appropriation rights over all later users of water from the same source.

Public supply Water withdrawn for all uses by public and private water suppliers and delivered to users that do not supply their own water. Water suppliers provide water for a variety of uses such as domestic, commercial, industrial, and public use. See also *Commercial water use, Domestic water use, Industrial water use,* and *Public water use*.

Public water use Water supplied from a public supply and used for fire fighting, street washing, and municipal parks and swimming pools. See also *Public supply*.

Recharge area An area in which water reaches the zone of saturation from surface infiltration.

Regolith A general term for the layer of unconsolidated (soillike) material of whatever origin that nearly everywhere forms the surface of the land and that overlies or covers the more coherent bedrock.

Return flow That part of irrigation water that is not consumed by evapotranspiration and that returns to its source or another body of water. The term is also applied to water that is discharged from industrial plants.

Riparian rights A concept of water law under which authorization to use water in a stream is based on ownership of the land adjacent to the stream.

Runoff The part of precipitation or snowmelt that appears in streams or surface water bodies.

Rural water use Water used in suburban or farm areas for domestic and livestock needs. The water generally is self-supplied and includes domestic use, drinking water for livestock, and other uses such as dairy sanitation, evaporation from stock-watering ponds, and cleaning and waste disposal. See also *Domestic water use, Livestock water use,* and *Self-supplied water*.

Safe yield (surface water) The amount of water than can be withdrawn or released from a reservoir on an ongoing basis with an acceptable small risk of supply interruption.

Saline water (general) Water that contains more than 1000 milligrams per liter (mg/L) of dissolved solids.

Saline water (specific) Water that generally contains 10,000 to 100,000 milligrams per liter (mg/L) of dissolved solids. See also *Brackish water* and *Brine*.

Saturated zone A subsurface zone in which all the interstices or voids are filled with water under a pressure greater than that of the atmosphere.

Self-supplied water Water withdrawn from a surface water or ground water source by a user and not obtained from a public supply. See also *Industrial water use* and *Rural water use*.

Sheet erosion Erosion in which relatively thin layers of surface material are gradually removed more or less evenly from an extensive area of gently sloping land by broad, continuous sheets of running water. See also *Sheet flow*.

Sheet flow An overland flow or downslope movement of water taking the form of a relatively thin continuous film flowing over relatively smooth soil or rock surface and not concentrated into channels.

Specific retention The ratio of the volume of water retained in a rock after gravity drainage to the volume of the rock.

Specific yield The ratio of the volume of water that will drain under the influence of gravity to the volume of saturated rock.

Storage coefficient The volume of water released from storage in a unit prism of an aquifer when the head is lowered a unit distance.

Transmissivity The capacity of an aquifer to transmit water; equal to hydraulic conductivity times the aquifer thickness.

Turbidity The state, condition, or quality of opaqueness or reduced clarity of a fluid due to the presence of a suspended matter.

Unconfined aquifer An aquifer that is unsaturated (i.e., is not full of water); has a water table.

Water budget An accounting of the inflow to, outflow from, and storage in a hydrologic unit such as a drainage basin, aquifer, soil zone, lake, or reservoir.

Water resources region Natural rainage basin or hydrologic area that contains either the drainage area of a major river or the combined drainage areas of a series of rivers. There are 21 regions, of which 18 are in the conterminous United States and one each is in Alaska, Hawaii, and the Caribbean. (See map on p. 39.)

Water table The level in the saturated zone at which the water is under a pressure equal to the atmospheric pressure.

Water year A continuous 12-month period arbitrarily selected to present data relative to hydrologic or meterologic phenomena during which a complete annual hydrologic cycle occurs. The water year selected by the U.S. Geological Survey is the period October 1 through September 30.

Withdrawal Water removed from the ground or diverted from a surface water source for use. See also *Offstream use*.

REFERENCES

R. C. Heath, 1984, *Ground-water Regions of the United States,* U.S. Geological Survey Water-Supply Paper 2242.

W. B. Solley, E. B. Chase, and W. B. Mann IV, 1983, *Estimated Use of Water in the United States in 1980,* U.S. Geological Survey Circular 1001.

U.S. Geological Survey, 1984, *National Water Summary 1983–Hydrologic Events and Issues,* U.S. Geological Survey Water-Supply Paper 2250.

APPENDIX B

Conversion Factors

Multiply	By	To Obtain
	AREA	
Acres	43,560	Square feet (ft^2)
	4047	Square meters (m^2)
	0.001562	Square miles (mi^2)
	FLOW	
Billion gallons per day (bgd)	1000	Million gallons per day (gal/day)
	1121	Thousand acre-feet (acre·ft) per year
	1.547	Thousand cubic feet per second (ft^3/s)
	694.4	Thousand gallons per minute (gal/min)
	3.785	Million cubic meters per day (m^3/day)
Million gallons per day (mgd)	0.001	Billion gallons per day
	1121	Acre-feet per year
	1.547	Cubic feet per second
	0.6944	Thousand gallons per minute
	0.003785	Million cubic meters per day
Thousand acre-feet per year	0.0008921	Billion gallons per day
	0.8921	Million gallons per day
	0.001380	Thousand cubic feet per second
	0.6195	Thousand gallons per minute
	0.003377	Million cubic meters per day

SELECTED WATER RELATIONSHIPS *(APPROXIMATIONS)*

$$
\begin{aligned}
1 \text{ gallon} &= 8.34 \text{ pounds (lb)} \\
1 \text{ million gallons} &= 3.07 \text{ acre-feet (acre·ft)} \\
1 \text{ cubic foot} &= 62.4 \text{ pounds} \\
&= 7.48 \text{ gallons (gal)} \\
\\
1 \text{ acre-foot} &= 325,851 \text{ gallons} \\
(1 \text{ acre covered by } 1 \text{ foot of water}) &= 43,560 \text{ cubic feet (ft}^3) \\
\\
1 \text{ cubic mile} &= 1.1 \text{ trillion gallons} \\
&= 3,379,200 \text{ acre-feet} \\
\\
1 \text{ inch of rain} &= 17.4 \text{ million gallons per square mile} \\
&= 27,200 \text{ gallons per acre} \\
&= 100 \text{ tons per acre}
\end{aligned}
$$

APPENDIX C
Additional Readings

PART I: Water: The Compound, the Resource

Balchin, W. G. V., ed. 1985. Special issue on water in the 1990's. In *The International Journal of Environmental Studies,* 25 (3), 139–211.

Jensen, Marvin E., and John D. Bredehoeft. 1983. *New efficiencies in water use vital for nation.* In *Yearbook of agriculture.* Washington, D.C.: U.S. Department of Agriculture.

L'vovich, M. I. 1979. *World water resources and their future.* Washington, D.C.: American Geophysical Union.

Mather, John R. 1984. *Water resources: Distribution, use, and management.* New York: John Wiley and Sons, Inc.

UNESCO. 1972. *Influences of man on the hydrologic cycle: Guidelines to policies for the safe development of land and water resources.* Paris: UNESCO

United States General Accounting Office. 1979. *Water resources and the nation's water supply.* CED-79-69. Washington, D.C.: GAO.

PART II: Water in the Environment

American Chemical Society. 1983. *Ground water: An information pamphlet.* Washington, D.C.: American Chemical Society.

Back, William, P. R. Seaber, and J. S. Rosenshein, eds. 1987. *Ground water hydrogeology of North America.* In *The Geology of North America,* Boulder, Colorado: Geological Society of America.

Hayes, R. T., K. A. Popko, and W. K. Johnson. 1980. *Guide manual for preparation of water balances.* Davis, Calif.: Hydrologic Engineering Center, U.S. Army Corps of Engineers

Hem, John D. 1985. *Study and interpretation of the chemical characteristics of natural water.* 3rd ed. U.S. Geological Survey Water-Supply Paper 2254.

Munger, J. William, and Steven J. Eisenreich. 1983. Continental-scale variations in precipitation chemistry. *Environmental Science and Technology,* 17 (1): 32a–42a.

U.S. Water Resources Council. 1980. *Essentials of ground water hydrology pertinent to water resource planning.* U.S. Water Resources Council Bulletin 16 (revised).

Wolman, M. Gordon, H. C. Riggs, Marshall Moss, and Vern Schneider, eds. 1987. *Surface water hydrology of North America.* In *The Geology of North America,* Boulder, Colorado: Geological Society of America.

PART III: Water Use

Abbey, David. 1979. Energy production and water resources in the Colorado River Basin. *Natural Resources Journal,* 19: 275–314.

Dendney, Daniel. 1981. Hydropower: An old technology for a new era. *Environment,* 23: 16–20, 37–45.

Engelbert, Ernest A., with Ann Foley Scheuring. 1982. *Competition for California Water: Alternative resolutions.* Berkeley, Calif.: University of California Press.

Kahrl, William L., ed. 1979. *The California water atlas.* Sacramento, Calif.: The State of California.

Keller, W. D. 1978. Drinking water: A geochemical factor in human health. *Geological Society of America Bulletin* 89: 334–336.

Lee, Terence. 1981. Rural drinking supply and sanitation in Latin America. *Natural Resources Forum* 5 (3): 282–290.

Pearl, Richard H., 1976. Hydrological problems associated with developing geothermal energy systems. *Ground water,* 14 (May–June): 128

PART IV: Problems and Hazards

Berk, R. A., C. J. LaCivita, K. Sredl, and T. F. Cooley. 1981. *Water shortage: Lessons in conservation from the great California drought 1976–1977.* Cambridge, Mass.: Abt Books.

Environmental Protection Agency. 1980. *Acid rain.* Washington, D.C.: EPA.

Griggs, Gary B., and Lance Paris. 1982. Flood control failure: San Lorenzo River, California. *Environmental management,* 6 (5): 407.

McFee, W. W., chairman. 1984. *Acid precipitation in relation to agriculture, forestry, and aquatic biology.* Council for Agricultural Science and Technology Report 100.

National Research Council. 1984. *Ground water contamination: Studies in geophysics.* Washington, D.C.: National Academy Press.

Oden, Svante. 1976. The acidity problem—An outline of concepts. *Water, Air, and Soil Pollution.* 6: 137–166.

Pye, Veronica I., Ruth Patrick, and John Quarles. 1983. *Ground water contamination in the United States.* Philadelphia: University of Pennsylvania Press.

Rahn, Perry H. 1974. Lessons learned from the June 9, 1972, flood in Rapid City, South Dakota. *Bulletin of the Association of Engineering Geologists,* XII: 83–97.

———. 1984. Flood-plain management nprogram in Rapid City, South Dakota. *Geological Society of America Bulletin,* 95: 8838–8843.

Turk, John T. 1983. *An evaluation of trends in the acidity of precipitation and the related acidification in North America.* U.S. Geological Survey Water-Supply Paper 2249.

PART V: Law, Economics, and Management of Water

Anderson, Terry L. 1983. *Water Crisis: Ending the policy drought.* Washington, D.C.: Cato Institute.

Back, William. 1981. Hydromythology and ethnohydrology in the New World. *Water Resources Research,* 17: 257–287.

Beattie, Bruce R. 1981. Irrigated agriculture and the Great Plains: Problems and policy alternatives. *Western Journal of Agricultural Economics,* 6 (2): 289–299.

Caponera, Dante A., and Dominique Alheritiere. 1978. Principles for international ground water law. *Natural Resources Forum,* 2: 279–290, 359–371.

Lindh, Gunnar. 1978. Socioeconomic aspects of urban hydrology. *AMBIO,* 7: 16–22.

Milliken, J. C., and G. C. Taylor. 1981. *Metropolitan water management.* Water Resources Monograph Series. Washington, D.C.: American Geophysical Union.

National Research Council. 1979. *Scientific basis of water-resource management: Studies in geophysics.* Washington, D.C.: National Academy Press.

New York State Department of Environmental Conservation. 1984. *Long Island ground water management program.* Albany, N.Y.: N.Y. Department of Environmental Conservation.

Postel, Sandra. 1985. *Conserving water: The untapped alternative.* Worldwatch Institute paper 67. Washington, D.C.; Worldwatch Institute.

Trelease, Frank J. 1978. Climatic change and water law. In *Climate, climatic change and water supply,* 70–84. Washington, D.C.: National Academy Press.